Prehistoric Modernization

Shengqian Chen

Prehistoric Modernization

From Hunting and Gathering to the Origins of Agriculture

三联书店

palgrave macmillan

Shengqian Chen
School of History
Renmin University of China
Beijing, China

Translated by
Huifang Li
School of Foreign Studies
Hebei Normal University
Shijiazhuang, China

ISBN 978-981-97-9422-5 ISBN 978-981-97-9423-2 (eBook)
https://doi.org/10.1007/978-981-97-9423-2

Jointly published with SDX Joint Publishing
The print edition is not for sale in China (Mainland). Customers from China (Mainland) please order the print book from: SDX Joint Publishing.
ISBN of the Co-Publisher's edition: 978-7-108-06876-7

This work was supported by B&R Book Program.

This Palgrave Macmillan imprint is published by the registered company Springer Nature Singapore Pte Ltd.
The registered company address is: 152 Beach Road, #21-01/04 Gateway East, Singapore 189721, Singapore

Preface

I always think if not for the purpose of promotion, the preface should be written by the author. It is the role of the author to tell his/her readers reasons, processes, crucial turning points, and perspectives of his work, and to provide some reading suggestions for a better understanding of his/her ideas. A book is like a friend, a person who opens a book and reads it is a person who has taken the author as his/her friend, and a preface is just like a self-introduction from the author. The best quality between friends is honesty. Therefore, a self-written preface should not be written like an advertisement. However, as Francis Bacon says: "There is no better person to flatter myself as me". People are tending to affirm themselves, so what I will say is just my point of view and its reliability depends on your judgement. I hope my readers could filter out something valuable from my rough writings.

The basis of this book is my Ph.D. dissertation *Adaptive Changes of Prehistoric Hunter-Gatherers during the Pleistocene-Holocene Transition* in China at Southern Methodist University in 2004 under the supervision of Dr. Fred Wendorf and Dr. Lewis R. Binford. Both are Fellows of the National Academy in the United States. Most Chinese archeologists have known Dr. Binford but know very little of Dr. Wendorf. He is an expert archaeologist on Prehistoric North Africa and an authority on the study of origins of agriculture in the Nile Valley. He was awarded his Ph.D. in Anthropology from Harvard in the 1950s and was elected Fellow of the National Academy in 1987, more than ten years earlier than Dr. Binford.

I consider it the greatest luck in my career to be taught by these two top archeologists in the world. Of course, talking about my supervisors' fame cannot prove my own learning. My own hope for my work is not to disgrace my supervisors' patient instructions.

For many people, doctoral dissertations are their most painstaking academic achievements, sometimes even peaks of their academic lives. I hope this is not my case, but I have to admit that my doctoral dissertation is by far my most systematic study. I still remember the day after my defence, I asked myself, "How was your writing?"—I was quite satisfied with its structure, and my arguments on the origins of agriculture in different culture-ecological zones were not bad. But I felt more work should be done in the collecting of archaeological materials, especially in the analysis of specific materials. When I was later working in the Institute of Vertebrate Paleontology and Paleoanthropology in the Chinese Academy of Sciences as a postdoc, I began to expand my research and strengthen my study on the human adaptations during the Upper Paleolithic in North China and the microblade technology. When I started my teaching at Jilin University, I opened a course called "Prehistory Since the Late Pleistocene" for graduate students, it was a follow-up course after my supervisor Binford's "Post-Pleistocene Adaptation". I tried to teach this course in two ways, one was from my own system, that is, the perspective of the origins of agriculture in China; the other was from the perspective of different regions in the world, with the aim to do some comparative studies. This course was taught four times, and some students sat there four times, maybe because I taught it differently each time. For me, teaching is a good way to develop and deepen this research. Every student in the class needed to ask questions, to learn, and to doubt, and I really appreciated their participation. Their thinking undoubtedly encouraged my study on the issue.

I started another three courses for graduates: "Contemporary Archeological Theories" "Lithic Analysis", and "Site Formation Process". I told the students in my class that it was not because I was academically qualified to teach these courses, what I expected was a two-way learning practice which would benefit both sides. These courses made up my knowledge deficiencies to some degree. The lithic analysis and the site formation research were very helpful for me to absorb more valuable information from those archaeological materials, and the theoretical exploration deepened my understanding of the study of the origins of agriculture. Meanwhile, I carried out specific projects on the origins of agriculture in China, and some of the results were published. The modelling hunter-gatherers of China in Chapter 3 was published in the *Anthropologica Sinica*, the study on the environment adaptation and diffusion of North China humans in the Upper Paleolithic was published in the *Quaternary Sciences*, the study on the microblade technology was published in *Archaeological Research*, the study on the Southern and Northern areas of Yanshan Mountain was published in the *Kaoguxuebao*. I have tailored some chapters in order to suit the overall structure of this book. These specific projects were more systematic and penetrating research developed from my Ph.D. dissertation. In recent years, a number of new findings on the origins of Chinese agriculture have appeared, which I too have integrated into this book. With the help of my teaching, my research and the new data, this book has become a much-improved version of my dissertation. They might share similar structures, but their contents are quite different.

My interest in the emergence of agriculture started with a course called "Study on the Neolithic in China" taught by Professor Yan Wenming for postgraduate students in 1993 at Beijing University. In 1991, when I was doing my undergraduate field practice excavating the site of Baiyinchanghan in Inner Mongolia, the well-preserved sites of housing, stone tools, and the beauty of archaeology impressed me a lot. Walking along the bank of the remotest Xila Mulun River, you couldn't help pondering what kind of life our ancestors were living, and whether they were the earliest agricultural producers. Twenty years later, when I returned to the Ningcheng Workstation, which is attached to the Institute of Cultural Relics and Archaeology of the Inner Mongolia Autonomous Region and resumed my study on the functions of stone tools of the Baiyinchanghan stone artifacts, the questions had become very specific and microscopic. The objects of my Master's thesis were the microcores in the Jijitan and Xishuidi of Nihewan Basin. It was a study on technology, and the age of these materials had reached the border of Neolithic. I don't remember whether it had occurred to me or not to connect these tools with the emergence of agriculture. Even if it had, there was nothing I could do at that moment. After six years' study in the U.S., with the patient guidance of Professor Binford and Professor Wendorf, especially the insightful instructions and direct help from Professor Binford, I eventually completed my doctoral dissertation and formed a rather comprehensive understanding of the origins of agriculture in China, although it was still a rough one. Since 2004, I have further developed my study on this issue.

What I present here is a steaming hot work, even if a somewhat over-brewed one. It was a result of my twenty years' efforts and a briefing of a phase of work.

The research structure of this book can be divided into four parts. The first part is a theoretical study, which aims to answer questions such as why agriculture emerged in some regions and why it did not happen in other regions. The second part is a modelling study, which is based on the culture-ecological principles of the hunter-gatherers. Through the modelling of the adaptive strategies of the hunter-gatherers with the data from modern weather stations, it predicts possible adaptive changes in different regions. Part three studies the initial conditions of the emergence of agriculture, including the cultural adaptations and the changes of environment during the Upper Paleolithic. The fourth part focuses on the appearance of agriculture and the adaptive changes in different regions. It answers two questions, one is the basic mode of the emergence of agriculture in China, the other is the adaptive changes of the hunter-gatherers who were living in the non-core areas of agriculture, and the process of their final acceptance of prehistoric agriculture. Each part composes of a series of rather independent studies. If you are interested in this book, I suggest you starting from the theoretical part, then move to your favourite region, or explore any regions you find interesting. The next step is the modelling study part and my analysis of the cultural adaptations during the Upper Paleolithic, both of which I assume a little confusing for common readers. The chapters on different regions are rather self-contained, and it would be fine for you to stop at the region that attracts you, and gain a glimpse of the others.

In my opinion, the highlights of this book include the following aspects:

First, the questions it addresses are why or why not agriculture emerged, how exactly it happened, what is the mode of it, and how many phases it experienced. Due to the fact that the contemporary Chinese archaeologists rarely consider these questions, successfully or not, this book tries to fill some of the blanks.

Second, compared with the hot topic of the origins of agriculture, the question of the non-emergence-of-agriculture in certain areas was rarely concerned. The book explains the cultural adaptation changes in these non-agriculture areas during the critical period, from the end of the Pleistocene to the beginning of the Holocene. Every region has its own adaptive strategies based on the local conditions. I also noticed the diversity in different regions when I was emphasising agriculture as a unified historical event.

Third, the standpoint of both unity and diversity is the study of the cultural adaptive mechanism of the hunter-gatherers. The essence of the emergence of agriculture is these hunter-gatherers replaced their mobile way of foraging with a settled way of agricultural production. Therefore, the research on the origins of agriculture must focus on the analysis of the cultural adaptation mechanism of the hunter-gatherers. The culture-ecological aspect of the hunter-gatherers is the most important perspective of this book, and it is also the key to understand this research.

Fourth, since our research started from the study of the hunter-gatherers, it is natural that the study of the origins of agriculture should start from the Paleolithic. The root of agriculture lies in the Paleolithic. In order to trace the originating process of agriculture, we should go back to the Paleolithic. Hence, the archaeological materials of this book are mostly from the Paleolithic, which is different from the similar studies in China that pay more attention to the Neolithic.

Fifth, in the study of archaeological materials, with reference to the research findings from the relevant disciplines, my focuses are the lithic analysis and the site formation research, which have become the new source for the study of the origins of agriculture. Therefore, rather than a summary of contemporary research, my study offers some new evidence on the issue.

Sixth, the emergence of agriculture is not a China-specific phenomenon, so I have no intention to isolate it. Instead, I have tried to compare it with materials from other agricultural cradles around the world, especially the Near East, so as to discover the basic mode of the origins of agriculture in China.

Seventh, it is not just research on an archaeological issue, it also attempts to reconstruct a prehistory through the relating of the origins of agriculture with the modernising process of Chinese society today. In light of the exploration of the origins of agriculture, i.e. the Prehistoric "modernisation", the great significance of the contemporary social changes in China could be understood.

Eighth, in the conclusion, there is a theoretical reflection on the archaeological research of the origins of agriculture, which adds a shade of deconstruction to the intended system construction. It might seem a little self-contradictory, but I regard it a necessary attitude for any scholar to reconsider his own work and be aware of his own limitations.

There are two further details to be noticed. The first is the issue of absolute age: the ages indicated as "Before Christ" "BC", or with a supplementary information "calibrated age" were all ages which were verified by the Tree Rings, and those indicated as "to present" or "BP" were unverified ages. The significance of the C14-dating technology in archaeology is extraordinary, but it has always been unsatisfactory in telling the exact dates of the formation of certain archaeological sites, especially when applied in the Upper Paleolithic. It is no strange to see the constant renewing of chronological data. Therefore, I took the absolute age as a reference, while at the same time, keeping some of my reservations. As to the question I am addressing, the requirement of the accuracy of the absolute date is not very high. In addition, I think, before we make clear of the forming process of the archaeological sites, the accuracy of age cannot prove anything, because we cannot tell whether it actually represents the time of the ancient people.

Another detail concerns the time division of the Neolithic. Basically, I followed the most traditional idea as taking the Cishan culture as the early culture of Neolithic without changing the chronological system for the newest discoveries. The reason is that we are not able to go back and revise those publications. My way is to add, such as adding "The Transition phase between the Paleolithic and the Neolithic" "Mesolithic" and "Post Paleolithic" before the "Early phase of Neolithic". I prefer an accumulative history rather than a modified one. This may result in the continuous use of some "incorrect" terms produced previously, but who can guarantee the correctness of our modifications? It might be modified by later scholars as well. Therefore, I am with the view to recognise the previous works and to do accumulative history.

My way of scholarship is largely a lonely one and I seldom consult people when I am writing. Even so, there is a full list of people to be thanked due to the long period of writing of this volume. From my supportive wife to my dear classmates and friends, from my encouraging colleagues to my most helpful supervisors, I sincerely thank them for the help and support they have provided! No matter what a careless person I am, there is one person I could never forget. Without him, there would be nothing of this book. It is my deceased supervisor: Dr. Lewis Binford. I don't argue in Heaven, nor do I argue in netherworld, but I argue in the continuity of ideas and spirit, and Dr. Binford's instructions and ideas are infused everywhere in this book. Because of this endless transmission, I argue in eternity. Every one of us is a link in the chain, and the real eternity lies in the ceaseless exploring spirit dwelling in human beings.

Beijing, China

Shengqian Chen

Introduction

Toute histoire digne de ce nom est histoire contemporaine

—*Benedetto Croce*

The Origin of the Question

On Oct. 31, 2011, a headline of *News for Reference* declared the birth of the 7 billionth citizen in the world in the Philippines. This is just a drop in the ocean of the hot discussions of world population, and previously the relevant reports had taken various media coverage in great quantity. The world population reached 6 billion in 1998, it was just 2.5 billion in 1900, and 1 billion in 1800, and 0.25 billion in 1000. If the current growth rate is maintained, the population will reach 15 billion in the year 2100. It is a number deserving our concern. Of course, there is another public opinion saying the growth of population will stop, because the population in some developed countries has suspended or even dropped in some cases. So, perhaps we should not worry about the increase in population, but its decrease. Ten thousand years ago, the population of the globe was around 5 million, a number closely related to this study. Undoubtedly, it is not an accurate number, but ever-after this, the human population has grown wildly and has expanded 1400 times by now. How could this happen? Cutis Marean published an article in the 9th issue of *Scientific American* in 2010, saying there was an Ice Age between one hundred and ninety-five thousand years ago and one hundred and twenty-three thousand years ago which deteriorated the living situation in Africa. He estimated there was no more than ten thousand people with fertility at most and was just several hundred at the least. Human beings were on the verge of extinction. Only a hundred thousand years later, human beings have taken over the whole earth. Should we be proud of the successful adaptation, or should we be concerned about the future of the world?

What is the relationship between the increase in population and the adaptive strategies of human beings? From the appearance of the hominid five or six million years before to the population crisis of the anatomical *Homo sapiens*, there was no substantial increase in population in this mobile, foraging group, but after that, there was an explosive increase in population. It is worth noting that, around ten thousand years ago, primitive agriculture appeared. Did the population crisis lead to the emergence of agriculture, or did the emergence of agriculture stimulate the population growth? Or was there a mutual encouragement between these two? In the twenty-first century, with the population is still growing, how will we survive? Against this background, the origins of agriculture became a question having to answer.

Compared with the foraging life, agriculture fed more people with the same area of land. Inevitably, there was a price to pay in sustaining the land, as agriculture took the space away from other species, the cost was actually taken by those other species. Agriculture greatly changed the landscape of the earth, which put an end to the long-lasting situation of "all the species are competing freely in the frosty days", the surface of the earth has become the colony of human beings. Then came the debates on global warming. Among them, the hot topics are in what degree human beings have interfered in the natural process of the earth, and what humans can do to reverse the ongoing direction towards a disastrous abyss. However, "Rome wasn't built in a day", how were human beings and their environment influencing each other within the past ten thousand years? In addition, how much did these interactions influence our life today?

Are these questions a little paranoid? Or are they large but impractical questions? Should the archaeologists keep clear from them? We should probe all the ins and outs when we are dealing with a tough issue. We should know the solutions to similar questions, and weigh the risks and opportunities it may bring about. Archaeologists study prehistory, but they are not living in it. They should also respond to the demands of their times.

Then, what kind of time I am living in? When I was young, I was living in a lakeside village in Southern China. The lake was huge and clear, and the swaying weeds and swimming fish could be seen when we were sitting in the fishing boat. There was a mystical forest behind the village where our ancestors were buried, and people were still living in the houses built by them. Just thirty years later, the lake is no longer clear, and numerous bamboo rods are stuck in the water to divide fishing farms. The old trees in the forest have been replaced by the more expensive Masson pines, and the old houses were demolished and there are fewer and fewer young men living in the village. Towns are emerging and spreading quickly in the places with convenient transportation. This is the urbanisation process that every Chinese today is very familiar with. This rapid social change has brought about an overwhelming uneasiness, even to its participants. For hundreds of years, the village had experienced many dramatic turbulences such as dynasty changes, wars and chaos, but its form never changed much. Why could this thirty-year peaceful development bring about such a huge change? How did this change of social structure occur?

It is an ongoing Industrial Revolution! It swept Europe and North America at the end of the eighteenth century and then spread throughout the rest of the world. The most familiar word describing the consequence of the Industrial Revolution is "smashing", for better or worse. Industrial production has replaced the agricultural economy which had lasted for several thousand years, and the major residential pattern has also changed from a rural mode into an urban one, which has transformed the core of social structure from a family-centred one into a citizen-centred one. Such epoch-making revolutions have penetrated every field of our life, including politics, economy, society, culture, military, and art.

As to the motivations of the revolution, some people attribute it to the commercial revolution resulted from overseas expansion; while institutional economists tend to argue the establishment of ownership cleared the way for capitalism and population growth was the impetus for institutional innovation; and others assert a comprehensive interpretation, arguing the industrial revolution was pushed by several wheels, the front two wheels being "capitalism" and "nation-state", and the rear ones including overseas expansion, world market, and technical innovations, etc (Dong 2009). Among all these reasons, the most fundamental is commerce or "market", because the ownership protection, overseas expansion, world market and technical innovations all derived from it. Without business there would be no industry. Otherwise, to whom would goods be sold? In addition, what was the use of money if there were no goods? The humiliations China had suffered in the early modern time push scholars of several generations to ponder the question of why the industrial revolution did not happen in China. The long-term constraint of "commerce" sculptured a highly stable agricultural society, and even today, the major restricting factor for the ongoing "industrious revolution" is still the constraint of "market".

Interestingly, what the scholars are discussing is not just the motivations of the industrial revolution, but also the existence of the "Industrious Revolution" itself. The debate is still hot between those who argue in catastrophe theory and those who argue in historical continuity. Since the invention of the term "Industrial Revolution" by Louis-Auguste Blanqui in the early nineteenth century, doubts about it have never ceased. With the popularity of relativism and pluralism in the postmodern philosophical discourse, the word "Revolution" has received more and more criticism and almost become a pejorative word. The "Grand Narrative" in history has become a derogatory term, and the searching for "universality" have been considered as equivalent to the simplifying "Reductionism", and thus is taken as a poison for modernism. However, if we deny "universality" completely, the advancement of modern science will be difficult to continue, regardless of whether it is natural sciences or social sciences.

The comparability between the origins of agriculture and the industrial revolution was already noticed in the early twenty century by V. Gordon Childe, the leading authority in the field of prehistoric research. He called it "Neolithic Revolution" (Childe 2008) in which the emergence of agriculture was undoubtedly the core issue. But how can these two issues be compared?

First, like the study of industrial revolution, the study of the emergence of agriculture also demands the clarification of some concepts, such as what is agriculture and what is the relationship between agriculture and domestication. Only after that, we could go further.

Second, the emergence of agriculture experienced a far longer process compared with the industrial revolution. It took several thousand years for agriculture to finish the journey from the early practice of domestication to the establishment of an agroecosystem, while the time span of industrial revolution was just several hundred years. But compared with the previous millions of years of hunting-gathering life, a thousand year is just a moment. A longer time scale is the feature of archaeological research.

Third, commerce is the foundation of industrial revolution, and it is the commercial revolution that motivated the industrial revolution. In this sense, the proper name of the industrial revolution should be the "Industrial and Commercial Revolution". If commerce (market) is the "invisible hand" controlling the direction of society, what is the foundation of agriculture? In addition, what is the "invisible hand" that pushed the development of agriculture?

Fourth, institutional economists hold that it is the population growth pushes the institutional innovation, and these institutions prepared the environment for industrial revolution. Therefore, it is still commerce. In terms of the origins of agriculture, was there also a kind of institutional constraint? Through his study of the hunting-gathering economics, Marshall Sahlins (2009) proposed that the innate equalitarianism in this economy killed all the enthusiasm for extra production. How was equalitarianism destroyed? What was its relationship with agriculture? And how could we find out an answer from archaeological materials?

Fifth, the industrious society is a civil society. The peasants left their family-centered village, went to cities, and became society-centred citizens. Compared with the hunting-gathering society, the village society was based on agriculture. It is agriculture that made the previously mobile, scattered hunting-gathering groups settle down, and their territories, possessions and blood lineage became clearer. During this process, a series of changes happened. How could these changes be seen from the archaeological materials?

Sixth, the industrial revolution started from coal and iron, the most needed and most vulnerable chain in the whole system. What did agriculture start from? Grain crops, or luxury goods like spices? And why?

Seventh, there are theories of rural origin and city origin when talking about the industrial revolution. Are there any other sources for the emergence of agriculture? Why did it only start in certain areas? Among those hunter-gatherers, who first encountered a food crisis? And who had the conditions to start agriculture?

Eighth, the industrial revolution brought about social transformation, changes in the religious system and the moral anomie. From a hunting-gathering society to an agricultural society, did the society also go through a social transformation? Could it be detected from the archaeological materials?

By putting these two issues together, my intention is to draw clues from the previous studies and find out whether the questions that existed in the process of the industrial revolution also happened in the process of agriculturalisation, rather than emphasise the importance of the emergence of agriculture. The goal of comparisons is not to reach the so-called historical unity or an inevitable law of cultural development, but the knowledge of the changing process itself. From the exploration of the initiative conditions, mechanism, phases, and various forms of agriculture, we can not only have a better understanding of the prehistory, but also achieve a better understanding of our own reality today. The industrialisation in China today is growing vigorously and the country itself is currently in the process of modernisation. This endows a special significance to the study of origins of agriculture, making it significant not just for Chinese archaeology but also for the study of contemporary China, and is the major reason why this book is entitled "Prehistoric Modernisation".

Li Hongzhang[1] once said that what the late Qing Dynasty faced was "Great changes not seen for three thousand years". Actually, the changes early modern China had experienced was not seen for just three thousand years, but for thousands of years. China was not only one of the earliest agriculturalised lands, but also the origins of two crops, the millet crop in the North and rice crop in the South, which thereby had a profound impact on human history. In addition, China is one of the major centres of human civilisation and has experienced a history of over five thousand years, for which there are varied reasons, but no doubt the stable development of agriculture is an essential one. However, as the saying goes "It is always the same person to be praised and blamed", this super-stable social structure based on small-scale peasant economy was severely criticised and was considered the reason for the poverty and weakness of modern China. The different historical paths between the East and the West can

[1] Li Hongzhang (1823-1901) was an important politician and diplomat in the late Qing Dynasty. He played a significant role in Chinese history as a key figure in China's modernisation process and in China's interactions and negotiations with foreign countries. His life spanned the tumultuous and challenging late Qing period, leaving a profound impact on Chinese history.

be traced back even earlier. In order to understand their different situations today, we need to trace back to their prehistories. Today's China is experiencing the deepest revolution during its thousands of years' history. The small-scale peasant economy is crumbled, and the whole country has rapidly modernised. Around ten thousand years ago, it was in China that the first agricultural revolution started, thus ushered in a cultural prosperity that lasted for nearly ten thousand years that was only surpassed by Westerners in modern times. Under this circumstance, it is very necessary to recapture that history, that is, the revolution which inaugurated the great course of "Prehistoric Modernisation". At that time, it seemed our ancestors had a better performance than us today.

Every era and every place have their own particular problems, and every era and every place have their own corresponding academic research. Even to the same question, answers from different eras and places will be different. As a scholar of this generation, I always feel dwarfed in front of my Chinese predecessor's broad learning and my Western counterparts' great achievements. But this does not mean there is nothing I can do. What is more, even the greatest sage cannot solve all the problems of his time, Western scholarship is deeply rooted in their own contextual problems. If we can strike home the problems of our own country and of our own time, and think independently, we could possibly achieve something as well. Although a little superficial it might be, this study tends to serve the time.

In his masterpiece *A History of Archaeological Thought*, Bruce G. Trigger (2010) states that the development of archaeology is influenced by the contemporary social-historical background, social thoughts and the conditions of the technological development. Our curiosity for certain questions does not come from nowhere, and what we want to know and what we could know cannot be divorced from our reality. This idea coincides with Croce's famous argument, "Toute histoire digne de ce nom est histoire contemporaine". What I want to emphasise here is that not just our understandings towards history, but also the questions we are dealing with are all derived from our concerns of reality, even those "ultimate" questions in every discipline.

I do not quite agree with the relativistic view of history. Although my prehistorical study is deeply influenced by the contemporary reality, it does not mean that our research is purely a relativistic one. Different understandings among researchers do not mean the facts are relativistic. I don't want to develop a discussion on philosophical ontology here which

I am not good at. What I want to state is that the research paradigm of this study is a scientific one. It may not be perfect, but it is still a mode we could rely on today. The essential materialistic view and the limited conceptual rationality frame this study within the category of processual archaeology.

What Kind of Question Is the Origin of Agriculture

In his new book *Origins and Revolutions: Human Identity in Earliest Prehistory*, Clive Gamble criticises Childe's concept of "Neolithic Revolution", for its lack of contents. When agriculture first started in the core zone of Southwest Asia, there was no agriculture at all in many other regions. But it does not mean there were no strategies to counter the survival crisis, that is why the archaeologists in America and Oceania prefer to use the concept "origins of food production" (Gamble 2007). The key issue is how to define "agriculture", which I will discuss in the next chapter. Because of the ambiguity of concepts, Gamble thinks that "Neolithic Revolution" is an imagined issue.

Another questioning on the origins of agriculture is from Kent Flannery (Flannery 1973). As an archaeologist, he shares many similar views with the biologist David Rindos. They both hold that the origins of agriculture are natural processes, and that the relationship between prehistoric human beings and plants and animals are closely symbiotic. Domestication happened naturally during this process, so there was no "revolution" at all (Rindos 1980), nor there is a need to explain it.

Does it mean that "the origin of agriculture" is a pseudo proposition? Maybe we can also ask why there is such a questioning now but not previously. I have mentioned the intellectual environment of postmodern era, namely, certainty is generally criticised and is replaced by uncertainty, and single path is replaced by multiple paths. The study on the origins of agriculture is no exception. What is agriculture exactly? Its definition has become so obscure, and its forms have become so diverse that it seems impossible to summarise or clarify. As a science thriving in the twenty-first century, ecology has brought more demanding challenges to the study of this issue, in which revolutionary changes are dissolved in the common historical relations between man and life. Therefore, even if there was an agriculture, it was just one link of the long chains of interactions between human beings and their domesticated objects.

I want to respond to this questioning from two aspects. The first is the background of these questions. As I mentioned, specific time and specific place produces specific questions. Gamble's questioning is based upon an academic background of Western archaeology and the postmodern situation of Western society, and it is a need from a post-industrial society. The question of prehistoric modernisation may not be a major concern for Western archaeologists, or they do not need a "modernisation" perspective at all when they are studying the origins of agriculture. But due to the fact that China's industrialisation is still at its height, the issue of prehistoric modernisation becomes a particular question to be answered for Chinese archaeology. It is in this context that we study the origins of agriculture from the perspective of prehistoric modernisation.

The other question is the perspective of archaeology. Archaeologists are good at long-term observations and their research scope covers 99% of the human history. Interestingly, archaeologists are also doing well in micro-specific studies, such as a historic site or an archaeological remain (the remain of an ancient plant or animal). From different time scales, the origins of agriculture changed into different questions. From the micro-perspective, the emergence of agriculture is a slow and transitional change; if we broaden the time scale, we will find just within several thousand years human beings have abandoned their long habitual way of life, that is, hunting-gathering and replaced it with a new one. It is undoubtedly revolutionary and is a milestone in human history. Although contemporary archaeologists have discovered many subtle changes which had been previously unnoticed from more multiple perspectives, the origin of agriculture is still one of the fundamental changes, and could be juxtaposed with the origin of humans and the origin of civilisation. Nevertheless, in the field of Chinese archaeology, the study of this topic is far from adequate.

The questioning of the revolutionary character of the emergence of agriculture from contemporary archaeologists does not reduce the great significance of it. On the contrary, it encourages contemporary archaeological study on this issue. Compared with past studies focusing on the newly found domesticated species and the continuous findings of "the earliest" dates, contemporary archaeologists are more concerned with the various forms of the origins of agriculture, providing explanations for why there were no such changes in other places and making comparisons of the different reactions afterwards in different regions. The emergence of agriculture is no longer a question about the findings of the earliest domesticated species or the appearance of domestication. It signifies an overall change of the human cultural system. Therefore, the understanding of this process rather than the date-confirmation should be the focus of our archaeological study.

Similarly, the opposition to causal theories of agriculture, especially the criticising of the single cause theory, does not deny the relational factors in the emergence of agriculture. On the contrary, it pays more attention to the interrelations among human, animals and plants and the environment, and gives priority to the study of the adaptive history of different groups. Therefore, in the study of agriculture, more attention should be paid to the environmental elements, the interactions between human beings and other species, the different adaptive histories, and the characteristics in different regions. This knowledge is critically important for the explanation of why agriculture emerged in some areas, but not in others.

In brief, the origin of agriculture is an issue with various forms and complicated correlations. It is not a simple question of listing the domesticated spices or finding some earliest dates.

The Origins of Agriculture in China as a Question

As a question, there are two layers of meaning of the origins of agriculture in China: the first is the current research situation on the issue, the second is the composition of the issue. With regard to the development in the studies of the origins of Chinese agriculture, Zhang Chi (Zhang 2008) has made a brief review from the establishment of the local origin theory to the exploration of the rice crop theory, and the South-North double-centre theory. While there has been progression in this field, there are still a number of gaps to be filled due to the short period of its research history (Zhang took the founding of the *Journal of Agricultural Archaeology* as a possible starting point).

During its brief history, the research of the origins of Chinese agriculture has focused on the discoveries of the domesticated species. Their major concerns are the tracing of the earliest dates and the definition of the domestication features. The archaeological world in China has realised the deficiencies in contemporary studies.

The origins of agriculture and its earliest forms were part of the systematic development of ancient culture. If we cannot understand them from the overall context of the prehistoric culture, this research project cannot be penetrating. Agriculture is a product of the later phase of prehistoric culture, and it was only part of the human survival practice at its beginning and developing phase. Only with a thorough knowledge of the overall structure and the evolution of the food economy in prehistoric society can we probe deep into the questions like the mechanism, the originating process, and the initial forms of primitive agriculture, and the regional differences and their impact upon society (Zhang 2008).

In China, it is usually people working in the Neolithic study the issue of the emergence of agriculture, however, if we want to probe its cultural origins, we have to examine the life of our pre-agricultural ancestors, that is, the hunter-gatherers during the Paleolithic. In China, there is a clear division between Neolithic Archaeology and Paleolithic Archaeology, and Paleolithic Archaeology belongs to the category of natural science and is considered a branch of Geoscience. This peculiar reality inevitably harms the development of both branches, especially the study of the origins of agriculture. Evidently, if we know nothing of the living situation of those Paleolithic ancestors, how can we explain the origins of agriculture?

In the late Paleolithic prehistoric men were still living on foraging, therefore, we must go deep into their hunting and gathering life to find out the ins and outs of the origins of agriculture. However, this Paleolithic way of life has long passed, what is left are some mute stone artifacts and limited amount of animal fossils. This does not mean there is no way for the archaeologists to know the cultural system of those hunter-gatherers. The abundant research materials from ethnology can provide a sound referential framework. From the simple analogy of tool functions to the studies of dozens of hunter-gatherers' cultural systems, some fundamental principles could be distilled and they can contribute to the study of the origins of agriculture, but until now, this field of study is largely ignored by Chinese archaeologists. It is not just because we do not possess a database similar to that of Western countries which were built during their colonial expansion, but also because archaeology is under the discipline of history in China, which always feels reluctant to accept the archaeological study with an anthropological tradition. These two factors led to the missing of the study of the cultural adaptive mechanism and principles of the hunter-gatherers. Consequently, when dealing with the origins of agriculture, Chinese archaeologists are more ready to answer the "When" "Where" and "What" questions, but not the "Why" and "How" questions.

Once viewed from the cultural adaptation perspective, the beginning of agriculture is not just embodied in the domesticated species, it also indicates an overall reaction of the hunter-gatherers' cultural system. In other words, it is interrelated with many other elements in a cultural system such as the settlement patterns, tool technology, social structure, and ideology, which means archaeologists can study it from multi-perspectives and multi-angles. There are many layers in the cultural system of the hunter-gatherers, and different anthropologists have different divisions (Wilson 1997), but the basic structures are very similar. The economy is usually at the bottom level, then comes the social structure, the top layer is ideology-related factors. The foundation of the cultural system is the environment. In terms of the function of the environment upon the cultural system, generally speaking, it starts from the bottom level, then moves upward to the higher levels. In the case of agriculture, the origins of agriculture came first, then emerged the agricultural society, a series of agricultural etiquette, and ideology. It cannot be in the opposite order.

When the emergence of agriculture is viewed from another aspect, the word "China" will be our first concern. It is a modern political concept today. In prehistoric times, there was neither such a concept nor the division of territories, hence here it is mainly a geographical term, especially in the sense of natural environment. Compared with the Levant in Southwest Asia and a centre in the origins of agriculture, China is a vast domain and with an area ten times bigger. Its scope covers various habitats: from the cold zone of the coniferous forest to tropical rainforest, from boundless grassland to narrow basins, and from the plateau known as the roof of the world to the vast continental shelf. It could hardly happen that all those habitats were responding to one environmental or cultural transition synchronously, not even in a modern industrial society. Therefore, the origins of agriculture in China must be a question that involves many regions and contains a number of differences. A contrastive study among many regions is also a major perspective of this book.

The second key element of the question is the forms of prehistoric agriculture. In terms of prehistoric China, the agricultural forms must be various, and each form was based on the local environmental conditions and developed into the corresponding species composition, cropping pattern, and tool technology. In addition, we can find that the hunting-gathering societies outside the agricultural area also underwent a transition, some of them accepted the agricultural pattern, and some developed their own way of adaptation.

The last layer of the question is about the word "origin". Here it means from when agriculture or the embryonic forms of agriculture first started in China. The spatiotemporal framework is of course a reference, but here I want to emphasise another dimension, that is, how we should understand the emergence of a new phenomenon. We should reflect the speed and the degree of a change. As a new phenomenon, how was agriculture able to grow from nothing? How could it grow from a sprout? From what phase could it be called "agriculture"? These questions were seldom asked and were overshadowed by the more general discussions on quantitative change or qualitative change. Contemporary science has made some explorations on these questions at the methodological level, and archaeologists are trying to apply those achievements in their studies (Bentley and Maschner 2003). This book is an attempt of this kind.

In brief, these questions above are the major concerns in the study of the origins of agriculture in China, and this research is trying to expand the study a little bit more, especially through its explanation of the appearance or non-appearance of agriculture in certain areas from the perspectives of Paleolithic archaeology and its exploration of the cultural adaptation mechanism of the hunter-gatherers; Correspondingly, it will pay more attention to the various forms of agriculture in different regions and the processes they each had undergone. The questions also affect the methods and perspectives I will apply in this study.

Methods and Perspectives

The question of the origins of agriculture in China is not just a question for Chinese archaeologists, it also draws constant attention from scholars in the neighbouring countries and the West, as a result, Chinese scholars working on this issue cannot ignore the studies abroad. From a broader perspective, the emergence of Chinese agriculture is part of global panorama of agriculture, so studies abroad are an indispensable angle for the study of Chinese issue. Thanks to the Opening-up policy, the economic prosperity and the convenience offered by the Internet, we can now get access to most of the advances in this field without stepping outdoors, and the rich literature we can possess is beyond the imagination of our predecessors. Meanwhile, with the efforts of Chinese archaeologists of several generations, we have achieved a significant accumulation in both materials and studies. Together with breaking the thinking barriers, all these conditions provide us an important guarantee to carry out our research. Under this circumstance, our quest for the origins of Chinese agriculture should be open and pluralistic, and I hope it could integrate the Western archaeological achievements as well.

To carry out our research on the origins of Chinese agriculture, as analysis of the question's composition showed previously, one aspect of this research is the study of the hunter-gatherers' cultural adaptation mechanism, which means a theoretical exploration aimed at providing an interpretation in theory. Since the uprising of early modern science, theoretical exploration has become a key part in every discipline, but it has been long ignored in traditional Chinese academia. Just like the manufacturing of guns, traditional Chinese craftsmen emphassed personal experience and comprehension while ignoring the quest for ballistic principles and material physics, which eventually prevented the advancement of this technology. Theoretical research in contemporary Chinese archaeology is weak and even considered dispensable. On one hand, compared with obtaining and studying of materials, theoretical studies are considered superficial and impractical; On the other hand, the existing theoretical exploration deviates from the key issues in archaeology and is limited to the discussions on the features of archaeological materials (Chen 2012). The theoretical exploration should go deep into the interrelations among human behavior, culture and society, and even further to include the epistemological basis of archaeological studies.

It is not easy to study the hunter-gatherers in the Paleolithic, because archaeologists cannot observe their life directly. However, with the help of ethnological studies, contemporary archaeologists can use the ethnological data on hunter-gatherers to find out their cultural mechanism. Based on this, they can go further with their studies, among which modeling is an effective approach. With this method, we can imagine the life of some hunter-gatherers according to the environment variables; then based on the prediction of their cultural adaptation mechanism, we can find out which regions would encounter the surviving crisis first with the changes of those variables. Finally, we can compare the results of this prediction with the actual archaeological materials and find out what we can learn from it.

The analysis of the archaeological materials is essential. The difference between my work and that of my predecessors might be my relating of the Neolithic with the Paleolithic because of my Paleolithic background. The origins of agriculture should be sought from the Paleolithic, and that is why the two Ages should be studied together. As the hunter-gatherers in the Paleolithic lived on natural rather than cultivated products, the environmental background thus becomes particularly important. Consequently, the analysis of the archaeological materials should include studying of both the cultural artifacts and the environmental ecofacts, and the relationship between them.

The emphasis on the relationship between culture and environment determines the major perspective of this book: Cultural Ecology. Cultural Ecology is one of the two branches of Anthro-ecology (the other branch is Human Bioecology). Human beings are related to the environment through two channels: culture and physical constitution. Cultural Ecology studies the cultural means of adaptation (Sutton and Anderson 2004) sometimes we call it "adaptive strategy". It represents human's choice of their way of living in face of new environmental conditions, it is not a passive one, and the final choice is decided by both the natural environment and the cultural system itself, which includes tool technology, some long-term structures as "habitus" (Bourdieu 2003) put forward by Pierre Bourdieu, social organisations and ideology. In this sense, it is not quite fair that the post-processual archaeologists accuse their processual counterparts of overlooking the study of the initiative power of human beings.

With the aim to discover the origins of agriculture, the focus of my study will be the cultural ecology of the hunter-gatherers and some simple agricultural forms (like slash-and-born farming and horticulture). Different cultural adaptation strategies relate to different environmental conditions and different populations, which means the relationship between human and environment changes with the change of culture. On the flip side, the changes of environment bring about the changes of cultural adaptive strategies, and this influence was far more powerful in the Paleolithic than in the Industrial age which we are living in today.

Above is a summary of my perspectives. Research perspectives usually represent the research strategies of a researcher and show the merits of his study, but they also reveal its limitations—the absence of others. With these perspectives, this research tries to make up for the deficiencies of previous research, but it does not mean research from other perspectives is unreasonable. The emergence of agriculture is not as simple as the domestication of animals and plants, it is an issue that affected the whole prehistoric cultural system. Therefore, multiple perspectives rather than a single one is the innate demand of this topic. It is also shirking from questions if we just take it as a natural outcome of the development of technology and culture. Like untying a chain sleeve, the study of the origins of agriculture in China needs to be tried at different levels and from different angles, and this study is just one of these many attempts.

References

B. G. Trigger. *A history of archaeological thought*[M]. Cambridge university press, 1989.

C. Gamble, *Origins and Revolutions: Human Identity in Earliest Prehistory.* Cambridge University Press, Cambridge, 2007.

Chen, Shengqian, *The hierarchical issue of archaeological theory*[J]. *Southeast Culture*, 2012 (4).

D. Rindos, Symbiosis, instability, and the origins and spread of agriculture: A new model *Current Anthropology*, 21: 751–772, 1980.

Dong Zhenghua, *Fifteen Lectures on Global Modernization.* Peking University Press, 2009.

For instance, R. A. Bentley and H. D. G. Maschner eds., *Complex Systems and Archaeology.* University of Utah Press, Salt Lake City, 2003.

Julian Steward, Marvin Harris, Rappaport, and others have proposed their own cultural systems models. David Wilson provides a synthesis in his book *Indigenous South Americans of the Past and Present: An Ecological Perspective.* Westview, Boulder, Colo., 1997.

K. V. Flannery, *The origins of agriculture. Annual Review of Anthropology*, 2: 271–310, 1973.

M. D. Sahlins, *Stone Age Economics. New York: Aldine de Gruyter*[J]. 1972

M. Q. Sutton and E. N. Anderson, Introduction to Cultural Ecology. Altamira, Walnut Creek, 2004.

P. Bourdieu, *Lesens pratique*, translated by Jiang Zihua, Yilin Press, 2003.

Same as the previous book, p. 110.

V. G. Childe, *Man Makes Himself*. Watts, London, 1936. The Chinese version can be found in 《人类创造了自身》, translated by An Jiayuan and Yu Jingdong, Shanghai Sanlian Bookstore Co.Ltd., 2008.

Zhang Chi, *Research on the Origin of Agriculture in China*[J], in *Century Review of Archaeological Research in China: Archaeological Volume of the Neolithic Age,* Science Press, 2008. 102–111.

Contents

List of Figures

List of Tables

CHAPTER 1

Theoretical Exploration on the Origins of Agriculture

Humans follow laws of the earth,
the earth follows laws of the heaven,
the heaven follows laws of the Dao,
the Dao follows the natural order.
—*Laozi*

1.1 Introduction

From whichever perspective one considers, emergence of agriculture is a monumental event in the human history. The advent of agriculture holds decisive significance for subsequent human evolution and cultural development, just like the emergence of bipedal locomotion, tool making, language, and the industrial revolution. For instance, cities cannot emerge without agriculture, as cities rely on agricultural surplus production, which poses a direct causal relationship. Moreover, the emergence of agriculture indirectly led to the invention of characters, as the development of writing is reliant on the agricultural surplus to sustain a class liberated from physical labour. In this case, agriculture impacts every facet of human life, from survival and reproduction to modes of dwelling, population distribution, kinship, social organisation, and ideology. Therefore, agriculture transcends being merely a livelihood revolution or a technological invention; it represents a comprehensive reorganisation of society and culture, giving rise to an entirely new cultural system within a new

S. Chen, *Prehistoric Modernization*,
https://doi.org/10.1007/978-981-97-9423-2_1

historical era. Within this transformation, agriculture acts as an "order parameter", which determines the direction of the cultural system (this concept will be elucidated in the section "New Theoretical Paradigm"). This is why, since the nineteenth century, the origin of agriculture has remained one of the three core questions in archaeological research. Over more than a century, vast materials and experiences have been amassed in this field. With the continuous emergence of new theories, methods, and materials, the exploration of the origin of agriculture continues to deepen.

Based on archaeologists, the study of the origin of agriculture is guided by three fundamental reasoning structures: analogy, induction, and deduction (Chen 2007). Analogy primarily emerges from experimental archaeology and ethnography. Classic works, such as *Prehistory of Agriculture: New Experimental and Ethnographic Approaches* (Anderson 1999), have proposed numerous insightful solutions to the issue of agricultural origins. Induction, which emphasises the discovery and analysis of archaeological materials, is the predominant research approach with the assistance of material evidence. Deriving from existing theories, the deduction could answer the questions about the origin of agriculture. When formulating theories in the deductive method, material support is unnecessarily required. In the history of agricultural origin research, classic theories include the Oasis Theory proposed by Childe, the Hilly Flanks Hypothesis by Robert Braidwood, the Marginal Zone Theory by Binford, the Population Pressure Theory by A. Belfer-Cohen, the Ecological Interaction Theory by Flannery, and Feasting Theory by Brian Hayden, which are not outcomes of inductive archaeological analysis. However, these theories lacked comprehensive factual support at their inception, thus, they possess the nature of being fact-free. Nevertheless, theoretical research could offer rich resources for reference, vital creativity, and the ability to maximise the subjectivity of archaeologists. This research method could also provide directional guidance for subsequent research, enabling a macroscopic understanding of issues and revealing principles and mechanisms of developmental processes. These theories, serve as a foundation for further studies in environmental archaeology, demographic archaeology, cultural ecology, and social differentiation, enriching perspectives on the study of agricultural origins. Building on previous research foundations, this chapter incorporates new advancements in theory, methodology, and materials to develop a novel theoretical model for explaining agricultural origins.

1.2 What is Agriculture?

Before defining agriculture, it might be essential to consider how to define and classify industrial and industrialised nations nowadays. Undoubtedly, precisely defining an industry is challenging. Generally, the industry refers to the large-scale machine production that emerged in the late eighteenth century, distinguishing it from traditional handicrafts. Throughout the twentieth century, there have been several new technological revolutions, including automation and the application of microelectronics, which further complicates the concept of industry. In the present world, even the least developed countries have some level of industry. Countries like China, with relatively developed industry, still cannot be regarded as industrialised nations, as a significant portion of their population is engaged in agriculture and their level of urbanisation has not reached that of industrialised nations. Therefore, an industrialised nation encompasses more than just industrial production and its output; instead, it involves transformations in the entire social structure, changes in the way of life, and revolutions in the realm of ideology.

In this case, how should we define agriculture? First, we need to clarify some fundamental conceptual distinctions among domestication, cultivation, food production, and agriculture. Figuring out these concepts is beneficial to identify the specific issues that need addressing. Regarding the study of agricultural origin, the difficulty often stems from the lack of clarity in relation to the object of explanation (Harris 1989, pp. 11–26).

Domestication primarily constitutes a biological process in which animals and plants increasingly rely on humans during their reproductive cycle, leading to genetic and physiological changes. Domestication is not limited to the human-animal and human-plant relationships; similar domestication relationships exist between organisms such as tropical leaf-cutter ants and fungi, as well as ants and aphids (Haim 2004). Domestication might not always be intentional; the prolonged interactions among humans, the wild ancestors of plants, and animals can unconsciously give rise to domestication. Taking plants as an example, the transition from wild to domesticated forms is a gradual process. Fully domesticated species, in the end, might lose their ability for self-seeding (Persall 1995, pp. 157–192) or they may lose traits, such as dormancy and physical and chemical defences against herbivores. For certain genetically labile plants, deliberate selection might not be necessary; through the repetitive cycle of harvest, cultivation, and harvest, a series of wild characteristics can naturally fade away, leading to dependence on humans (Blumer and Byrne 1991). From a human perspective, domestication can be viewed as the initial phase of controlling the reproduction of animals and plants. As long as humans stably exploit a species with genetically mutable traits, which can be inherited, it can result in the domestication of the species.

In contrast, cultivation represents a distinct cultural phenomenon, wherein plant cultivation involves intentional practices, including soil preparation, sowing, weeding, irrigation, bird and animal repellent, harvesting, and storage. The process of cultivation requires a range of technical conditions and labour inputs. Similar to animal husbandry, cultivation necessitates the deliberate alteration of the human-animal relationship, that is, involving human control over the scope of animal activities. Correspondingly, adjustments to human living space and temporal patterns are required. The cultivation of both animals and plants, which includes animal husbandry, does not necessarily indicate the completion of domestication; rather, it might still be in progress. Cultivation constitutes a defined human action characterised by purpose and organisation. While cultivation can potentially lead to domestication, it does not inevitably result in species domestication.

Food production represents the delineation of human livelihood states, involving changes in land usage and shifts in human social structures and organisations. Agriculture, in turn, represents a further process of matured food production, encompassing not only changes associated with food production but also the formation of a new ecosystem (Rindos 1980) and the emergence of novel social organisation and structure. It is often referred to as subsistence methods, which involve utilising a small number of domesticated plants and animals, as food production, or sometimes as "low-level food production" (Smith 2001), which could be represented by the Formative phase in North America. Slash-and-burn horticulture and simple root crop cultivation can also be included in the realm of food production. In both livelihood models, the range of domesticated plants and animals is limited; compared to the domestication of grains like wheat, rice, and maize, less labour input is required. Classic examples include the Dani people in the Grand Valley of the Baliem River in the highlands of New Guinea and the Lozi people in western Zambia (Sutton and Anderson 2004). In these societies, hunting and gathering remain important means of acquiring food.

Possessing various types, agriculture commonly refers to forms primarily centred around cereal crop production and animal husbandry, which marks as a significant variant. Animal husbandry heavily relies on animal rearing and encompasses livelihoods based on animal meat and by-products such as milk, blood, fur, hide, and even dung. Known ancient agricultural forms include terraced agriculture, which covers ancient Saudi Arabian bunded terraces, Peruvian terraces, etc.; raised field agriculture in South America; raised-bed planting, which includes chinampa in Mesoamerican agriculture, coral reef fields in the Pacific; dryland farming or rainfed agriculture; and irrigation agriculture that supports various major centres of civilisation. Animal husbandry could be seen as the pastoralism in the Eurasian grasslands that centred around rearing horses, cattle, and sheep; Arabian Peninsula's pastoralism focused on camels and sheep; and Africa's pastoralism centred around cattle rearing. In the colder regions of the Northern Hemisphere, populations primarily rely on reindeer herding for livelihoods. However, since reindeer are perpetually in a semi-domesticated state, it cannot be categorised as a distinct form of animal husbandry.

Drawing upon the historical and ethnographic evidence of agricultural and food production practices, these subsistence methods exhibit significant distinctions from hunting and gathering. The former involves the production and consumption of domesticated animals and plants, while the latter revolves around the hunting and gathering of wild animals and plants. The differences manifest in settlement patterns, kinship relationships, social systems, and ideologies, representing fundamental distinctions within the entire cultural framework. Thus, food production and agriculture could pertain to the classification of cultural systems. In comparison, cultivation is a cultural behaviour, whereas domestication is a biological one. It is undoubtedly that food production and agriculture rely on cultivation, which also represents an evolutionary phase towards food production, that is, food production invariably starts with the cultivation of a certain species (Harris 1996, pp. 1–9). Therefore, through conceptual analysis, at least three main questions need to be addressed:

First, the formation of biological characteristics through domestication, primarily studied by archaeologists who are specialised in animal and plant remains.

Second, the driving forces and environmental conditions that led humans to alter their relationship with wild animals and plants, initiating purposeful behaviours aimed at influencing the growth of animals and plants.

Third, the process of reconfiguring cultural systems involved in the production and consumption of domesticated animals and plants.

For archaeologists who mainly focus on the origin of agriculture, the second issue is the most valuable, which involves explanations of human's attempts at altering their relationship with animals and plants and tracing the entire process of changes. In essence, regarding the origin of agriculture, the fundamental theoretical issue is the origin of cultivation (Keeley 1995, pp. 243–272).

Occasionally, hunter-gatherers also tend to care for young animals, clear weeds around frequently gathered plants, and even engage in water irrigation for certain plants (Pringle 1998). Cultivation behaviour has long existed within the cultural systems of hunter-gatherers, and cultivation is not unfamiliar to them. Just as market exchange behaviour has existed within various cultural systems and phases of cultural development, but only became the foundation of Western modern societies after the Renaissance. When studying industrial and industrialised nations, one cannot isolate the study to a specific indicator; rather, one must examine changes across various levels of the entire social system. Similarly, regarding the origin of agriculture, the fundamental issue is to discover and explain the holistic adaptive changes in the cultural system, instead of merely seeking occasional evidence of animal and plant cultivation. How did the behaviour of animal and plant cultivation evolve within the cultural system? How did the behaviour influence the structure of the entire cultural system? Why could animal and plant cultivation lead to agriculture in certain regions but not in others?

In its origin phase is not fully developed, thus agriculture is closer to food production. The terms "food production" and "primitive agriculture" are often interchangeable in the text, which should be distinguished.

1.3 Classical Theories on the Origins of Agriculture

The Oasis Theory: The theoretical study of the agricultural original could trace to the Oasis Theory proposed by Gordon Childe. Previously, Childe was not only regarded as the earliest archaeologist to propose the theory of agricultural origins but also the founder of the Oasis Theory. In current academia, archaeologists recognise that the origins of this theory are attributed to another individual (Watson 1995, pp. 21–38), namely Rafael Pumpelly, the leader of the Central Asiatic Expeditions in 1904. In 1908, Pumpelly suggests that the arid conditions at end of the Late Pleistocene drove people to seek refugia in oasis regions to utilise the remaining plant resources, consequently leading to plant domestication (Pumpelly 1908). There are three points could be derived from Pumpelly's theory: firstly, environmental aridity (drastic change or deterioration) resulted in reduced natural resources; secondly, the proximity of humans and species, which is a prerequisite for domestication, thus his theory is sometimes referred to as the "propinquity theory"; thirdly, due to the insufficient natural resources, people needed to increase the yield of natural species, by utilising human's desires.

Childe's Oasis Theory initially appeared in his work *The Most Ancient East* (Childe 1928); while its dissemination primarily occurred through his well-known book *Man Makes Himself* (Childe 1936). Childe proposes that West Asia experienced arid climatic conditions during the transition from the Late Pleistocene to the Early Holocene, leading both humans and animals to inhabit areas around limited water sources, that is, river valleys and oasis zones. Eventually, through utilising grains and straws, certain animals were domesticated. Childe did not explicitly mention that humans also domesticated plants in this environment, but later researchers attributed this aspect to Childe's theory. As a result, subsequent scholars associated the Oasis Theory with Childe, suggesting that both animal and plant domestication began in "oasis" regions. From his perspective, it is evident that plant domestication preceded animal domestication. Childe argued that plant domestication possibly originated in the Nile Valley or Palestine, considering the creators of the Natufian Culture as the earliest plant cultivators.

Childe and Pumpelly share similar key points, that is, environmental driving forces compose the fundament of domestication. Human-species proximity and the undeniable human capacity and willingness resulted in domesticating animals and plants, thereby favouring agriculture over hunting and gathering. Childe's works further indicated that agriculture would rapidly disseminate once it emerged, and other groups would readily adopt it accordingly.

In the early twentieth century, when Childe introduced his theory, the origin of agriculture was not a primary concern for archaeologists. Thus, there were scarce archaeological materials related to agricultural origins, resulting in his research primarily theoretical deduction. Childe's contribution lies in his theoretical framing of the issue and proposing potential answers, thereby offering a conceptual framework for subsequent researchers to reference.

Hilly Flanks Hypothesis Proposed by Robert Braidwood: Braidwood stands as the pioneer in multidisciplinary research on the origin of agriculture. His investigations into agricultural origins began with the excavation project at the Jarmo site, which is located in the Zagros Mountains in Iraq, under the auspices of the Oriental Institute of the University of Chicago from 1950 to 1951. The University of Chicago is the birthplace of radiocarbon dating. Willard Libby, who was awarded the Nobel Prize in Chemistry, facilitated Braidwood with limited radiocarbon dating data. This enabled Braidwood to recognise that the earliest agriculture might have originated in elevated areas with favourable water conditions, which was termed by Braidwood as "hilly flanks of Fertile Crescent", instead of valleys or oases, as Childe had suggested. Braidwood proposed that in this zone, evidence of domestication of crops like wheat, barley, legumes, along with sheep, goats, and cattle, would likely be found. Moreover, he anticipated the presence of village sites like Jarmo, showcasing early food production practices (Braidwood 1960).

The years of 1954 and 1955 marked the official initiation of multidisciplinary collaborative research led by Braidwood, involving zoologists, botanists, and geologists. Geologist Herbert E. Wright Jr. made a significant finding during this period, determining that glaciers existed in the Zagros Mountains during the Late Glacial period, where the Jarmo site was located, contrary to earlier assumptions of early Holocene aridity. Moreover, no major climate changes between the inception of early food production and the present can be found. The research conducted by Herbert E. Wright Jr. compelled Braidwood to seek explanations beyond environmental shifts. As a consequence, cultural factors emerged as the primary explanation for the origin of agriculture. The continuous advancement of human knowledge and technology led to a gradual realisation of the potential for domestication of plants and animals in the hilly flanks region. This realisation eventually gave rise to the invention of food production practices.

Braidwood's Hilly Flanks Hypothesis shares a commonality with Childe's Oasis Theory, as both propose that agriculture is a progression and an inevitable outcome in human history, rooted in humanity's capacity to this technology invention. However, Braidwood's perspective diverges by considering agriculture as a leap in human understanding and accumulation over an extended period, giving rise to emergent cultural features. In contrast, Childe views agriculture as a technological innovation that is fundamentally driven by the need to address food scarcity. The key distinction lies in Braidwood's belief that agriculture would emerge in the optimal zones of survival for the wild ancestors of domesticated species, rather than Childe's notion of refuges like valleys and oases, serving as shelters for both humans and species.

The Marginal Zone Theory proposed by Binford: In 1968, Lewis Binford published a paper titled *Post-Pleistocene Adaptation*, introducing his theory of agricultural origins in peripheral zones (Binford 1968, pp. 331–341). He contested the notions proposed by Childe and Braidwood about the birthplace of agriculture, arguing that it was not the valleys and oases as Childe suggested, nor the hilly flanks as Braidwood proposed, but rather certain periphery areas. Drawing upon ethnography and cultural ecology, Binford posited that hunter-gatherer populations inhabiting optimal zones like hilly flanks could maintain a dynamic equilibrium between natural resource availability and population growth. However, this balance might be disrupted by environmental deterioration or population growth, such as when one group encroached upon another's territory. The disrupted hunter-gatherer group might splinter, with some of its population moving away from the optimal zone to settle in peripheral environments. This arrangement ensured that the optimal zones remained within the limits of environmental carrying capacity. Meanwhile, the hunter-gatherer groups relocating to the periphery areas could easily exert pressure on food resources, resulting in an imbalance that eventually led people to intensify the use of certain promising species, triggering the domestication of animals and plants.

Later, Kent Flannery further refined this theory by introducing the concept of the "broad spectrum revolution" (Flannery 1986). Flannery observed a trend exists in the characteristics of archaeological materials during the Upper Paleolithic and early Holocene periods, where there was an increase in the diversity of species exploited by humans, especially various small animals that were challenging to capture, as well as plant seeds that required more human labour to process. Flannery regarded this as a pre-adaptation to agriculture origins. Flannery endorsed Binford's theory and proposed that plant domestication might have begun with conscious cultivation activities among hunter-gatherers in peripheral regions. He also suggested that these hunter-gatherers aimed to enhance grain density in these areas, attempting to achieve distribution levels comparable to those in optimal zones.

Binford's theoretical exploration of agricultural origins undoubtedly goes further, delving into the mechanisms of origin and discussing population growth as a major variable. In 2001, Binford published the book *Constructing Frames of Reference*, providing a comprehensive discussion of the impact of population factors on different types of hunter-gatherer groups (Binford 2001). On the other hand, Flannery's focus lies in studying the process of agricultural origins, instead of considering agricultural origins as a black box but instead revealing each step of its occurrence. Their theories challenge the stereotypical perspectives that agriculture is an inevitable and progressive development, while proving that agriculture originated as products of pressures and adaptations.

Theory of Coevolution Proposed by Rindos: The perspective of David Rindos differs from the aforementioned archaeologists in that Rindos focuses on explaining the co-evolutionary mechanism between humans and plants; rather than pays attention to the reason of original of agriculture in certain regions at the end of the Late Pleistocene. Based on Rindos, it lacks true significance to probe why humans established a co-evolutionary relationship with plants, much like questioning why certain ants co-evolved with aphids or why certain bird species co-evolved with specific fruits. Such co-evolutionary relationships represent outcomes of long-term adaptations, allowing species to achieve maximum fitness under specific spatiotemporal conditions. The occurrence of domestication is neither inevitable nor driven by human subjective desires; it simply occurred (Rindos 1984).

Rindos categorises this co-evolution into three phases: The first phase is incidental domestication, where humans bring about morphological changes in certain species through harvesting, protection, or accidental seed dispersal. In archaeological remains, the evidence of certain species might be more abundant than others or their distribution extending beyond their native habitats. The second phase is specific domestication, where humans become the primary dispersal agents for certain plants. Through practices, such as burning, clearing, and cultivation, humans create anthropogenic locales, resulting in further morphological changes in these species, which could be seen as larger seeds, thinner husks, and altered dispersal mechanisms. Archaeological materials often reveal coexistence of wild and domesticated species. The third phase is agricultural domestication, characterised by a significant reduction in species diversity and stabilisation of species morphology. As sedentism intensifies, changes occur in settlement patterns, eventually giving rise to agricultural ecosystems. With continued population growth or encountering famine, agricultural populations may expand outward again (Rindos 1984).

Flannery partly supports the viewpoint held by Rindos, emphasising that agriculture is an extension of early broad spectrum adaptation. Flannery argues that searching for specific reasons for the origins of agriculture is unnecessary (Flannery 1986). Deborah Pearsall, a plant archaeologist, also aligns with the coevolution theory held by Rindos (Persall 1995, pp. 157–192). However, there are also critics who question Rindos for lacking exploration of the mechanisms behind the origin of agriculture. They raise concerns about the reasons for coevolution only occurring in certain places and the means of its spread (Redding 1988). Others criticise Rindos for overlooking artificial selection, which indicates the purposeful cultivation practices launched by humans, suggesting that cultivation should precede domestication rather than the reverse (Blumer and Byrne 1991).

Rindos contributes to our understanding of evolutionary mechanisms, emphasising an ecologically reciprocal perspective on the relationship between humans and plants. Within this mutualistic mechanism, plant domestication takes place. Humans are not intentionally domesticating specific crops, instead, their role is to foster the growth of certain crops.

Feasting Theory Proposed by Hayden: Brian Hayden's theory is undoubtedly one of the most unconventional. He critiques all theories suggesting that agriculture originated due to population pressure or resource scarcity in certain areas (Hayden 1995, pp. 273–299), suggesting that agriculture should have emerged in regions with relatively advanced technology capable of generating abundant and stable resources. In such areas, the accumulation of individual awareness, including the concept of personal wealth, would lead to competitive feasting, thus, people would use this practice to develop, expand, or consolidate their power and prestige. Within this environment, the earliest domesticated species, such as dogs, gourds, chilli peppers, and avocados, appeared rapidly and spread (Hayden 1990). Afterwards, Hayden extended his theory to other technologies. Hayden argued that practices, which include metalworking, pottery, arts, and animal and plant domestication, all began as techniques aimed at establishing prestige and authority, transforming into practical technologies over the past 30,000 years (Hayden 1998). This perspective seems to stem from observations of contemporary society, where many practical technologies have originated from non-practical domains, such as the military. The purpose of labour is not merely survival but achieving success.

The pursuit of prestige is crucial for the success of reproduction, including procreation. In recent years, Hayden has further posited that competitive feasting not only holds significance in terms of reproduction but also of production. After analysing risk-reduction strategies among hunter-gatherers, Hayden contends that complex hunter-gatherers differ significantly in their strategies from simpler ones. Due to the possession of certain surplus production, complex hunter-gatherers develop a need for storage; otherwise, their surplus would rot unused. One of their methods involves animal husbandry, converting surplus into future meat sources. Another method is the establishment of social relationships through storage, transforming excess food into useful commodities, services, debts, or relationships. Taking Southeast Asia and Polynesia as examples, Hayden illustrates how feasting can create "social security nets". This approach enables individuals to mitigate risks related to livelihood, reproduction, and social conflicts, thereby enhancing overall fitness (Hayden 2009).

The Feasting hypothesis is predicated on the emergence of non-equalitarian hunter-gatherer societies, sometimes referred to as complex hunter-gatherers. As Sahlins states, "affluent foragers" store resources in nature and lack the concept of personal ownership (Sahlins 2009)[1] whereas non-equalitarian hunter-gatherers disrupt this structure. Hayden does not elucidate the reason for the disruption of the social structure, which persisted for millions of years. Nevertheless, he does offer a novel perspective on the study of agricultural origins from a developmental standpoint.

1.4 Theories on the Origins of Agriculture: A Critical Thinking

There are numerous theories regarding the origin of agriculture, leading to diverse and inconclusive viewpoints. How should we approach these theories? What are the foundations or premises upon which these theories are built? What is their logical reasoning? How should we ultimately comprehend the issue of agricultural origins?

Firstly, the debate revolves around whether it was the economic foundation or the superstructure that led to the origin of agriculture. This is a focal point of contention. While most theories lean towards examining the economic foundation, a minority of theories focus on the superstructure, exemplified by scholars like Hayden and Barbara Bender (Bender 1978). The question of why ancient hunter-gatherers engaged in agriculture raises the query of whether it was for immediate subsistence or for societal relationships. Upon closer analysis of these theories, it becomes evident that they stem from fundamental differences in philosophical backgrounds, that is, the dichotomy between materialism and idealism. Materialism accentuates external constraints, with the external world serving as the starting point for the problem. In contrast, idealism underscores human agency, with subjective human desires constituting the key for the problem. This philosophical dichotomy influences the perspectives of archaeologists, thus, resolving this issue requires more than the insights of archaeologists alone.

Secondly, the question arises whether the origin of agriculture was a natural transition or a revolution. Different from other scholars, Rindos and Flannery evidently argue that the origin of agriculture is less revolutionary. It is readily observed that significant social disparities between the long-standing hunter-gatherer subsistence and the agricultural societies formed over thousands of years. However, Childe, who introduced the notion of the "Agricultural Revolution", failed to explain the revolutionary significance of agriculture as a transformative change in a mechanistic perspective. On the other hand, transition theorists struggle to explain why, after millions of years of continuous hunting and gathering, agriculture emerged much later, around 10,000 years ago, in regions like Western Asia, China, and even later in the New World. Hayden, who emphasises socio-economic reasons, is essentially a transition theorist, employing the accumulation of population and technological knowledge to pinpoint the specific timing of agricultural origins (Hayden 1995, pp. 273–299). Conversely, understanding the multi-millennial process of agricultural origins remains a challenge for revolution theorists. As a completely new human adaptation, why did agriculture emerge?

Thirdly, the debate revolves around whether the driving forces behind the origin of agriculture were internal or external factors. Early studies by scholars such as Pumpelly and Childe leaned towards external factors, like environmental climate change. Subsequent researchers have strived to avoid a deterministic environmental stance, favouring the notion of environmental possibility. However, since the twenty-first century, a resurgence in emphasis on environmental factors has emerged. P. J. Richardson et al. noticed that agriculture did not emerge during the Upper Paleolithic but rather abruptly appeared in the Holocene, despite the advanced technological culture emerging in the Upper Paleolithic. As a consequence, scholars argue that the climate of the Late Pleistocene was entirely unsuitable for agriculture. New climate data also indicate that the climate during the Last Glacial Maximum (LGM) was arid, with low carbon dioxide concentrations and highly unstable conditions, rendering it practically impossible for ancient populations to successfully develop agriculture (Richerson et al. 2001). In contrast, the rapid maturation and spread of early agriculture around 8000 years ago coincided almost precisely with the stable climate of the Holocene Climatic Optimum (HCO). Corresponding to the external environmental deterministic view, the research stems from internal factors. For instance, while Braidwood considers agriculture as an outcome of the accumulation of knowledge and technology, Binford and other scholars view it as a result of the accumulation of population; and Hayden regards it as a product of socio-economic development.

Fourthly, the question arises whether agriculture emerged as a result of favourable environmental conditions or as a response to challenging circumstances. Braidwood supports the view that agriculture emerged within favourable environmental situations, positing that the origin of agriculture should be rooted in the optimal habitats of wild progenitors of domesticated species. Similarly, Hayden suggests that only regions with abundant resources could have facilitated the accumulation of surplus production by foraging societies, thereby giving rise to nonegalitarian complex hunter-gatherer societies. On the other hand, leaning towards Thucydides' style, scholars, who advocate asserting that agriculture emerged as a response to challenging circumstances, argue that moderate challenges are most conducive to cultural innovation and technological invention. Consequently, technologies emerge in response to threats. Childe and Binford, for instance, argue that agriculture emerged as a response to population pressures and the need to address resource scarcity. In relation to this issue, whether agriculture originated in the core habitats of wild domestication progenitors or in peripheral regions should be addressed. Additionally, should the origin of agriculture be linked to the optimal habitation areas of hunter-gatherer societies, or does it align with the edge habitats?

Fifthly, is a universal explanation regarding the origin of agriculture attainable? The models presented by scholars from Childe to Hayden offer different kinds of generalised framework, attempting to reveal the mechanisms and reasons behind the emergence of agriculture. Drawing upon Popper's concept, P. J. Watson refers to the agricultural origin as a covering law (Watson 1995, pp. 21–38). Can scholars discover a covering law akin to those found in the natural sciences when it comes to studying the origin of agriculture? Clearly, regular laws cannot be applicable in human societies, which are composed of complex phenomena. Post-processual archaeologists emphasise interpretation, which is defined as an understanding developed from time, geographical, and gender perspectives. This suggests that the origin of agriculture might not await discovery of a law but rather entails interpretation from the standpoint of modern society. Furthermore, based on scholars who specialise in archaeological research of specific regions, there exists a considerable gap between universal theories and regional archaeological evidence. A universal theory that encompasses archaeological material from various regions is essentially absent, as diverse timelines, species variations, and environmental and climatic differences associated with the agricultural origins vary greatly, seemingly rendering universal theories inadequate to explain regional particularities. Notably, emphasising particularities does not necessarily negate the exploration of universality; otherwise, it is easy to fail to comprehend the adaptive mechanisms of prehistoric human cultures.

Sixthly, a challenge exists in explaining the origin of agriculture lies in how to perceive different factors and their interrelationships. Pumpelly and Childe regard climate change as the fundamental driving force of agricultural origin; Richardson also consider climate conditions as determinative factors. Binford and other scholars see population as the most crucial factor, while Hayden notes socio-economic elements and emphasises singular driving causes. Rindos, among other researchers, incorporates the coevolutionary relationship between humans and plants to explain the origin of agriculture. It seems that Rindos considers the role of ecosystems; however, he is still confined to a biological perspective, regarding it as a primary approach to analyse agricultural origins. In the past decades, an increasing number of studies have embraced an ecological systemic approach, striving to comprehensively consider various factors. Binford and Rindos have already embedded the perceptions on ecological systems in their scholarly works. Meanwhile, the developments in system philosophy, which compasses synergy, chaos theory, and self-organisation, offers new perspectives to contribute to the understanding of the origin of agriculture. For instance, Binford employed self-organisation theory in his recent research (Binford 2001).

Lastly, divergent viewpoints exist in research methodologies. Scholars, like Binford, advocate for scientific objectivity, highlighting explicit presuppositions to emphasise central issues within the paradigm of processual archaeology. Similarly, Ofer Bar-Yosef, who also investigates agricultural origins, suggests that the historical narrative method is equally effective, as providing detailed accounts of events serves as a valid form of explanation (Bar-Yosef 1998). Within archaeological research methodologies, the most significant division currently lies between the scientific and humanistic approaches, prominently manifesting in the paradigms of processual and post-processual archaeology. Thus, two questions arise: Can a scientifically objective analysis of the agricultural origins be conducted? What constitutes a valid and reasonable analysis? Studies rooted in the natural sciences, like those of Rindos and paleoethnobotanist D. M. Pearsall, posits that human subjective intent is a constant, thus no need to be considered; instead, the focus should be emphasised on the gradual formation of domesticated species. In contrast, archaeologists emphasise changes in human intentions. They contend that domestication was bound to happen sooner or later, once deliberate cultivation commenced. The crux lies in the reasons for humans abandoning long-standing hunting-gathering practices but in favour of food production afterwards. How can this immense gap in archaeological epistemology be bridged?

Processual archaeology emphasises that archaeological research should evolve into a science akin to geology, gradually moving beyond empirical materials. In this case, archaeological research could recognise the developmental patterns of human culture and society through the exploration of natural laws (Binford 1972). However, the past two to three decades of archaeological development have proven this path to be untenable. Archaeological research deals with human societal phenomena, traversing a dual interpretive process from the formation of archaeological materials to their discovery by archaeologists, and further to the archaeologists' examination of these materials to derive insights about human society. In contrast, only a singular interpretive process is involved in the interpretation of natural phenomena leading to natural laws. The uniqueness of human societal phenomena constitutes the foundation of modern philosophical hermeneutics. Creativity, which composes the nature of human activities, unbounded necessitates an "interpretation from within" or a "re-experience" to comprehend past behaviours, being similar to the archaeological research advocated by phenomenological archaeology (Tilley 1994). Building upon this foundation, post-processual archaeology emphasises "reading the past" (Hodder 1986). In this case, archaeological materials are regarded as "texts", which are used to repeatedly read and interpret, fundamentally rejecting the idea that human history possesses any inherent patterns.

However, situated between the polarities of seeking patterns and cultivating understanding, an intermediate state exists, that is, comprehension of cultural evolutionary mechanisms. A certain mechanism is entailed in the development of any entity, from the simplest bacterial division to the most intricate human societal phenomena. During human societal evolution, mechanisms are not equivalent to patterns, as strong contextual conditions are involved in complex social phenomena. In this case, due to differing contextual backgrounds, vastly divergent outcomes will be attributed under the same mechanism. Nevertheless, it is undeniable of the existence of mechanisms within social evolution merely because of dissimilar outcomes. This situation is analogous to warfare. For instance, *The Art of War* written by Sun Tzu elucidates some fundamental mechanisms governing the conduct of warfare (though some people might prefer "mechanisms" as "patterns", these terms fundamentally differ from universally applicable natural laws). Yet, rigidly employing mechanisms can often lead to defeat, just as deviating from them can. Victory belongs to those who both adhere to the mechanisms of warfare and possess the ability to read the timing and situation.

The study of the agricultural origins involves the adaptive mechanisms of hunter-gatherers. Our analysis encompasses three main aspects:

Firstly, the mechanism of how hunter-gatherers respond to pressures, such as population growth, resource risks, and environmental changes.

Secondly, the evolution of adaptive mechanisms among ancient hunter-gatherers, reaching a point where agricultural origins become plausible, and the interrelation between these mechanisms and the general coping strategies of hunter-gatherers.

Thirdly, the initial conditions that led to the emergence of agriculture and how these advantageous conditions interact. The prevailing circumstances should be perceptive in this case.

The interconnectedness of these three aspects is crucial, that is, understanding the transition from hunter-gatherer coping mechanisms to the ones that gave rise to agriculture, and how the initial conditions for agricultural emergence influence these mechanisms. Clarifying these aspects is essential for achieving a comprehensive understanding of the mechanisms underlying the emergence of agriculture.

Understanding the mechanisms behind the emergence of agriculture equips us with the ability to predict the trajectory of cultural development, whether in prehistoric societies or our contemporary culture, much like accurately anticipating the events based on the understanding of natural laws. Comprehending these mechanisms is similar to mastering *The Art of War*, as it becomes a vital source of guiding future actions, particularly insights into the ecological relationship between hunter-gatherers and other organisms, their responses to environmental changes, and the impact of population pressures on hunter-gatherer societies, which all provide numerous cautionary lessons in this industrial or post-industrial society. Bacon's adage "histories make men wise" holds true because history presents a myriad of possibilities. Yet the ultimate direction we take depends on our own actions as humans.

1.5 Seeking a Universal Explanation

1.5.1 A New Theoretical Foundation

The establishment of theories is grounded in foundational assumptions of significance. For processual archaeology, explicit premises, which stand as one of the fundamental requirements, clarifies the basis upon which subsequent conclusions are drawn. Simultaneously, the process of reasoning needs to be lucid, facilitating a clear understanding of the research and averting both overly generalised assertions and the inherent pitfalls of evading critique.

At the end of the twentieth century, the development of science presented a new panorama characterised by emerging holistic perspectives that are represented by ecological science and complexity science. The new complex theories cover dissipative structure theory, synergy theory, chaos theory, and hypercycle theory, replacing conventional control theory, information theory, and general systems theory. This shift provided a deeper understanding of the ways of alterations. The application of complex systems theory to social science research is no longer a novelty; within the field of archaeology, this trend is still burgeoning. For instance, the publication of *Complex Systems and Archaeology*, which was edited by R. A. Bentley and H. D. Maschner in 2003, exemplifies this NEW trend (Blumer 1996, pp. 25–50). From three perspectives, the following discussion delves into the relationship between complex systems theory and the origin of agriculture.

(1) Non-Equilibrium Nature and Discontinuous Evolution

Undoubtedly, the origin of agriculture constitutes an evolutionary phenomenon within the human ecological system, whether through transition or revolution, or the cumulative development of cultural evolution. Within the research on the evolutionary framework of agricultural origins, whether cultural development is a purposeful transformation by humans or a process of natural selection constitutes a central question. The former represents a Spencerian perspective, deeply rooted in notions of progress; while the latter aligns with Darwinian views, emphasising natural selection propelling cultural change (Rosenberg 1990). Teleology and progressivism are evidently problematic, as they inherently require other theories to substantiate them, and the history of evolution or human practice cannot validate their rationality. The Theory of Evolution proposed by Darwin underscores the randomness of evolution, critiquing the notion of an undisturbed nature. Based on Darwin, in contrast to cultural change, nature is not entirely passive and static in the face of transformation; human interference in nature is not necessarily linear or continuously deteriorating (Blumer 1996, pp. 25–50). This could lead us to understand that pre-agricultural hunter-gatherers did not exist in an untouched purely natural state. Instead, hunter-gatherers significantly influenced their environment. Some scholars argue that hunter-gatherers engaged in "move to produce" (Politis 1996), during which process they continuously altered landscapes through the mobile foraging and dispersing of plant seeds. Particularly after the adoption of fire, hunter-gatherers frequently employed it to clear living areas (Pyne 2006), resulting in considerable impact on the natural environment.

For a long time, archaeologists have regarded the origins of agriculture as a cumulative interaction process between plants, animals, and human populations. This process is perceived as linear, devoid of revolutions or abrupt changes. Similar viewpoints are evident in the aforementioned theories proposed by Rindos and Flannery, as well as in the research of scholars like MacNeish (MacNeish 1992). Based on their research, adaptation is a slow, gradual accumulation process, thus, agriculture naturally emerged without further explanation. In contrast, the new evolutionary perspective emphasises the punctuated equilibrium during the evolutionary process, which alters between slow and steady evolution and rapid and explosive growth, resulting in an evolutionary trajectory resembling stair steps (Gould 2008). This new evolutionary perspective does not simply depict the origin of agriculture as a single revolution or a linear gradual transition. Instead, scholars, who hold this viewpoint, suggest that the origin of agriculture might involve several revolutionary changes, each potentially marked by leaps of development. This macro-evolutionary view finds support in the context of the Neolithic Revolution in the Near East, where the origin of agriculture is seen as a product of the interplay between long-term evolution and revolutionary leaps (Zeder 2009).

(2) Synergetics and Broad spectrum adaptation

Possessing distinct characteristics, synergetic research investigates various systems, which are composed of numerous subsystems. This research analyses how these subsystems cooperate to create spatial, temporal, or functional structures on a macroscale. Its focus lies in the loss of stability, which indicates identifying the time, conditions, and domains of stability loss, and determining the role of order parameters and their driving principles (also referred to as servomechanism or governing principles) in the process. A classic example is the formation of lasers, where a system undergoes a phase transition far from equilibrium, termed non-equilibrium phase transition. Subsystems exhibit competitive and cooperative traits, which Hermann Haken termed "synergetics" (Haken 2001).

An order parameter is a crucial concept within synergetics. Regardless of the system, Haken argues that the level of order, which is termed an order parameter, in new structures could be reflected, if a parameter undergoes a transition from absence to presence during the system's evolution and indicates the emergence of new structures. On one hand, the order parameter emerges as a result of collective motion of numerous subsystems within the system, being characterised by both competition and cooperation. On the other hand, once the order parameter forms, it assumes a governing or directive role over the subsystems, dominating the overall evolutionary process of the system.

For the cultural systems of hunter-gatherers, sustenance has historically relied on the growth of plants and animals, with the process of food production in early agriculture undergoing a transition from absence to presence. The emergence of agriculture fundamentally impacted the entire human societal system, as it facilitated settled living, altering the nomadic lifeway of millions of years, while also giving rise to new settlement patterns and the predominance of new technologies, such as pottery and ground stone tools. Within agricultural societies, significant changes in kinship structures, social organisations, religious rituals, and ideologies took place. From an organisational perspective, agriculture supported the emergence of complex civilisations. Ideologically, while hunter-gatherer societies often held animistic beliefs, agricultural societies saw the rise of organised religions. With the establishment of agriculture, the subsequent developments of writing systems, states, and metalworking followed suit. Thus, for a hunter-gatherer cultural system existing in a critical and non-equilibrium state, food production serves as its order parameter. Hunting shaped the formation of hunter-gatherer societies; industry and commerce contributed to the modern societies. Similar to these, under the dominion of the order parameter, significant shifts in social structure occurred. Those living in contemporary China can profoundly grasp the tangible experience of these profound shifts in times.

Complex systems exhibit synergistic reactions at critical states. According to the principles of synergetics, once a particular decision is adopted, it implies the exclusion of other choices. Various possibilities would be tested through dramatic changes within systems in critical states. As asserted by P. Bak "The critical state is a very good state… because, in general, this is when we perform best" (Bak et al. 1996). A system, which is situated at a critical state, undergoes a break in symmetries, leading to critical fluctuations. At this point, numerous possibilities emerge, and the system is the most vibrant during this phase.

In the process of agricultural origins, Flannery identifies a phase known as the "broad spectrum revolution", which he deems a pre-adaptation process to the emergence of agriculture. In fact, the broad spectrum revolution reflects the response of hunter-gatherer cultural systems at a critical state. hunter-gatherers expanded their diet, incorporating plants with lower yield returns, maximising the utilisation of all available natural foods within a limited area. From this perspective, the broad spectrum revolution represents a hunter-gatherer strategy of maximising utilisation.

However, it must be acknowledged that broad spectrum adaptation represents an unstable system state, as it possesses relatively low levels of adaptation. Generally, higher levels of adaptation indicate a narrower and more specialised diet (Dobzhansky 1968, pp. 109–122). After the broad spectrum revolution, numerous choices were eliminated, and certain species began to be intensively utilised. Simultaneously, the diet narrowed, and the level of adaptation increased. For instance, at the Abu Hureyra sites, over 160 plant species were identified from the end of the Upper Paleolithic, whereas in the Pre-Pottery Neolithic period, only around eight to nine plant species were utilised, with over 90% of plants being discarded (Hillman and Harris 1989, pp. 240–269).

(3) Self-Organisation Theory and the Emergence of Agriculture Origins

The cultural system of hunter-gatherers, as the precursor to agriculture, is a self-organising system. A self-organising system refers to a system that can spontaneously organise, generate, and evolve without specific external instructions. It has the capacity to autonomously transition from disorder to order, creating self-structured systems (Wu 2001). For self-organising systems, which are situated at a critical state, even minor variations in initial conditions can lead to significant changes in its states, resulting in the emergence of new structures and forms of organisation. A commonly used analogy is the "butterfly effect", which implies that a butterfly flapping its wings in the Amazon rainforest could trigger a hurricane in North America. As P. Bak pointed out, instead of gradually forming, things emerge through revolutions, as dynamic systems constantly generate in critical states (Bak et al. 1996). Similarly, in human's evolutionary process, complex adaptations are rarely generated from a single leap, but often require multiple intermittent breakthroughs, with accumulated changes eventually leading to revolutionary events. The same pattern applies to the emergence of agriculture, involving numerous factors such as species, environmental climate, tool technology, social organisation, ideology, etc. Conditions for its development require an extended period. Additionally, the full establishment of agriculture also spanned thousands of years, encompassing several significant leaps.

The process of agricultural origins operates as a cyclical driving force. The reduction in mobility among hunter-gatherers necessitates the diversification of food sources or the intensification of food production, both of which contribute to expanded regional population and land coverage, thereby promoting sedentism. During this cyclical process, various factors promote iterative development, continuously elevating the level of agriculture. This progression evolves from rudimentary food production to complex agricultural ecosystems, displaying a spiral ascent structure, often referred to as "super-cycles" (Eigen and Schuster 1990). The cyclical progression not only continuously ascends but also enhances complexity. Moreover, multiple cycles interact and mutually reinforce each other. For instance, amid the end of the Last Ice Age, the extinction of large animals rendered hunting unsustainable, compelling a heightened emphasis on plant gathering and a subsequent reduction in mobility for hunter-gatherers. Concurrently, advancements in technology and the accumulation of knowledge facilitated intensified plant utilisation. Meanwhile, the accumulation of Late Palaeolithic population and increased social complexity bolstered the regional concept of land ownership, making mobility increasingly challenging. Numerous interconnected cycles perpetuate the emergence and escalation of food production, ultimately leading to the establishment of agriculture.

According to the Chaos Theory, the systematic evolutionary trajectory displays diversity, encompassing at least three distinct paths: Firstly, the evolutionary path that passes through critical points or critical regions, leading to phenomena occurring along this trajectory are the most intricate and complex, akin to turbulent rapids. Thus, the outcome of evolution is exceptionally hard to predict, with minor perturbations potentially causing significant fluctuations. Secondly, intermittent evolutionary paths feature substantial peaks and valleys, often accompanied by abrupt changes. While most evolutionary pathways within this context are predictable, merely certain areas or structures remain unpredictable. Thirdly, the gradual evolutionary path is adopted by the majority of entities in evolution. The evolution process involves smooth transitions without significant changes, and it is largely predictable (Wu 2001). All three of these evolutionary paths aforestated can be observed in the process of agricultural origins. The first scenario is evident in the initial emergence and development of food production, where the transition from food production to agriculture is uncertain and could either materialise or be abandoned. The second scenario is observed during the gradual formation of agriculture. The third path characterises the ongoing refinement and enhancement of agriculture once it reaches a mature phase. When investigating the origins of agriculture, particular attention is directed towards the first path. To comprehend this scenario, it is essential to explore the adaptive strategies of hunter-gatherers, both short-term and long-term mechanisms, thus facilitating a better understanding of the inevitability and contingency of agricultural origins. This, in turn, contributes to a deeper grasp of the mechanisms underlying the emergence of agriculture.

1.5.2 Long-term Trends in Hunter-gatherer Adaptation

Archaeology, as a discipline benefiting from temporal depth, holds the advantage of analysing extended periods. This allows us to identify variables, which act slowly yet endure significantly, and detect the crucial implications for human evolutionary history. Without a holistic perspective, relying solely on materials from a specific site or period makes the detection of variables challenging. Furthermore, the unveiling of long-term trends assists us in viewing agricultural origins from a perspective of self-organising development. That is, agricultural origins can be understood as novel structures that emerged from cultural systems breaking their critical states, potentially arising without explaining through external factors. Furthermore, accumulated long-term trends could explain why the earliest instances of agricultural origins occurred around ten thousand years ago rather than earlier or later. Over the course of millions of years, humanity underwent a lifeway of hunting and gathering. This extensive temporal process grants the necessary temporal depth to examine long-term trends, thereby facilitating a more comprehensive understanding of the subject.

(1) Human Food History: Inherent Strategy of Energy Maximisation

Undoubtedly, for humans, suitable food is a product of long-term evolution. Regardless of the differences or preferences that may exist, humans share the same physiological mechanism, forming the basis of this discussion. The more humans adapt to a certain food, the more it is required, and the longer its history of consumption may exist. Tracing the evolution in physiology and psychology can assist in reconstructing the fundamental history of human diet. Following contemporary nutritional recommendations, the sequence of foods required by humans, from most to least, is fruits and vegetables, proteins, which include fish, lean meat, poultry, and eggs, grains, fats, and sugars (Fig. 1.1). The history of human ancestors consuming natural fruits and vegetables can be traced back to earlier primate ancestors. Variations exist in human adaptation to different types of proteins: adaptation to low-density lipoproteins, which could be found in red meat, is less favourable than to high-density lipoproteins, aligning with human evolutionary history. Relatively speaking, the hunting of large animals occurred later than the utilisation of certain aquatic resources, like shellfish and turtles, due to their slow mobility, thus ease of capture. Existing archaeological evidence indicates that the human population may have been astonishingly reduced to less than 10,000 in the early phases of modern human origin (around 200,000 years ago) (Marean 2010). The findings from the PP13B cave in South Africa suggest that humans might have relied on shellfish to survive this significant crisis. Similar evidence is present in other regions of Africa, such as the discoveries of harpoons and shells in Katanda (Brooks et al. 1995). However, the peak of large animal hunting occurred much later in the Upper Paleolithic, which is well after the human survival crisis, thus, explaining why human adaptation to low-density lipoproteins is better than to high-density lipoproteins.

Compared to protein-rich food, energy density of grains, as the origin of agriculture, is increased through human processing. The popular high-protein diet is founded on this premise. Domesticated animals offer great fat content, while wild terrestrial animals typically have lower fat content. High-energy food, such as sugar, represents innovations in historical periods. Thus, contemporary nutrition support practices gradually advocate minimal consumption of sugar. The widespread obesity in modern society is closely associated with excessive intake of high-energy

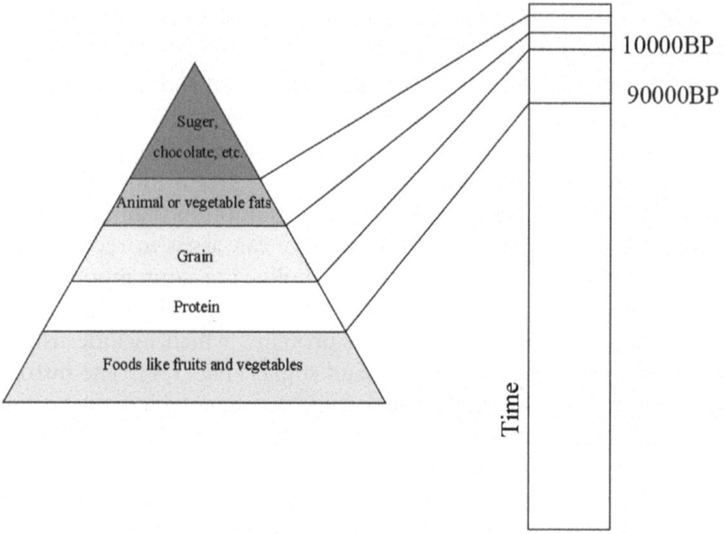

Fig. 1.1 The evolution history of human food

food, which contradicts human physiological adaptations to such dietary patterns.

Correspondingly, from a psychological standpoint, humans have consistently been inclined to seek high-energy foods. Clearly, higher energy density can reduce time of seeking food and enhance reproductive opportunities. A glimpse into the evolutionary history of human diets reveals an enduring pursuit of energy maximisation. From this perspective, the origin of agriculture can be viewed as the outcome of the human strategy for energy maximisation.

(2) Mobility—Unlocking the Understanding of hunter-gatherers

Hunter-gatherers rely on hunting animals, gathering plants, and utilising aquatic resources for sustenance. The emergence of hunter-gatherer societies represents a certain phase of human evolution. The advent of hunting is of paramount importance, as it might be a parameter for the formation of prehistoric societal systems and the structure of the entire cultural system, determining people's technological tools, modes of habitation, social organisation, division of labour, and ideology. It is notably that the division of hunting as solely a male activity and gathering as a female activity is an oversimplification. In a broader sense of hunting, beyond the act of animal hunting, women also participate in all phases, from preparation and trapping to butchering and processing, endowing females with the nature of hunters as well (Brumbach and Jarvenpa 1997).

It is easily inclined to perceive hunter-gatherers through the lens of the agricultural era, assuming that they always lived in one place, much like settled communities. However, the most pronounced characteristic of the hunter-gatherer lifeway is mobility, as searching for prey and gathering plants are necessary. Simultaneously, a specific location typically can merely support one hunter-gatherer group during certain seasons, while it is rarely capable of sustaining the population year-round. Over-exploitation might lead to the inability of recovery of flora and fauna to their original densities, potentially resulting in the collapse of the entire ecosystem. Furthermore, sedentism implies that a group resides in one place throughout the entire year, although certain members might venture out for hunting (Rice 1975), entailing transporting all food back to the residential camp. Nevertheless, hunter-gatherers lack draft animals for carrying loads. Therefore, instead of expending substantial effort to carry back game, it becomes more practical for humans to settle down to where the food source is located. This constitutes another reason why hunter-gatherers must maintain a mobile lifeway.

The mobility of hunter-gatherers serves not only to acquire food but also to gather information about food sources. Clearly, food resources are not consistently available, prompting hunter-gatherers to accumulate information as a reserve, much like a bank, so that when necessary, they can move to locations with food rather than waiting until hunger compels them to seek sustenance. According to a conventional understanding, hunter-gatherers lack the notion of property, thus, do not accumulate reserves. From the perspective of hunter-gatherers, all food is stored in the natural environment and is accessed as needed. Mobility, which is the process of collecting such information, enables them to acquire advanced knowledge of the distribution of food. Thus, mobility represents a risk mitigation strategy employed by hunter-gatherers.

During the process of mobility, whenever hunter-gatherers arrive at a new location, they engage in practices, which include tree felling, surface clearing, burning of vegetation, waste accumulation, and discarding of fruit pits. In such cleared spaces, the sunlight and nutrients resulting from the clearing, coupled with the absence of competitors, allow seeds to rapidly germinate and flourish in the soil. Over time, the diversity of accessible plant species increases, gradually altering the vegetation landscape through inadvertent human intervention. Hence, it can be argued that mobility also constitutes a form of production (Politis 1996).

Mobility holds significant implications for hunter-gatherers, including the aspect of reproduction. Through mobility, individuals are able to come into contact with various groups spread across vast regions, facilitating the exchange of members and the sharing of genetic material, thereby preventing inbreeding.

Furthermore, mobility brings about other ancillary benefits, such as avoiding the accumulation of waste at sedentisms, preventing the proliferation of mosquitoes and water source contamination. These issues mentioned often lead to the spread of infectious diseases, which constitute significant challenges for sedentary agricultural societies. In essence, for hunter-gatherer communities, survival is intrinsically tied to mobility. That is, mobility, as the fundamental survival strategy, serves as the key to understanding the way of life of hunter-gatherers.

Certainly, mobility also carries evident negative consequences, with the most prominent impact being population limitation. High mobility does not permit a fertile woman to bear too many children. Typically, she can only carry one child while on the move, leaving the other children to walk on their own, necessitating a significant age gap among children. In this case, contraception is achieved by prolonging breastfeeding. Moreover, there are other means of population control that lead to an average fertility rate of around 4 children per fertile woman, despite an actual reproductive potential of 12 to 15 children. Coupled with various accidents and illnesses, the number of individuals who can survive is quite limited. By contemporary standards, population growth occurs at an exceedingly slow pace (Divale 1972).

Once mobility is restricted or diminished, the population growth can rapidly increase, posing significant challenges to the sustenance of hunter-gatherers. Deprived of mobility, hunter-gatherers experience a substantial reduction in the spatial range of available resources, leading to a rapid overexploitation of resources and significantly heightening the vulnerability of their livelihoods.

(3) Hunter-Gatherers' Concept of Territory—A Gradual Clarification of Ownership

Conventionally, archaeologists consider primitive hunter-gatherer societies to be highly egalitarian, devoid of hierarchical distinctions and property concepts, with the characteristic of extensive sharing. Marshall Sahlins proclaimed hunter-gatherers to be the "original affluent society", which indicates hunter-gatherers held possessions in common and avoided individual ownership, instead of the possession of a great deal. In this case, this absence of ownership meant no surplus production and, consequently, no incentive for surplus production's development (Sahlins 2009). In the hunter-gatherer societies, the absence of ownership went throughout the practices of hunting-gathering and agricultural production. The cultural structure of hunter-gatherers constrained the origin of agriculture, which is featured with surplus production.

However, the egalitarianism of hunter-gatherer societies is more mythical than factual. The idea of collective ownership within a group does not imply an endless capacity to share with any and all groups. Groups have their territories, and under certain circumstances, they might allow related or closely associated groups to utilise these territories. The concept of territory is not unique to hunter-gatherers; many animals exhibit distinct territorial behaviour, especially among predators. Higher primates also exhibit clear territorial behaviours. The presence of territories is crucial for the survival of individuals and populations, and territorial defence is one of the main reasons behind conflicts.

The concept of territory among hunter-gatherers is evolving. In the Early to Middle Paleolithic s, at least in terms of lithic assemblage styles, the concept of territory was not particularly apparent. However, by the Upper Paleolithic, lithic assemblage styles began to exhibit distinct regional trends. H. Martin Wobst proposed that the increased population in the Upper Paleolithic prompted the need to restrict the scope of reciprocity, transitioning from general to constrained reciprocity. This evolution led to the gradual formation of territorial concepts, manifested in the regional stylistic variations of lithic materials (Wobst 1976).

The awareness of territory would inevitably limit the mobility of hunter-gatherers. With a continuous increase in population, it is evident that the territorial range of each group might diminish. Typically, when a hunter-gatherer group surpasses a certain size, it would splinter, with the newly formed group relocating elsewhere. As the groups keep expanding, some might be compelled to move into more peripheral environments. This phenomenon led Binford to propose The Marginal Zone Theory agriculture origin. In reality, reduced mobility is a more direct and widespread outcome. Specific strategies must be adopted for hunter-gatherers, who reduced their mobility, to address their subsistence challenges.

(4) Development of Technology and Livelihood Strategies

Human cultural adaptation begins with technological advancements, which are characterised by an accumulative nature. In the process of technological transmission, each generation contributes creativity and innovation, resulting in technological progress. Therefore, excluding cases of isolated populations over prolonged periods, the subsequent generation's technology tends to be more advanced and intricate than the previous one. Currently, based on Grahame Clark's framework, known lithic technologies can be categorised into five groups: Oldowan technology, Acheulean technology (represented by handaxes), Mousterian technology (represented by flakes), blade technology, and microlithic technology (Clark 1977). Advancements in lithic technology are evident in the increase of high-quality stone material availability, efficiency in stone utilisation and blade production; moreover, it also could be seen in the emergence of composite tools, enhanced portability of lithic tools, and the broadening of their adaptability range. The evolution of lithic technology showcases a trend favouring the mobility of hunter-gatherers. Heightened mobility implies that hunter-gatherers can exploit resources over a wider range, particularly animal resources.

The adaptive strategies of hunter-gatherers also exhibit accumulative development. In the earliest phases, hunter-gatherers survived by actively seeking out food resources, allowing people to seek the food rather than vice versa. Subsequently, another strategy emerged: allowing food to come to people. This involved establishing relatively fixed residential camps where multiple groups were dispatched for hunting and gathering, followed by bringing the food back to the camp. The former is referred to as the "forager", while the latter is known as the "collector" (Binford 1980, pp. 1–17). The distinction between these two groups marks a landmark and holds environmental contextual relevance. The forager strategy tends to prevail in regions with evenly distributed resources, whereas the collector strategy is more prevalent in areas with uneven resource distribution. In comparison to the forager strategy, the collector strategy exhibit lower mobility rates. In addition, collectors necessitate and invest more in constructing durable dwellings and tools. Concurrently, reduced mobility leads to higher birth rates, larger group sizes, and sedentism dimensions among collectors. However, these phenomena lead to increased population pressure or a greater potential labour force.

From the perspective of livelihood, hunter-gatherers could be differentiated into three main forms: hunting-oriented, gathering-oriented, and fishing-hunting-oriented. This distinction is closely related to local resource conditions and possesses cultural evolutionary significance. Generally, gathering economies are the earliest, followed by hunting economies, while fishing-hunting economies are the latest to emerge. Hunting can be categorised into opportunistic hunting and effective hunting, with the latter potentially appearing in the Late Upper Paleolithic (LUP) (M. H. Nitecki and D. V. Nitecki 2005). Fishing, excluding shellfish collection and slow-moving animals (such as turtles) hunting, requires more complex technologies, including boats, harpoons, and fishing nets. As a result, fishing emerged the latest. Rarely does any hunter-gatherer group solely rely on one mode of subsistence; rather, economies tend to be mixed, with varying degrees of emphasis. Different subsistence focuses correspond to different tool technologies and social organisations. Fishing economies could sustain the highest population density, followed by gathering economies, and then hunting economies with the lowest. According to Binford's theory, when population pressure predominantly affects hunting-oriented groups, hunter-gatherers might undergo a transition towards gathering economies. Similarly, gathering-oriented groups under population pressure might shift to fishing economies; if in the absence of fishing-hunting resources, they would need to intensify the use of plant resources (Binford 2001). Consequently, agriculture, which is centred on crop cultivation could only have originated from gathering-oriented hunter-gatherer groups.

(5) Sustained Population Growth

According to contemporary studies on human evolution, the phenomenon of sustained population growth happened relatively recently. Modern humans once faced a population bottleneck, coming close to extinction. However, thereafter, populations began to increase steadily: modern humans spread out from Africa, exploring the vast landmasses of Eurasia, as well as the unexplored territories of Oceania and the Americas. Whether this expansion was driven by population pressure or the successful radiation of adaptation, a continuous increase in the total population could be seen. As humans continue to enter new spaces that have never been utilised by humans before, population density can remain constant or even dilute. However, with all continents suitable for human habitation, two potential outcomes of population growth are presented: an increase in population density or the migration of hunter-gatherers to the peripheral areas of their subsistence territories.

Archaeological evidence from the Upper Paleolithic already demonstrates a significant rise in population density, leading to localised resource pressures. The research on multiple sites in the Mediterranean region during the LUP conducted by Mary Stiner indicates a shift in human hunting targets. hunter-gatherers transitioned from collecting slow-growing and sluggish animals, like shellfish and turtles, to rapidly growing and agile ones, like rabbits and grouse. It is evident that the labour investigation required for hunting the former is much smaller than for the latter. However, overexploitation of the former is easy to achieve, compelling a shift towards more challenging food sources (Stiner 2002). Population pressures could be a major factor driving overexploitation.

The expansion of populations into peripheral zones encompasses migrations to cold regions, such as Siberia and Alaska, utilisation of desert fringe areas, and the extension into extreme environments. Current archaeological evidence also indicates human presence in the marginal areas of the Qinghai-Tibet Plateau during the LUP (Goebel 1999; Brantingham and Gao 2006). Humans might have also ventured into tropical rainforest regions. Nevertheless, some anthropologists raise doubts about the adaptive capacity of hunter-gatherers in such harsh environments. The expansion of humans into extreme environments likely stemmed from population pressures; even if other reasonable explanations exist, this behaviour further contributed to population growth. If unfavourable conditions arise, populations tend to concentrate in regions with relatively abundant resources, much like the inhabitants of the western deserts of Egypt migrating into the Nile Valley.

For humans, factors that have historically constrained population growth include predation by hunters, limitations on total food availability, diseases, and the nomadic lifeway. These factors bear resemblance to the ones limiting the expansion of other animal populations. By the LUP, or the earlier era when modern humans left Africa, the development of technological culture, language, and social organisation conferred upon humans a comprehensive adaptive advantage over other species, positioning them at the top of the food chain. The threat posed by predators to humans was largely eradicated, and the only predators humans had to contend with were themselves in the form of warfare and conflict. Compared to contemporary human societies, population growth rates for early hunter-gatherer societies were markedly slower and possibly fluctuating; the enduring trend was one of gradual accumulation and growth.

(6) Agency: The Development of Human Spirit

In contemporary archaeological research, agency constitutes a prominent area. Archaeologists are increasingly dissatisfied with the fact that humans are solely portrayed as animals. They argue that humans are not mere appendages of systems, slaves to structures, or passive adaptors to their surroundings; rather, humans are autonomous agents. Closely aligned with the essence of humanity, agency offers a more profound understanding of intrinsic human attributes. Devoid of agency undoubtedly overlooks the very essence of humanity. However, the domain of agency presents formidable challenges for research, thus leading to its long-standing state of suspension.

Human agency arises from the evolution of the human spiritual realm. Humans' spiritual realm represents a transcendence, transformation, or even rebellion against the tangible world, propelling individuals to challenge natural laws. As a result, humans acquire knowledge of, adapt to, and utilise these laws. Without the formation of the spiritual realm, humans would have remained akin to other animals, passively conformed to nature and lacked the capacity for comprehension and mastery. Thus, the human spiritual realm bestows autonomy upon individuals, fostering the development of creativity. The process through which the human spiritual realm develops remains relatively obscure. Steven Mithen introduced the analogy, suggesting that distinct functional brain modules, akin to the various chambers of a cathedral, contribute to the formation of the human spiritual realm once complete interconnectivity between these "chambers" is achieved. Thus, each brain module possesses its unique developmental history (Mithen 1996).

A pivotal piece of evidence for the emergence of modern humans lies in the appearance of adornments and artworks, notably flourishing during the LUP. In this case, scholars have postulated an emergence of the "LUP Revolution" (Bar-Yosef 2002). During this period, in terms of physicality and cognitive abilities, humans exhibited little disparity from contemporary counterparts. The primary manifestation of the LUP revolution resided in the development of the spiritual domain, encompassing cognitive advancement and the origination of rudimentary religious concepts. The advent of agriculture would have been improbable without cognitive progression, the rupture of dietary habits, and the exploration of novel food sources. Thus, the origin of agriculture also represents a breakthrough in human cognition and the spiritual domain, underscoring the manifestation of human agency.

All these trends mentioned are interconnected. If they synergistically collaborate, the cumulative impacts could undergo magnification, resulting in a phenomenon known as the "avalanche effect". Nevertheless, there are instances where these trends counterbalance one another, for instance, advancements in technology and societal organisational levels might offset a portion of population pressures. Notably, not all trends play identical roles in the process of agricultural emergence. The mechanisms through which could contribute to the genesis of agriculture will be further explored in subsequent discussions.

1.5.3 Coping Mechanisms of Hunter-Gatherer Pressures

Following the discussion on the long-term adaptive trends of hunter-gatherers, the emergence of agriculture can be regarded as a result of self-organised development within cultural systems. Conversely, numerous archaeologists consider the origin of agriculture as a response of cultural systems to pressures, particularly population or resource pressures, which might be induced by environmental changes, with this response occurring around 10,000 years ago.

Throughout history, when archaeologists have deliberated the pressures confronting hunter-gatherers, the focus often rests on the temporary resource uncertainty brought about by unpredictable environmental and climatic changes, that is, risks. Additionally, the other aspect is constituted by population pressures, which are resulted from demographic growth, leading to sustained resource scarcity.

Regarding resource uncertainty, P. Weissner categorised four risk reduction strategies among hunter-gatherers (Weissner 1982), an elaboration later expanded into eight categories by Hayden:

a. Control of resource loss (e.g. burning weeds, transplantation, and off-season harvesting), which increases actual yields or establishes exclusive territories to ensure group provisioning;
b. Sharing and redistribution within the group, including reciprocal sharing, intergroup or intra-group feasting based on reciprocity, and redistribution through authoritative figures or institutions, which is akin to tribal chieftains.
c. Establishment of regional mutual aid alliances, allowing movement to ally territories for sustenance. The ally establishments are based on kinship networks, shared totemic ancestors, frequently visited relationships, shared names, meat-sharing relationships, and exchange of ornaments and prestige goods;
d. Creation of adversaries, expelling them when necessary;
e. Storage, although its utility is relatively limited in highly mobile livelihoods.
f. Simple food exchanges;
g. Development of risk-reduction techniques, such as microblade technology;
h. Diversification of livelihoods and broadening of the diet (Hayden 2009).

The composition of population pressures is intertwined with the natural growth of the population, and it is additionally associated with environmental degradation, which manifests as a decline in the carrying capacity of the environment, a reduced population that a given area can support, or in other words, a diminished spatial extent conducive to human habitation. In comparison to transient resource uncertainties, these changes persist over a longer timeframe. For instance, as the LGM concluded, the adaptive resources of both flora and fauna diminished, leading to their depletion and extinction.

Animal ecology has extensively studied the mechanisms underlying population pressure. If the population of a given animal species continues to increase, there is an inevitable potential for surpassing the carrying capacity of the environment, subsequently necessitating the consideration of population regulation. Animal ecologists identify two modes of regulation: density-independent and density-dependent. The former, labelled as "catastrophic", pertains to factors such as climate conditions, pollutants, and changes in concentration of hydrogen ions that influence mortality rates regardless of population density. The latter, which entails changes in population size, links to increasing population density, encompassing factors, including inter- and intra-species competition, predation, parasitism, and disease. Animal ecological theories expound a range of mechanisms underlying density-dependent regulation, including inter-specific interactions, intra-specific factors, and food availability. Inter-specific factors encompass predation, parasitism, and inter-species competition for shared resources. Intra-specific factors involve regulatory behaviours, endocrine regulation, and genetic modulation. For instance, the genetic regulation theory posits that genetic traits decline when species density becomes excessive, leading to persistently elevated mortality rates, even in resource-abundant conditions. Food regulation places food availability as the most pivotal limiting factor (Sun 2001). For humanity, the population pressure induced by external environmental changes is a case of density-independent regulation, while population pressure resulting from natural population growth represents density-dependent regulation.

Animal ecology benefits from robust empirical evidence for testing various theories, while the study of mechanisms behind population pressure in human populations lacks such convenience. At present, the primary evidence is derived from anthropological research, as historical hunter-gatherer groups left valuable yet limited materials.

Based on the cultural ecology among hunter-gatherer societies, the following mechanisms for coping with population pressure can be inferred:

Firstly, the intensification of mobility and relocation. In situations of inadequate food resources, a primary response involves frequent hunting and gathering expeditions, or migration to regions with relatively abundant food resources. Hunter-gatherers possess minimal belongings, facilitating ease of mobility and migration.

Secondly, broadening the diet by consuming foods that are not frequently eaten, often consisting of lower quality or more challenging-to-process items.

Thirdly, population control. In ethnology, behaviours such as infanticide and abandoning the elderly are commonly observed among hunter-gatherer societies. These practices aim to reduce the consuming population and overall population size.

Fourthly, competition or cooperation. The management of limited resources can occur through either competitive or cooperative means. For humans, this might involve conflicts within groups and wars between different groups. Alternatively, resources could be managed by exchanging goods, relying on allies, or seeking assistance from kin-related groups to overcome challenges.

These four mechanisms are not unique to humans and are also found among animals. To realise population control, certain fish species, for instance, consume smaller individuals of the same species to deal with limited food resources.

Fifthly, the environmental adaptation by humans is fundamentally a cultural adaptation, most conspicuously manifesting as a series of technologies that address challenges encountered. These technologies include the manufacturing of stone tools, the utilisation of fire, and the construction of shelters. Technological innovations serve to expand the scope and efficiency of resource utilisation. In this sense, the emergence of agriculture can be viewed as a technological revolution.

It is necessary to discuss the hierarchical aspects of cultural adaptation to further delve into this issue. Technological transformation stands as the foundational variable within the human cultural system. Subsequent to this, alterations at the level of social organisation follow, encompassing kinship systems, division of labour, and societal complexity, which is exemplified by exchange as it can temporarily alleviate resource scarcity stemming from population pressure. Concurrently, social networks serve as pivotal sources of resource-related information and technological knowledge, thereby mitigating risks arising from population pressure. Finally, transformations manifest at the spiritual level, encompassing religious rituals, ideologies, and belief systems.

Evidently, a pending question arises: population pressure does not inevitably propel technological transformation. Ester Boserup once asserted that population growth stimulated technological advancement (Boserup 1965). However, this statement is proved unsubstantiated both in reality and across history. This implies that population pressure only potentially drives technological innovation under specific circumstances.

These circumstances delineate why population pressure impacts certain cultural systems but not others.

The contextual conditions underpinning the effects of population pressure generally encompass the following aspects:

Firstly, the nature of the environment, whether open or closed. Specifically, in open environments, surplus population can disperse; whereas in closed or regionally bounded environments, which is referred to as "packing" by Binford, excess population cannot diffuse and is constrained to address issues within a limited area.

Secondly, technological conditions. The presence of technological reserves available for utilisation. For instance, groups, which solely rely on hunting, face challenges in developing plant cultivation.

Thirdly, social organisational conditions that influence diverse human groups' impetus to expand production. Sahlins questioned whether egalitarian distribution systems utilised by certain hunter-gatherer societies nullify surplus production.

Fourthly, ideological conditions. Although it remains unclear which ideological factors precisely favour or hinder agricultural origins, it can be inferred that ideologies conducive to generating surplus production will favour the emergence of agriculture. Conversely, notions of high-level egalitarianism would hinder the occurrence of agricultural origins.

Fifthly, environmental conditions serve as explicit external limiting factors, such as environments lacking domesticate-able species.

Reserves, which are more closely associated with seasonal environmental considerations, should not be perceived as a response to population pressure. For hunter-gatherers in cold regions, reserves are essential. However, reserves do not inherently link with the occurrence of agriculture, even though agricultural societies often exhibit the habit of reserves.

Summarising the mechanisms underlying the effects of population pressure, two sets of contradictions are evident, although further discoveries may emerge in the future. One set pertains to human intrinsic attributes and characteristics as individuals (Marxist termed it as "social attributes"). Human intrinsics determine how humans respond to intra-group population growth, much like the fundamental approaches seen in animal ecology research. Meanwhile, human nature encompasses at least three forms of responses: technological, social, and ideological. Humans respond to population pressure on these three levels.

Another set of contradictions involves the opposition between habit and agency. Bourdieu, a cultural structuralist, refers to habit as "habitus", which is an inherent structural feature of culture that influences people's action modes (Bourdieu 2003). Agency, which indicates human spirituality or consciousness, pertains to the capacity of individuals to overcome environmental and cultural constraints, engaging in creative activities that defy conventions (Johnson 1989). This dichotomy encapsulates a new perspective within contemporary archaeology, which undoubtedly can be employed to analyse the process of agricultural origins. However, research in this realm remains limited at present.

1.5.4 New Insights into the Mechanism of Agricultural Origins

Based on novel scientific theories and the long-term trends and short-term adaptive mechanism of hunter-gatherer adaptation, a new model for the mechanism of agricultural origins can be constructed (Fig. 1.2). This model aims to elucidate the general process and conditions underlying the emergence of agriculture. This model seeks to explain the rise of food production, which is a parameter; how various factors within the hunter-gatherer cultural system interact with one another; and under what circumstances such interactions might lead to the emergence of food production.

This model primarily emphasises that agriculture, which is represented by food production, views an entirely new factor within the hunter-gatherer cultural system, manifested as an emergence, or a sudden appearance. The most conspicuous distinction between farmers and hunter-gatherers lies in the fact that the former engage in food production and tend towards sedentism (pastoralism is a derived form of agriculture, branching off from settled agriculture); while the latter rely on naturally occurring resources and exhibit a mobile lifeway. More significantly, food production functions as a parameter, decisively influencing the trajectory of cultural system development. This parameter governs the coordinated changes of other variables, subsequently giving rise to a series of Neolithic cultural features, such as sedentism, ground stone tool, and pottery. In summary, the emergence of food production characterises the nature and features of agricultural origins.

The initiation of food production signifies the departure of hunter-gatherers from their former livelihood characterised by a nomadic existence centred around hunting, gathering, and fishing economies. Why

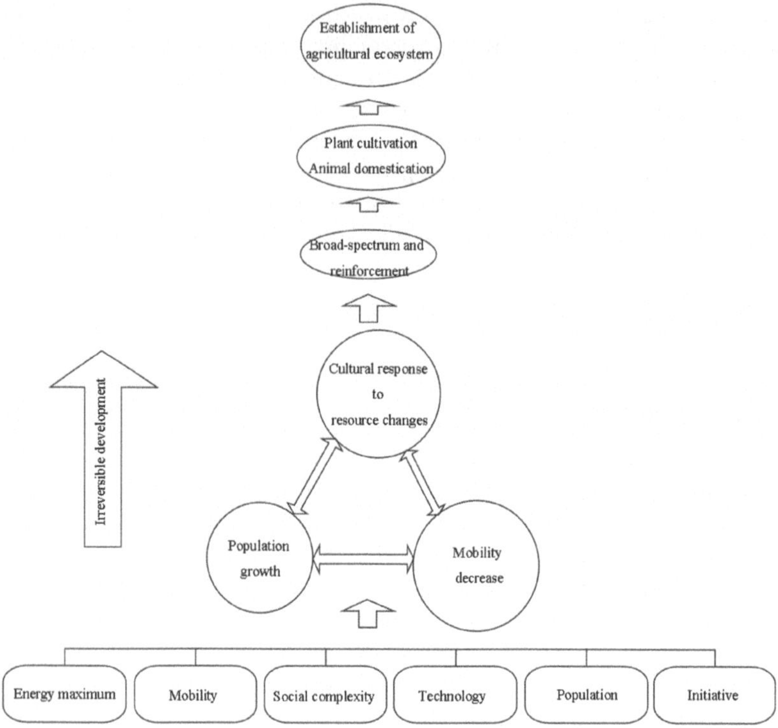

Fig. 1.2 The hypercycle model of agricultural origins

would hunter-gatherers abandon their previously well-adapted way of life? Hunter-gatherer societies are structured around hunting, which serves as the parameter for cultural system formation, being defined by mobility, which is a primary adaptive trait. Hunting and mobility are intrinsically linked: a decline in hunting practices leads to a decrease in mobility. As a consequence, reduced mobility results in a narrower range of life resources for the group. It is established that humans had already become proficient hunters by at least the LUP. After tens of thousands of years of hunting, hunting resources began to experience strain. Simultaneously, climatic and environmental shifts influenced the quantity of huntable animals. The combination of these factors undoubtedly exacerbated the situation, rendering the sustenance of an economy dependent on hunting increasingly unsustainable.

The loss of mobility is not solely linked to hunting but also directly connected to population density and the likelihood of migration. The increase in population density not only intensifies resource pressure, forcing a shift from larger to smaller animals, from slow-moving to agile prey, but it also reduces the per capita territorial domain and diminishes resource acquisition possibilities. For hunter-gatherers, mobility is the fundamental strategy to mitigate livelihood risks; a contraction in the scope of mobility significantly augments livelihood risks. Generally, when the population of hunter-gatherer groups reaches a certain threshold, a split happens, with the new faction migrating to new territories to alleviate population density and resource pressure. However, when suitable new territories for migration become scarce, population density inexorably rises. The potential for migration is not solely determined by the availability of new lands but is also associated with the sensitivity of individuals to territorial boundaries. Evidently, higher population densities and greater social complexity correspond to heightened sensitivity to the boundaries of productive areas, rendering migration increasingly improbable.

From the perspective of species, agriculture involves a selective focus on only a few species through the cycles of cultivation, maintenance, harvesting, processing, and then restarting the cycle. The number of species typically utilised by hunter-gatherers far surpasses that of farmers. That is, the process of agricultural development represents the reduction or concentration of a few specific species. This phenomenon is called "intensification", which refers to enhancing the output, accessible components, and efficiency of processing for certain species. Intensification is the fundament of domestication, and domestication is the outcome of ongoing intensification. The absence of intensification implies the absence of domestication. Thus, in the study of agricultural origins, the exploration of intensification should be the initial focus.

How does intensification occur? There are two fundamental pathways for hunter-gatherers to enhance resource acquisition. Firstly, they expand the geographical range of resource exploration, a process termed "between-patch mobility" due to the patchy distribution of resources, involving movement between different patches. The second pathway involves elevating resource utilisation within a specific patch or area, referred to as "within-patch mobility". The second path entails improving the efficiency of resource extraction within a given patch, broadening the range and quantity of resources utilised. Regarding between-patch mobility, hunter-gatherers need to develop techniques that enhance mobility, such as the microblade technology, which involves the creation of lightweight and versatile tools that enable individuals to easily transport equipment between different areas. Furthermore, the usage of watercraft allows people to traverse regions that are challenging to access on foot. As for the within-patch mobility, to enhance resource utilisation within a patch, there are two pathways: Firstly is the broad spectrum adaptation, which involves expanding the range of resources utilised; The second way focuses on intensification, aiming to increase the yield of specific resource categories.

The relationship between broad spectrum adaptation and intensification is both parallel and developmental in nature. Broad spectrum adaptation represents a critical ebb and flow within a cultural system, exemplifying the highest efficiency within a hunter-gatherer economy. hunter-gatherers utilise all available species, yet such a scenario is unsustainable in the long term. Ultimately, certain species are selected for intensification. Due to the constraints of limited time and resources, once the intensified utilisation of specific species occurs, hunter-gatherers are compelled to narrow their focus, gradually phasing out the utilisation of other species. Thus, from this perspective, intensification can be regarded as a progression stemming from broad spectrum adaptation (Table 1.1).

The establishment of agriculture is a relatively long-term process, referred to in this model as the "hypercycle". Due to the demands of tending to plants, following the initiation of intensification, the mobility of hunter-gatherers inevitably decreases. This reduction in mobility implies a smaller utilisation area and a decreased total resource volume, necessitating further intensification of resource utilisation. Meanwhile, the reduction in mobility contributes to population growth, fostering a strengthened territorial mindset among hunter-gatherers, which in turn further limits their mobility. Amid the improvement of human hunting capabilities and the occurrence of environmental changes, the diminishing availability of prey introduces this change into the cycle, further propelling the development of intensification. The defining characteristic of hypercycle is that the incorporation of new variables does not disrupt the cycle process, but rather drives its enhancements. As an effective mechanism, the hypercycle could explain the gradual maturation of agriculture, aligning with phenomena observed in archaeological materials.

Table 1.1 Impact of broad spectrum adaptation and intensification on archaeological evidence

Broad-Spectrum	*Intensification*
Diverse tools with adaptability, such as standardised techniques	Specialised and formal tools
Tools with high mobility, with features of lightweight and easy maintenance	Tools with reduced mobility and durability
Wide range of food sources, including large, medium, and small animals, especially: ① Agile and predatory animals with high costs and low returns; ② Slow-growing and depletable species, such as turtles	Narrow food sources, focused utilisation of specific species, especially: ① Complexly processed items, such as small plant seeds and toxic nuts; ② Items requiring sophisticated tools to process, like hunting marine mammals
Extensive spatial utilisation encompassing diverse habitats, leading hunter-gatherers to expand from optimal to suboptimal habitats	Limited habitat range, extended occupation within a specific habitat type

Agriculture emerged gradually from the Upper Paleolithic to reach full maturity in the Middle Neolithic, subsequently rapidly spreading. As a driving force, the hypercycle enabled humanity to progressively transition into an agricultural system. This process can be divided into at least three phases:

In the initial phase, hunter-gatherer mobility decreases, with the emergence of intensification, marking the embryonic phase of agriculture. Changes occur in the tool assemblages and residential structures of the Upper Paleolithic.

The second phase witnesses the emergence of settled communities and the establishment of domesticated species. During the Early Neolithic, the tool assemblages formed, constituting the early agricultural phase.

In the third phase, during the Middle Neolithic, an agricultural ecosystem is established, marking the true onset of the agricultural era. The progression from inception to full realisation spans thousands of years.

With the understanding of the mechanisms of agricultural origins, it is imperative to delve further into the reason for the capability of certain cultural systems to initiate the mechanisms of agricultural origins, that is, the reasons why hunter-gatherer cultural systems could reach critical states. Once a cultural system attains a critical state, even slight variations in initial conditions can lead to significant shifts in system structure.

Undoubtedly, the interplay of human societal and environmental factors brings cultural systems to critical states, yet concretely, which factors are precisely relevant?

Based on an environmental perspective, initial conditions primarily encompass species availability, climatic conditions, animal extinctions, and unique geographical conditions. Notably, environmental conditions incorporate both fundamental limiting factors and favourable conditions. Species availability stands as the most fundamental limiting factor, as the regions where agricultural domestication occurs must host corresponding wild species. For instance, the region of Western Asia boasts a rich variety of wild large-seeded plant species. The earliest domesticated crops encompassed wheat, barley, several legumes, and flax, which could provide oil and fibre, collectively offering comprehensive nutritional sustenance. Subsequently, the domestication of goats and sheep contributed to a holistic agricultural ecosystem. The native habitat of these species aforementioned lies within the Western Asian region. The Old World also saw the domestication of animals that provided draft power, like cattle, horses, and camels. In comparison, the Americas lacked domesticable large animals, as only limited options were available, such as turkeys, llamas, and guinea pigs, rendering the establishment of a complete agricultural ecosystem unfeasible. Species availability dictates that agriculture can only occur in regions with suitable domesticable species, as not all species are fit for domestication. The majority of crops cultivated today were domesticated during the Early Neolithic, which historical period witnessed minimal additions to the list. Although human domestication techniques have advanced significantly, the limits set by species availability are still constrained.

Due to the limitations posed by climatic conditions, the fluctuating and unpredictable climate of the Pleistocene made it challenging to achieve successful yields from cultivation; whereas the stable climate conditions of the Holocene provided the necessary climatic environment for agriculture. For a cultural system that is situated in a critical state, the transition in climatic conditions is undeniably pivotal. To some extent, climatic conditions determine that the emergence of the earliest agriculture happened around 10,000 years ago, at the end of the Late Pleistocene and the beginning of the Holocene, rather than earlier.

As previously mentioned, the cultural systems of hunter-gatherers are structured around hunting as the prevailing parameter. If the hunted prey species were to go extinct, hunting would become unsustainable, and the far-reaching consequences are evident. Due to the extinction of numerous large herbivores animals towards the end of the Pleistocene, intensified human hunting practices already emerged in the Upper Paleolithic era. As the decline and extinction of prey species directly compelled hunter-gatherers to explore and exploit new resources. The hunting difficulties in the hunter-gatherer cultural system extends beyond this. The decrease in available prey not only lowers mobility, leading to reconsiderations of social organisation and labour division, but also triggers shifts in societal prestige evaluation norms. That is, the entire social fabric faces challenges and necessitates adjustments. Similar to the transition from the agricultural era to the industrial and commercial era, every facet of a cultural system is imbued with challenges.

Regarding the geographical context of the origins of agriculture, certain regions exhibit greater sensitivity to changes, particularly environmental changes. This phenomenon is often referred to as an ecotone, with the classic examples being the transitional zones between forests and grasslands, as well as the coastal zones where land meets the ocean. As an advantage, ecotones lie in the diversity of resources, allowing individuals to access resources from two distinct ecological zones; it is particularly beneficial in reducing the risk associated with food acquisition due to the temporal differences in resource availability across these zones. However, every opportunity comes with costs. The drawback of ecotones is their resource instability. Taking the example of the temperate forest-grassland ecotone in northern China, the location of this boundary has shifted repeatedly throughout history. During colder periods, grasslands have encroached southwards, prompting the migration of northern grassland societies, which posed significant threats to the political powers that were centred in central plains. Similarly, considering coastal environments, the timespan of 10,000 from the LGM to the Holocene climatic optimum, sea levels have risen by over 100 metres, with intermittent fluctuations causing instability in coastal environments. Nevertheless, compared with the example in ancient northern China illustrated, coastal ecotones exhibit relatively higher levels of stability.

Due to the resource richness, ecotones often attract larger populations. In addition, due to evolutionary psychology, regions where forests and grasslands intersect, such as the forest-grassland ecotone, align with the psychological predispositions that emerged from humans' early adaptations to "tropical sparse-wooded grasslands". Thus, humans are more inclined to inhabit these zones where trees and grasslands intertwine.

Moreover, these areas tend to host a greater diversity of herbivorous animals that humans prefer to hunt, shaping them conducive to a livelihood centred around hunting. However, given the sensitivity of these regions to climatic and environmental changes as well as resource utilisation, these areas are more prone to experiencing population and resource pressures. Thus, the significance of ecotones in the context of the origins of agriculture requires careful consideration.

From the perceptions of human society, changes in initial conditions should encompass aspects of population growth, societal complexity, and technological development.

The accumulation of population growth of all species presents a significant challenge for adaptation, as well as humans. After establishing comprehensive adaptive advantages over other predators, there have been few factors that could restrict the population growth of prehistoric hunter-gatherers, apart from human-specific and resource-related limitations. Population pressure varies under different resource conditions. Livelihoods primarily encountered popular pressure under hunting, followed by gathering-centric livelihoods, and finally fishing-based livelihoods. Population pressure does not necessarily lead to technological advancement; while technological development could happen when popular pressure interacts with other factors. For instance, the population exerting pressure on the livelihood of hunter-gatherers might become indispensable human resources for agricultural societies, as clearing new land requires a considerable workforce. Therefore, population pressure is relative to a certain structured cultural system. The upgrades of cultural system structure could contribute to the transformation of surplus population to resources.

Environmental factors have a decisive impact on animal adaptation, but it is not necessarily applicable in the case of human societies. Fundamentally as a form of cultural adaptation, human adaptation characterised by the dominance of technology, social organisation, and ideology. Technological constraints encompass food processing techniques, grinding plant seeds with the utilisation of grinding tools, and baking or cooking through using hearths and pottery. The technological limitations of the Upper Paleolithic did not fundamentally constrain early agricultural production, with diverse technological pathways. For instance, regarding plant seeds as food does not necessarily require pottery; and grinding tools can include grinding slabs, handstones, mortars, stone grinders.

Relatively speaking, the constraints of social organisation are more significant than technological conditions. In comparison to primarily foraging strategies of hunter-gatherers, those who adopt predominantly collector strategies typically possess more complex technologies. As hunter-gatherers who utilised collector strategies residented at the central camps longer, enabling investments in durable tools, such as heavy grinding stones and ground stone axes, as more solid constructions. Correspondingly, extended periods of habitation lead individuals to become more familiar with the plants in their vicinity and pay greater attention to them. With lower mobility, collector strategies are associated with higher population densities, resulting in a greater labour force compared to foraging strategies.

In comparison to social organisation, the constraints imposed by ideological aspects are more pronounced. It is known that hunter-gatherer societies establish prestige standards centred around hunting. However, in early agricultural societies, the assessment of prestige shifts to agricultural production capabilities. When the prestige gained from hunting outweighs that from food production, the dominance of a prestige assessment system rooted in hunting becomes a barrier to the origin of agriculture. The disruption of ideologies involves both internal struggles within a society and external influences. Conflicts within hunter-gatherer societies primarily manifest at the levels of gender, age, and competition for prestige. The transition to agriculture in relation to these conflicts remains yet unclear.

The emergence of agriculture entails comprehensive changes within cultural systems, involving a multitude of interconnected factors. A monocausal explanation is insufficient to account for the origins of agriculture. In addition, overlooking the interrelations between factors cannot understand the mechanisms underlying agricultural origins. In light of the current theoretical paradigm, drawing from insights provided by cultural ecology in the context of hunter-gatherer adaptations, the understanding of the mechanisms behind the origin of agriculture has evolved as outlined above. The following chapter will focus on explaining why agriculture did not develop in certain regions.

Note

1. According to Sahlins, the notion of "affluence" can be approached from two perspectives. Firstly, it involves the augmentation of supply and the possession of more abundant resources. The second avenue lies in the reduction of desires, bearing resemblance to the principles found in Zen Buddhism. Hunter-gatherer society is commonly characterized as egalitarian communities, where affluence predominantly refers to the latter perspective. In this sense, it denotes that the natural resources are collectively owned, thereby rendering every individual inherently "affluent".

References

B. Bender, Gatherer-hunter to farmer: A social perspective. *World Archaeology*, 10: 204–222, 1978.

B. D. Smith, Low-level food production. *Journal of Archaeological Research*, 9: 1–43, 2001.

B. Hayden, A new overview of domestication. In T. D. Price and A. B. Gebauer eds., *Last Hunters-First Farmers*, pp. 273–299. School of American Research Press, Santa Fe, New Mexico, 1995.

B. Hayden, Nimrods, piscators, pluckers, and planters: The emergence of food production. *Journal of Anthropological Archaeology*, 9: 31–69, 1990.

B. Hayden, Practical and prestige technologies: The evolution of material systems. *Journal of Archaeological Method and Theory*, 5: 1–53, 1998.

B. Hayden, The proof is in the pudding: feasting and the origins of domestication. *Current Anthropology*, 50: 597–601, 2009.

Bar-Yosef, the Natufian culture in the Levant, threshold to the origins of agriculture. *Evolutionary Anthropology*, 7: 159–177, 1998.

Blumer, A. A., and R. Byrne, The ecological genetics of domestication and the origins of agriculture. *Current Anthropology*, 32: 23–54, 1991.

M. A. Blumer, Ecology, evolutionary theory and agricultural origins. In D. R. Harris ed., *The Origins and Spread of Agriculture and Pastoralism in Eurasia*, pp. 25–50. Smithsonian Institution Press, Washington, 1996.

C. Tilley, *A Phenomenology of Landscape*. Berg, Oxford, 1994.

C. W. Marean, *Humanity on the Brink of Extinction. Scientific American* Chinese version, 2010 (9).

Chen Shengqian, *The structure of archaeological reasoning*. *Archaeology*, 2007 (10).

D. A. Brooks et al., Dating and context of three Middle Stone Age sites with bone points in the Upper Semliki Valley, Zaire. *Science*, 268: 548–553, 1995.

D. M. Persall, domestication and agriculture in the New World tropics. In T. D. Price and A. B. Gebauer eds., *Last Hunters-First Farmers*, pp. 157–192. School of American Research Press, Santa Fe, New Mexico, 1995.

D. R. Harris, An evolutionary continuum of people-plant interaction. In D. R. Harris and G. C. Hillman eds., *Foraging and Farming: The Evolution of Plant Exploitation*, pp. 11–26. Unwin Hyman, London, 1989.

D. R. Harris, Introduction: Themes and concept in the study of early agriculture. In D. R. Harris ed., *The Origins and Spread of Agriculture and Pastoralism in Eurasia*, pp. 1–9. Smithsonian Institution Press, Washington, 1996.

D. Rindos, Symbiosis, instability, and the origins and spread of agriculture: A new model. *Current Anthropology*, 21: 751–772, 1980.

D. Rindos, *The Origins of Agriculture: An Evolutionary Perspective.* Academic Press, Orlando, 1984.

E. Boserup, T*he Conditions of Agricultural Growth: The Economics of Agrarian Change under Population Pressure.* Aldine, Chicago, 1965.

G. C. Hillman, S. M. Colledge and D. R. Harris, Plant-food economy during the EpiPaleolithic at Tell Abu Hureyra, Syria: dietary diversity, seasonality and modes of exploitation. In D. R. Harris and G. C. Hillman eds., *Foraging and Farming:The Evolution of Plant Exploitation*, pp. 240–269. Unwin Hyman, London, 1989.

G. Clark, *World Prehistory in New Perspective.* Cambridge University Press, Cambridge, 1977.

G. G. Politis, Moving to produce: Nukak mobility and settlement patterns in Amazonia. *World Archaeology*, 27: 492–511, 1996.

G. Rice, A systematic explanation of a change in Mogollon settlement patterns. Ph.D.dissertation, University of Washington, 1975.

H. Haken, *Synergetics: An Introduction to the Order and Complexity of Nature*, translated by Ling Fuhua. Shanghai Translation Publishing House, 2001.

H. J. Brumbach and R. Jarvenpa, Ethnoarchaeology of subsistence space and gender: a subarctic Dene case. *American Antiquity*, 62: 414–436, 1997.

H. M. Wobst, Locational relationships in Paleolithic society. *Journal of Human Evolution*, 5: 49–58, 1976.

H. Pringle, The slow birth of agriculture. *Science*, 282: 1446–1450, 1998.

Hodder, *Reading the Past.* Cambridge University Press, Cambridge, 1986.

K. V. Flannery ed., Guilá Naquitz: A*rchaic Foraging and Early Agriculture in Oaxaca, Mexico.* Academic Press, Orlando, 1986.

L. H. Keeley, Protoagricultural practices among hunter-gatherers: A cross-cultural survey. In T. D. Price and A. B. Gebauer eds., *Last Hunters-First Farmers*, pp. 243–272. School of American Research Press, Santa Fe, New Mexico, 1995.

L. R. Binford, *An Archaeological Perspective*. Seminar Press, New York, 1972.

L. R. Binford, *Constructing Frames of Reference: An Analytical Method for Archaeological Theory Building Using Ethnographic and Environmental Data Sets*. University of California Press, Berkeley, 2001.

L. R. Binford, Post-Pleistocene Adaptation. In S. Binford and L. Binford eds., *New Perspectives in Archaeology*, pp. 313–341. Aldine, Chicago, 1968.

L. R. Binford, Willow smoke and dogs' tails: hunter-gatherer sedentism systems and archaeological site formation. *American Antiquity*, 45: 1–17, 1980.

M. A. Zeder, The Neolithic Macro-(R)evolution: Macroevolutionary theory and the study of culture change. *Journal of Archaeological Research*, 17: 1–63, 2009.

M. C. Stiner, Carnivory, coevolution, and the geographic spread of the genus *Homo*. *Journal of Archaeological Research*, 10: 1–63, 2002.

M. D. Sahlins, Stone Age Economics, translated by Zhang Jingwei, Zheng Shaoxiong, and Zhang Fan, SDX Joint Publishing Company, 2009.

M. Eigen, P. Schuster, *The Hypercycle: A Principal of Natural Self-Organization*, Shanghai Translation Publishing House, 1990.

M. H. Nitecki and D. V. Nitecki eds., The Evolution of Human Hunting. Plenum Press, New York, 1987. Chen Shengqian, *The Hunting of Prehistoric Humans*. *Fossils*, 2005 (2)

M. Johnson, Conceptions of agency in archaeological interpretation. Journal of Anthropological Archaeology 8: 189–211, 1989.

M. Q. Sutton and E. N. Anderson eds., *Introduction to Cultural Ecology*. Altamira, Walnut Creek, 2004.

M. Rosenberg, The mother of invention: Evolutionary theory, territoriality and the origins of agriculture. *American Anthropologist*, 92: 399–415, 1990.

O. Bar-Yosef, The Upper Paleolithic Revolution. *American Review of Anthropology*, 31: 363–393, 2002.

O. Haim, *Second Nature: Economic Origins of Human Evolution*, translated by Zhang Dunmin, China Social Sciences Press, 2004.

P. Bak, How Nature Works. Copernicus, New York, 1996.《大自然如何工作》, Translated by Li Wei and Cai Lu. Central China Normal University Press, 2001.

P. Bourdieu, Lesens pratique, translated by Jiang Zihua, Yilin Press, 2003.

P. C. Anderson, *Prehistory of Agriculture: New Experimental and Ethnographic Approaches*. The Institute of Archaeology, University of California, Los Angeles, 1999.

P. J. Richerson, R. Boyd and R. L. Bettinger, Was agriculture impossible during the Pleistocene but mandatory during the Holocene? A climate change hypothesis. *American Antiquity*, 66: 387–411, 2001.

P. J. Watson, Explaining the transition to agriculture. In T. D. Price and A. B. Gebauer eds., *Last Hunters-First Farmers*, pp. 21–38. School of American Research Press, Santa Fe, New Mexico, 1995.

P. Weissner, Beyond willow smoke and dogs tails. *American Antiquity*, 47: 171–178, 1982.

R. A. Bentley and H. D. G. Maschner eds., Complex Systems and Archaeology. University of Utah Press, Salt Lake City, 2003.

R. J. Braidwood, The agricultural revolution. *Scientific American*, 203: 130–141, 1960.

R. Pumpelly ed., *Explorations in Turkestan. Expedition of 1904. Prehistoric Civilizations of Anau: Origins, Growth, and Influence of Environment.* Publication no.73. Carnegie Institution, Washington, 1908.

R. S. MacNeish, *The Origins of Agriculture.* University of Oklahoma Press, Norman, 1992.

R. W. Redding, A general explanation of subsistence change: From hunting and gathering to food production. *Journal of Anthropological Archaeology*, 7: 56–97, 1988.

S. J. Gould. *The Panda's Thumb: More Reflections in Natural History*, translated by Tian Ming. Hainan Publishing House, 2008.

S. J. Pyne, *A Brief History of Fire*, translated by Mei Xueqin, Niu Ruihua, Jia Jun, and others. SDX Joint Publishing Company, 2006.

S. Mithen, *The Prehistory of the Mind: A Search for the Origin of Art*, Religion and Science. Orion, London, 1996.

Sun Ruyong, *Principles of Animal Ecology*, Beijing Normal University Publishing Group, 2001.

T. Dobzhansky, Adaptedness and fitness. In R. C. Lewontin ed., *Population Biology and Evolution*, pp. 109–122. Syracuse University Press, Syracuse, 1968.

T. Goebel, Pleistocene hunter colonization of Siberia and peopling of the Americas: an ecological approach. *Evolutionary Anthropology*, 8: 208–227, 1999.

P. J. Brantingham and Gao Xing, Peopling of the northern Tibertan Plateau. *World Archaeology*, 38: 387–414, 2006.

V. G. Childe, *Man Makes Himself*. Watts, Library of Science and Culture, London, 1936. 《人类创造了自身》, translated by An Jiayuan and Yu Jingdong, Shanghai Sanlian Bookstore Co.Ltd., 2008.

V. G. Childe, *The Most Ancient East.* Routledge and Kegan Paul, London, 1928.

W. T. Divale, Systemic population control in the Middle and Upper Paleolithic: inferences based on contemporary hunter-gatherers. *World Archaeology*, 4: 222–243, 1972.

Wu Tong, *Research on Methodology of Self-organization*, Tsinghua university press, 2001.

CHAPTER 2

Why Agriculture Did Not Happen

> Let people return to using knotted ropes for writing,
> to eating their food as enjoyable
> to viewing their clothes as beautiful,
> to being content with their residences,
> to finding joy in their customs.
> People of neighbouring countries reside within sight of each other;
> Dogs barking and cocks crowing can be heard across the border,
> Yet they do not have to deal with each other even when old and dying.
> If such a time existed, then the ultimate state of order was already achieved.
>
> —*Laozi*

2.1 Introduction

The issue "Why agriculture happened" is always hotly debated, whereas "why agriculture did not happen" is not even considered a question. Contemporary archaeological research rarely delves into the explanation of stability of cultural systems. When discussing the origins of the Industrial Revolution, not only its development in Europe is examined but also the absence of industrialisation in other regions is explored. This is particularly relevant for Chinese scholars who ponder over issues like the "Needham's Grand Question", constituting a fundamental component of modern Chinese academic research. Prehistoric agriculture originated in only a few regions across the world, with its introduction to other areas occurring later. This holds true even for China. Why did these regions

S. Chen, *Prehistoric Modernization*,
https://doi.org/10.1007/978-981-97-9423-2_2

lack the emergence of agriculture? With exhibition of varying conditions, different regions are different. However, are there shared factors among various regions? In other words, which factors are pivotal? This chapter will focus on discussing this issue. It is also the other facet of prehistoric modernisation, a dimension often overlooked. It must be acknowledged that the issue of "why agriculture did not happen" derives from the perspective of agriculture's origins, as agricultural societies had a significant impact on hunting and gathering societies. Just as the idealism of an agricultural era, as depicted by Laozi, faltered in the face of industrialisation, hunting and gathering societies either embraced agriculture, retreated to more marginal environments, or devised alternative strategies to effectively respond to the impacts of agricultural societies.

2.2 Environmental Constraints

What is the definition of environmental constraints? Theoretically, climates and lighting can be artificially manipulated with the assistance of technological advancements, and soil can be altered or transported. Thus, agricultural production could be achieved anywhere. Limitations lie in certain regions that are unable to engage in agricultural production within specific periods, giving existing technological and societal conditions. Furthermore, the species, which are suitable for domestication, possess biological limitations. For plants, the limits are confined by temperature and precipitation. With ongoing selections made by humans, some species, such as rice, have expanded from subtropical regions to areas as far north as the Heilongjiang Province at 45°N latitude. That is, environmental constraints refer to the environmental conditions that contemporary cultural adaptability cannot overcome.

Emilio F. Moran categorised several extreme environmental challenges that factor into human adaptation (Table 2.1) (Moran 2000), which cannot be overcome with modern technology. For prehistoric hunting and gathering societies, surmounting these challenges was impossible. Thus, only with certain cultural and environmental conditions can humans survive in these extreme environments. The Arctic-dwelling Inuit serves as a classic example. They have developed more sophisticated and reliable tools to exploit marine mammal resources. High-latitude regions usually face difficulties in plant growth, with low primary biological productivity. Inhabitants of such areas cannot sustain themselves solely on plants; instead, they must rely on hunting, thus adapting to highly mobile livelihoods. Similar constraints apply to high-altitude regions. Due to the low density of oxygen, the mobility of hunters is restricted. Inhabitants in these areas must rely on additional resources to survive. Grassland areas host immense and high-mobile animal populations, requiring humans to possess great mobility to effectively utilise these animal resources. In this case, mobility contradicts the need for agricultural sedentism, and high mobility presents a primary barrier to agricultural development.

Regarding agricultural origins, environmental constraints do not solely have negative impacts but also yield positive influences, which is frequently overlooked. Hunter-gatherer populations, who inhabited temperate, subarctic, and Arctic environments, often experience food scarcity and inadequate caloric intake during late winter and early spring. During such periods, they tend to rely on carbohydrate resources, such as storable plants (Speth and Spielmann 1983). During their long-term adaptational history, hunter-gatherers tend to pay more attention to the supply of carbohydrate resources. The success of their adaptation is determined not by the most abundant periods of food availability, but rather by their ability to secure sustenance during the leanest seasons. This emphasis on guaranteeing supply during times of scarcity, even among groups primarily reliant on animal-based diets, necessitates consideration of acquiring reserve food resources. Thus, such environmental constraints effectively promote a transition of hunter-gatherer populations towards agricultural production, rather than vice versa. Undoubtedly, as aforestated, negative effects also exist, with all environmental constraints being issues of cultural adaptability. Negative effects require corresponding cultural development to be overcome. The greater the

Table 2.1 Constraint factors of extreme habitats (environment)

Habitat	*Characteristics*
Arctic Region	Long and extremely low temperatures; Alternating seasons of polar night and polar day; low biological productivity
High-Latitude Regions	Low oxygen content; Large diurnal temperature variations; Low biological productivity; High neonatal mortality rate
Arid Regions	Erratic rainfall; High evaporation; Low biological productivity
Grassland Regions	Lengthy dry seasons; Periodic droughts; Large and well-organised animal populations
Humid Tropics	Extremely diverse species; High precipitation; Intense solar radiation; Rapid secondary succession

difficulty, the more time is needed for its resolution. This also composes the reason for agricultural infeasibility and the delayed agricultural occurrence within certain extreme environments. Thus, regions, which are most conducive to agricultural origins, indicate the ones where the positive effects of environmental constraints are maximised and the negative effects are minimised.

2.3 Desert and Tropical Forest Adaptation

Desert and tropical rainforest zones were not considered as possible origins of agriculture by archaeologists for an extended period. However, in the past two to three decades, research conducted suggests that certain regions, such as the western desert of Egypt in North Africa, the highlands of New Guinea, and even the Amazon rainforest, may have independently developed primitive forms of agriculture. Nevertheless, the impact of these early agricultural practices on the subsequent cultural development in these areas was minimal and even absent. The later agricultural systems in these regions were introduced from other regions. This stark contrast underscores that extreme environmental conditions pose formidable barriers to agricultural production that are difficult to overcome. Nevertheless, the positive effects of environmental constraints have supported the emergence of initial attempts at agriculture in these regions.

In North Africa, the earliest form of agriculture was the domestication of cattle, instead of plant cultivation. The earliest evidence of cattle domestication practice can be found at the Nabta site in the western desert of Egypt, dating back approximately 8000 years (Wendorf and Schild 1998). The inability to access water within a 24-hour period could lead to cattle's death by thirst. However, even during the warm and humid Holocene Optimum, the Nabta site had very limited surface water resources apart from a few wells. Naturally, cattle could not penetrate deep into the western desert area unless driven there by humans during certain seasons. Materials discovered from Nabta suggest that people had begun to domesticate cattle at that time. Nevertheless, existing archaeological evidence indicates that prehistoric agriculture in Egypt originated from West Asia rather than being locally developed. This implies that the western desert region of Egypt was completely abandoned by people after the Holocene Optimum, rendering the Nabta experiment inconclusive.

Through the study of ethnographic materials, John. E. Yellen proposed that cultural adaptation among hunter-gatherers in desert regions emphasises flexibility rather than stability due to the typically rapid and unpredictable resource fluctuations in such areas (Yellen 1976). Agricultural production, as a constant strategy for food acquisition, requires a relatively stable environmental condition. Sudden changes can disrupt the rhythm of agricultural production, delaying planting seasons or impacting crop growth. Extreme droughts can entirely undermine the foundation of agriculture. This is why the western desert region of Egypt, despite attempts, ultimately could not sustain agriculture. Desert regions undeniably require stable food resources provided by agriculture, but supporting the consistent environmental conditions necessary for agricultural endeavours is hard to achieve.

Regarding adaptation to tropical rainforests, the primary concern is not whether early agricultural production could be sustained here, but rather whether it could support hunter-gatherers, or in the other word, human survival itself. Until now, there is no reliable ethnographic evidence suggesting that hunter-gatherers could subsist independently within tropical rainforests without relying on exchanges with neighbouring agricultural groups (Bailey 1989; Bailey and Headland 1991). Consequently, some scholars propose that hunter-gatherers in the Amazon rainforest transitioned from groups engaged in horticulture, with their adaptation to the rainforest being secondary, which resulted from pressure imposed by colonisers and neighbouring agricultural communities. Undoubtedly, colonisers and agricultural groups brought advanced tools and opportunities for trade, rendering adaptation to the rainforest feasible (Lee and Daly 1999). This is analogous to how, after the introduction of horses to North America, Indigenous peoples who had practised primitive agriculture reverted to being hunter-gatherers.

Meanwhile, agricultural practices in tropical rainforest regions are also constrained by their impoverished soils: high temperatures lead to rapid organic matter decomposition into inorganic ones; and the frequent and intense rainfall leaches these minerals from the soil, rendering the land extremely infertile. The fertility generated from burning vegetation is insufficient for crop growth; if vegetation regenerates, the robust cycling capabilities of the tropical rainforest retain or recycle most essential nutrients, with only a minimal portion being lost to the river systems. However, early Amazonian rainforest horticulturist seemed to have known how to solve this issue. They created fertile black soil known as terra preta, covering approximately 10% of the Amazon region, equivalent to the size of France. Unlike the approach of "slash and burn" practised in tropical areas, Amazonian rainforest gardeners employed "slash and char" technique. In contrast to those who incinerate all organic matter into ash, the vegetation was partially burned, leaving behind substantial charcoal. This charcoal is mixed into soil and then fertilisers are applied, much like adding yeast to bread dough nowadays. As a result of microbial activity, the soil here became significantly more fertile (Mann 2000b). Similarly, in the Beni region of Bolivia, evidence of mounds, water channels, and irrigation canals dating back 2000 years has been found (Mann 2000a). In Xingu region in Brazil, being perceived as "primitive" forest and grassland, had been anthropogenically modified 1000 years ago, with high-density agricultural populations residing in a network of villages (Heckenberger et al. 2003).

Agriculture in the highlands of New Guinea may present a more illustrative case. Three distinct phases of plant utilisation have been identified: early utilisation (8270 BC to 7970 BC), mounding cultivation (5050 BC to 4490 BC), and ditch cultivation (2400 BC to 2030 BC). Evidence from plant pollen, phytoliths, and starch residues suggests that bananas were cultivated in the highland region, despite their original growth habitat not being in the highlands (Denham et al. 2003; Neumann 2003). If these findings are further validated, it appears that prehistoric hunter-gatherers possessed the capability to overcome survival constraints in tropical rainforest zones. Based on existing ethnographic records, numerous hunter-gatherers and some horticulturalists inhabited tropical rainforest regions, yet very few examples could be found of fully developed agricultural societies. Why did the incipient agriculture in this region not evolve into intensive agriculture similar to what transpired in Western Asia and China?

Tropical rainforests exhibit remarkably high primary productivity, mostly concentrated in tree trunks and leaves. The most valuable resources are situated in the canopy, suitable for avian exploitation and beyond the reach of larger-bodied humans (Binford 2001). Another constraint lies in the scarcity of terrestrial animals, inversely correlated with primary productivity (Eisenberg 1980, p.35–55). Although resources in tropical rainforest regions lack pronounced seasonality, the overall availability of utilisable resources, especially preferred animals and aquatic species, are limited. Thus, these regions are akin to extreme environments, akin to deserts, albeit with relatively stable resource conditions.

An additional major challenge of tropical rainforest adaptation is disease pressure, particularly for settled communities. The accumulation of waste in settled areas can easily contaminate water sources and food. Deforestation associated with land cultivation provides a suitable environment for mosquito proliferation, vectors of diseases such as malaria, the most prevalent and perilous ailment in tropical regions (Mithell 2003). Crop cultivation entails prolonged labour in fields and proximity to rivers, increasing susceptibility to malaria. Hence, compared to hunter-gatherer lifeways, settled agriculture in tropical rainforest regions is much riskier. Elevated disease pressure leads to shortened average lifespans, necessitating higher birth rates to maintain basic population growth. Consequently, each family must support more dependent children, leading to a decrease in the labour force and increase in household burdens. Correspondingly, each child receives less technical training and educational resources (Sachs et al. 2001). In comparison to temperate regions, the smaller labour population in the tropical rainforest areas limits their ability to clear rapidly growing weeds and trees. The aforementioned factors hinder labour-intensive agricultural production activities. In the tropical rainforest regions, sedentism entails risk, while mobility equates to safety. Thus, developing agriculture as a primary livelihood faces more formidable obstacles in this area.

2.4 Mobility: A Double-Edged Sword

Mobility serves as a safety measure for hunter-gatherers. Human mobility not only contributes to food acquisition; it also represents a means of gathering information about food distribution, timing of availability, and others (Binford 2009). Agricultural communities store grain in granaries, while hunter-gatherers "store" food in the natural environment, making foraging accessible when needed, thus eliminating the risk for temporary searches. In addition, mobility allows hunter-gatherers to encounter other groups, ensuring exogamous marriages. In short, mobility determines both the production and reproduction of the life of hunter-gatherers, constituting their fundamental adaptive strategy. It is imperative to recognise that mobility is not the sole environmental adaptation strategy (Minc and Smith 1989). Foraging mobility constitutes the hunter-gatherers' mode of production (Sahlins 1968), much like agriculture in traditional societies and market economies in contemporary societies, serving as the cornerstone upon which their societies function.

How could mobility be lost? There must be factors impeding the mobility of hunter-gatherers. The Optimal Foraging Theory offers some insights into this issue. Hunter-gatherer adaptation is not a uniformly simple strategy. As delineated by Robert. Bettinger, two distinctive strategies can be discerned: processing, which indicates individuals invest more time in food preparation through activities like grinding and filtering to render inaccessible items, such as plant seeds; and foraging, characterised by investing more time in the direct collection of accessible items (Binford 1980; Bettinger and Bumhoff 1985).[1] The varying emphasis on these strategies results in divergent mobility patterns. When hunter-gatherers lean more towards a "processing" strategy, they need to reside in one location for longer periods to prepare food, consequently diminishing their mobility.

However, as resource patches become scarcer, hunter-gatherers inevitably increase mobility, encompassing both inter-patch and intra-patch movement (Chapter One). The augmentation of mobility has its limits. Without the aid of animals (like horses) or watercraft, the coverage of food-seeking territory by foot is limited by walking speed and carrying capacity, and remains constrained. This implies that inter-patch movement cannot infinitely expand. According to E. S. Higgs, a hunter-gatherer group typically operates within a resource domain with a radius approximating two hours of travel distance (Higgs 1975). When the degree of resource scarcity extends beyond this range and exceeds the utilisation scope of hunter-gatherers, the option of intra-patch movement might be favoured. However, such a strategy is also unsustainable, as no single patch can continuously supply naturally available food resources over the long term. To prolong residence within a resource patch, two alternatives emerge: broadening the diet to incorporate typically unused food sources; or adopting a "processor" strategy to transform inaccessible resources into accessible ones. In accordance with the Optimal Foraging Theory, hunter-gatherers would opt to prolong their stay in high-quality resource patches, unless the overall resource conditions are unfavourable, or until a comprehensive understanding of resource distribution is processed, which would guide the decision to relocate from optimal patches.

As long as some groups choose to stay, the overall mobility of hunter-gatherers might be influenced. Here, a "dance floor theory" is employed. When people dance, although the space is limited, collisions are rare because everyone is in motion, allowing them to avoid collisions promptly. However, if someone within the dance floor suddenly stops for some reason, three possibilities can be anticipated: Firstly, colliding with others, resulting in discomfort and necessitating the departure of both parties from the dance floor to not obstruct others; Secondly, the stationary individual chooses not to leave, causing adjacent dancers to slow down to avert collisions. As their speed decreases, it inadvertently affects the pace of others. If this situation persists, more individuals might stop. The more people stop, the less viable the dancing becomes within the dance floor, ultimately compelling everyone to halt; The third possibility involves adopting a differentiation strategy. That is, the stationary individual remains at the centre the dance floor, while those intending to continue dancing revolve around the periphery, refraining from entering the central region. There remains a defined distance between the two groups (Fig. 2.1).

This model can explain the repercussions of lost mobility. There are three scenarios could be predicted as certain hunter-gatherer's groups reduce their mobility or initiate sedentism due to certain factors:

a. Escalated conflicts between settled and mobile groups could be seen;
b. Expansion of settled livelihoods, even without distinct attractions;
c. Hunter-gatherers seek to preserve mobility, while avoiding conflicts with settlers. This requires relocation to peripheral regions.

The least plausible eventuality is the departure of settlers, similarly implausible is the transformation of all hunting groups into settlers. In case conflicts arise between settled and mobile groups, mobile groups possess more advantageous manoeuvrability. In this case, to ensure greater safety, settled groups developed larger communities. A buffer space is needed to maintain peaceful coexistence. However, as the size of the settled population grows, the options for mobile individuals diminish. Consequently, mobile hunter-gatherers either become settlers or continue migrating towards peripheral environments. As recorded in ethnographic materials, hunter-gatherer communities often inhabit marginal environments; while no hunter-gatherer community can coexist within settled communities, as they cannot share the same territory. Moreover, these environments are often characterised by extreme low biological productivity, such as deserts, or unsuitable for human habitation, such as tropical rainforests. Survival in such settings demands higher mobility, which is contrary to the origin of agriculture.

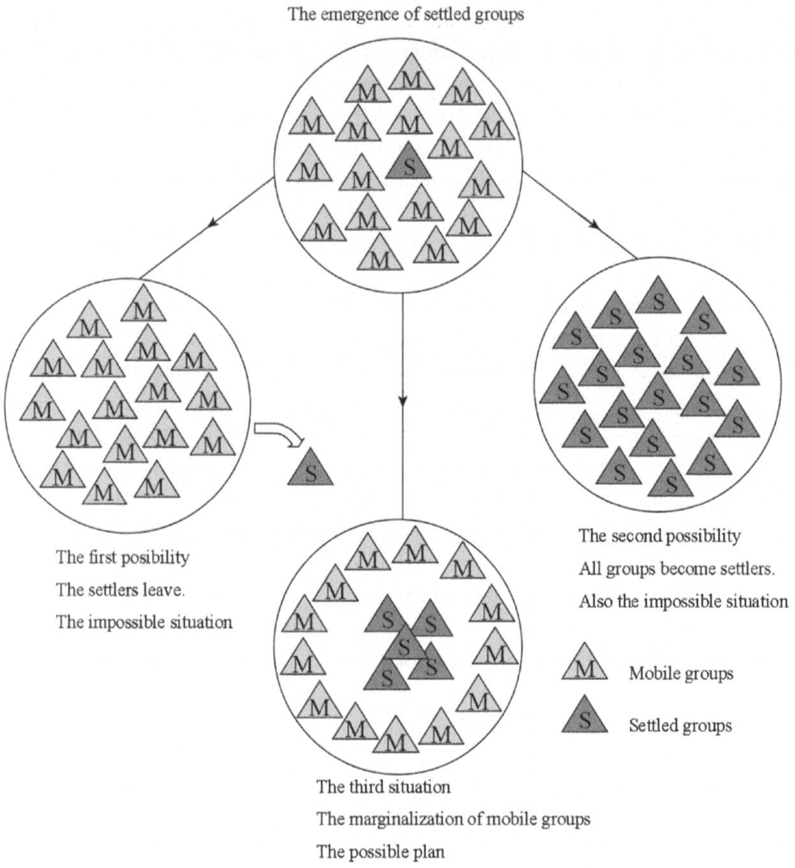

Fig. 2.1 The "dance floor model" of settlement formation

In summary, mobility is the fundamental survival strategy for hunter-gatherers. Additionally, sedentism and hunting-gathering are undeniably contradictory, except in specific circumstances. Hunter-gatherers who lose their mobility cannot sustain the hunting-gathering subsistence strategy. To obtain essential livelihood resources within limited living space, the choice of agriculture becomes inevitable, in whatever form it may take. Without adopting agriculture, the alternative is to abandon agriculturally viable land and relocate to more inhospitable environments in pursuit of survival, which necessitates the continuation of heightened mobility. Consequently, the path to the origin of agriculture deviates further from the trajectory of these mobile hunter-gatherers.

2.5 The History of Adaptation: Essential Endowments Structure

The juxtaposition of the origins of agriculture and the Industrial Revolution in the present discourse serves not only the purposes outlined in the introduction but also due to a significant reason, which is their potential explication through a unified theory. The profound historical significance stemming from both the origins of agriculture and the Industrial Revolution is beyond dispute. These transformations entail not only profound shifts in subsistence strategies but also changes in settlement patterns, social organisations, and even ideological frameworks. Such internal consistency compels us to question why some hunter-gatherer societies failed to invent or adopt agriculture, just as we inquire why certain societies did not experience an Industrial Revolution. Similarly, the modernisation of developing nations to catch up with developed ones constitutes a core question in modern economics. Economists universally acknowledged that the development of advanced industry, particularly in high-tech sectors, has been the foundation of development in developed countries. Over the past half-century, economists have proposed numerous developmental strategies aimed at propelling developing nations to surpass developed ones. However, most of these strategies culminated in failure, with only a handful of countries and regions achieving success. The promising beginnings of the experiments undertaken by South American nations in the 1970s eventually led to catastrophic consequences, resulting in prolonged economic stagnation. The divergence in outcomes prompts the question: Why did these two scenarios yield such disparate conclusions?

Yifu Lin and other economists have employed the economic theory of "comparative advantage" to explain this divergence, underscoring the role of disparate "endowment structures" in shaping different paths of development. This concept encompasses the three key factors of economics: resources, labour, and capital. All developing countries face scarcities in capital and skilled labour, both of which necessitate prolonged periods of cumulative development to acquire. Western developed countries have undergone centuries of capital accumulation and societal advancement, which marks a process developing nations lack. To catch up with developed countries, South America resorted to borrowing to procure the capital required for technology-intensive advanced industries (heavy industries), resulting in the higher cost of this capital compared to Western developed countries. Simultaneously, due to the absence of skilled workers, the products produced lacked competitiveness in the market, leading to the failure of catch-up initiatives and the accumulation of significant debt, which subsequently hampered further development. In contrast, the "Four Asian Tigers" leveraged their comparative advantage, utilising their abundant labour resources to progress from raw material processing to labour-light industries with low demands for technological capital accumulation. By amassing capital and nurturing skilled labour, the "Four Asian Tigers" transitioned to capital and technology-intensive industries only when conditions were ripe, successfully achieving industrial upgrading and becoming developed countries and regions (Lin 2004).

The endowment theory can also be employed to explain the variations in the timing of agricultural origins across different regions; and the reasons why some societies, even when surrounded by agricultural communities, did not adopt agriculture as livelihood. Similar to the Industrial Revolution, the emergence of agriculture is not a sudden intellectual revelation, instead, it necessitates through cultural accumulation and suitable environmental conditions. Fundamental prerequisites of agricultural origins include domesticable species, favourable climates, and suitable soils. For hunter-gatherers, the realisation of agricultural origins requires a series of preparatory elements. Binford categorises the mobility strategies of hunter-gatherers into two types: collectors and foragers. Collectors possess relatively fixed central camps, accompanied by temporary camps and specialised activity sites such as slaughter areas, storage points, and hunting blinds. Foragers, on the other hand, lack such differentiated dwelling types and spend comparable amounts of time at each location (Binford 1980). The difference between the two strategies is obvious, that is, the collectors employ the strategy of food approaching people; while the foragers seek for food. Based on Binford, these two strategies represent the extremes of mobility. Many hunting and gathering societies lie somewhere in between and may concurrently employ both approaches. Generally, the forager strategy can meet needs in areas with evenly distributed resources. Conversely, in regions characterised by highly patch resource distribution, where food resources are concentrated in specific areas where each with distinct concentrations of different food resources, the collector strategy is more advantageous. In essence, these strategies also exhibit regional variations (Table 2.2).

Different mobility strategies have significantly affected various other factors. For collectors, the necessity to reside in one place for longer durations implies a value in investing in long-lasting habitation construction, making durable tools, and developing storage technologies and facilities. In addition, to avoid missing critical resource acquisition opportunities, the patch distribution of resources suggests that these foraging groups need improved time planning and more intricate labour division and organisation. For instance, the annual migration of mahseer fish to inland rivers occurs within specific timeframes, and the success of their capture depends heavily on accurate timing judgments, as being too early or too late can result in missed opportunities (Arima and DeWhirst 1990).These

Table 2.2 Endowment theory differences between the two foraging strategies

Strategy	*Distribution area*	*Sedentism structure*	*Endowment structure*
Foragers	Regions with relatively even resource distribution, such as tropical rainforest	Lack of distinct differentiation in sedentism types with strong mobility	Pragmatic tools; smaller social group sizes; higher mobility; loose social organisation; shorter time for depth considerations; lower fertility rates
Collectors	Regions with patch resource distribution, varying concentrations in different areas, common in temperate environments	Clear differentiation in sedentism types, presence of central camps with potential for longer stays	Potential investment in robust infrastructure, durable tools, storage technologies, and facilities; development of intricate labour division and social organisation; longer time for depth considerations; higher fertility rates, possibly leading to higher population density and larger social scale; greater potential for surplus production and increased social complexity

adaptive practices are essential for the origins of agriculture. Furthermore, relatively stable sedentisms and more refined food processing might lead to higher fertility rates, potentially resulting in larger population sizes for the group (Hillman 1989).[2] Agriculture, as a labour-intensive livelihood, demands increased labour input, especially from males. A society, with larger population and more skilled labour organisations, could realise agriculture as the mode of subsistence. With more male labourers devoted to agricultural production, a reduction in hunting time would be seen. Consequently, the overall mobility of the group would decrease.

Comparatively, hunting and gathering societies inclined towards the forager strategy cannot realise the agricultural origin, as the lack of the investments could be seen in sedentisms, tools, storage technologies, and facilities, which are characteristic of collector strategies. Similarly, foragers exhibit a deficit in complex labour division, social organisation, and population size. Furthermore, collector strategy often prevail in regions characterised by highly patch resource distribution. These foraging groups tend to focus on limited kinds but abundant resources in size such as migrating animal herds, mature fruit-bearing trees, and migratory fish schools. Through this strategy, foragers could gain a better understanding of the growth patterns of these food resources, enabling more manageable human intervention and the likelihood of obtaining surpluses that cannot be consumed all at once. These surpluses, whether plant fruits or hunted animals, serve as foundational conditions for domestication (Chapter 1, Sect. 5). Also, surplus production fosters the growth of social complexity; the expansion of social complexity and surplus production mutually reinforce each other. The absence of surplus products is also complementary to the dynamics of an egalitarian society. Unless this equilibrium is disrupted, hunting and gathering communities lack the impetus to expand their production.

Based on the analysis above, a region with a greater adoption of the collector strategy is more likely to adopt agriculture as a substance under consistent conditions. Compared to those with fewer adherents to this strategy, hunting and gathering societies that emphasise the collector strategy are more inclined to embrace agricultural livelihoods. The mobile strategies produced by two distinct endowment theories are products of long-term cultural adaptation processes rather than deliberate designs. Archaeological research benefits from its capacity for investigating processes on extended timescales. Through archaeological research, a comparative cultural system across diverse cultural ecologies and discern variations in their element endowments could be examined, thus, subsequently determining the relative timing of agricultural emergence in different regions.

2.6 New Friends: Symbiotic Relationship

If hunter-gatherers desire to preserve their mobility while reaping the benefits of agricultural production, what actions should they take? A straightforward approach is exchange, specifically exchanging the products from hunting and gathering practices with neighbouring agricultural societies. Such a relationship allows hunting and gathering communities to maintain their mobile lifeway while securing a more reliable food source. Durable storage crops, textiles, tools, and other products from agricultural societies enhance the ability of hunter-gatherers to withstand adaptive challenges. Simultaneously, agricultural societies can acquire animal hides, meat, and other forest resources from hunting and gathering communities, allowing them to focus on agricultural production without dispersing labour. This mutually beneficial arrangement constitutes a symbiotic relationship. Ethnographic materials show that hunter-gatherers rarely exist in isolation but possess varying degrees of exchange with neighbouring agricultural societies, spanning from tropical rainforests to frigid regions. Despite often being perceived as isolated entities, these hunter-gatherers are not as secluded as commonly assumed (Bailey and Headland 1991; Headland and Reid 1989). For instance, the Oroqen people in the Daxing'anling Mountains, who rely on hunting for subsistence, engage in extensive exchange relationships with nearby agricultural groups (Editorial Committee for the Revised Edition of Collection of Historical Survey Materials on Chinese Ethnic Minority Societies, 2009).[3] In tropical rainforest regions, hunter-gatherers often foster reciprocal relationships with neighbouring agricultural and pastoralist groups, engaging in work with agricultural communities during busy farming seasons or exchanging forest resources (Binford 2001). Similar phenomena are observed in archaeological materials, such as in prehistoric Northern Europe (Zvelebil 1996).

Such reciprocal relationships are particularly applicable in regions with uneven resource distribution. For instance, hunter-gatherers can hunt by exploiting forested areas; while agricultural groups engage in farming activities in river valley regions. Sparse woodland grasslands, which are surrounded in tropical rainforests, are struggling for agricultural communities to utilise resources effectively. Hunting and gathering communities, in turn, address this deficiency. During the Neolithic, reciprocal relationships persisted in Central Europe where agricultural communities farmed in the loess-covered hills; while valley and swamp areas were inhabited by hunting and gathering groups akin to the Mesolithic period (Wittle 1996). They coexisted in close proximity, engaging in a flow of goods and services. With undeveloped early agricultural systems, the synergy between cereal cultivation and animal husbandry had not developed optimally. Limited usage of draft animals for ploughing and transportation, as well as inadequate utilisation of products from animals, like milk, wool, hides, and even dung, leading to these incipient agricultural systems. For hunter-gatherers, their living spaces were limited due to the continuous encroachment and partition by agricultural communities. Consequently, sustaining an entirely independent lifeway became challenging, making products from agricultural communities indispensable. This interaction partially elucidates human successful adaptations in tropical rainforest regions (Baily et al. 1989). Areas, which were once unsuitable for human habitation, have been transformed due to the support of agricultural communities. As a result, certain groups have been able to pioneer the tropical rainforest as a new habitat, employing hunting and gathering practices.

Generally, interactions between the two types of communities are more advantageous for agricultural groups than for hunting and gathering societies. Agricultural communities exhibit higher population density and greater food production per unit of land area. In contrast, the population of hunting and gathering societies continually drains into agricultural communities through means such as marriage, dowry, wage labour, and adoption. As the population of agricultural communities grows and their technological capabilities improve, they gradually expand into previously untapped territories. For instance, people in agricultural groups can drain swamps for cultivation or develop crop varieties suitable for mountainous regions. Thus, the living space of hunter-gatherers is further encroached upon: they face a choice of either integrating into agricultural societies or migrating to even more marginal environments. The reciprocal relationship between the two groups eventually dissolves as agricultural societies mature. However, during the initial phases of agricultural origins, this mutual dependence ensures the early agriculturists' maintenance of a fragile mode of production. Simultaneously, the mutual relationship enables certain hunter-gatherers to sustain their substance without adopting agriculture.

2.7 Garden of Eden: Aquatic Resources Dependents

As aforementioned, mobility, as the fundamental adaptive strategy of hunter-gatherers, sustains a hunting and gathering livelihood. However, this assertion is conditional, as it primarily applies to the utilisation of terrestrial resources by hunter-gatherers. Ethnographic and archaeological evidence both suggest that those hunter-gatherers who heavily depend on aquatic resources tend to exhibit higher levels of sedentism (remaining in one place for over a year), some degree of social stratification (departing from egalitarianism), and population densities resembling those of agricultural communities (Yesner 1980). Those reliant on aquatic resources do not engage in agricultural production and solely depend on natural resources. Apparently, their cultural adaptation strategies distinctly differ from those of other hunting and gathering societies, challenging the cultural adaptation mechanisms discussed in previous sections.

The apparent exception to this is due to the fact that the scope of resources obtained by aquatic resource-dependent societies is much broader than that of terrestrial resource-dependent ones. For instance, marine fish are not subject to the same constraints as land animals. Compared to land resources, aquatic resources are dynamic and their sources encompass a vast expanse, thus, hunter-gatherers cannot easily deplete resources in a specific aquatic area. For instance, although species like the Oncorhynchus aggregate in inland rivers during their reproductive season, their habitat spans oceans. Although hunter-gatherers exploit resources from localised points, their resource catchment includes the boundless marine habitat where such species thrive. In contrast, terrestrial resources, whether animal herds or plant species, possess narrower distribution ranges often restricted by geographical features like mountains, rivers, and deserts that hinder biological mobility. Consequently, at a given resource collection position, terrestrial resources lack continuous replenishment, diminishing their capacity to sustain hunter-gatherer livelihoods in comparison to aquatic resources.

Furthermore, the seasonality of aquatic resources differs from that of terrestrial resources, allowing aquatic resources to compensate for the seasonal scarcities of land resources. Thus, enhancing the utilisation of aquatic resources can mitigate the seasonal risks faced by hunter-gatherers, resembling the advantages of agriculture. In regions where huntable animal resources available are relatively scarce, such as the Pacific Northwest coast of North America, the exceedingly damp climate is also unfavourable for storing plant-based food. Hence, aquatic resources become an attractive alternative (Arima and DeWhirst 1990, pp. 393–411). In Arctic regions where plant-based food is scarce and terrestrial animals are few, relying on marine resources becomes the optimal choice.[4]

Additionally, aquatic resources, like fish, shellfish, aquatic plants, and marine mammals, can provide abundant sources of protein, fat, minerals, and carbohydrates; offer raw materials for clothing and tools (Editorial Committee for the Revised Edition of Collection of Historical Survey Materials on Chinese Ethnic Minority Societies 2009). These aquatic resources can even serve as substitutes for animal hunting and partial plant gathering. Some resources, such as shellfish collection, can be managed by women, elderly individuals, and children. Rendering protein acquisition is easier than hunting animals. This effectively expands the labour force population and contributes to the survival of the community.

Certainly, the utilisation of aquatic resources requires certain prerequisites, often involving more intricate technological preparations, such as boats, fishing nets, harpoons, and spears. Complex technologies can enhance the reliability of resource acquisition. In areas lacking alternative resources, survival could be jeopardised without stable and reliable technological support. This also explains why Arctic hunter-gatherers possess the most sophisticated technologies. Similarly, it can be argued that technology serves as one of the limiting constraints on people's utilisation of aquatic resources. Take the prehistoric hunter-gatherers along the Nile River in Egypt as an example: their ability to fish in deeper waters emerged as late as the Epipaleolithic (Wendorf and Schild 1989), while earlier hunter-gatherers could only catch fish in shallower waters.

Apart from the Native Americans of the Pacific Northwest coast and the Eskimos of Alaska, the Jomon Age (ca. 10,500–ca. 300 B. C.) in Japan and the Mesolithic in Europe also exemplify archaeological instances of dependence on aquatic resources (Aikens and Higuchi 1982; Price 1987). These societies share the common traits of enduring over extended periods, along with possessing several common features of the Neolithic, such as ceramic vessels, fixed sedentisms, and ground stone tools. However, there are no signs of agricultural production found in these cultures. It was not until fairly mature agricultural systems emerged, marked by improved domesticated species and more efficient tools like ploughs and iron tools, that these regions gradually transitioned to accepting agriculture.

For all hunter-gatherers, the utilisation of aquatic resources might not necessarily represent a utopia for maintaining their existing lifeway. However, this distinct mode of resource utilisation does indeed set them apart from the vast majority of pure terrestrial resource-dependent hunter-gatherers. It becomes an additional factor, which is beyond the general cultural adaptations of hunter-gatherer societies, to explain why agriculture failed to emerge in certain regions.

2.8 Conclusion

Chinese people might wonder, would China have developed an industrial revolution on our own without Western intervention? This is possible, as in the economic domain, significant sprouts of capitalism appeared in the southeastern parts of China by the late Ming Dynasty; in the realm of thought and culture, a group of Enlightenment-inspired pioneers had emerged. Even during the Qing dynasty, which saw a reversion to traditional values, the market economy was slowly advancing, as witnessed by the prosperity of *qian zhuang*/unincorporated bank. However, China's path to a mature market economy was interrupted by the Western naval power and artillery, just as it was on the brink of such a transformation.

Similar issues arise in research on the origins of agriculture. If there had been ample time and no interference from agricultural societies, could those hunter-gatherer groups in ethnographic records have initiated agriculture? Agriculture, as a livelihood technology, has been reinvented multiple times. It could be confirmed that the Old and New Worlds independently developed their own agricultural practices without mutual influence. In fact, within the Old World, agriculture emerged in different regions, in near isolation from one another. There is no evidence to suggest that these regions, including China, Western Asia, New Guinea, and West Africa, experienced external influences during the initial phases of agricultural development. Although the agricultural origin indeed facilitated and stimulated population growth, it cannot be denied that agriculture provides a solution to the population crises. To address humans' sustenance challenges, agriculture represents the inevitable and the sole effective method. Among the various strategies employed by hunter-gatherers to adapt to environmental pressures, only the strengthening of resource utilisation could lead to food production initiation, ultimately emancipating them from the constraints of natural resources on population growth.

Agriculture is a product of inherent human societal needs, a culmination of cultural adaptations over a prolonged period. Simultaneously, agriculture represents the response that hunter-gatherers were compelled to adopt under the pressures of subsistence. There were no more suitable alternatives available. Therefore, the essence of the question "why did agriculture not originate" lies in understanding "why did agriculture not originate at that particular time" instead of asserting that agriculture never has originated. In summary, the reasons encompassed the lack of necessity, insurmountable environmental constraints, inadequate preparation, and the presence of alternative options. In regions without population pressure, there was little impetus for the origin of agriculture. Since agriculture is a labour-intensive production activity, there was no incentive to adopt increased labour in circumstances where hunter-gatherers could subsist on natural resources. The population crisis not only manifested in the scarcity of food resources but also in competitive allocation. This situation promoted the development of social complexity, constituting an additional impetus for the advancement of agricultural production. Even possessing this impetus, the agricultural origins remain challenging without environmental and technological conditions and social organisational preparedness, which belong to the aforementioned endowment theory. Finally, if the adaptive pressures faced by hunter-gatherers could be addressed through other less costly channels, such as dependence on other agricultural groups or utilisation of aquatic resources, agriculture would hold little appeal for them.

The question of "why did agriculture not originate" remains a component of the overarching issue of the origin of agriculture, albeit addressing the aspect of its delayed inception. The subsequent chapter will address, under ideal conditions, which regions were the earliest to witness the emergence of agriculture and which regions experienced this transition at a later phase.

Notes

1. In contrast to Binford's collector-forager model, which primarily focuses on the composition of archaeological sites and emphasizes different levels of mobility, Bettinger's model takes a different perspective. It directly stems from the collector -forager theory in animal ecology and provides a more helpful explanation for the loss of mobility.
2. The consumption of accessible aquatic plant rhizomes facilitates early weaning of infants, leading to a shortened reproductive interval and population growth.
3. For example, the Hezhen people use fish skin to make clothing.
4. The population size of terrestrial animals is contingent upon the availability of plants. Despite the substantial population size of North American caribou, their distribution range is exceptionally extensive, resulting in a persistently low secondary productivity, that is, animal production, per unit area.

References

A. Whittle, *Europe in the Neolithic*: The Creation of New Worlds. Cambridge University Press, Cambridge, 1996.

C. C. Mann, Earthmovers of the Amazon. *Science*, 297: 786–789, 2000.

C. C. Mann, The real dirt on rainforest fertility. *Science*, 297: 920–923, 2000.

C. M. Aikens and Takayasu Higuchi, Prehistory of Japan. Academic Press, New York, 1982.

D. R. Yesner, Maritime hunter-gatherers: ecology and prehistory. *Current Anthropology*, 21: 727–750, 1980.

E. Arima and J. DeWhirst, Nootkans of Vancouver Island. In W. Suttles ed., *Handbook of North American Indians*, Vol. 7, Northwest Coast, pp. 391–411. Smithsonian Institution Press, Washington DC, 1990.

E. F. Moran, *Human Adaptability*. Westview, Boulder, 2000.

E. S. Higgs ed., *Palaeoeconomy*. Cambridge University Press, Cambridge, 1975.

Editorial Committee for the Revised Edition of *Collection of Historical Survey Materials on Chinese Ethnic Minority Societies: A Historical Survey of the Oroqen Ethnic Group* (Revised Edition). The Ethnic Publishing House, 2009

F. Wendorf and R. Schild, Summary and synthesis. In F. Wendorf et al eds., *The Prehistory of Wadi Kubbaniya*, Vol. 3, Late Paleolithic Archaeology, pp. 119–161. Southern Methodist University Press, Dallas, 1989.

F. Wendorf and R. Schild, Are the early Holocene cattle in the eastern Sahara domestic or wild? *Evolutionary Anthropology*, 3: 118–128, 1994. Nabta Playa and its role in Northeastern African Prehistory. *Journal of Anthropological Archaeology*, 17: 97–123, 1998.

Editorial Committee for the Revised Edition of Collection of Historical Survey Materials on Chinese Ethnic Minority Societies: A Historical Survey of the Oroqen Ethnic Group (Revised Edition). The Ethnic Publishing House, 2009.

For example, the Nuu-Chah-Nulth people of the Northwest Coast of North America. Reference: E. Arima and J. DeWhirst, Nootkans of Vancouver Island. In W. Suttles ed., *Handbook of North American Indians*, Vol. 7, Northwest Coast, pp. 391–411. Smithsonian Institution Press, Washington DC, 1990.

G. Hillman, Wild plant foods and diet at Late Paleolithic Wadi Kubbaniya: the evidence from charred remains. In F. Wendorf eds., *The Prehistory of Wadi Kubbaniya*, Vol. 2, Stratigraphy, Paleoeconomy and Environment, pp. 119–161. Southern Methodist University Press, Dallas, 1989.

J. D. Sachs et al., The geography of poverty and wealth. *Scientific American*, 284 (3): 70–75, 2001.

J. D. Speth and K. A. Spielmann, Engergy source, protein metabolism, and hunter-gatherer subsistence strategies. *Journal of Anthropological Archaeology*, 2: 1–31, 1983.

J. E. Eisenberg, The density and biomass of tropical animals. In M. E. Soule and B. A. Wilcox eds., *Conservation Biology: An Evolutionary-Ecological Perspective*, pp. 35–55. Sinauer Associates, Sunderland, 1980.

J. E. Yellen, Long term hunter-gatherer adaptation to desert environment: a biogeographical perspective. *World Archaeology*, 8: 262–274, 1976.

K. Neumann, New Guinea: A cradle of agriculture. *Science*, 301: 180–181, 2003.

L. Minc and K. Smith, The spirit of survival: cultural responses to resource variability in North Alaska. In P. Halstead and J. O'Shea eds., *Bad Year Economics*, pp. 8–39. Cambridge University Press, Cambridge, 1989.

L. R. Binford, *Constructing Frames of Reference: An Analytical Method for Archaeological Theory Building Using Hunter-Gatherer and Environmental Data Sets*. University of California Press, Berkeley, 2001.

L. R. Binford, *In Pursuit of the Past*, translated by Chen Shengqian, Shanghai Sanlian Bookstore Co.Ltd., 2009.

L. R. Binford, Willow smoke and dog's tails: Hunter-gatherer sedentism systems and archaeological site formation. *American Antiquity*, 45: 1–17, 1980.

Lin, Yifu, *China's Miracle: Development Strategy and Economic Reform* (Revised and Enlarged Edition), Shanghai Sanlian Bookstore Co.Ltd., 2004.

M. D. Sahlins, Notes on the original affluent society. In R. B. Lee and I. DeVore eds., *Man the Hunter*, pp. 85–89. Aldine, Chicago, 1968.

M. J. Heckenberger et al., Amazonia 1492: Pristine forest or cultural parkland. *Science*, 301: 1710–1714, 2003.

M. Zvelebil, The agricultural frontier and the transition to farming the in the CircumBaltic Region. In D. R. Harris ed., *The Origins and Spread of Agriculture and Pastoralism in Eurasia*, pp. 323–345. Smithsonian Institution Press, Washington DC, 1996.

P. Mithell, The archaeological study of epidemic and infectious disease. *World Archaeology*, 35: 171–179, 2003.

R. B. Lee and R. Daly, *Cambridge Encyclopedia of Hunter and Gatherers*. Cambridge University Press, Cambridge, 1999.

R. C. Bailey and G. Headland, The Tropical Rainforest: Is it a productive environment for human foragers? *Human Ecology*, 19: 261–285, 1991.

T. N. Headland and L. A. Reid, hunter-gatherers and their neighbors from prehistory to the present. *Current Anthropology*, 30: 43–66, 1989.

R. C. Baily et al., Hunting and gathering in Tropical Rain Forest: Is it possible? *American Anthropologist*, 91: 59–82, 1989.

R. L. Bettinger and M. A. Bumhoff, The Numic spread: Great Basin cultures in competition. *American Antiquity*, 47: 485–503, 1985.

T. D. Price, The Mesolithic of Western Europe. *Journal of World Prehistory*, 1: 225–306, 1987.

T. P. Denham et al., Origins of agriculture at Kuk Swamp in the Highlands of New Guinea. *Science*, 301: 189–193, 2003.

CHAPTER 3

Modelling Hunter-Gatherers of China

> The heavens and the earth are not benevolent; they treat all things as straw dogs.
>
> —— Laozi

3.1 Introduction

In the era of industrialisation, it appears that humans have seemingly forgotten their origins in nature and the constraints imposed upon them by the natural world, which does not inherently favour humanity. For over 99% of human evolutionary history, our ancestors relied on hunting and gathering as their primary means of subsistence. Even in some marginal environments, such as polars, tropical jungles, and deserts, the hunting and gathering lifeway persisted until modern times. All food resources for these hunter-gatherers came from naturally growing plants and animals, upon which their survival depended. Apart from tool usage, there was no fundamental distinction between human foraging behaviour and that of other animals. However, from the Late Pleistocene to the Early Holocene, this hunting-gathering mode of subsistence was gradually substituted by agricultural production in the northern regions of China and the middle and lower reaches of the Yangtze River. Humans no longer relied on naturally occurring plants and animals for survival, but instead started to artificially control the growth of plants and animals

S. Chen, *Prehistoric Modernization*,
https://doi.org/10.1007/978-981-97-9423-2_3

through domestication and food production, thereby no longer restricted by food constraints of the natural state. As a consequence, pastoralism, which marked another form of food production, emerged in regions beyond the Great Wall where were less suitable for grain agriculture. Despite this, only a few hunter-gatherer groups survived in some special ecological environments. Today, when we seek to understand prehistoric hunter-gatherers, the most common approach is to rely on archaeological discoveries and research.

With the continuous development of archaeology, the exploration of human history has evolved from simple artifact studies to the realm of interpreting human cultural behaviour. However, a considerable gap between the material evidence found from archaeological discoveries and the interpretation of human cultural behaviour still remains, requiring a process of inference. Archaeologists could bridge this gap through deductive reasoning based on archaeological theory, inductive reasoning by multidisciplinary analysis, and analogical reasoning by ethnoarchaeology and experimental archaeology (Chen 2007). Among these approaches, research based on ethnoarchaeology often faces a dilemma where individual cases lack sufficient persuasive power despite the wealth of ethnographic material available. Consequently, archaeologists make efforts to distil applicable frameworks from this abundant material, primarily by combining ecological simulations and comprehensive analysis of ethnographic data, following a pioneering approach like that of Binford (Binford 2001). The present simulated research is conducted on this basis.

The advantage of simulation research lies in its nature to simplify investigations. This approach is widely used in Natural Sciences and also Social Sciences, such as Economics. Simulation research proves to be a valuable method to analyse phenomena that cannot be directly observed, especially those related to vanished processes in human history. The goal of simulation research is not to replicate the most accurate processes. As Robert L. Kelly puts it, simulating the behaviour of hunter-gatherers does not entail duplicating all actual circumstances but rather emulating the genuine aspects of hunter-gatherer actions based on predefined models and conditions within specific contexts (Kelly 1995). Simulations can inspire further research, particularly when dealing with uncertain situations. By intentionally designing certain inaccurate models and then examining deviations from expected outcomes, researchers can uncover unknown aspects and identify directions for future investigations (Binford 2001). Simulated research on hunter-gatherers often focuses on environmental variables, simulating their livelihood behaviours (Jochim 1976; Winterhalder 1986; Mithen 1989), dwelling forms (Pickering 2003), or comparing the effectiveness and exploring certain phenomena through utilising multifaceted variable simulations incorporating economic, cultural, and environmental elements (Ebert 2004). The simulation approach proposed by Binford not only uses environmental variables but also draws from abundant ethnographic data, making the simulations more realistic and reliable. This method is valuable for understanding adaptive behaviours of prehistoric ancestors.

3.2 The Method of Ecological Modelling for the Prehistoric Hunter-Gatherers

All hunter-gatherers live within specific ecological environments, occupying defined geographical areas, and rely on the available food resources within these regions for sustenance. Their adaptive behavioural patterns are deeply influenced by the characteristics of the ecological environment and its fluctuations. Ecological simulation involves calculating and simulating the environmental characteristics and food resource conditions upon which hunter-gatherers depend for survival. By establishing a livelihood model based on the aforementioned ecological factors, researchers can consequently deduce and predict how hunter-gatherers would respond to changeable resource conditions.

Remarkable regional differences and species diversity exhibit in the Earth's biosphere, and which variations are connected through the circulation of energy and matter. Particularly, energy, which serves as the "currency" in comparative ecological research, holds commonality and enables comparisons between different ecosystems, acting like currency in Economics, therefore, facilitating the comparison of values between various economic activities. The primary source of biological energy is derived from plants through photosynthesis, followed by the ecological cycle of energy, such as from plant-eating animals to carnivores that feed on herbivores, and finally to decomposers like microorganisms.

Two ecological concepts could be commonly employed to describe energy: production and productivity. Production encompasses primary production and secondary production. Primary production refers to the conversion of solar energy by plants in an ecosystem into chemical energy, which takes the form of organic matter that can be utilised by other consumers and decomposers. Secondary production mainly refers to the assimilation and reproduction of animals using the organic matter produced by primary productions. Productivity represents the production quantity and refers to the amount of organic matter produced per unit area and unit time, indicating the rate of production. It is typically calculated as the mass of new cell production per square metre per year. For example, in this simulated research, the concept of "Net Above-Ground Productivity" (NAGP) is utilised, developed by Michael L. Rosenzweig (2001) and adjusted by Binford:

$$\log_{10} \text{NAGP} = \{[1.0 + (1.66 \pm 0.27)] \times [\log_{10} \text{AE}]\} - (1.66 \pm 0.07)$$

In this case, AE represents the annual actual evapotranspiration rate (in millimetres).

The actual annual evapotranspiration rate is not only related to water resource conditions but also to solar radiation energy. The factors that affect plant photosynthesis include solar radiation, water, and carbon dioxide. Since carbon dioxide is essentially constant, thus, in our simulation research, only the conditions of water and heat will be considered.

Regarding the detailed process of the simulation method, Binford supplies a comprehensive introduction, which will not be reiterated here. In this simulation research, to simulate the impact of ecosystem characteristics on hunter-gatherers, the variables utilised are derived from 431 weather stations in China, including monthly average temperature, monthly average precipitation, latitude and longitude, elevation, and distance from the ocean (the Climatic Data Center, National Meteorological Information Center, China Meteorological Administration 1985).

3.3 Basic Features of the Ecological System of Hunter-Gatherers in Prehistoric China

Undoubtedly, human behavioural characteristics are profoundly influenced by the characteristics of the ecosystems they inhabit. China boasts a vast territory with diverse and complex topography. The horizontal distribution determined by latitude and longitude, the vertical distribution determined by altitude, and the land-sea distribution of the Eurasian continent all significantly affect the structure and distribution of human ecosystems in China.

The productivity of an ecosystem primarily depends on the length and quality of the growing season, such as the temperature. H. P. Bailey considers 8 °C as a constant for the beginning and end of the growing season (Bailey 1960). In China, when considering the length of the growing season (Fig. 3.1), regions where plants can grow throughout the year could be only found south of the Tropic of Cancer, where food resources are available year-round. Within the Sichuan Basin, the length of the growing season is similar to that in South China, as the surrounding high mountains block cold air currents. Additionally, potentially due to more abundant rainfall, the length of the growing season along the Qinling Mountains, the Funiu Mountain, and the Dabieshan Mountains is greater than that in the middle and lower reaches of the Yangtze River. On the other hand, China's northwest, southwest, and northeast regions are influenced by altitude and the continental climate, resulting in a decrease in the length of the growing season with higher altitude, latitude, and inland distance. In the hinterland of the Qinghai-Tibet Plateau, the length of the growing season tends towards zero, with almost zero plant growth in this region.

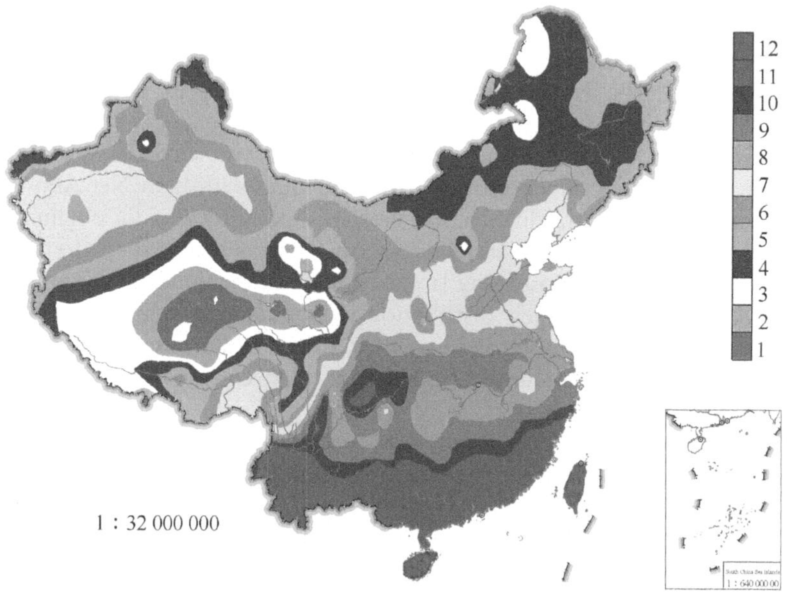

Fig. 3.1 Distribution of the length of growing season

Despite the elements aforementioned, plant growth requires water. Thus, the true productivity of an ecosystem demands both water and thermal conditions. Based on Fig. 3.2, it can be observed that the distribution of Net Above-Ground Productivity does not entirely coincide with the distribution of the length of the growing seasons. However, the high Net Above-Ground Productivity does not necessarily represent abundant food resources for hunter-gatherers. In areas with high Net Above-Ground Productivity, plants often invest most of their energy in elongating to compete for sunlight; with a significant portion of this energy distributed to branches, which are found at treetops and ends of branches, which are either inaccessible or difficult to gather. In contrast, in regions with lower Net Above-Ground Productivity, plants normally invest more of their energy in reproduction, such as reproducing seeds. Additionally, in arid and semi-arid areas, plants commonly develop extensive rhizomes to adapt to frequent droughts and wildfires. Thus, for hunter-gatherers, the most abundant terrestrial plant-based food resources are found in relatively arid areas (aquatic resources are excluded), rather than in lush tropical and subtropical forest regions, or in plant-scarce desert and arid regions. If hunter-gatherers in these areas are solely reliant on terrestrial plants for sustenance, larger areas should be covered in their search for sufficient energy. In other words, their mobility would be higher.

Corresponding to Net Above-Ground Productivity, secondary productivity also exhibits similar characteristics. In areas with high Net Above-Ground Productivity, due to the scarcity and difficulty in obtaining accessible plant parts, animals tend to be smaller to facilitate access to food in the forest canopy. Possessing larger body sizes, the numbers of animals are often lower, and their distribution is sparse, as seen in the case of elephants. In regions, such as Lingnan and the middle and lower reaches of the Yangtze River, animal resources are relatively scarce compared to surrounding areas. In contrast, in semi-arid grassland and forest, large groups of large herbivores can be observed. As depicted in Fig. 3.3, the regions with the most abundant animal resources are located in the transitional zone between the Inner Mongolian grasslands and the temperate deciduous broad-leaved forests of North China and Liaoning (coinciding with the 400-millimetre isohyet of precipitation), the eastern edge of the Qinghai-Tibet Plateau (also a transitional zone between grasslands and forests), and certain localised areas in the southwest. Additionally, there

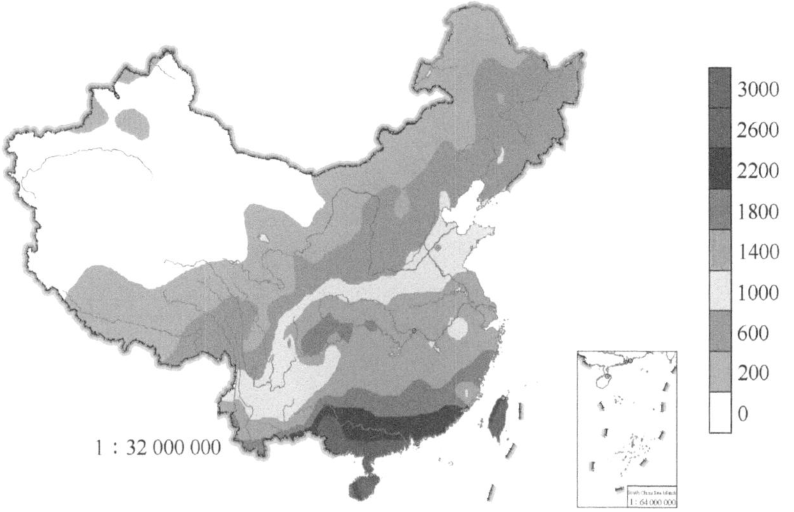

Fig. 3.2 Density distribution of the net surface productivity

is an unusual high-abundance zone distributed in the Hengduan Mountains, which could be related to significant elevation differences in the layout of meteorological stations in this region. However, despite the fact that we may occasionally encounter groups of wild herbivores, the animal resources in the vast northwestern region are extremely sparse. Nevertheless, their activity range is vast, resulting in very low actual resource density.

3.4 Minimalist Terrestrial Model

The Minimalist Terrestrial Model will disregard the roles of rivers, lakes, and coastlines in the habitat of hunter-gatherers, assuming that humans primarily utilise terrestrial resources. In other words, humans rely solely on the plants and animals produced on land as food resources. This model, which serves as the baseline, observes the extent to which the real world deviates.

Simulating the Quantity of Accessible Plant Foods: As mentioned earlier, high Net Above-Ground Productivity does not necessarily result in more accessible food for hunter-gatherers, as plant growth structures

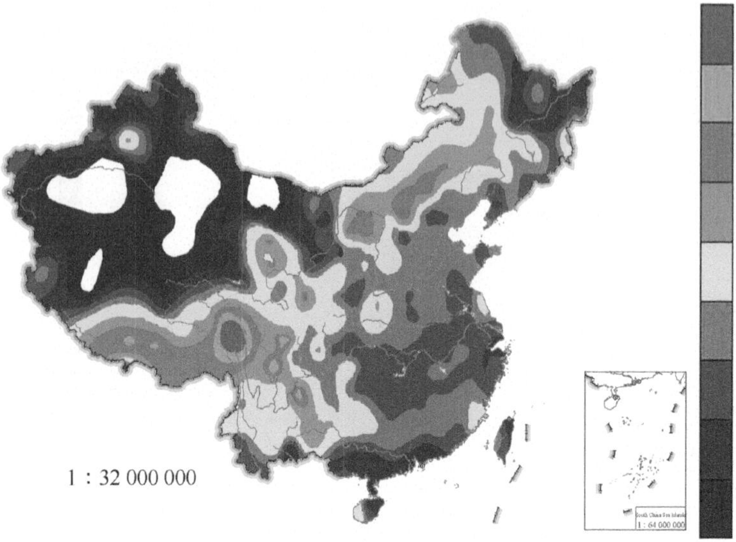

Fig. 3.3 Density distribution of secondary productivity (animals)

consume a significant portion of the energy for sunlight. Therefore, when simulating the quantity of accessible plant foods, this energy allocation must be taken into account. The specific utilisation pathways are not considered here; only the total quantity is addressed. For instance, in regions with high Net Above-Ground Productivity, most accessible food grows in the forest canopy, making it hard to access. Thus, the remaining quantity of accessible food is still relatively high, after accounting for the energy in the inaccessible part (Fig. 3.4). From Fig. 3.4, the regions with the highest quantity of accessible plant foods are located in the southern part of southwest China, the Sichuan Basin, and some areas in South China; while the regions with the lowest quantity are in the Qinghai-Tibet Plateau. The quantity of accessible plant foods in the North China and the middle and lower reaches of the Yangtze River regions is moderate, with North China surpassing the latter, despite the higher Net Above-Ground Productivity in the middle and lower reaches of the Yangtze River.

Simulating the Quantity of Accessible Animal Foods: Hunter-gatherers predominantly hunt ungulate animals (the mammal-lacking continent of Oceania is excluded). Therefore, Binford develops another

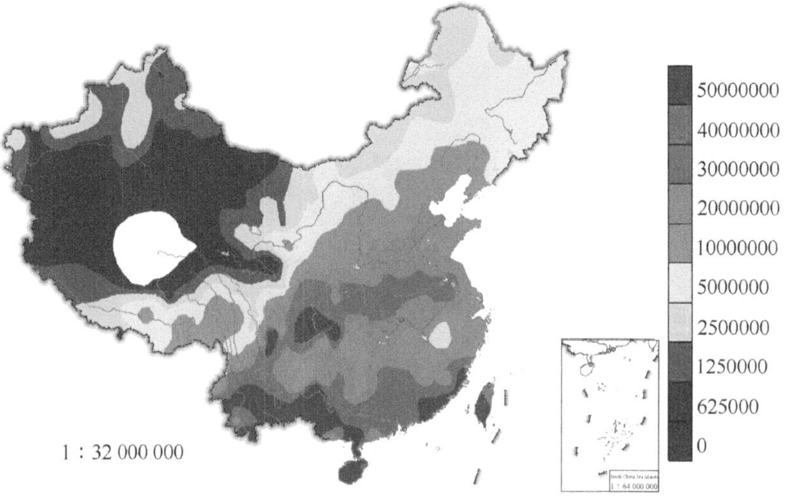

Fig. 3.4 Distribution of edible plant biomass

variable RXPREY, which is used to simulate the quantity of ungulate animals and viewed as a representative value for the quantity of accessible animal foods. As depicted in Fig. 3.5, the transitional zones, which cross the northeast to the southwest, and certain localised areas in southwest China exhibit relatively high quantities of accessible animal foods. In contrast, the middle and lower reaches of the Yangtze River and South China show lower quantities of accessible animal foods, while the vast northwest and most of the Qinghai-Tibet Plateau exhibit even lower quantities. Ungulate animals are widely dispersed in these regions, with a high degree of mobility exceeding what hunter-gatherers can achieve on foot. Without the aid of horses and long-range weapons, hunting such animals would be impractical. Hunter-gatherers tend to target animals near forest edges rather than those in grasslands, as the forest regions offer richer and more diverse food resources. The character of the forest results in these animals possessing well-defined territories without migratory natures. This explains why the highest quantity of accessible animal foods comes from the transitional zone between grasslands and forests, rather than from purely grassland or forest.

Simulation of Hunter-Gatherer Population Density: According to the biological principle of minimum subsistence level, an organism can

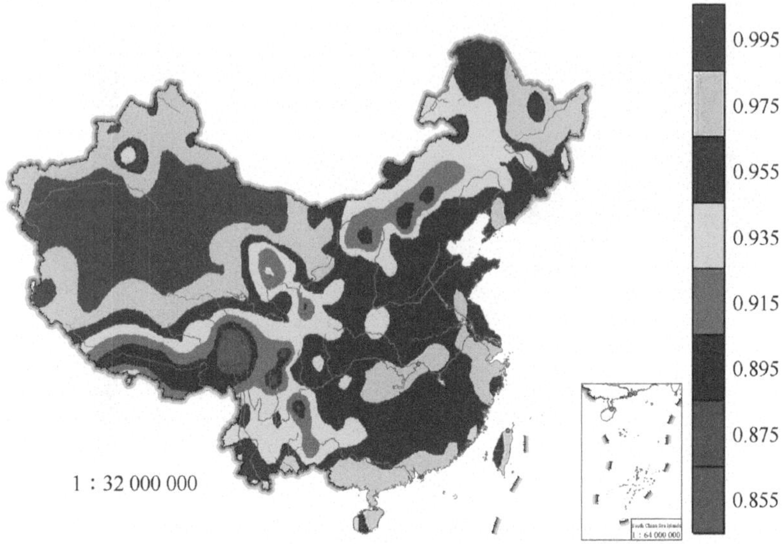

Fig. 3.5 Distribution of edible animal biomass

survive only when all conditions are met. The population of hunter-gatherers within an environment is determined by the period with the scarcest food supply during the year. If relying solely on plant-based food, the population density of hunter-gatherers would peak in certain regions of the southwest and South China. On the other hand, if relying solely on animal-based food, the population density peak would be limited to the transitional zone from the northeast to the southwest. Certainly, hunter-gatherers cannot solely rely on either plant-based or animal-based food. Considering both elements, we can conduct a population density model based on the complete reliance on terrestrial plants and animals (Fig. 3.6). From this model, it can be observed that if hunter-gatherers were completely dependent on hunting and gathering instead of food production, under the current climate conditions in China, the highest population density should be distributed in the northern segment of the transitional zone from the northeast to the southwest and in the southwestern region of China. However, the densely populated regions of China nowadays, such as the middle and lower reaches of the Yangtze River and the Bohai Sea region, would actually be relatively sparsely populated. In other words, these areas can only support much smaller population densities. For example, when the population density reaches 5 people per 100 square kilometres, the lifeway of hunter-gatherer could still be sustained in the northern segment of the Yanshan Mountains and the southwestern region. However, it would not be practical in the middle and lower reaches of the Yangtze River and the Bohai Sea region.

Simulation of Subsistence Diversity: By dividing the population density supported by terrestrial plants by the total population density, the proportion of hunter-gatherers relying on plant-based food is obtained, known as the hunter-gatherer economy ratio. Similarly, we can calculate the ratio of relying on hunting and fishing. The most developed hunter-gatherer economy is expected in the southwestern region of China, including the Sichuan Basin, followed by mountainous regions in the south. Conversely, the least developed foraging economies should be found in the northeastern and central areas of the Qinghai-Tibet Plateau. Comparatively, the developed hunting economy is mainly observed in the northern parts of the Northeast, Northwest, and Inner Mongolia.

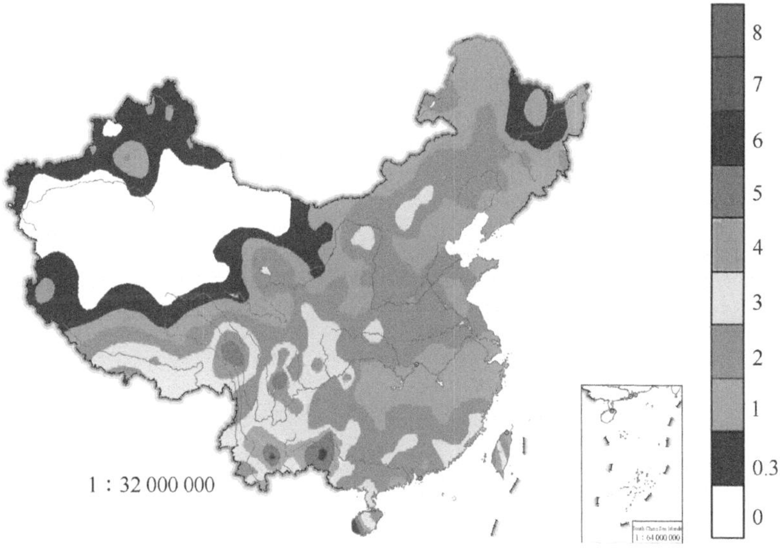

Fig. 3.6 Population density distribution of fully dependent on terrestrial animal and plant foods

Fishing-centred subsistence is primarily located in coastal areas, and there is also a considerable potential for a fishing economy in most parts of Northeast and North China. Historically, North China was once a region abundant in marshes, and the decline of these ecosystems was more related to agricultural activities rather than climate changes. Additionally, there is a fishing developed area in the Qinghai-Tibet Plateau, where numerous lakes offer abundant fish resources. However, local populations in this region do not consume fish, making it an exceptional case. The least developed fishing economies are found in the southwestern and northwestern regions of China, where rapid river flows and water scarcity naturally limit fishing opportunities. Taking these factors into account, it is possible to map the distribution of subsistence diversity (Fig. 3.7).

Diversity in subsistence represents the dietary breadth of hunter-gatherers to some extent. Based on Fig. 3.7, the highest diversity in subsistence could be found in the southern region of China, surprisingly followed by the western parts of Xinjiang and Tibet. The areas with the lowest subsistence diversity are in the northern parts of Northeast China and the eastern part of the Qinghai-Tibet Plateau. Most regions in

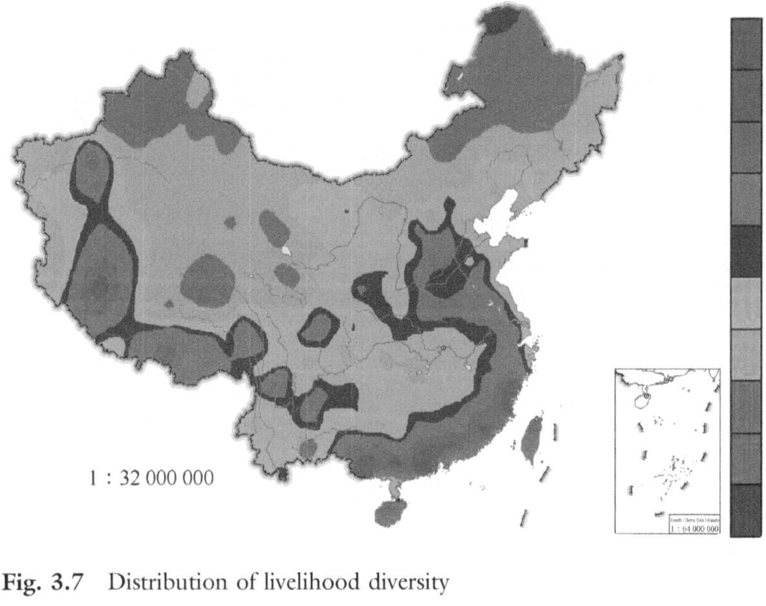

Fig. 3.7 Distribution of livelihood diversity

China fall within a moderate level of subsistence diversity, with the notable exception of the lower subsistence diversity in the middle reaches of the Yangtze River. Lower subsistence diversity implies less adaptability and increased vulnerability to livelihood crises. Another noteworthy finding is the relatively low subsistence diversity in the northern coastal areas around the Bohai Sea.

A significant limitation in the simulation of subsistence diversity could be observed, that is, the mode fails to distinguish the unique geographical environments of the Qinghai-Tibet Plateau and the Northwestern regions. To address this issue, Binford developed another variable, which is SUBSPX, to represent the primary forms of subsistence (Fig. 3.8). From Fig. 3.8, it is clear that most areas in Xinjiang and the Qinghai-Tibet Plateau are not conducive to the survival of hunter-gatherers. Gathering-dominated subsistence mainly exists in Southern China, the southern regions of Southwest China, Sichuan Basin, Jianghuai, and the middle and lower reaches of the Yangtze River. Hunting-dominated subsistence is prevalent in Northeast China, Northwest China, North China, the eastern part of the Qinghai-Tibet Plateau, and the mountainous regions of Southern China. In the transitional zones where hunting and gathering are the main subsistence strategies, some hybrid subsistence patterns can be observed.

Simulating the Cultural Ecological Zones of Hunter-Gatherers: Based on the SUBSPX, the basic ecological zones for Chinese hunter-gatherer ecosystems could be established (Fig. 3.9). Within ecologically similar systems, hunter-gatherers would face comparable resource conditions and environmental challenges. This zoning could serve as a reference framework for future analyses and comparisons of prehistoric hunter-gatherers. Notably, the simulated ecological environments are under modern climatic conditions but without the disturbances caused by agriculture and industry.

(1) Northeast Forest Zone: This zone is characterised by forest vegetation, with a growing season of less than six months. Rivers in this region offer relatively abundant fish resources.
(2) Inner Mongolia Grassland Zone: This zone possesses the richest animal resources, especially in the areas adjacent to the North China temperate deciduous broad-leaved forest. The transitional area has relatively plentiful plant resources available for utilisation.

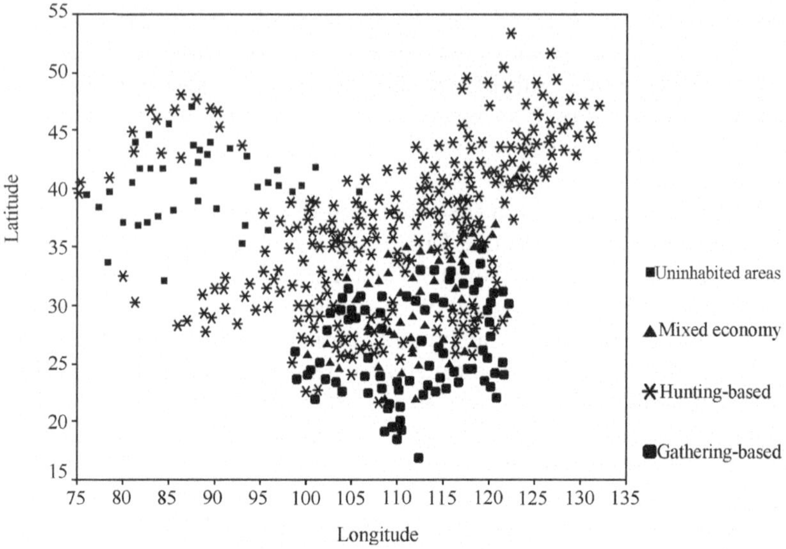

Fig. 3.8 Distribution of major livelihood types

(3) Northwest Desert and Gobi Zone: In this region, both plant and animal resources are extremely scarce, same as water resources. Humans with hunter-gatherer reliance found it challenging in survival.
(4) Qinghai-Tibet Plateau Alpine Zone: Both primary and secondary productivity are low, except for the eastern edge where the environment is relatively suitable for hunter-gatherers.
(5) North China Deciduous Broad-leaved Forest Zone: This zone has relatively abundant animal and accessible plant resources, making it more suitable for hunter-gatherers, particularly in the drier northern and western regions.
(6) Yangtze River Middle and Lower Reaches Subtropical Evergreen Broad-leaved Forest Zone: Although this zone exhibits higher primary productivity, the actual availability of terrestrial food resources is deficient, and animal resources are relatively scarce. However, rivers and lakes cover this area, making ample aquatic resources a compensation for the limited terrestrial resources.

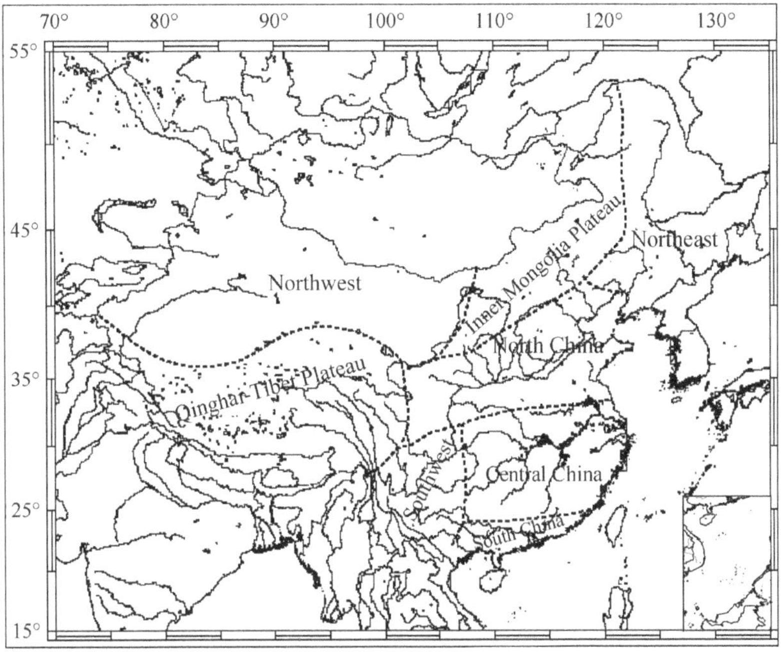

Fig. 3.9 Cultural ecological regions of hunter-gatherers

However, utilising aquatic resources requires certain technological capabilities.

(7) Southwest Valleys and Basins Zone: The most distinctive feature of this zone is the diversity of ecosystems within a limited space. Due to topographical separation and vertical distribution, numerous relatively independent small ecosystems have formed here, creating conditions for a rich and diverse local culture.

(8) South China Tropical Monsoon Rainforest Zone: This zone has high primary productivity but limited accessible plant resources and scarce animal resources. Moreover, this zone faces significant disease pressure, presenting a challenge for hunter-gatherers. For food producers, this challenge is even more severe.

Three transitional zones are particularly noteworthy: the east–west transition zone from northeast to southwest, the north–south transition zone along the Qinling Mountains and the Huai River, and the coastal areas and islands. The east–west transition zone from northeast to southwest is roughly located around the 400 mm isohyet, especially in its northern section, which has traditionally served as a boundary between pastoral and agricultural regions. Its southern section is influenced by the high altitudes of the Qinghai-Tibet Plateau, leading to an eastern shift in its position. The Late Neolithic and historical archaeological cultures in this transitional zone exhibit considerable similarities (Tong 1986). The north–south transition zone along the Qinling Mountains and the Huai River also exhibits distinct characteristics, with complex terrain and greater sensitivity of ecosystems to climate changes. Although this simulation study does not include coastal areas, it is crucial to consider these regions when differentiating hunter-gatherer ecosystems, as these areas display exceptional stability in human ecosystems.

3.5 Inference and Predictions

After completing the Minimalist Terrestrial Model, the possible adaptive transitions of hunter-gatherers could be deduced and predicted based on existing knowledge and the model. Generally speaking, the adaptive form for hunter-gatherers in different habitats are diverse, but this does not imply the absence of commonalities. An example could be seen as the Indians, who settled on the North American Great Plains, engaged in hunting and gathering, while others were involved in agricultural production, before the arrival of European colonists. However, after European colonists introduced horses to the region, horseback hunting and gathering, as a common adaptation, emerged among the Native Americans. What are the reasons for different groups of Native Americans with various backgrounds developing similar adaptations after the horses' introduction? This convergence is undoubtedly related to key variables, which is mobility for hunter-gatherers, in their cultural adaptive systems, leading to changes. The introduction of horses greatly increased the mobility of hunter-gatherers. In other words, with the speed and load capacity of horses, the range of resource utilisation for hunter-gatherers was significantly expanded; while, some groups, which had abandoned hunting and gathering, returned to a nomadic lifeway to forage for food.

Based on different criteria, the adaptive forms of hunter-gatherers can be classified into various categories. However, not all classifications hold equal significance when considering the reasons for transitions occurring in their primary modes of subsistence. Some prehistoric hunter-gatherers developed food production, while others did not. Among those who did not invent food production, some quickly learned this new way of life from the ones who invented, as a consequence, they could engage in food production. In contrast, other hunter-gatherers, despite knowing the existence of food producers around them, retained their original lifeway; and some were completely unaware of the food production. These differences cannot simply be attributed to the intelligence or talents of individuals or certain groups. In fact, a hunter-gatherer group, which is hypothesised named Group A that does not adopt or invent food production, cannot imply the nature of Group A, but rather because groups similar to Group A are unable to invent or adopt food production. This is the pattern we will explore, that is, the underlying essence of hunter-gatherer adaptation. In this regard, three factors, which play crucial roles in understanding the regularities of hunter-gatherer adaptations, will be emphasised: mobility, population, and environmental characteristics.

Within resource constraints, whether it arises from population growth, changes in resources, or other factors, the change in mobility is the first observable adaptation among hunter-gatherers, as mobility is the fundamental means through which hunter-gatherers address their resource supply. A group's daily mobility is limited when travelling on foot. An increasing mobility primarily manifests as higher migration frequencies, implying shorter stays at each foraging location. However, if population growth occurs simultaneously, resulting in smaller territories for each group, hunter-gatherers will have to search for food within these limited areas. As a result, certain food items, which were not typically consumed regularly, might be included in their diet, leading to "broad spectrum adaptation" (Flannery 1969). Meanwhile, some measures, including human intervention, would be employed to increase the density and yield of certain species with growth potential, known as intensification. Based on this model, it could be inferred that hunter-gatherers' subsistence strategies would continue to evolve in a particular direction if they faced prolonged resource constraints during the late Pleistocene.

From an ecological perspective, resources typically exhibit a patchy distribution. Under favourable resource conditions, the distances between different resource patches are relatively close. Conversely, under unfavourable resource conditions, the distances between resource patches increase. For hunter-gatherers, unfavourable resource conditions mean they have to travel greater distances to obtain necessary food, which is known as inter-patch mobility. However, hunter-gatherers are capable of accumulating and sharing information, as they are intimately familiar with their living territories. Moreover, when resources become scarce, competition among groups intensifies. In such situations, hunter-gatherers may choose to spend more time in patches with richer resources. Additionally, when the benefits outweigh the costs within resource patches, hunter-gatherers will stay within a single patch to expand the variety and quantity of available resources, which is viewed as intra-patch mobility. Archaeologists could observe the following patterns: ① evidence of frequent activities within a limited area, such as site clusters; ② assemblages of portable tools; ③ utilisation of small animals; ④ increased reliance on a greater variety of foods, especially aquatic resources; ⑤ dominance of certain food types in the hunter-gatherers' diet; ⑥ the discovery of diverse and abundant artifacts at certain sites; ⑦ a reduction in the average body size of hunter-gatherers (Binford 2001).

As foretasted, the mobility of hunter-gatherers can be classified into two types: foragers and collectors. Typical foragers exhibit high mobility, with less pronounced site differentiation, and they rarely stay in one place for extended periods. Consequently, their site structures are relatively simple, mainly consisting of camps and other places for outdoor activities. Due to their short residence, the variety of artifacts is relatively limited. Collectors need to store food for at least one period of the year. Meanwhile, collectors often establish a central base camp and send out groups to collect large quantities of food from resource-rich areas. As their site structures are complex, collectors have relatively fixed camps, leading to the construction of more elaborate dwellings, food storage facilities, large-sized durable tools, which do not need frequent transport, and a greater diversity of artifacts. Based on the analysis of available anthropological record from 395 hunter-gatherer societies, it is observed that in all subarctic and colder regions where water-based resources are crucial, hunter-gatherers tend to adopt the collector strategies (Generalisation 8.17). Collector strategies are also commonly found among animal resource-dependent societies in polar and northern forest climatic zones and are prevalent among temperate plant resource-dependent societies. Conversely, hunter-gatherers, who rely on animal resources in colder temperature and warmer regions, and those expend on plant resources in all environments, tend to employ the forager strategies (Generalisations 8.15 & 8.16). Therefore, if the hunter-gatherers in the North China and Yangtze River Basin regions heavily rely on animal resources, they are more likely to primarily employ the forager strategies. Alternatively, if they depend on plant resources, they are more inclined to adopt collector strategies.

Binford proposed that population density imposes fundamental constraints on subsistence types. If hunting terrestrial animals is the primary means of subsistence, population density of 1.57 people per 100 square kilometres will lead to pressure from packing pressure. As population density continues to increase, the proportion of hunting as a subsistence strategy will decrease. When population density reaches 9.098 people per 100 square kilometres, the main food resources should shift towards terrestrial plants or aquatic resources, depending on the local environmental conditions (Proposition 10.14). Population growth does not automatically lead to adaptive changes, but if this transition does not occur, the population is destined to decrease due to malnutrition and diseases. Based on this model, it could be inferred that the previous animal resource-reliance hunter-gatherer groups will shift towards terrestrial plants or aquatic resources once population density surpasses a certain limit. The aquatic resources primarily refer to coastal aquatic resources, as terrestrial aquatic resources and terrestrial plant resources face similar climate constraints, such as aridity, while marine resources are relatively more stable. Excluding the hunter-gatherers in coastal areas, the ones in both northern and southern regions shift towards plant dependence after reaching population saturation. Regarding archaeological remains, it might be discovered a shift in the focus of resource dependence, that is, transition from an economy primarily based on hunting to one primarily based on plant gathering.

Not only does population growth have an impact on the subsistence types of hunter-gatherers, but the environments in which they inhabit also impose some fundamental constraints on their subsistence strategies. Binford identified four thresholds that frequently lead to adaptations (Fig. 3.10):

1. Growing Season Threshold: This threshold is defined by an effective temperature of 18 °C, indicating a 12-month growing season. If temperatures exceed 18 °C, plants can grow year-round; if temperatures are below 18 °C, a year-round growing season is not possible.

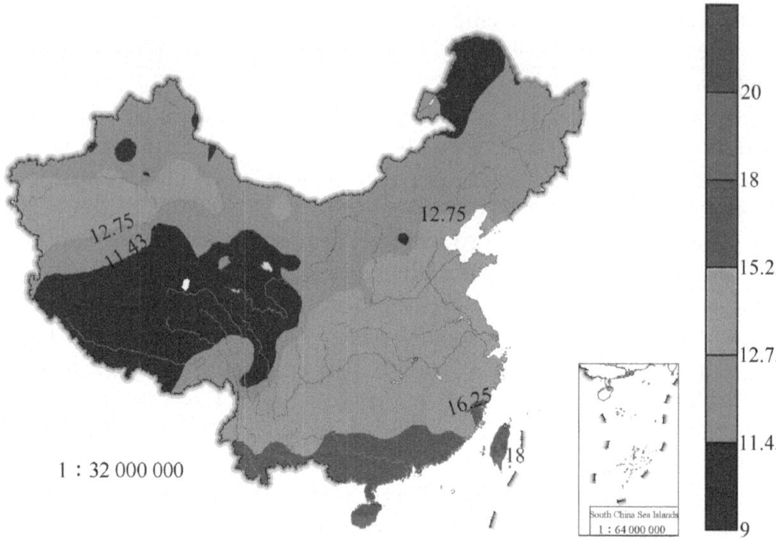

Fig. 3.10 Distribution of effective temperature

2. Storage Threshold: This threshold becomes essential above latitude 35° or with an effective temperature of 15.25 °C or below, as storage becomes necessary for survival.
3. Terrestrial Plant Threshold: This threshold requires an effective temperature of above 12.75 °C, making the plant-based subsistence strategy feasible. Below 12.75 °C, such strategies are only viable if the climate is uniformly mild throughout the year.
4. Subpolar Bottleneck: This threshold corresponds to an effective temperature of 11.53 °C or above, representing the basic requirement for global Net Above-Ground Productivity growth (Generalisation 8.12). If hunter-gatherers live in regions with lower temperatures than the subpolar bottleneck, the availability of plant-based food becomes severely limited. In such cases, hunter-gatherers must rely on hunting large herbivores or aquatic resources, as seen in the example of Native Americans hunting caribou and Eskimos relying on marine animal hunting for sustenance.

For hunter-gatherers in China within the simulated model, nearly the entire Qinghai-Tibet Plateau falls below the subpolar bottleneck's effective temperature, as well as the northern regions of Northeast China. For these areas, the utilisation of plant-based and aquatic resources is not viable; hunting remains the only feasible subsistence strategy. An ethnographic example of such a lifeway is found in the Evenki people living in the Daxing'anling (Greater Khingan Range), who have long term relied on hunting as their means of sustenance. On the Qinghai-Tibet Plateau, hunting is practically impossible for nomadic hunter-gatherers who rely on foot for travelling, not only due to the scarcity of animals and high mobility but also because of the thin oxygen at high altitudes. As for the Northwest desert region, with scarce water resources, only some oases could be utilised by populations, engaging in agriculture or animal husbandry. The analysis of ethnographic materials from hunter-gatherer societies worldwide proposed by Binford reveals that there is no record of hunter-gatherers in true deserts and high-altitude tundra regions. Similarly, very few traces of hunter-gatherers are found in the world's second driest semi-arid shrubland regions. These barren regions are typically inhabited by populations engaged in agriculture or animal husbandry. Therefore, it is considered that the unsuitable habitation areas for hunter-gatherers include the Qinghai-Tibet Plateau, Northwest desert, and semi-arid deserts. If the prehistoric climate was similar, these areas would have been uninhabited.

Regarding the terrestrial plant threshold, which requires an effective temperature of 12.75°C, it corresponds to the regions just beyond the Great Wall in North China and coincides with the southern edge of the Qinghai-Tibet Plateau. This means that hunter-gatherers who rely primarily on plant resources can only survive in North China and the southern regions of China. The vast Northwest and Northeast, with their distinct continental climates, cannot support a plant-based subsistence but rather favour a hunting-based way of substance. Another boundary is the storage threshold, which requires an effective temperature of 15.25°C. The need for winter food storage is significant north of this line; while to the south, the need is less pronounced. At the same time, storage is a fundamental characteristic of gatherer strategies, especially for the domestication of annual plants. Without stored seeds, planting for the following year would be impossible. Therefore, it could be inferred that agricultural practices centred around the domestication of cereals are only viable in the regions between these two thresholds, specifically in North China and the middle and lower reaches of the Yangtze River.

References

B. Winterhalder, Diet choice, risk and food sharing in a stochastic environment. *Journal of Ethnobiology*, 6: 205–223, 1986.

Chen Shengqian, The structure of archaeological reasoning. *Archaeology*, 2007 (10).

Compiled by the Climatic Data Center, National Meteorological Information Center, China Meteorological Administration): *China Average Temperature Data (1951–1980)*. China Meteorological Press, 1985.

D. Ebert, *Predictive Modeling and the Ecology of hunter-gatherers of the Boreal Forest of Manitoba*. Archaeopress BAR International, 2004.

H. P. Bailey, A method of determining the warmth and temperateness of climate. *Geografiska Annaler*, 42: 1–16, 1960.

K. V. Flannery, Origins and ecological effects of early Domestication in Iran and the Near East. In Ucko P. J. and Dimbleby G. W. eds., *The Domestication and Exploitation of Plants and Animals*. Aldine, Chicago, 1969.

L. R. Binford, *Constructing Frames of Reference: An Analytical Method for Archaeological Theory Building Using Hunter-Gatherer and Environmental Data Sets*. University of California Press, Berkley, 2001.

M. A. Jochim, *Hunter-Gatherer Subsistence and sedentism: A Predictive Model*. Academic Press, New York, 1976.

M. L. Rosenzweig, Net primary production of terrestrial communities: prediction from climatological data. *American Naturalist*, 102 (923): 67–74, 1968.

M. Pickering, Modelling *Hunter-Gatherer settlement patterns: An Australian Case Study*. Archaeopress BAR S1103, 2003.

R. Kelly, *The Foraging Spectrum: Diversity in Hunter-Gatherer Lifeways*. Smithsonian Institution Press, Washington, D C, 1995.

S. J. Mithen, Modeling hunter-gatherer decision making: complementing optimal foraging theory. *Human Ecology*, 17: 59–83, 1989.

Tong Enzheng, *Discussion on the Cultural Transmission Belt in a Crescent Shape from Northeast to Southwest China*. can be found in the book *Collected Works on Cultural Relics and Archaeology: Commemorating the 30th Anniversary of the Cultural Relics Press*. Cultural Relics Press, 1986.

CHAPTER 4

Accumulation of Resource Endowments in the Birthplaces of Agriculture

A tree that fills one's embrace grows from a tiny sprout;
A tower of nine stories begins with a pile of earth;
A journey of a thousand miles commences beneath one's feet.
——*Laozi*

4.1 Introduction

The preceding three chapters cover theoretical discussions and simulation studies to predict the conditions and regions where the origins of agriculture might have occurred. However, the validation of these viewpoints requires proving through archaeological evidence. From this chapter onward, archaeological materials will become the primary focus of our research. Archaeological evidence, on its own, does not speak for itself; it demands archaeologists' meticulous examination and research through eliminating irrelevant information and connecting relevant data, thus, a comprehensive archaeological reasoning could be constructed. Undoubtedly, archaeological materials possess multiple attributes; they not only provide temporal and spatial frameworks but also reflect the outcomes of ancient human behaviours, including the structure and social context of these behaviours. This enables archaeologists to explore the mechanisms and fundamental processes of prehistoric human cultural evolution.

In Chinese archaeology, scholars typically focus on the origins of agriculture within the context of the Neolithic. However, the roots

S. Chen, *Prehistoric Modernization*,
https://doi.org/10.1007/978-981-97-9423-2_4

of agriculture lie in the Paleolithic, which poses a necessary period to understand the complete process and formation mechanism of Chinese agriculture. What happened during the Late Paleolithic that led certain regions' hunter-gatherers to embark on the path of agricultural origins? Why did other hunter-gatherer groups in different regions not choose this path? The origin of agriculture, as an overall transformation in the cultural adaptation system of hunter-gatherers, akin to the industrial revolution, necessitates the accumulation of resource endowment factors, all of which were shaped during the Paleolithic. In the current academia, studies of agricultural origins are extremely limited, which is the significance of this chapter.

Based on the predictions from theoretical and simulation studies, as well as the current archaeological findings, it is considered that the North China and the Yangtze River middle-lower region, which could be viewed as the core areas for the origins of agriculture, will be analysed initially. In addition, when issuing other regions, research on the Late Paleolithic will still be referenced, although not as extensively as in this chapter. The North China and the Yangtze River middle-lower region possess the richest and longest history of archaeological materials from the Paleolithic, which provides a crucial foundation for the discussion on the adaptive transformations of the Late Paleolithic hunter-gatherers in this chapter.

4.2 Early Upper Paleolithic Adaptive Changes in North China

4.2.1 *Evolution of the Upper Paleolithic Cultural Adaptations Worldwide*

Since the Late Pleistocene, anatomically modern humans have gradually dominated in the Earth's ecosystem, spreading to all continents except Antarctica. Evidence of human existence has been found in diverse habitats ranging from plateaus to islands, from the equator to subarctic regions. Human evolution indicates critical changes in both biological and cultural behaviours. Mary Stiner analysed the entire evolution process of human cultural adaptations, considering eight dimensions, which include hunting efficiency, consumption efficiency, and types of hunting. She argued that human ecological space expanded twice, around 500,000 and 250,000 years ago, with significant achievements in the successful use of fire, potential hunting of larger animals, and utilisation of aquatic resources. Additionally, multiple great progresses of human ecological space expansion occurred between 50,000 to 10,000 years ago (Stiner 2002), but the archaeological evidence for these events is still not entirely clear.

It is clear that human behavioural adaptations underwent at least two revolutionary changes during the period between 50,000 and 10,000 years ago. The first was the Upper Paleolithic Revolution (Bar-Yosef 2002), during which humans created a series of new cultural and technological features. Thus, Richard Klein suggests that modern humans must have experienced a biological genetic mutation during this time, resulting in unprecedented competitive advantages such as language ability. Subsequently, humans migrated out of Africa and spread across the world (Klein 1992; Feinman and Price 2001). The second revolutionary change was the Food Production Revolution, which occurred during the Late Pleistocene. At least two regions in the world, West Asia and China, saw the emergence of food production practices. Based on this, humans not only began cultivating crops and domesticating animals but also developed entirely new settlement patterns, which is sedentism, and social organisational forms, thus laying the foundation for later civilisations, including the development of writing, cities, and states.

During the early phase of the Late Pleistocene (i.e. earlier than 50,000 years ago), the emergence of new cultural features was primarily observed on the African continent, corresponding to the Middle Stone Age (MSA). Examples of such features include bone and antler tools from sites like Howiesons Poort and Blombos (Henshiwood et al. 2002), as well as harpoons discovered at Katanda (Brooks et al. 1995). However, these discoveries in Africa are not present in all contemporaneous sites, and their dating remains controversial. Moving towards the late phase of the Late Pleistocene, particularly during the Upper Paleolithic (i.e. earlier than 40,000 years ago), the richest records are found in Europe. Southeastern and southwestern Europe demonstrate some continuity of cultural traits, meanwhile, certain characteristics of the Late Paleolithic emerge (Straus 1995). Other regions of the Late Paleolithic are established based on the European chronology, including West Asia (Bar-Yosef 2004), and China. The features of the Late Paleolithic in West Asia were initially evidenced at sites such as Boker Tachtit, which dating back to approximately 47,000 to 46,000 years ago, and Ksar' Akil, which dating back to around 45,000 to 43,000 years ago, predating European examples by 7,000 to 10,000 years (Bar-Yosef 2002).

John Hoffecker subdivided the Upper Paleolithic in Europe into three phases: the Early Upper Paleolithic (EUP), the Middle Upper Paleolithic (MUP), and the Late Upper Paleolithic (LUP) (Hoffecker 2002).The boundary between EUP and MUP coincides with the disappearance of Neanderthals. Technological innovations in the Upper Paleolithic (UP) of Europe cannot be found in the Middle Paleolithic cultures but are common in the MSA of Africa, with evidence dating back approximately 30,000 to 40,000 years earlier in Africa.[1] This imported technological advancements includes ground bone points, bone awls, and perforated shells. The differences between Neanderthals and modern humans may stem from a Brain Mutation, which cannot be discerned from anatomical features in the skull (D'errico et al. 2003). Moreover, solely anatomical features cannot explain the changes in human behaviours. Additionally, while Neanderthals and modern humans may have used similar lithic assemblage in the Middle Paleolithic, there might have been significant differences in their usage methods (Nitecki 1987, pp. 1–9): only modern humans demonstrate the ability to hafting composite tools and show the social cohesion needed for hunting large animals (Niewoehner 2001). Some scholars suggest that hafting technology may have been utilised before the emergence of modern humans, while modern humans show a distinct advantage in terms of frequency and precision of tool usage (Churchill 2001).

During the Late Pleistocene, human cultures evolved from the Middle to the Late Paleolithic. Bar-Yosef, referred to this development as the "Late Paleolithic Revolution", which spread outward from core areas. The triggering factors for this revolution could be biological evolution or socio-economic (techno-cultural) (Bar-Yosef 2002). Some archaeologists argue that actual revolution did not exist, but rather the accumulation of cultural developments led to the formation of new features (McBrearty and Brooks 2000; Clark 1997, pp. 60–76). Additionally, some scholars hold different views, believing that the Late Paleolithic cultures appeared roughly simultaneously in West Asia, Central Asia, and Europe. While some attribute these cultural features to local populations (Derev'anko 1998), others suggest they were brought by migrating modern humans.

The essence of the Upper Paleolithic Revolution ultimately lies in the fundamental aspect of subsistence strategies. Based on the overall trend of human evolution, the revolution includes continuous improvement of the success rate of hunting large ungulates,[2] expansion of the breadth of the diet (such as utilising small animals), and enhancement of the efficiency of hunting and food processing through technological advancements, and extension of the realm of symbolic behaviour in human societies (Stine et al. 1999). During the Middle Paleolithic and its earlier periods, although significant amounts of large animal remain have been found in archaeological sites, few artifacts can be unequivocally identified as hunting tools. A few exceptions, such as the early wooden spear discovered in Germany (Thieme 1997), fall short in comparison to the Late Paleolithic stone spears regarding the effectiveness in hunting larger animals. Hunting large animals is normally considered a phenomenon of the Late Upper Paleolithic (Binford 1987, pp. 47–105), although some archaeologists disagree on this point. Bar-Yosef argued that there was no animal archaeological evidence supporting a change in adaptive strategies from the Middle to the Late Upper Paleolithic. The lithic assemblages, which represent hunting and gathering strategies in the two periods, were determined by the available resources in the local regions (Bar-Yosef 2004).For instance, in the Levantine region of West Asia, the hunting strategies of Neanderthals and Late Paleolithic modern humans cannot be distinguished (Speth and Tchernov 2001, pp. 52–72).

Theoretically, around the same time as the onset of the Upper Paleolithic in various regions of Eurasia, anatomically modern humans dispersed from Africa and occupied all continents except Antarctica. From this, it could be inferred that when culturally adaptive modern humans arrived, the large mammals in these regions were still unfamiliar to modern humans, who were not yet adapted to hunting mammals. Particularly at the time humans entered the Americas, where the local animals had never been hunted by humans. It is unreasonable to assume that humans would equally hunt different-sized animals, as human behavioural mechanisms and interests favoured hunting large animals. Based on the Expensive Signalling Theory and the Central Place Foraging Theory utilised by D. N. Grimstead, the profitability of hunting large animals is demonstrated, even with long-distance transportation involved (Grimstead 2010). During the Upper Paleolithic, humans already possessed long-range hunting tools and could utilise various facilities, including traps and snares, for hunting. Humans also employed techniques such as smoking and drying to preserve food. Therefore, the only hindrance for human hunting of large animals was whether the animal resources were sufficient and sustainable. It can even be inferred that the regions where agriculture originated were likely areas where hunting of large animals was prevalent, but the prey resources were easily depleted and suitable for agriculture.

Overall, it can be argued that a crucial development in human hunting capabilities during the Upper Paleolithic has grown (Table 4.1). The lithic assemblage included refined stone spearheads, arrowheads, points, wooden and bone spears with standardised flakes or blades (used for hunting), end scrapers (for processing hide), and burins (for working bone and antler tools), and others. The artifacts exhibited more standardised shapes, specialised functions, and higher-quality materials. However, during the Upper Paleolithic, the most fundamental elaboration of the significant advancement in human hunting capabilities might be attributed to the strengthening of social organisation, planning, cooperation, and the exchange of hunting-related information, all of which depended on the development of language, which marks as a phenomenon that could be referred to as the "First Information Revolution". During the Upper Paleolithic, the significant improvement in hunting ability among humans can account for the widespread population expansion, as well as partially explain the phenomenon of animal extinctions at the end of the Late Pleistocene.

4.2.2 Cultural divisions of the Upper Paleolithic in North China: Chronology and Materials

Regarding the entirety of China, North China possesses the most abundant archaeological discoveries and has the longest archaeological research history. Ever since the discovery of the Salawusu site in the 1920s, other significant the Upper Paleolithic sites in North China include the Shandingdong Cave, Xiaonanshan, Shiyu, Xiachuan, etc. The North China lithic industry in the Upper Paleolithic traditionally divides into two technological types: the large flake chopping tool, that is, triangular large points tradition (also known as the "Kehe-Dingccn series") and the boat-shaped scrapers tradition (also known as the "Zhoukoudian Locality 1—Shiyu Fm") (Jia and Wei 1976). The materials from the Middle Paleolithic (approximately 200,000/300,000 to 40,000 years ago) are scarce and often present dating challenges, leading some scholars, such as Xing Gao, to suggest abandoning the Middle Paleolithic (Gao 1999). Currently, the Xujiayao site boasts the best material conditions in North China

Table 4.1 Manifestations of advancements in human hunting capabilities during the upper paleolithic and their corresponding archaeological records

Progress in human hunting capabilities	*Archaeological records*
Extended shooting range and improved accuracy of projectile tools, such as the use of throwing devices, bows and arrows	Stone arrowheads, spearheads, standardised flakes, blades, microblades, and points
Enhanced lethality (or effective stopping power) of projectile tools, which contain more lethal arrowheads, spearheads, and poisons; increased efficiency in processing hunted game and manufacturing and maintaining tools	Development of composite tools, higher-quality stones, organic materials; more standardised and orderly forms of artifacts; specific tool types include burins (used for engraving composite tools), end scrapers (for processing fur), awls, and grinding tools
Growth in hunting skills, knowledge, and experience, relying on longer periods of adolescent growth (later age of sexual maturity) and extended lifespans (accumulation of knowledge)	Increased age at death and later onset of sexual maturity; Improved overall health conditions
Invention of untended facilities, such as traps, deer pits, fish weirs, ambush hunting, etc., depending on group organisation and cooperation	Presence of larger sedentisms; more complex site layouts, including residential areas, hearths, graves, butchering areas, lithic manufacturing sites, and more evident seasonal resource utilisation
Food storage techniques, including smoking, drying, wind-drying, oil immersion, and storage rooms	Construction of storage pits
Exchange of hunting information, crucial for seasonally hunting animal herds	Creation of ornaments, art, symbolic objects, and exotic items

(Jia 1979), with distinctive characteristics of its lithic assemblage, which include a large number of stone balls, mostly small-size stone tools, and some new tool types, such as thumbnail-shaped scrapers, round-headed scrapers, concave scrapers, primitive prismatic core, and possibly bone tools with intentionally cut animal antlers. Furthermore, the scale of the site is considerable. These features represent a relatively typical the Upper Paleolithic cultural profile. However, studies on these human fossils indicate similarities with European Neanderthals while also possessing some primitive features akin to the Peking Man (Wu 1980). Initially estimated to date back 60,000 to 30,000 years ago by Jia Lanpo and Wei Qi (Jia and Wei 1976), following uranium-series dating pushed the chronology to approximately 120,000 to 100,000 years ago (Wu 1980; Chen et al. 1984; Chen et al. 1982), with some radiometric data suggesting even more recent dates of around 50,000 years ago.[3] The ambiguous nature of China's Middle Paleolithic cultural characteristics highlights the emergent changes in North China's Upper Paleolithic cultural landscape.

The Upper Paleolithic in North China was established with the European one as a model, thus its starting date is set at approximately 40,000 years ago (Tang and Gai 1986). In the book *Ancient Humans in China*, which was published in 1989, the Salawusu site was regarded as representative of the small tool tradition in North China, potentially indicating an early date and the existence of pressure-flaking retouch technology (Huang et al. 2004; Yin and Huang 2004). Infrared stimulated thermoluminescence dating pushed its chronology back to the Middle Paleolithic, no later than approximately 70,000 years ago (Huang et al. 2004; Yin and Huang 2004), significantly earlier than the 14C radiocarbon date of around 35,000 years ago (Li et al. 1984), and the Uranium-series dating of approximately 50,000 to 37,000 years ago (Yuan et al. 1983). However, this dating is still later than the thermoluminescence date of approximately 124,000 to 93,000 years ago (Dong et al. 1998). Two other sites, Liujiacha (Gansu Provincial Museum 1982) and Xiachuan Fuyihe, are considered as representatives of the starting phase of the Upper Paleolithic in North China but lack specific dating (Wang 1978). Currently, the most reliable dates are from Shandingdong Cave and Shuidonggou sites. The Shandingdong Cave site underwent the most systematic dating research (Chen et al. 1984; Chen et al. 1989, Chen et al. 1992) with accelerator mass spectrometry (AMS) dating indicating dates of approximately 28,000 years ago (with a half-life of 5730 years), and the lower layer dated to around 34,000 years ago, which aligns well with the faunal group (Yuan 1993). The Shuidonggou site, with a long research history, underwent more in-depth investigations in recent years through new methods and international cooperation, and its chronology is estimated to be approximately 29,000 to 24,000 years ago (Madsen et al. 2001, pp. 706–716; Gao et al. 2002). Another representative Shiyu site in the UP, yields a new AMS date of approximately 32,000 years ago (Yuan 1993). Based on the current evidence, it is appropriate to designate the starting date of the Upper Paleolithic in North China as approximately 30,000 to 35,000 years ago (Li et al. 2012).

The stone tool industries of the Upper Paleolithic in North China are commonly divided into different cultural traditions (Zhang 1987, 1990), such as the "small tool-dominant tradition" and the "long flake-microlithic",[4] or the "small tool tradition", "blade technology", and "microlithic technology" (Huang 1989). Another classification scheme includes the "culture series dominated by blades", "culture series dominated by microblades", and "culture series dominated by flakes" (Li et al. 1993). Shenglong Lin (Clark 1977) also introduced the classification system of five "lithic modes" or types of stone tools published by Grahame Clark (Clark 1977) which could infer that North China consists of at least three types of industries in the Upper Paleolithic: flake industry, blade industry, and microblade industry. However, such a classification is based on relatively limited materials and lacks sufficient theoretical support from related disciplines. As a consequence, the following questions should be thoroughly explored and supported by robust theoretical and interdisciplinary research. Does the classification relate to the characteristics and changes in the environment? What kind of adaptive behavioural patterns do these classifications imply? If there are behavioural adaptive pattern distinctions, how are they related to each other?

Based on the analysis of the technological features, spread process, and specific ecological connections of the microblade technology, its origin can be dated to approximately 24,000 to 21,000 years ago, making it a product of the Last Glacial Maximum (LGM) (as discussed in the following section). The distribution range of the microblade technology, represented by the Shuidonggou site, is likely confined to the Ningxia Basin and the Hetao region, with some discoveries reported in the Liaoning Benxi area.[5] The predominant industry in North China is the blade industry or "small tool tradition", which is interconnected with the microblade-dominant industry. The blade industry gradually gave way to the microblade technology, and the latest occurrence of the blade industry is around 18,000 years ago based on the uranium disequilibrium dating method, as seen in the Xibaimaying site (Xie and Yu 1989), and potentially even earlier. As for the Shandingdong Cave site, whose chronology has been rectified due to newer data. In the Xiaonanshan site, the assemblage of the upper Layer (layers 2–3), dated to 11,000 ± 500 years ago, exhibits no significant differences from the assemblage of the lower layer (layer 6), dated to 24,000–19,000 years ago, according to the research conducted by Zhimin An. Consequently, they are considered putting together (An 1965). Conversely, it is highly likely that the Upper Layer's assemblage at the Shandingdong Cave site may have been disturbed and mixed from the lower layer. Based on this, a clearer archaeological culture sequence for the Late Pleistocene in North China could be established (Table 4.2).

Table 4.2 Archaeological culture sequence and representative sites of the Late Pleistocene in North China

Time (kya BP)	*Phase*	*Representative sites*
Approx. 20–10	LLUP	**Microblade technology**: Xiachuan, Shizitan (Yuan 1998; Zhang 1990; Shizitan Archaeological Team 2002), Lingjing (Zhou 1974), Dagang (Zhang 1996), Jijitan (Hebei Provincial Institute of Cultural Relics and Archaeology, 1993), Hutouliang (Gai and Wei 1977), Tingsijian (Sollberger and Patterson 1976) Hebei Provincial Institute of Cultural Relics et al. 1992), Yujiagou, Xueguan (Zhang et al. 1985), Chaisi (Wang et al. 1994), Daxianzhuang (Ge and Lin 1985a, 1985b), Heilongtan (Cultural Relics Management Committee and Tancheng County Library 1986), Ang'angxi (Huang et al. 1984)
Approx. 35–20	EUP	**Microblade technology**: Shuidonggou (Jia 1964) **Blade Industry**: Shiyu (Jia et al. 1972), Shandingdong Cave (Pei 1940, pp. 1–100), Wangfujing (Li et al. 1998, 2000), Xiaonanhai, Xiaogushan (Zhang et al. 1985), Liujiacha, Upper Cave of Xiaokongshan (71) Xiaokongshan Joint Excavation Team 1988), Zhijidong (Zhang and Liu 2003), Tashuihe (Chen 1989), Xibaimaying, Mengjiaquan (Hebei Provincial Institute of Cultural Relics, et al. 1991)
Approx. 120–35	MP	Xujiayau, Xigou (Liu 1986), Longyadong (Wang et al. 2004), Gezidong (77) (Gezidong Excavation Team 1975). Salawusu

4.2.3 The Upper Paleolithic Revolution in North China

From the ambiguous Middle Paleolithic to the Upper Paleolithic with a distinct style, the cultural landscape of ancient humans in North China underwent a revolutionary transformation, giving rise to a series of new characteristics. This transformation shares certain synchronicity with the Upper Paleolithic of West Asia and Europe on the Eurasian continent (Table 4.3). Compared to the Upper Paleolithic revolution in the western parts of West Asia and Europe, the revolution in North China appears to be more abrupt. However, the development of lithic technology occurred later in North China, mainly during the LUP phase.

Compared to stone tool technologies, which are easier to preserve, the Upper Paleolithic may have also existed other techniques for which archaeological evidence has not been found yet, such as traps, fishing nets, or the usage of poison for hunting and fishing. The development of

Table 4.3 Contrasting late Pleistocene innovations in North China with West Asia and Europe on the Eurasian Continent

West Asia and Europe (Mellars 2005, pp. 12–27)	*North China*
indirect percussion used on blades and microblades	Tools became smaller and materials more refined, introducing blade technology EUP, as seen in sites like Shuidonggou
Emergence of new types of end scrapers and burins	New types of tools appeared, such as end scrapers, thumbnail-shaped scrapers, burins, and backed knives
Emergence of tools with standardisation and regional characteristics	The appearance of microblade LUP products
Complex bone, antler, and ivory tools	Elaborate tools like harpoons from the Xiaogushan site, bone awls from the Shuidonggou site, and bone needles from the Shandingdong site
The appearance of personal ornaments, such as perforated teeth, seashells, polished stone ornaments, and ivory beads	There are 9 sites in North China, including the Shandingdong, Xiaonanhai, and Shiyu. Besides, red iron ore powders is found in the Dongfang square in Wangfujing, Shandingdong, and Shuidonggou, etc
Abundance of art: Rich artistic expressions such as sculptures and fresco	Antler rods from the Shandingdong site and frescos from the Shizitan site
Emergence of symbolic systems	Specimens with incised marks in the Dongfang Square; and hundreds of specimens with marks in the Shiyu site (which require further research to verify whether they are symbolic or not)
Musical instruments: bird bone flutes	
Long-distance exchange networks	Examples include seashells from the Shandingdong site
Long-distance projectile technology	For instance, arrowheads from the Shiyu site and refined points
Rapid changes in technological patterns	Examples include pressure retouch and indirect percussion, as seen in the emergence of the microblade technology
Increased population density	Larger sites and more location clusters
Clearer sedentism structures	Shuidonggou site includes areas for tool production and habitation; LUP sites like Mahanshan in the Nihewan Basin
Specialised animal exploitation	Abundance of sheep bones at the Shiyu site

composite tools marks a significant advancement in technological innovation during the Upper Paleolithic. Composite tools, which combined the flexibility of organic materials such as bone, antler, and wood with the rigidity of stone, forming modular combinations (cutting, scraping, carving, throwing, digging), exchanged blades or spearheads to achieve greater adaptability and efficiency. During the LUP, it can be determined that hunter-gatherers in North China had already begun to domesticate dogs. Ancient DNA evidence indicates that dog domestication may have occurred as early as 15,000 years ago in East Asia, and the history of human utilisation of dog ancestors' dates back even earlier (Savolainen et al. 2002). Additionally, bone and antler tools thrived during the Upper Paleolithic. Examples could be found as the bone awls at the Shuidonggou site and the bone needles at the Shandingdong site, which clearly reflect the beginning of leather working during this period. Furthermore, discoveries like the harpoons at the Xiaogushan site indicate that humans began utilising aquatic resources.

In the Upper Paleolithic sites of the western Eurasian continent, there is often evidence of prehistoric humans utilising the broad spectrum of food resources, including small animals such as rabbits; and starting to exploit aquatic resources, such as fishing and gathering shellfish (Stiner et al. 1999; Stiner 2002). In the Upper Paleolithic sites of North China, a general trend can be observed where individual animal figures became smaller, the degree of fragmentation of animal bones increased, and the variety of animal species decreased from early to late sites. This trend is particularly obvious when comparing the Upper Paleolithic sites, like Shiyu, Shandingdong, and Xiaonanhai in the EUP.

Beyond the technological perspective, a further comparison of the overall efficiency of the cultural system reveals a more remarkable revolutionary change during the Upper Paleolithic. The overall efficiency of a cultural system consists of four aspects: technology, social organisation, information transmission, and belief systems. For the Upper Paleolithic humans, social organisation is characterised by clear individual identities, the development of self-awareness, and the formation of social exchange networks. The widespread presence of ornaments and exotic objects indicates progress in human social organisation during this period. In terms of information transmission, the key aspect is the development of language, and its material manifestation can be seen in symbolic notations. Although this aspect is not yet entirely clear in North China, the incised marks on animal bones found at the Dongfang Square at Wangfujing site,

Xiaogushan site, and Shiyu site may be related to symbolic notations. Regarding belief systems, archaeological evidence is manifested in artistic artifacts. Currently, the most developed the Upper Paleolithic art is found in Western Europe, especially in France. The sophisticated European Paleolithic art disappeared with the end of the Last Ice Age and was a product of specific environmental conditions (Straus 1995). In North China, the evidence of belief systems is limited to the antler rods at the Shandingdong site and the fresco at the Shizitan site, but their functions and dates remain uncertain.

During the EUP in North China, there were at least four adaptive patterns observed: the Shuidonggou pattern, the Shiyu pattern, the Shandingdong—the Dongfang Square at Wangfujing—Xiaonanhai pattern, and the Xiaogushan pattern (Fig. 4.1). These patterns are distinguished based on the characteristics of the ecological environment and the composition of technological culture. The Shuidonggou pattern corresponds to an environment at the junction of grasslands, forests, and deserts, where the productivity of the ecosystem is relatively low. Thus, human subsistence, which primarily relies on hunting, and plant-based foods, are relatively limited. Consequently, the corresponding lithic assemblage emphasises the highly mobile blade industry. The Shiyu pattern corresponds to an environment with higher productivity, located mainly at the border between grasslands and forests, with abundant hunting resources and more gathering resources compared to the Shuidonggou pattern. The lithic assemblage is more diverse in the Shiyu pattern. On the other hand, the Shandingdong—the Dongfang Square at Wangfujing—Xiaonanhai pattern is an adaptation to a temperate forest environment, characterised by even more diverse and less standardised lithic assemblage, possibly due to a wider range of choices for organic tools. In comparison, the Xiaogushan pattern reflects an adaptation to a waterside environment, with the presence of artifacts such as harpoons, providing better stability and allowing for longer term residence in a specific area (Binford 2001). Examining the distribution of these four patterns, a trend towards increasing humidity in the environment could be observed, leading to a transition from a primary focus on animal hunting to a more diversified approach encompassing hunting and gathering (with the Shiyu pattern emphasising hunting), and eventually incorporating the utilisation of aquatic resources for hunting and gathering activities.

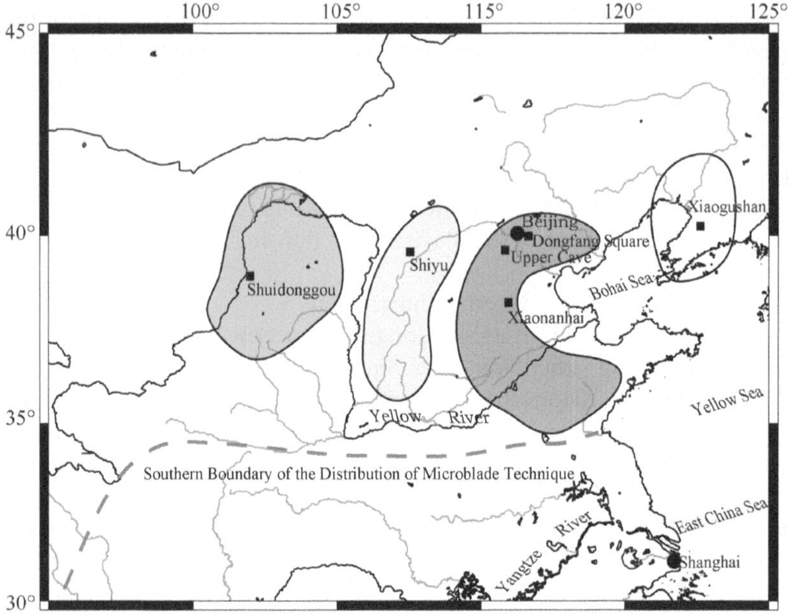

Fig. 4.1 Four regional adaptation patterns in the early phase of the late Paleolithic of North China

4.3 Microblade and Late Upper Paleolithic Adaptation

4.3.1 Macroscopic Distribution Characteristics of Microblade in China

When considering the distribution of microblade in China, it is necessary to differentiate from various perspectives. From a macroscopic standpoint, the relationship between microblade and ancient natural geography must be scrutinised, as well as its correlation with human subsistence strategies. From a mesoscopic perspective, distribution patterns within specific regions must be explored, such as the distinct distribution characteristics observed in Northern Jiangsu, Southwest Shandong, Shanxi, and Hebei. At a microscopic level, the distribution characteristics of microblade within archaeological sites must be examined, particularly the differences among various sites within a site complex. For this discussion, a macroscopic viewpoint will be primarily employed.

Time serves as an essential dimension to understand the distribution of microblade technology. Generally, microblade products can be divided into two phases: the Paleolithic and the Neolithic. After entering the Neolithic, microblade products rapidly disappeared from the North China region and was limited to areas outside the Great Wall. Alternatively, similar to the emergence of microblade technology, preliminary phases, such as those observed at the Xujiayao and Shiyu sites (Ge and Lin 1985), can be regarded as distinct phases and be identified. However, strictly speaking, the Xujiayao and Shiyu lithic assemblages cannot be definitively labelled and defined as possessing microblade technology. When solely considering the Paleolithic microblade technology, Chun Chen and Xiangqian Wang suggest that two phases, represented by the Xiachuan and Xueguan sites, can be differentiated due to the earlier appearance of conical cores compared to wedge-shaped cores (Chen and Wang 1989). However, based on materials explored from the Nihewan Basin (e.g. the Toumafang site) and Siberia (e.g. the Sumnagin culture) (Hoffecker 2005), true conical cores, which are commonly referred to as "pencil-shaped microcore", only emerged in the mid-Holocene. Currently, the earliest dates for Chinese microblade come from the Xiachuan site, with six radiocarbon dates ranging from 23,900 ± 1000 BP to 16,400 ± 900 BP (Wu 1984). Even earlier materials, such as those found at 77:01 of the Chaisi site (also known as the Dingjiagou site) in Shanxi Province, yielded a Carbon-14 dating of 26,400 ± 800 BP (Wang 1989). Recent discoveries in Henan Province and Shanxi Province have also surpassed 20,000 years.[6]

Regarding the distribution of microblade in the Late Pleistocene in North China, Fei Xie proposed a horseshoe-shaped distribution belt that stretches from the Yanshan Mountains westward and then southward along the Taihang Mountains and Funiu Mountain, extending further east to the Malingshan in northern Jiangsu and southwestern Shandong. This region is divided into five sub-regions. The two eastern sub-regions, along the Yanshan Mountains and in the northern Jiangsu and southwestern Shandong areas, are dominated by boat-shaped cores (broadly considered a variation of the wedge-shaped cores, but with a wider platform and shorter flakes).[7] On the other hand, the three western sub-regions are primarily characterised by wedge-shaped cores

(Xie 2000). However, this regional division does not include the materials discovered from the Shayan site in Dali, Shaanxi Province (An and Wu 1957; Xi'an Banpo Museum, Dali County Cultural Museum 1983), whose semi-conical cores are similar to those from Xiachuan (Wang et al. 1978) wedge-shaped cores are similar to those from the Hutouliang site (Gai and Wei 1977). Therefore, the Shayan site could be categorised among the three western sub-regions. Notably, the discoveries in the Nihewan Basin significantly differ from those in the Xiachuan site. While Xiachuan is characterised by finely crafted columnar-conical microblade cores, the Nihewan Basin is dominated by wedge-shaped cores (Hebei Provincial Institute of Cultural Relics 1993). Thus, further evidence is needed to verify the division into these five sub-regions, particularly in the eastern and western sides. In addition, further excavation and evidence are required to understand the significance of this distinction. It has been observed that the size of microcores differs between the eastern and western regions, with microcores found in the western regions generally larger than those in the east. Especially, microcores discovered in the sites of Tingsiji and Malingshan have reached the most advanced technology for microblade technology. Although Xiachuan site has also yielded some exceptionally small-size microcores, they are far less common compared to the east (see Fig. 4.2).

The coexistence of microblade and small tools in North China during the Upper Paleolithic requires further examination. It could be argued that no conclusive evidence could infer the extensive coexistence of the microblade and small-sized tool industry during this epoch. The representative site of the typical small-sized tool industry in North China, the Shandingdong site, dates back earlier than the origin of microblade (approximately 21,000 years ago). The dating of the Xiaonanhai site, which is situated in Anyang, spans several tens of thousands of years. Based on the reports, no significant changes in artifacts could be observed in the Xiaonanhai site, thus, these artifacts were regarded for relative research (An 1965). It is astonishing for highly mobile prehistoric hunter-gatherers to repeatedly use a site for such an extended period without any changes. Moreover, it is notably that prehistoric hunter-gatherers were not limited to only using microblade technology; therefore, the microblade assemblages in different regions of North China include various components, ranging from exquisitely crafted back knives and shoulder-shaped points to ordinary scrapers. Based on the nature of different raw materials, users of microblade differentiated their tool for usage,

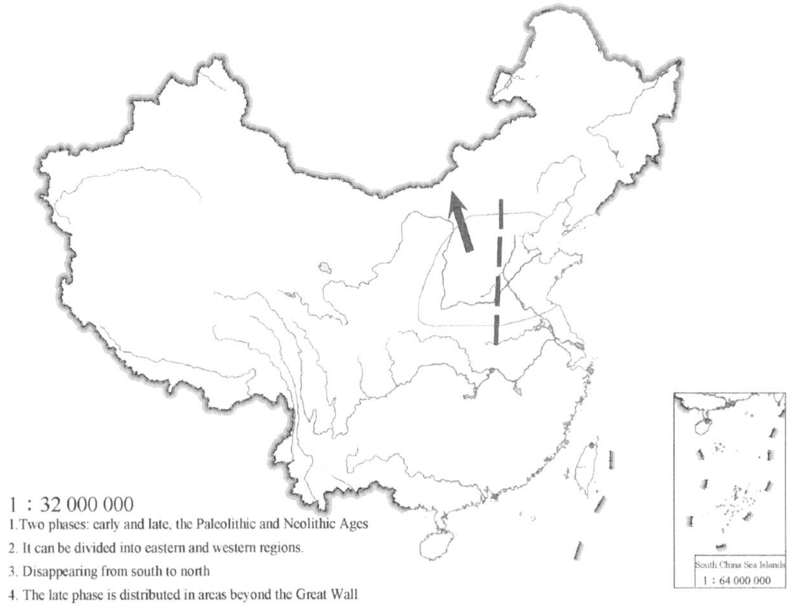

Fig. 4.2 Spatial and temporal distribution characteristics of microblade technology in China[8]

for example, producing expedient tools from lower-quality stones. Additionally, as mobile hunter-gatherers, users of microblade utilise resources seasonally. It is not necessary for users to employ microblade at every sedentism site, as a consequence, the presence of so-called small-sized tool industries in some sites within a region where microblade was prevalent should be reconsidered. At the initial phases of the emergence of microblade technology, the coexistence with small-sized tool industries might have been possible. The discovery of small-sized tool components in assemblages using microblade is also normal (as seen in sites like Mengjiaquan in Yutian [Hebei Provincial Institute of Cultural Relics, et al. 1991]). artifacts assemblages have never been solely composed of microblade technology, satisfying all their tool needs.

4.3.2 *What Is the Nature of Microblade Technology?*

(1) Theoretical Framework

Establishing theoretical explanations and expanding research materials, as well as developing research methods, are essential means of investigation in all academic disciplines. However, in our past research on the origins of microblade technology, the development of theoretical explanations has been relatively rare, as the research has heavily relied on archaeological findings. Clearly, this approach is insufficient. In comparison, the Western (Paleolithic Archaeology of the Western Hemisphere) has made more attempts in this regard. For example:

1. M. Shott used ethnographic materials to analyse the relationship between lithic technology and site mobility, emphasising site mobility can affect the composition of tool assemblages (Shott 1986).
2. M. C. Nelson proposed the importance of analysing tool morphology and distribution, which serves as a powerful means to understand the characteristics of activity areas, leading to discussions on technological strategies, societal and economic strategies of a group, and their adaptation to the environment (Nelson 1991).
3. D. E. Young and other scholars attempted to develop a cognitive approach from the study of lithic technology, in order to extract more information about human behaviour (Young et al. 1994). B. Hayden devised comprehensive criteria for evaluating stone tool design and usage strategies (Hayden et al. 1996).
4. P. Bleed analysed the impacts of reliability and maintainability on lithic technology (Bleed 1986). S. Kuhn conducted cost–benefit analyses to understand how weight and functionality influence lithic technology (Kuhn 1994), with each researcher developing their theoretical perspectives from a specific aspect.
5. B. P. Winterhalder and other researchers developed theoretical explanations based on the "Optimal Foraging Theory" (Winterhalder et al. 1999).

All the aforementioned studies revolve around a crucial factor, which is the mobility of hunter-gatherers. Mobility is a vital strategy for populations relying primarily on hunting and gathering as their livelihood, enabling them to adapt to their environment effectively (Kelly 1983). In other words, they engage in "mobile production" (Politis 1996). Through mobility, these hunter-gatherers disperse accessible fruits around their dwellings, while using other waste materials as fertiliser. Clearing trees around the sedentism allows sunlight to reach these seeds, inadvertently transforming the environment and producing their food. Additionally, their mobility facilitates the discovery of potential food resources, ensuring a constant supply, akin to an "information bank" (Binford 2009). For these hunter-gatherers, mobility, which is the essential means of securing food resources, is akin to food production in agricultural societies or market economies in modern times.

For hunter-gatherers, technological features must be suitable for their nomadic lifeway. Unlike pastoral societies with domesticated horses and oxen pulling carts, hunter-gatherers primarily rely on human labour for transportation, except for Arctic populations utilising sleds and communities residing near rivers or lakes employing boats or canoes. Simultaneously, hunter-gatherers require a diverse set of tools to address the ever-changing tasks in their mobile existence. As a result, hunter-gatherers face a dilemma: they must strive to minimise the costs of transportation while maximising the durability and functionality of their tools.

Certainly, the mobile toolkits used in the hunter-gatherer subsistence are just one part of their entire toolkit. Also, the role of mobile toolkits in the livelihood of hunter-gatherers is also variable. Generally, the toolkit of hunter-gatherers can consist of three components: expendable tools, including choppers and flakes; durable tools, like grinding slabs, handstones, which are occasionally referred to as "site furniture" and may include partially finished handaxes; the mobile tools, i.e. tools that are carried by humans, with products in the microblade as the most typical example, which greatly facilitate mobility. Therefore, to understand the emergence of microblade technology, it is essential to grasp the temporal and spatial characteristics of the mobility of hunter-gatherers. Kelly and Binford have provided detailed discussions on the mobility of hunter-gatherers, emphasising its close connection with the environmental features of resources. Binford further emphasises that the mobility is related to the types of resource utilisation, such as animal hunting, plant gathering, aquatic resource exploitation, and population density, all of which can lead to different levels of mobility and technological complexity (Kelly 1995; Binford 2001).

Once the key variable related to the mobility of hunter-gatherers, which is relevant to the analysis of microblade technology, has been established, constructing a theoretical framework for lithic analysis to determine the position of microblade within the lithic manufacturing techniques should be further considered. Microblade is a product of a historical phase, and it has relevant historical origins. Kuhn's model of cost and benefit, which envisions an optimal design, is theoretically valid, but it fails to adequately estimate non-utilitarian utility, such as the role of gender selection in handaxe production. Similarly, the model is not well-suited for analysing the production of exquisitely crafted stone artifacts used as ceremonial objects in Mesoamerica. The analytical frameworks conducted by Young and Hayden are highly detailed but lack operational simplicity, especially when compared to early stone tool industries. Bleed's approach is highly valuable, but it is not yet comprehensive enough. Moreover, reliability and maintainability are not necessarily conflicting, and the concept of efficiency or lethality seems more specific. The most crucial issue is that Bleed overlooks two key variables in the technological organisation of hunter-gatherers: mobility and durability, which form a contracting set, as durable tools are often less portable; an assembly of durable tools indicates reduced mobility among the hunter-gatherer population. Additionally, the adaptability breadth of tools can serve as an indicator for comparing different lithic technologies.

Through utilising the five indicators mentioned above to assess the main lithic technological patterns that humans have possessed (Table 4.4), some basic trends can be observed. Clark summarised five technological patterns (Clark 1977), which are frequently cited in the Chinese archaeological academia. Although Clark's classification is solely considered applicable to the Western Hemisphere (Kuhn and Elston 2002, pp. 1–8), it can still be used for a broad technological analysis. Additionally, another technological pattern could be included, namely grinding technology, representing the most typical design for increasing durability and reducing mobility. In reality, grinding technology represents a redundant design for accomplishing stone tool tasks. In settled conditions, the transportation cost of this technology is minimal, and its utility is significantly enhanced.

Throughout human hunting and gathering adaptation history, two basic trends can be summarised: first is the maximisation of energy

Table 4.4 Comparison of six human lithic technological patterns

	Mobility	*Adaptability*	*Maintainability*	*Efficiency*	*Durability*
Pattern I Choppers	Low	Medium	Medium	Lowest	Medium
Pattern II Handaxes	Low	Medium	Medium	Medium	Medium
Pattern III Flakes	High	Medium	Medium	Medium	Low
Pattern IV blades	High	High	High	High	Low
Pattern V microblades	Highest	Highest	Highest	Highest	Lowest
Pattern VI Ground stone tools	Lowest	Lowest	Lowest	High	Highest

strategy, which means that hunters and gatherers seek food with the highest energy within the same foraging time. Of course, humans, like all living beings, also consider the issue of nutritional balance, thus intentionally collecting some low-energy food, but this is not the primary contradiction in hunting and gathering economies. Compared to gathering plant-based foods, hunting represents a strategy of energy maximisation. Similarly, the domestication of animals and plants is also such a strategy. For hunter-gatherers, they aim to increase the proportion of their hunting economy, meaning they pursue hunting tools with higher effectiveness. The microblade technology can be considered the pinnacle of chipped lithic technology and one of the most advanced stone tool-making techniques in the Paleolithic. The second trend is the maximisation of mobility, which means that hunter-gatherers seek to cover as large area as possible in the same amount of time during their foraging process. A larger area represents more hunting opportunities and access to other food sources, as well as a more secure food supply. Therefore, increasing mobility is another goal pursued by hunter-gatherers. To enhance mobility, the burden of carrying tools must be reduced, and microblade technology precisely serves this purpose (Fig. 4.3).

From Mode I to Mode V, each successive mode exhibits a certain degree of inclusiveness towards the previous mode, much like software upgrades. For hunter-gatherers who can produce hand axes, making choppers is easy; similarly, for those who can produce microblades, creating larger blades is also easily manageable. From the perspective of mobility, moving from Mode I to Mode V, the lithic technology of hunter-gatherers becomes increasingly conducive to enhancing mobility. The adoption of microlithic technology represents the peak of mobility,

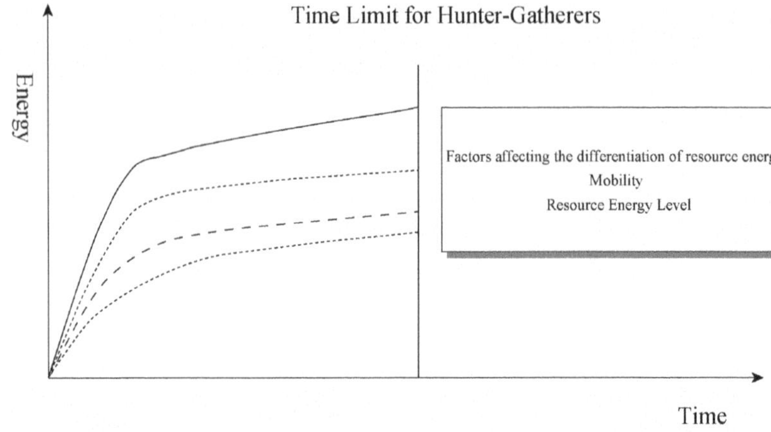

Fig. 4.3 Factors influencing the efficiency of hunter-gatherer livelihoods[9]

but with the widespread adoption of grinding technology (Mode VI), the mobility represented by lithic technology quickly declines. Microlithic technology not only represents the pinnacle of mobility but also embodies the widest adaptability, highest maintainability, and highest lethality (or efficiency). However, microlithic technology exhibits the lowest durability, which is related to the small and sharp edges used in this technology. In sharp contrast, grinding technology emphasises durability, as settled communities do not require high mobility or broad adaptability and prioritise developing stone tools with increased durability to reduce maintenance costs. From Mode I to Mode V, the progression reflects humanity's continuous efforts to expand the utilisation of resources, including increasing foraging areas, diversifying food resources, and pursuing the maximisation of food resource energy (i.e. seeking high-energy foods). However, before the advent of rapid transportation like horses, human mobility inevitably had its limits.

Additionally, it is noticeable that different system states of technology yield different results. Directly investing more labour in lithic technology as a system state does not proportionally increase the utility of stone tools. Instead, the utility and production of compound tools marks a rapid improvement and other higher levels. In other words, when a system state reaches its limit, there is a possibility of a system state transition, such as from microlithic technology to grinding technology. If humans cannot further enhance mobility to obtain essential resources, they may have to transition their way of life from hunting and gathering to food production. Furthermore, lithic technology is just one part of the entire technological repertoire of humans. If a group heavily relies on non-lithic technologies, it can lead to variations in the nature of their technological system. For instance, the prevalence of choppers in lithic technology in Southern China and Southeast Asia may be related to non-lithic technologies employed in the region.

As the nature of microblade technology is already known, it is crucial to emphasise that the microblade technology is built upon existing lithic technological traditions, including Patterns I, II, III, and IV. The lithic assemblages associated with microblade technology should also include expedient tools, such as large choppers, which can be used alongside microlithic tools to fulfil specific functions. Thus, it is natural to find a small number of coarse choppers, blade tools, and microblade tools within lithic assemblages containing microblade products. Additionally, there may be partially finished pieces with signs of repair, indicating reserve materials. Solely viewing microblade as a single lithic assemblage would fail to explain the co-occurrence of large lithic tools and the diverse subsistence activities of the hunting and gathering communities. Consequently, it is more appropriate to consider a complete lithic assemblage containing microblade products as a unified entity, rather than splitting it into several lithic assemblages or even different industrial traditions.

Lastly, microblade technology is not merely a continuation of traditional technological elements; rather, it represents a new technological invention and a completely novel way of tool usage. Currently, there is no definitive archaeological evidence to prove that bows and arrows appeared in the preliminary phases of the LUP. The emergence of microblade industry indirectly suggests an inseparable connection between bow and arrow technology and microblade technology. The utilisation of bow and arrow technology allowed small and delicate microblades to exert significant killing power, enabling the hunting of large animals. Nevertheless, the products of microblade were not limited to being arrowheads with embedded blades, despite the invention of the bow and arrow may have been a crucial prerequisite for the development of microblade technology.

(2) Technological Element Analysis

Understanding the microblade technology, which include its constituent elements, technological origins, and typology, is the premise of comprehending its origin. Subsequently, a thorough investigation into its origins and processes can be pursued. In addition, understanding the composition of microblade requires delving into archaeological remains. Currently, typical microblade technology is associated with at least ten types of remains: bifacial cores, debitages flakes, retouched bifacial flakes, heat-treated gloss, pre-knapping abrasion marks, ski spall, lame à crête, retouched platform fragments, microblades and microcores. Certainly, the production of microblades involves using typical bifacial cores as blanks, which are then transformed into platforms through either flaking or retouching. Another method involves utilising the natural flat surfaces or naturally fractured planes of raw materials as platforms; however, even in this case, the blank itself still undergoes modifications similar to typical bifacial cores, indicating that this technique may have derived from the simplification by means of flaking and preparing cores.[10] Thus, the production technique of bifacial tools constitutes the primary element of microblade technology.

The bifacial technology itself represents a pursuit of tool standardisation and versatility, making it particularly suitable for mobile hunter-gatherers (Odell 1998). Bifaces can serve as durable and versatile tools and may also serve as by-products in the production of points (Kelly 1988). It's functional adaptability and morphological standardisation align with the objectives of the microblade technology. From a technological origin perspective, bifacial technology is one of the sources of microblade technology; another source is the carving technique, which is marked as a method to produce flakes. Andrei Tabarev also argues that microblade combines both bifacial and carving techniques (Tabarev 1997). Additionally, soft hammer percussion, pressure retouch, heat treatment, and friction techniques are crucial components of the microblade technology (Table 4.5). Among them, pressure retouch can be employed in edge retouch before producing the primary flake, as well as in the production of microblades, thus reflecting another source of the microblade technology. Madsen et al. argue that bipolar technology and the

Table 4.5 Microblade components

Microblade components
Bifacial technology—Preparation of the cores
Carving technique—Chipping of removal flakes
Prismatic core technology—Preparation of the cores
Pressure retouch—Retouching of the cores, flaking of the blades
Soft hammer percussion—Retouching of the cores
Heat treatment—Treatment of the cores
Friction technique—Processing the platforms

origin of microblade are intricately linked (Elston et al. 1997). However, bipolar technology is not an indispensable component in microblade technology.

Particularly, before and during the development of microblade technology, there existed the "Prismatic blade technology", which was present in the microblade of the Upper Paleolithic Shuidonggou site in the northern region and was also utilised by ancient Native Americans for producing blades (Sollberger and Patterson 1976). In North China, the original prismatic blade technology can be traced back to the Early Pleistocene remains in the Nihewan Basin (Hou 2000), and it is also found at the Early Upper Pleistocene Xujiayao site (Jia and Wei 1976). This prismatic blade technology is closely related to the "boat-shaped core technology" in the microblade technology. The implementation process of the microblade can be divided into two methods (Fig. 4.4): one involves processing bifacial blanks and then producing microblades, while the other involves retouching block-like core blanks and then flaking microblades (Kobayoshi 1970). Although typical blanks of boat-shaped microblade cores adopt the retouching technique of bifacial tools, their morphology is closer to the original prismatic blade cores. The prismatic blade technology also served as the precursor of the flaking technology of blades, and it's derived from double platform blade flaking technology, as observed in the Shuidongou site (Ningxia Institute of Cultural Relics and Archaeology 2003). Microblade technology evolved from blade technology. In artifacts assemblage using the microblade technology, microblades are frequently present, providing evidence that the prismatic blade technology constitutes another source for the microblade technology.

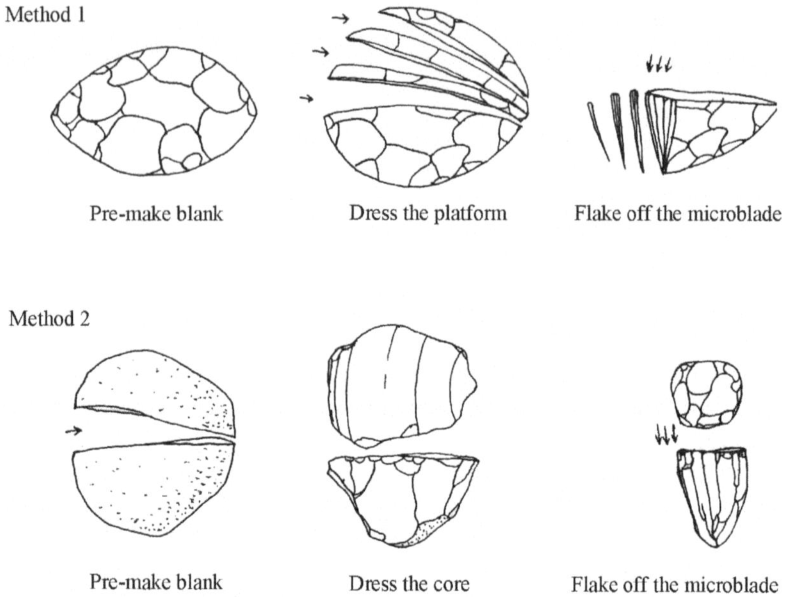

Fig. 4.4 Two methods of microblade technology[11]

One significant characteristic of the microblade is the production of bifaces and boat-shaped core blanks, which is followed by making microblades. Kobayashi suggests that this process originates from pre-existing mental templates in the human mind, where the forms and processes of microblade cores and microblades already exist, and the production process aims to achieve these pre-existing goals. On the other hand, J. J. Flanniken proposes that the process is determined by the cultural type and the nature of the raw materials (Flanniken 1987). However, the microblade technology, which only represents a manifestation of cultural types, is not entirely determined by functionality. Therefore, the microblade can represent the regional forms of cultural types.

Based on the above illustrations, the microblade and handaxe technology represented by the Mode II share similarities in seeking formal consistency beyond functionality and exhibit more complex and diverse production processes. The aforementioned technology, that is, bifacial technology features standardisation and multifactoriality. While further processing bifaces into microblade cores and then flaking microblades represent a specialised extension of the technology but cannot be fully explained based on functionality. Based on J. J. Flanniken, the total length of the edge flakes struck from bifaces exceeds the total length of the edge flakes produced by flaking microblades. Although microblade production conserves raw materials compared to biface production, it requires twice as much time consumption. Additionally, if waste flakes are considered during the process, microblade production does not excel in providing cutting edges compared to bifacial tools, but its primary significance lies in offering standardised products. Even though bifaces can serve as versatile, long-lasting tools, their form is not essential for producing microblades. Combining the bifacial technology with a specific blade production technology resulted in the typical microblade technology, leading to the possibility of the blade production technology equaling prismatic blade technology.

4.3.3 Causes of the Emergence of Microblade Technology

The purpose of technology is to address specific problems, microblade technology emerged in the Upper Paleolithic, during which epoch hunter-gatherers faced challenges fundamentally similar to earlier periods, as these challenges involved addressing the uncertainty and risks associated with resources. The stability of livelihoods depended not on the abundance of resources during the peak season but on the minimum available during the lean season, known as the "tolerable lower intake level" principle in biology. While the problems remained unchanged, the initial conditions for hunter-gatherers to solve underwent changes. Firstly, the sustained population growth led to a corresponding increase in resource demands; however, natural resource production did not increase accordingly, and disruptions in ecological processes could lead to a decline in resource productivity, presenting a significant disadvantage. Secondly, human culture evolved through accumulation, providing hunter-gatherers with the potential to find innovative solutions through cultural advancements. Finally, changes in the natural environment posed challenges for human adaptation. During the Upper Paleolithic, the Earth entered the LGM, reducing the available living space for humans in high latitudes; however, in mid to low latitudes, the reduction of tropical climate zones and the lowering of sea levels expanded human living space. In this case, the LGM did not necessarily signify a deterioration in the human living environment; instead, the rapid end of this glacial period and climate fluctuations presented greater challenges for human adaptation.

During the Upper Paleolithic, several new characteristics can be observed in the hunting and gathering adaptations across the Eurasian continent. Firstly, there was a shift towards hunting small-sized animals, as evidenced by the discovery of humans hunting small mammals, such as rabbits, during the Middle Upper Paleolithic in West Asia (Stiner et al. 1999), possibly in response to the reduction in large animal populations. Consequently, hunting tools also became smaller. Secondly, stable isotope analysis of human remains in the Gravettian phase in Europe, including Siberia, indicated that humans were utilising a broader range of terrestrial and freshwater food resources (Hoffecker and Elias 2003; Goebel 1999). This expansion in resource utilisation demanded more versatile and flexible tools. Thirdly, the rapid human population expansion and occupation of all continents even venturing into polar regions except Antarctica (Derev'anko 1998). This unprecedented migration rate demonstrated a prominent level of mobility, necessitating a set of tools suitable for mobility. The global trend of reducing the size of human stone tools was evident, and microblade technology represented one of its forms. Lastly, as aforestated, one fundamental trend in human adaptation was to maximise energy intake. Simultaneously, humans sought stable resources with minimal risks. During the hunting and gathering phase, this involved discovering more available resources and hunting the most rewarding prey, primarily large- and medium-sized herbivores, especially deer species. Therefore, increasing mobility can be regarded as a fundamental trend in human adaptation during the hunting and gathering phase.

Of course, it is essential to consider that microblade technology is built upon the technological foundations of the Upper Paleolithic and earlier periods. In addition to the aforementioned techniques, such as chopping, blade, and flake technology, the microblade technology basis should also encompass the following: human's understanding and utilisation of high-quality stone materials; the usage of mechanical devices, which cover bows, throwing implements, or blowpipes, to enhance the distance and accuracy of projectile tools; the possible employment of chemical poisons to increase the lethality of tools. Moreover, other supportive technologies are engaged, that is, various tended and untended facilities, including traps and nets. Building upon these technological foundations, microblade technology fully utilises its characteristics in creating mobile and disposable tools, without considering temporary tasks such as chopping or cutting. Due to the assistance of these technologies, the products of microblade technology can be remarkably small in size.

Microblade technology aims to produce composite tools with embedded blades, but archaeologists have limited knowledge about their specific uses. Current archaeological findings mainly consist of embedded-blade knives from the Neolithic sites, often using flat animal bones with a slot on one side to insert the blade. These tools can serve as daggers or small knives for personal carry, while spears with embedded blades are quite rare, with some discoveries in Siberia (Derev'anko 1998) (Fig. 4.5), but not as prevalent as embedded-blade knives. The question arises whether microblade technology was primarily used for manufacturing embedded-blade knives rather than spears. In the case of Aboriginal Australian lithic technology, spears with regular blades embedded are highly effective due to their uneven edges, resulting in larger wounds upon piercing prey (Mulvaney and Kamminga 1999). In this case, it could be questioned the reason for producing specially microblade embedded in spears. The scarcity of spears with embedded blades also suggests that the main purpose of microblade technology was not for making spears but for manufacturing embedded-blade knives, as only such tools require even edges. Another piece of evidence is that even in lithic assemblage containing microblade products, there are specifically retouched projectile points, found in sites like the Nihewan Basin and various Northeastern Neolithic sites, as well as Siberia. These assemblages indicate the presence of specialised tools for hunting large animals. Elston and Brantingham argue that embedded tools are more advantageous compared to spears with embedded blades, which are not easily maintainable and more likely to be discarded once damaged (Elston and Brantingham 2002). In fact, regarding hafting tools, ethnographic evidence shows that manufacturing the arrow shafts is more labour-intensive than making the arrowheads (Kelly 1982). In conclusion, for hunting large animals, spears with embedded blades are not essential and economical.

Although microblades are more likely to have been used for making embedded-blade knives, however, no the Upper Paleolithic embedded-blade knives have been discovered in China yet. Discoveries from Siberia during this period can serve as a reference. microblades used for embedded-blade knives generally have a width of no less than 5 millimetres and a length of no less than 20 millimetres, as they need to be large enough to be hafted. However, the microblades found in North China

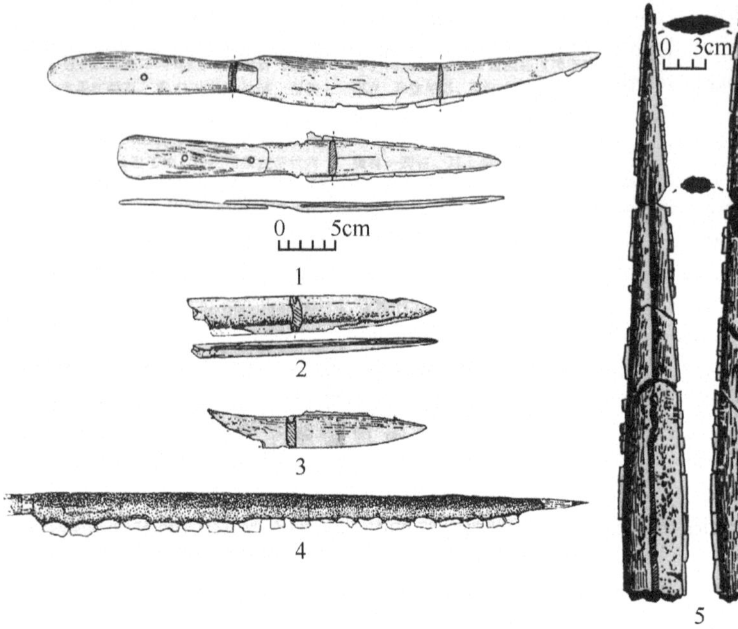

Fig. 4.5 Composite tools inlaid with microblade in archaeological materials and composite tools inlaid with stone flakes in ethnology[12]

in the Upper Paleolithic, especially materials from Tingsijian, Northern Jiangsu, Western Shandong, and parts of Xiachuan, are exceptionally small and differ from the typical microblades used for embedded-blade knives. These microblades seem unsuitable for usage as cutting edges in embedded-blade knives but are more appropriate for some form of barbs in hunting tools. However, this is only a plausible conjecture, and archaeological evidence is needed to substantiate.

Bone and antler tools were common in both the Upper Paleolithic and Neolithic sites, especially in the Neolithic sites, composite tools with embedded blades disappeared in the North China region where grinding techniques prevailed but were still present in areas outside the Great Wall (Gansu Provincial Museum Cultural Relics Work Team, Wuwei Regional Cultural Relics Survey Team 1974; Chinese Academy of Social Sciences Institute of Archaeology Inner Mongolia Work Team 1985, 1997; Shenyang Municipal Cultural Relics Management Office,

Shenyang Imperial Palace Museum 1985). Instead, numerous ground bone arrowheads took their place, which might be related to differences in hunting targets. Modern hunting and gathering communities often employ different arrowheads when hunting different prey. Therefore, microblade technology neither replaced bone and antler tools nor purely stone tools, such as projectile points. On the contrary, during the Neolithic in North China, microblade technology was replaced by grinding techniques. Microblade technology can be seen as a product to satisfy specific tasks and needs; while it would lose its advantage once these particular demands disappeared. Lightweight, mobility, and adaptability of microblade technology constitutes its specific advantages. Thus, the microblade technology was suitable for manufacturing embedded-blade knives, embedded-blade spears, and barbs for hunting tools (Fig. 4.5). Residue analysis of microblade tools with embedded blades indicates that they were in contact with sturgeon, deer, bears, felines, rodents, and even human blood (Gerlach et al. 1996).

The purpose of microblade technology is to maximise the production of standardised, sharp cutting edges that could be used with the minimum amount of lithic material possible. In this sense, microblade technology represents a way to optimise the utilisation of high-quality lithic resources or to minimise the burden of carrying raw materials. Technically, waste can be reduced significantly. The concept of "non-hafting" proposed by Flenniken indicates that waste is challenging, which might be encountered by modern replicators. Based on archaeological materials, evidence suggests that there is no significant issue with hafting, which could be demonstrated through renewal of platforms, the creation of new platforms during production, and the existence of extremely small microcores in regions such as Tingsiji and Malingshan.

For highly mobile hunter-gatherers, their tools must be lightweight and efficient. Coincidently, microblade technology precisely fulfils these requirements. Understanding the purpose of microblade technology leads us to investigate when and where humans had a strong need for this technology, ultimately providing a comprehensive explanation for the emergence of microblade technology.

Notably, as a form of technology, microblade technology can be considered from a cost-effectiveness perspective. However, it must also be recognised that microblade technology exists as a cultural form, and its emergence has an element of randomness. When microblade technology becomes widespread, its cultural significance gradually strengthens, while its utilitarian significance may diminish. Similar to pottery, which initially served as containers and cooking vessels, gradually acquired cultural significance and became a symbol of different archaeological cultures, without any loss of its utilitarian value. The same applies to microblade technology. A typical example is the microblade technology on the Japanese archipelago, where high-quality stone materials suitable for making microblades, such as obsidian, are not rare due to volcanic activity. Additionally, the livelihood of hunter-gatherers on these islands is not characterised by hunting large herds of herbivores like in Eurasia. Moreover, the region boasts abundant aquatic resources, and the utilisation of these resources allows hunter-gatherers to reduce mobility. Therefore, microblade technology is not essential to the subsistence strategy of hunter-gatherers in the Japanese archipelago; rather, it was culturally absorbed. However, when tracing the origin of microblade technology, it is necessary to consider from a utilitarian perspective. This technology inevitably appeared in the places where it was most needed and where the relevant technological foundations were most developed.

4.3.4 Region and Chronology of the Origin of Microblade Technology

Zhimin An presented four hypotheses regarding the origin of microblade technology: the European hypothesis, Siberian hypothesis, Mongolian hypothesis, and North China hypothesis. He argued that microblade technology originated in North China, primarily based on the presence of sites like Shiyu and Xiaonanhai, which exhibit similarities to microblade technology from the Upper Paleolithic in North China. Another evidence is the discovery of blade technology of microblade technology at the Shuidonggou site in Ningxia (An 1978). However, the lithic technology at Shiyu site and Xiaonanhai site does not include a crucial element of microblade technology, that is, the bifacial technique. Moreover, the lithic assemblages at the Xiaonanhai site are a mixture of materials from different periods, and there has been no reliable dating to ascertain their chronological sequence. Similarly, lithic assemblage in the Shuidonggou site also lacks the bifacial technique. Shuisheng Du proposed that the Xiachuan site represent the origin of microblade technology in North China, while the Hutouliang site represents its origin in Siberia (Du 2004). However, in terms of technological elements, there is no fundamental distinction between these two types of techniques.

Even though there were microliths resembling "pseudo-microblades", which are similar to those found at the Xiaonanhai site, during the early phase of the Upper Paleolithic in Japan, however, Kobayashi argues that these artifacts are not related to microblade technology. Instead, he proposes that Japanese microblade technology has an independent origin. Prior to the emergence of microblade technology, there existed a "backed blade industry" with the width of such blades often exceeding 11 millimetres in the Japanese archipelago. However, microblade technology is not simply a reduced version of these bladelets; rather, it emerged because of the local lithic industry "adopting" new stimuli, leading to the disappearance of the bladelet. If this "adoption" refers to the influence of microblade technology, it becomes difficult to argue that Japanese microblade technology has a purely indigenous origin. Because of the uniqueness of microblade technology in the Japanese archipelago, it is unlikely for its independent origin.

A. V. Tabarev argues that the core region and origin of microblade technology lies in the western Baikal Lake area, Upper Amur Basin, and northeastern Mongolia (Tabarev 1997,). Similarly, Ted. Goebel suggests that microblade technology might have originated in central and eastern Mongolia (Goebel 2002) mainly considering that the securely dated microlithic materials are around 17,500 BP, predating the less reliable materials dated to 18,000 BP. Additionally, humans had left the Siberian region during the LGM, and the crucial technological element of microblade technology, that is, the bifacial technology, is present in the early phases of the Upper Paleolithic in Siberia, and this technology is found in earlier Paleolithic cultures on the western side of the Eurasian continent.

Simply judging the date and region of the origin of microblade technology based on archaeological findings is insufficient, and it is overly simplistic to regard the central distribution area of microblade technology as its place of origin. As previously discussed, microblade technology, as a technique advantageous for its highly mobile livelihoods, cannot be considered a risk-minimisation strategy. For hunter-gatherers, risk minimisation involves increasing mobility, and microblade technology serves as a means to achieve this. In ethnographic materials, the most mobile groups are often those whose primary livelihood is hunting, residing either in grasslands or near deserts where prey resources are not densely concentrated. It would be challenging for nomadic hunter-gatherers to solely survive in grassland regions. It was also demanding for modern groups who are marginalised to edge regions, as they must inhabit transitional zones between grasslands and forests to avoid over-reliance on a single resource zone (Binford 2001); also, they can hunt and seek shelter in the forested areas during cold and snowy seasons. Thus, these transition zones between grasslands and forests are the most suitable areas for hunter-gatherer livelihoods. Based on modern meteorological data (Fig. 4.6), a model regarding hunter-gatherers reveals that the most suitable regions for a hunting-gathering economy are the natural transition zones between forests and grasslands from northeast to southwest, extending northwards to eastern Mongolia and east of Lake Baikal in Russia (mainly determined based on the distribution of primary productivity in the biosphere (Leith 1972) under current climatic conditions in China. In other words, microblade technology should have originated in the area aforementioned if it were to originate under modern climate conditions. This also explains the reason for the consistent usage of microblade technology in this region. However, the time of origin of microblade technology coincided with the period around the LGM, during which the climatic zones in the Northern Hemisphere shifted southward by 3–5 degrees of latitude (Wei et al. 1997), with greater changes at mid to high latitudes. Therefore, if microblade technology originated during the LGM, it should have originated in the inland regions of North China, rather than the transitional zone from northeast to southwest, and it is even less likely to have originated in the regions some scholars have suggested, such as the Baikal Lake area or eastern Mongolia (Goebel 2002, pp. 117–132).

During the period around the LGM, the main characteristic of environmental changes was the fragmentation of existing resources into patches, and the distances between these patches continued to expand until they were entirely replaced by new environments. As the distances between resource patches increased, hunter-gatherers had to enhance their mobility to access the resources they were accustomed to utilising (Fig. 4.7). Once their mobility reached its limit, they had no choice but to explore and exploit untapped resources, diversify their resource usage, which is known as broad spectrum adaptation, and intensify the utilisation of previously unused resources, such as processing toxic plants to detox them.

During the LGM, the tundra extended in Northeast China up to around 45°N, while currently, tundra distribution is absent within China's territory; and the southern boundary of coniferous forests have shifted from approximately 50°N to around 23°N. The forest-grassland transition zone, which exists from Northeast to Southwest China, appeared

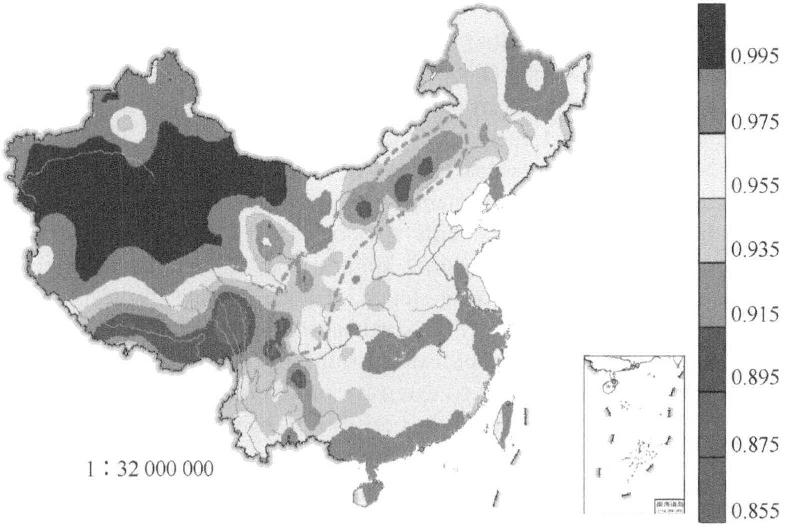

Fig. 4.6 Relationship between animal resource distribution and hunting simulated by modern weather station data[13]

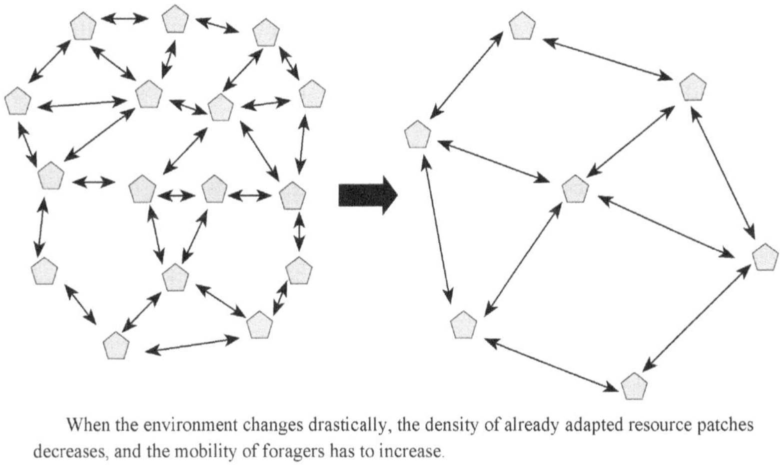

Fig. 4.7 The impact of changes in resource patch density on the mobility of hunter-gatherer[14]

in the core region of North China during that period, while vast areas of Northwest, Qinghai-Tibet Plateau, and Inner Mongolia where are covered by dry grasslands (Figs. 4.8 and 4.9) (Winker and Wang 1993, pp. 221–261). Northwestern parts of Siberia were covered by ice sheets during the LGM, and there were glaciers in the Siberian Plateau, the Aldan Mountains, the eastern region of Lake Baikal, and the western Altai Mountains, with the remaining areas mainly consisting of tundra and mammoth-steppe ecosystems (Goebel 1999). However, Goebel's paleoenvironmental reconstructions suggest that the southern fringe of Siberia was a zone of pine forest-steppe. However, some studies propose that the adjacent northern part of Northeast China was tundra. It is more plausible that the forest-steppe existed only in localised areas along the southern edge of Siberia due to extensive mountain glaciers, followed by vast tundra and dry grasslands as transitional zones. Therefore, during the LGM, the extensive forest-grassland transition zone was distributed around the core area of North China, which is the most likely region where hunter-gatherers migrated southwards from Siberia. In historical times, with each cooling period, there have been phenomena of northern populations migrating southwards. The southern boundary of this forest-grassland transition zone lies around 33°N, which aligns with the southern boundary of the distribution of the microblade technology. It also corresponds to the distribution of the mammoth-rhinoceros fauna, not exceeding 33°N. This intriguing coincidence highlights the adaptability of the microblade technology to a specific environmental context.

Nowadays, Siberia's environment transitions from north to south, ranging from tundra and taiga forests to grasslands and deserts. The grassland region is distributed between 52°N and 53°N latitude (approximately 300 kilometres wide), and the forest-grassland transition zone possesses abundant animal resources, making it highly suitable for human habitation. Humans likely inhabited the resource-rich mountainous areas of southern Siberia during the Early Late Pleistocene (approximately 120,000 to 110,000 BP). Evidence indicates that humans reached this region to 55°N latitude during the Early Late Pleistocene and expanded further north to around 60°N latitude during the Middle Upper Pleistocene (Clark 1997). However, with the onset of the LGM during the Late Upper Pleistocene, humans departed from Siberia, with only a small population remaining in the southern mountainous regions of Siberia, while the majority migrated to North China.

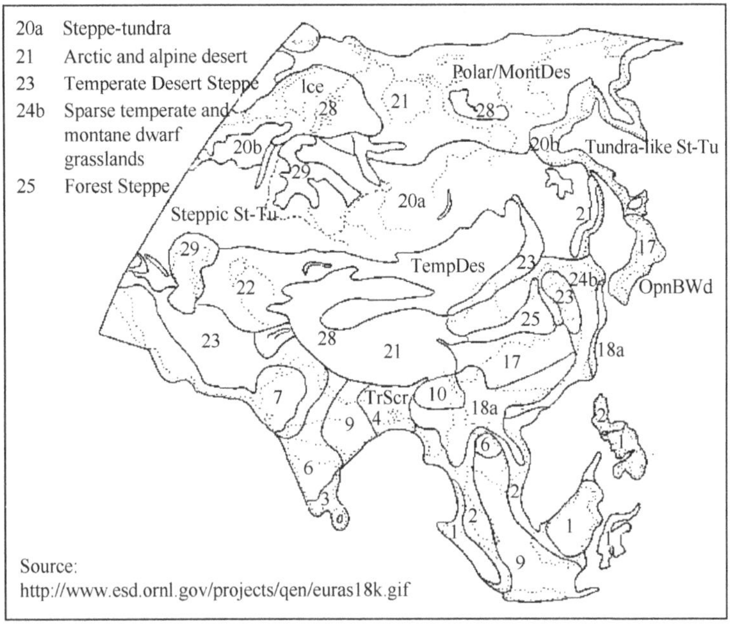

Fig. 4.8 18,000-year-old reconstruction of vegetation in Asia

It is now established that the origin of the microblade technology lies in an ecological transition zone, which also served as a cultural transition zone. The southward migration of the Siberian Paleolithic culture occurred even before the LGM, as evident in the Shuidonggou culture in Ningxia. Elston et al. also proposed that the spread of microblade technology in Eurasia was from north to south, with a dating range of 29,000 to 24,000 BP. The discovery of bipolar bladelet like microblades in certain sites possibly indicated preliminary phases of microblades (Elston et al. 1997). However, these findings are more likely to be related to repair activities rather than genuine microblade production. For example, during the excavation season at the Shuidonggou site in 2005, numerous small flakes resembling microblades were found in the second location of the first excavation area. Yet, the analysis of technological elements cannot prove that bipolar technology was an essential aspect of microblade technology. During the LGM, as human populations migrated southward, this cultural transition zone also moved southward. In the western regions,

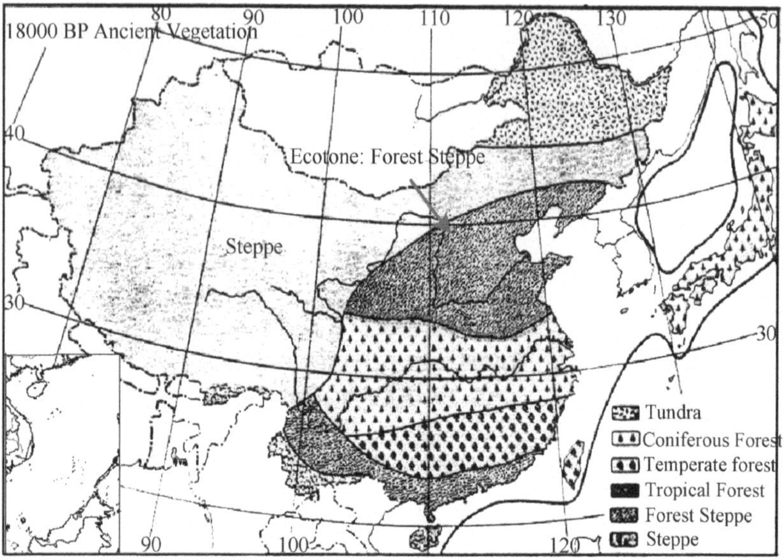

Fig. 4.9 The relationship between the change of ancient vegetation and the process of microblade[15]

significant changes were observed in the appearance of all Paleolithic cultures in the Gansu-Ningxia and southern Shanxi areas (Ji et al. 2005). Generally, new cultural elements are often facilitated through cultural exchanges. It is evident that the microblade technology, which is distinct from geometric microblade technologies prevalent in North Africa, West Asia, and Europe, emphasised the performance of the preparation of bifacial cores and prismatic core technology. Its appearance in North China can even be traced back to the Early Pleistocene (Fig. 4.9), indicating that it was the result of cultural exchanges and interactions.

After establishing that the microblade technology originated in a cultural-ecological transition zone, it becomes possible to trace the chronology of this zone, which represents the earliest period when microblade technology could have emerged. With four site ages exceeding 20,000 BP, Flanniken proposed that the microblade technology at the Dyuktai site could date as early as 36,000 BP, with the latest dating around 11,000 BP. However, subsequent investigations have refuted the early dating of those sites. Tabarev suggests that the origin of microblade technology lies between 25,000 and 22,700 BP, whereas the confirmed dating of microblade technology within the Russian territory currently falls around 17,500 BP, as discovered in the South Baikal Lake region (Goebel 1999). In the North China region, the earliest evidence of microblade technology was originally found in the Xiachuan site and the Locality 77:01 at the Dingcun site, but the dating has remained disputable. However, through detailed excavations and new dating methods, the microblade technology at sites like the Xishi site in Henan Province and the Shizitan site in Shanxi Province has been pushed back to approximately 20,000 BP (Shizitan Archaeological Team 2010, 2002; Zhao and Huo 2007).

With microblade technology discovered, the Jinsitai site located in the eastern part of Inner Mongolia is one of the most abundant Paleolithic sites in recent years. It contains three cultural layers with approximate ages of 30,000, 20,000, and 10,000 BP (Wang et al. 2010). The first two cultural layers do not show any evidence of microblade technology. The middle layer exhibits a blade industry similar to the one found in the Shuidonggou site, characterised by the typical Levallois technique. Evidence of microblade technology was merely found in the latest layer. If microblade technology originated in the eastern region of Mongolia, its dating should not be later than that in North China.

The origin of the microblade technology is linked to the date of humans first entering the Americas. In turn, the earliest Native Americans undoubtedly originated from Asia, which allows us to infer the timeframe for the origin of microblades. The molecular biology analysis suggests that the first human migration to the Americas occurred around 20,000 years ago, through a coastal route from the central and southern regions of Siberia. Subsequently, humans reached the Americas via an inland route from the same area, influencing the genetic makeup of North and Central America. A third migration, occurring later, impacted the Aleutian Islands, Eskimo populations, and Na-Dene Native Americans (Schurr and Sherry 2004). Another study indicates that the genetic distance between Native Americans and indigenous populations of Southeast and Northeast Siberia is the closest (Uinuk-ool et al. 2003). If humans were already proficient in using microblade technology before their migration to the Americas, they would have continued using it upon arrival. The transition from microblade technology to Clovis point technology would not have occurred in a brief period. In fact, the dissemination of microblade technology to North America took place during the third wave of human migration to the continent. It is highly likely that microblade technology matured after the earliest Asian populations migrated to North America as its spread occurred from south to north and from west to east. This implies that the timeframe for the origin of microblade technology is likely to be around the LGM, which aligns closely with the first human migration to the Americas. After a significant cooling of the climate, the forest-grassland zone moved southward, resulting in the scarcity of previously adapted resource patches. Meanwhile, the Beringia began to form. Humans started moving away from the frigid regions of Siberia, migrating southward and eastward to find familiar resources and unexploited ones, respectively. Consequently, migration occurred predominantly in these two directions and not towards the north or west. Thus, the theoretical timeframe for the origin of microblade technology could be placed around the LGM, approximately 24,000 to 21,000 BP.

4.3.5 *The Upper Paleolithic Adaptation Represented by Microblade Technology*

In summary, the following conclusions can be drawn regarding microblade technology:

1. Microblade technology represents a type of lithic technology that is advantageous for highly mobile hunting and gathering populations.
2. Microblade technology is the result of combining the traditions of the bifacial technique and the prismatic core technique.
3. Microblade technology reflects the adaptations of hunting and gathering populations to the changing resources during the period before and after the LGM.
4. Microblade technology marks the pinnacle of development in the mobile hunting and gathering way of life.
5. The origin of microblade technology can be traced back to the region of North China during the period before and after the LGM.

Microblade technology was the predominant technological feature in the late phase of the Upper Paleolithic in North China. During this period, two main subsistence patterns can be roughly identified: The first pattern is characterised by the production of larger microblade technologies, represented by the sites of Xiachuan and Hutouliang. This pattern corresponds to what Xie Fei referred to as the "distribution area dominated by the wedge-shaped core technique". The second pattern involves the production of smaller microblade technologies, represented by sites in the Tingsijian and Malingshan regions, corresponding to the "distribution area dominated by the boat-shaped core technique" (Xie 2000). The distinction lies in the possibility that techniques were used to process several types of resources: The former pattern may have been more focused on hunting and gathering plant-based foods; while the latter pattern may have been more related to the utilisation of aquatic resources. However, this distinction is not mutually exclusive but merely reflects a macroscopic difference.

Compared to the early and middle phases of the Paleolithic, the techno-cultural complex is more diverse and complex in the late phase. Through a comparison along five dimensions: mobility, efficiency, maintainability, adaptability, and durability, the microblade technology products demonstrate excellent mobility, as they are easily portable. They also exhibit high efficiency, with the capability of being easily maintained by replacing the blade edge when necessary. Moreover, these microblade technologies can serve multiple purposes, showcasing broad adaptability. Considering cost and efficiency, the high efficiency of the microblade technology comes at the expense of lower durability. This also reflects that humans emphasised portability, utility, maintainability, and broad adaptability of tools in the late phase of the Upper Paleolithic. In summary, in the Upper Paleolithic of North China, certain trends can be observed in technological advancements:

1. Hunter-gatherers achieved unprecedented levels of mobility, aiming to expand their food collection territory or frequency of utilisation. This is driven by concerns related to resource density, population pressure, and the need for certainty in resource availability to avoid sudden scarcity of food.
2. Standardisation of technology was adopted to accommodate various tasks, as hunter-gatherers needed to undertake a wider variety of tasks with the expansion of the range of food collection. In addition, the uncertainty and unpredictability of resource availability necessitated the preparation of tools with broader adaptability, aligning to higher mobility.
3. An increased emphasis on tool maintainability, consistent with the standardisation and multi-functionality of tools.

4. Utilisation of the highest quality stone raw materials, indicating a pursuit of tool efficiency. This trend might be related to the seasonal and sporadic nature of resources, as people needed to gather as many resources as possible within limited time frames.
5. A relatively lower priority on tool durability reflected the strong mobility of the population. Lightweight small tools were preferred, and they inherently had lower durability compared to exquisitely polished large tools.

The production of blade and microblade technology is oriented towards manufacturing composite tools. Microblade technology rapidly became prevalent in the late phase of the Paleolithic in North China, and it spread to the Japanese archipelago and Alaska in North America. The practical use of productions is attributed to their widespread adoption. The emergence of industries, which is primarily based on microblade technology and signifies the expansion of human living environments, enabling the utilisation of more marginal environments. The continued existence of this industry from northeastern to southwestern China, across the natural transition zones, further demonstrates its adaptability to marginal environments and its highly suitable nature for a mobile hunter-gatherer lifeway.

4.3.6 *Microblade Technology and the Origins of Agriculture*

After a detailed discussion on the nature and origin of microblade technology, it might be wondered how this technology relates to the origins of agriculture. The first chapter covers the coping mechanisms of hunter-gatherers, which include increasing mobility, expansion of diet, population growth control, enhancement of competition or cooperation, and technological innovations. Among these mechanisms, increasing mobility is the most basic and straightforward approach with the lowest cost, without need to expand the diet or change well-established diet, nor is there a requirement for extensive labour or the development of specific technologies to process challenging food items (such as processing nuts or other small plant seeds, which would require grinding slabs, handstones, or filtration tools). Controlling population growth is difficult and painful for hunter-gatherers. Similarly, competition enhancement is limited since it is challenging to find suitable partners to assist on a large scale of adaption pressures. Moreover, seeking assistance carries its own costs and can lead to a decrease in the overall social status of the group. Technological innovations require more favourable conditions and involve a long-term process.

The microblade technology represents an innovation in lithic technology domain that catered to the needs of mobile hunter-gatherers by providing lightweight tools with versatile functionalities. In this sense, microblade technology, which can be viewed as a fundamental response to adaptation pressures for hunter-gatherers since the LGM, served as the initial step in the origin of agriculture. It seems that the increased mobility of hunter-gatherers is contradictory to the sedentary nature of agricultural production. However, agricultural production can be seen as a comprehensive cultural response to adaptation pressures for hunter-gatherers, as its first step involves strengthening the utilisation of resources within their habitat. Enhancing mobility is the initial phase of this strategy, whereby hunter-gatherers gain a deeper understanding of the types, distribution, and quantities of resources in their habitat and reinforce their sense of ownership. Evolving from increased mobility to the intensified utilisation of certain food resources, species domestication and the emergence of agriculture becomes the eventual outcomes. In this case, without this initial step, the subsequent development of agriculture would not have been possible.

Certainly, mobility is not infinitely scalable. Initially, increasing the frequency of resource searches within their habitat can indeed boost the yield of resources for hunter-gatherers. However, continuous escalation would lead to diminishing marginal returns. The primary productivity within a habitat is ultimately determined by factors such as temperature brought by solar radiation and precipitation. While the primary productivity exhibits annual fluctuations, it does have an upper limit. That is, regardless of the degree of mobility of hunter-gatherers, the resources they can acquire are not boundless. Furthermore, once hunter-gatherers disrupt the ecological balance, the productivity of the entire habitat may rapidly deteriorate. Additionally, hunter-gatherers relying on terrestrial resources can only cover a limited range through walking while hunting and gathering. Even when technology becomes advanced, it is still impossible to endlessly increase the quantity of resources obtained.

In terms of mobility, hunter-gatherers have only two choices. One option is to maintain a highly mobile way of life, which could be achieved through the following methods: (1) Finding accessible tools, such as boats or horses, to facilitate movement and expand the coverage of their habitat; (2) Controlling the mobility of resources through the domestication of herbivores, to establish a pastoral economy; (3) Engaging in symbiosis with agricultural communities and exchange for agricultural products; (4) Developing a diversified economy, with some members focusing on hunting and gathering while others abandon these practices. In summary, before species domestication, the only means of enhancing mobility is through water transportation, such as boats and canoes. The other option is to reduce mobility until complete sedentism, which requires obtaining necessary resources within a limited area. If there are no sustainable natural resources available, food production becomes the only viable strategy. Within regions lacking aquatic resources and suitable conditions for water transportation, the transition towards intensified resource utilisation becomes inevitable once hunter-gatherers face adaptive pressures.

Existing archaeological and ethnographic evidence indicates that not all hunter-gatherer groups using the microblade technology transitioned to agricultural production. This observation is not limited to the natural transition zones from northeastern to southwestern China; it is also evident in various regions, including Northeast China, Northwest China, the Mongolian Plateau, vast areas of Siberia, and the Japanese archipelago. Hunter-gatherer groups in these regions, utilising microblade technology, were able to acquire the necessary resources for their livelihoods. There is a noticeable differentiation process: Hunter-gatherer groups employing microblade technology followed distinct developmental paths. Some, like those in North China, facing higher population pressures and specific resource conditions (this will be elaborated in the next chapter), chose to reduce mobility and move towards intensified resource utilisation. On the other hand, other hunter-gatherer groups, leveraging the advantages of mobility offered by microblade technology, continuously expanded into marginal habitats, such as regions in Siberia, the Pacific Northwest of North America, the Mongolian Plateau, or developed aquatic resource utilisation as seen in the Japanese archipelago. Consequently, they did not face the same adaptive pressures and maintained their hunting and gathering livelihoods over the long term.

Previously, the adaptive patterns of hunter-gatherers in the Upper Paleolithic in North China were roughly classified into four models:

Shuidonggou, Shiyu, Shandingdong Cave—Dongfang Square—Xiaonanhai, and Xiaogushan. In the Late Pleistocene, only the microblade technology mode was observed. However, this static categorisation fails to fully capture the variations in prehistoric human adaptive strategies, and drawing a clear boundary between the Upper Paleolithic and the Neolithic poses challenges, as there exists a transitional period between them. The current chronological methods consider the boundary between the Pleistocene and Holocene as the demarcation between the Paleolithic and Neolithic periods. However, it is important to note that geological divisions and prehistoric cultural periods cannot be strictly equivalent. The existing chronological system reflects the tradition of Chinese prehistoric research, where Paleolithic archaeology was pioneered by scholars with geological backgrounds, while Neolithic archaeology was established by scholars with historical backgrounds. The transition from the Late Pleistocene's later phase to the typical features of the Neolithic, when agriculture was established, spanning from approximately 15,000 to 8000 years ago, encompasses a period of nearly 7000 years. During this transitional phase, the prominence of the microblade technology in North China eventually declined and disappeared entirely from the Central Plains region.

Archaeological findings in North China clearly demonstrate the changes in archaeological remains, which is associated with microblade technology. Regarding the artifact assemblage, the most notable change is the appearance of durable tools, such as adze-shaped tools and site furniture like grinding slabs. The abundance of such artifacts increases as sites approaches the Holocene period. Subsequently, pottery and ground-stone tools emerged, marking the beginning of a new era. In terms of site structures, sites containing microblade technology typically manifest as location clusters, often referred to as site groups. This change reflects the strengthening of mobility among hunter-gatherer groups within their habitats, indicating a noticeable increase in the frequency of food resource exploration within limited areas. However, in the later phase of the Upper Paleolithic (approximately 15,000 years ago), intriguing differentiations appear within these location clusters. Some sites contain particularly rich and diverse types of remains, which usually signify longer periods of residence and carry more convincing evidence. The longer a group resides in a particular location, the broader the range of activities they engage in, leading to a more diverse assemblage of artifacts. A detailed discussion of specific archaeological materials will be provided in Sect. 2 of Chapter 6.

In summary, it is clear that the regional differentiation in adaptation strategies have already appeared during the Early Late Pleistocene in North China. With the onset of the LGM, the climate became colder and drier, leading to a reduction in surface productivity. During the late phase of this period, microblade technology became widespread in North China, and hunter-gatherers significantly increased their mobility to cope with the scattered and mobile food resources. As the LGM ended, significant changes occurred in human habitats and resource distribution, with the microblade technology reaching its peak in North China. Subsequently, with further climatic warming, the prominence of microblade technology declined, and new adaptation strategies emerged. The existing archaeological evidence records the fundamental process of these changes.

4.4 Late Upper Paleolithic Adaptive Transition in the Middle and Lower Reaches of the Yangtze River

4.4.1 *Materials and Research*

When *Ancient Humans in China* was published in 1989, no significant archaeological materials had been found from the Upper Paleolithic in the middle and lower Yangtze River region. However, since the 1990s, numerous Paleolithic sites and locations have been discovered in provinces such as Hunan, Hubei, Anhui, and Jiangsu. Even the historically under explored regions of Jiangxi and Zhejiang have seen breakthroughs in archaeological findings. Particularly, large-scale cultural surveys and excavations resulting from the Three Gorges Dam Project and South-to-North Water Diversion Project have led to a concentration of archaeological materials in western Hubei and its surrounding areas. However, despite these locations, the discoveries and research on the Upper Paleolithic in the middle and lower Yangtze River region still remain limited and fragmentary, as the lack of absolute dating[16] due to poor organic preservation in the soil, making it difficult to discover datable materials and establish a chronological framework. In addition, the absence of long-term systematic research on representative sites hinders a comprehensive understanding of the cultural landscape during the Late Pleistocene. Furthermore, the research method, which mainly focuses on typology of lithic technology, has resulted in a shortage of archaeological materials from diverse perspectives, which is unfavourable for exploring the changes in ancient human cultural adaptation.

Based on the available archaeological materials, Hunan Province stands out due to its relatively sustained research efforts, making a relatively abundant and systematic collection of materials available that thus forms the basic cultural framework for the middle and lower Yangtze River region. There are over 200 Paleolithic sites distributed across four river basins: Xiangjiang, Zijiang, Yuanjiang, and Lishui. The Yuanjiang and Lishui river basins, in particular, have yielded particularly rich findings. Based on terrace stratigraphy, the dating is primarily classified as follows: the first terrace is assigned to the Upper Pleistocene, the second terrace to the Middle Pleistocene, and the third and fourth terraces to the Early Pleistocene. This dating is predicated on the preference of ancient people to live near water sources for easier access to water and hunting as animals also need to drink water. The Upper Paleolithic deposits are also found in the surface layer of weathering crust with reticulate degeneration, possibly formed during the LGM. The lithic industry in Hunan Province is overall classified as southern conglomerate stone technology, and its internal variations are first manifested in regional differences. In comparison, the eastern Xiangjiang and Zijiang river basins have fewer discoveries, which characteristics remain less clear; while the Yuanjiang River middle reaches and Lishui River basin, which are located in the western parts of Hunan Province, formed distinct cultural regions known as the "Wushui Culture Group" and the "Lishui Culture Group". These two culture groups exhibit distinct stages of development, with the Wushui culture group represented by the Erluanshi, Yanwutan, and Changping sites; and the Lishui Culture Group represented by the Huzhaoshan, Jigongdang, Wuyashan, and Shiligang typical sites. The small flint lithic industry found at the Yan'erdong site and the Shiligang site shows a continuity with the small flint lithic industry discovered in the Lixian Plain in the early to middle Neolithic cultures. The Wushui Culture Group shares closer affinities with the Paleolithic industries in Guizhou and Lingnan regions, serving as a local cultural variant within a small area. However, no miniaturisation phenomenon of stone tools appears in the late phase of the Wushui Culture Group, making it distinctive from the Lishui River basin (Yuan 1996).

The Lishui Culture Group is mainly distributed in the Lishui River basin and the western bank of the Dongting Lake, including the hilly plains of the lower Yuan River. There are over 100 known sites, and among those that have been excavated are Huzhaoshan site in Jinshi City, Jigongdang and Wuyashan in Lixian County, Yan'ershan, Dashengmiao, Shijiaping, Hujiabao, Xiejiabao, Wangjiashan, and Linlizhuma. The Lishui Culture Group shows clear evidence of two distinct phases: The early phase is characterised by a typical conglomerate lithic industry; while the late phase shows the appearance of a small flint lithic industry,[17] which can be further divided into two phases: the earlier phase represented by Wuyashan, showing a trend towards miniaturisation in the lithic industry, and the later phase represented by Yan'erdong, Shiligang, and Zhuma sites (Table 4.6).

As one of the most well-preserved sites in Hubei Province, the Jigongshan site exhibits two distinct cultural phases with different lithic industry characteristics. The lower cultural layer represents a typical gravel lithic industry, while the upper cultural layer shows a trend towards miniaturisation of lithic tools (Liu and Wang 2001). The Zhangnaodong site in the northwestern part of Hubei Province has been radiocarbon dated to approximately 13,490 ± 150 years ago (Huang et al. 1987), marking the final phase of the Paleolithic. The investigation and excavation of Paleolithic sites in the Hanjiang River basin, in conjunction with the South-to-North Water Diversion Project, revealed evident stylistic differences in lithic industries between an earlier period characterised by gravel lithic industry and a later period featuring miniaturisation of lithic tools, predominantly flakes, found at locations such as the rear slope Shigushan

Table 4.6 Cultural sequence of the lower and middle Yangtze River regions in the paleolithic (Wu Maolin 1980)

Time Period (kya BP)	*Phase*	*Typical sites*	
		Middle Yangtze River Region	*Lower Yangtze River Region*
Approx. 35–10	LUP	Zhuma, Yan'erdong, Shiligang, Zhangnaodong	Sanshandao, Xianrendong
	EUP	Wuyashan, Upper Stratum of Jigongshan, Zhangjiaying	Daguling
Before 35	The Early to Middle Paleolithic	Shilongtou, Jiudaohe, Lower Stratum of Jigongshan, etc	Maozhushan, Shuiyangjiang, Guanshan, etc

(Hubei Provincial Museum 1987), Danjiangkou City Museum and Yuzui (Chen et al.). The clear distinction between the two periods aligns with the findings in the Hunan Province. However, due to the limited available materials, precisely determining the exact dates to distinguish the late and early periods of the lithic industry remains challenging. Thus, for the time being, it is categorised as the Upper Paleolithic lithic industry, representing the Upper Paleolithic.

The cultural remains from the Upper Paleolithic in the lower Yangtze River region is relatively scarce. The area of the Daguling site in the Yuanhe River basin, Jiangxi Province, is argued to have been occupied during the Upper Paleolithic. Yingsan Fang and other scholars applied optically stimulated luminescence (OSL) to date to two sediment samples from middle and upper layers of the Daguling site, yielding dates of approximately 36,100 ± 2,600 years ago and 64,200 ± 4,900 years ago, respectively (Fang et al. 2003). Therefore, the lithic assemblage at this site has already undergone miniaturisation, with stone flakes being predominant. Regarding the more recent period, the archaeological material is limited to the Shenxiandong site in Lishui, Jiangsu Province, with an absolute age of approximately 11,200 ± 1,000 years (Ge 1990). The presence of excavated pottery shards suggests that, like the Wanxian Xianrendong site and the Yuchanyan site in Dao County, the Daguling site belongs to the earliest phase of the Neolithic (or the transitional period between the Paleolithic and Neolithic). In addition, the lower layers of Wanxianxianrendong site and Diaotonghuan site, which exceed 15,000 years, are more strongly associated with the Upper Paleolithic (MacNeish 1995), characterised by small-size flake industries. The upper layers of the Qiliting site, Zhejiang Province, are likely belong to the Upper Paleolithic as they display miniaturisation of the lithic industry (Xu et al. 2006). Similar sites include the Hualongdong site in Dongzhi, Anhui Province (Anhui Provincial Institute of Cultural Relics and Archaeology, Jilin University Border Archaeology Research Center 2012).

4.4.2 Characteristics of the Upper Paleolithic Culture

Due to the lack of materials, the current insights of the characteristics of the Upper Paleolithic culture in the middle and lower Yangtze River region primarily relies on the limited lithic tools. One prominent feature of these materials is the miniaturisation of stone tools. The gravel industry tradition during the early to middle Paleolithic witnessed the emergence of flake industries, marking a significant reduction in the size of stone tools. Notably, the western part of the middle and lower Yangtze River region, particularly the areas bordering the southwestern region, such as the Three Gorges area and the Wushui River basin, exhibits distinct regional characteristics and is covered by high mountains and dense forests with complex terrain, displaying lithic cultural traits more akin to those of the southwestern region and thus, does not demonstrate the same miniaturisation phenomenon observed in the central and lower Yangtze River region. The Xiniudong site in the Shennongjia Forestry District yielded fossils dated to approximately 100,000 years ago through uranium-series dating. The stone tools predominantly consist of large and medium-sized implements, with 90% made of flint raw material (Wu 1998). Researchers consider the lithic characteristics of the Xiniudong site to be more similar to the Guanyindong site in Guizhou Province, while distinct from the typical South China Paleolithic industries (Yuan 2004). The stone tools from the Wushui culture group primarily consist of choppers and cutting tools, a small number of scraping tools, and few points. The dominant manufacturing technique involves percussion, particularly hammering, supplemented by the ridged hammer bipolar flaking technology (RHBF) commonly seen in the southwestern region.

The miniaturisation of stone tools also affected the coexisting gravel stone tools. In various sites across Hunan, the lithic assemblages predominantly consist of small-size stone tools, but also include larger stone tools. However, the proportion of large stone tools at the Wuyashan site is less than 30%, and the proportion only occupies 10% at the Shiligang site, indicating that gravel tools have undergone miniaturisation. Therefore, it could be observed that the Upper Paleolithic lithic culture in the middle and lower Yangtze River region has distinct developmental characteristics, as neither of these regions exhibited a trend towards lithic miniaturisation, differing from both the southwestern and South China regions (which will be illustrated in the Chapter 8). Instead, the middle and lower Yangtze River region shares similarities with the North China region, as it witnessed the emergence of more refined and smaller microblade technology during the Upper Paleolithic.

The miniaturisation of stone tools is accompanied by the changes in selection of raw materials of stone tools during the Upper Paleolithic in the middle and lower Yangtze River region. Siliceous rocks, such as flint, became more commonly used at sites like the Wuyashan, Yan'erdong, Shiligang, Zhangnaodong, and Hualongdong sites. Meanwhile, sites in the Daguoling area and the middle reaches of the Hanjiang River utilised high-quality quartz. In the upper layer of Jigongshan, both flint and quartz were employed, with flint accounting for 20% of the raw materials.[18] Notably, some of these high-quality raw materials were not naturally available locally, indicating the need for long-distance exchange or transportation. The refinement of raw materials directly resulted in the sharper cutting edges.

Regarding lithic assemblage, the early phase of the Paleolithic in the middle and lower Yangtze River region belongs to the Southern gravel tool industry tradition. However, there are regional variations. For instance, in the Hanjiang River basin of Hubei Province and the Lishui River basin of Hunan Province, the differences are generally minor. In these two regions, representative artifacts include various choppers, large triangular points[19] (referred to as handaxes in the Hanshui River basin), proto-handaxes (or adze-like handaxes), cleaver (primarily found in the Hanshui River basin), spheroid[20] (mainly found in the Lishui River basin), and large flakes, all produced by direct percussion using large gravels as cores. The assemblages also contain small tools such as scrapers and points, but their proportions are relatively small. The representative lithic assemblage of the Upper Paleolithic shows a reversal in the proportion of these two components, with scrapers occupying a dominant position, followed by points and choppers. Even among choppers, the proportion of those from the late period are noticeably smaller than those from the initial period. Manufacturing techniques of stone tools also became more refined, with the discovery of shallow and small processing scars on some tiny blade portions at the Shiligang site, indicating the usage of the pressure techniques (Feng 1999; Feng et al. 2012).[21]

The differences in stone tool size, raw materials, and type assemblages are associated with variations in their functional uses. Tools made from large gravels, such as choppers, large triangular points, and cleaver, are evidently more suitable for activities, which include felling, pounding, and digging, but less effective for cutting and scraping tasks (Chen and Chen 2012). The Upper Paleolithic tool assemblages, which are crafted from fine materials like flint, are predominantly scrapers, being suited for cutting and scraping. Considering stone tool weight, larger tools, which possess the nature of cumbersome, are less portable, thus limiting their usage within a confined area. On the contrary, small stone tools made from fine raw materials are highly portable, allowing people to utilise them over a broader range. Moreover, small groups are more likely to combine with other materials, forming composite tools and thus enhancing efficiency in various tasks, such as hafting or inlaying.

Apart from the differences reflected in the lithic assemblage, there have been significant changes in the site structure and location during the Upper Paleolithic. The richness of lithic assemblage generally represents the types of human activities, as the longer people reside in a place, the more diverse activities they engage in. Thus, the richness of lithic assemblage indicates the duration of human habitation in a particular area. Compared to the Early Paleolithic lithic assemblage, the Upper Paleolithic assemblages not only include the content of gravel tools but also feature various forms of scrapers. The abundance of artifacts in the Upper Paleolithic sites is much greater than those in early sites, reflecting extended periods of human habitation. Taking the example of Hunan Province, typical the Upper Paleolithic sites, which include Wuyashan and Shiligang sites are located on the left bank of rivers, the sunny side; while early typical sites, such as Huzhaoshan and Jigongdan are on the right bank, the sunny side (Tan 1999, Yuan 1990). This difference may reflect adaptive strategies under different climatic conditions; and these sites could surpass the constraints of terraces and disperse to different terraces in the Upper Paleolithic, unlike early sites, which were predominantly close to water. This expansion of human habitat reflects a broader utilisation of resources. The differences between the Early and the Upper Paleolithic periods can be summarised as shown in the table (Table 4.7).

4.4.3 Changes in the Adaptation Strategies of the Upper Paleolithic Hunter-Gatherers

In the Upper Paleolithic of the Yangtze River region, significant changes occurred in human adaptation strategies, clearly evidenced by the size, raw materials, types, and functions of stone tool materials. During the Early and Middle Paleolithic, humans were likely to rely on plant-based foods, emphasising gathering activities; while the Upper Paleolithic saw an increased emphasis on hunting, evidenced by the development of cutting and scraping tools. The LGM led to the northward migration of flora and fauna, resulting in open grasslands replacing dense forests in certain areas (Chapter 5). This unfamiliar environment attracted herbivorous animals, making hunting a viable means of subsistence. However,

Table 4.7 A comparison of cultural characteristics between the early and the upper paleolithic periods in the middle and lower Yangtze River Region

	Lower Paleolithic	*Upper Paleolithic*
Stone Tool Size	Relatively large gravel tools	Evident trend towards miniaturisation
Raw Material Selection	River pebbles with less strict selection standards	Flint, high-quality quartz, and other refined materials
Tool Assemblage	Dominance of coarse stone tools, such as choppers, handaxes, cleavers, spheroid, etc	Dominance of stone tools, such as scrappers, points, etc
Functional Use	More suitable for activities such as cutting, chopping, and digging for processing plants	More suitable for activities such as cutting and scraping for processing animal products
Structures of Archaeological Remains	Structure relatively simple, with a prevalence of gravel tools	Richer and more diverse archaeological remains
Site Location	Caves and near water terraces, with relatively fixed locations	Diverse habitat types

the Middle and Lower Yangtze River Region, which is located on the edge of the northern animal habitat, akin to an ecological transition zone, features with resource diversity and instability, thus, resulting in higher risks in a hunting-focused subsistence strategy. With climate changes and the northern fauna retreats, the region may face the depletion of hunting resources accordingly. Moreover, compared to a gathering strategy, a hunting strategy can support a smaller population density. This is since animals, as secondary consumers, can only supply a fraction of the energy obtained to humans. As hunters, humans occupy a similar position to apex predators, such as tigers and lions, requiring extensive living areas to sustain their survival. Consequently, it can be inferred that the same population density may not pose a population pressure for groups reliant on plant gathering. However, the population of groups, which are dependent on hunting for survival, has already exceeded its limit (Chapter Three), as a consequence, a transition in livelihood became necessary. This constitutes one of the significant reasons why early agriculture first emerged in the Middle and Lower Yangtze River region.

Based on the distribution of the archaeological sites, the influence of the LGM can be observed. People chose to reside on the sunny side of the rivers (northern or left bank) for warmth, showing distinct differences from the preliminary period sites. Additionally, with the increase in grasslands, the hunting economy grew, leading to an expansion of human activities and greater long-distance mobility, which, in turn, gave rise to the emergence of small stone tool industries. With the transgression of river terraces, people in the Upper Paleolithic expanded their habitats and utilised a wider range of resources, as reflected in the diverse types and varied blade forms of stone tools. To deal with the high-risk environment in an ecological transition zone, a more flexible adaptation strategy was needed to adopt. Ancient people retained the previous subsistence factor, which emphasised plant gathering in the Yangtze River region and remained the tradition of the gravel industry, also added components of a hunting economy. The proportion of hunting in the Yangtze River region might be slightly lower compared to North China, where it featured standardised and specialised tool assemblages of the Upper Paleolithic. This mixed strategy represents an adaptive response to encounter a dynamic environment; it is not a specialised adaptation nor an evolutionarily stable strategy (Sun 2001).[22] This mixed strategy is fragile, as hunting resources in the Yangtze River region are unstable.

In the middle and lower reaches of the Yangtze River in the early and middle phases of the Paleolithic, archaeological sites and locations were widely and sporadically distributed. A systematic survey conducted by Jiarong Yuan in the Lishui River basin revealed a relatively high density of Paleolithic sites with over 70 sites having been discovered, with stone tool materials found in almost all the net-patterned red soil terraces along the river. However, large areas with concentrated stone tools are rare (Yuan 1996). Due to the bulky and heavy nature of gravel tools and the availability of convenient raw materials, many stone tools were made and used on-site and discarded afterwards. As a consequence, the types of lithic assemblage found at these sites are relatively simple, which might be due to repeated visits. Additionally, the simple tool assemblages might reflect limited human activities and short periods of residence at the site. In contrast, the Upper Paleolithic sites show a more diverse range of artifacts, indicating longer periods of occupation. According to the research conducted by Binford on the mobility of hunter-gatherers (Binford 1980), the subsistence strategy, which was utilised by the Paleolithic hunter-gatherers in the middle and lower reaches of the Yangtze

River in the early and middle phases, was closely aligned with that of foragers who did not have distinct central base camps. Additionally, there was little functional differentiation between sites, leading to the migration according to the distribution and movement of food resources. In contrast, the contemporary hunter-gatherers during the Upper Paleolithic were more likely to adopt the strategy of collectors. In this case, central base camps emerged, and functional distinctions might exist between different sites. For instance, at the sites of Wuyashan and Shiligang, the remains are abundant; while some other sites were only temporarily utilised, as people only brought back food, which were needed to restore, to the central base camp. During the LGM, the average temperature in the middle and lower reaches of the Yangtze River decreased significantly, leading to a shortened growing season. Thus, ancient people had to rely on more food reserves to survive the food scarcity in winter and spring. As mentioned in the Chapters 1 and 2, the strategy of collectors composed a necessary adaptation for Paleolithic hunter-gatherers to evolve towards the origin of agriculture. Only through the strategy of collectors could hunter-gatherers develop the durable tools, architectural facilities, storage techniques, and even social organisation required for agricultural production. The shift in human adaptive strategies from foragers to collectors during the Upper Paleolithic in the middle and lower reaches of the Yangtze River was a crucial prerequisite for the emergence of agriculture, as it represented the accumulation of resource endowments.

4.5 Conclusion

Tracing back to the Upper Paleolithic, the agricultural origins mark a significant research topic. However, due to limitations in research systems and theoretical studies, this topic has not been extensively explored. In this chapter, a detailed analysis of the Upper Paleolithic materials in both North China and the middle and lower reaches of the Yangtze River has been conducted, thus exploring the human adaptive changes that occurred in these two regions before the emergence of agriculture and their connections. North China boasts the richest archaeological materials in China. The current material analysis indicates that the Upper Paleolithic in North China can be divided into at least two phases. In the early phase of the Upper Paleolithic, human adaptive strategies can be divided into four patterns: Shuidonggou, Shiyu, Shandongdong—Dongfang Square—Xiaonanhai, and Xiaogushan. In the late phase, specifically after the LGM, the microblade technology became prevalent. Aimed at increasing mobility, this technology represented the initial step of the utilisation of intensified resources by prehistoric hunter-gatherers. Around 15,000 years ago, the microblade technology reached its peak, coinciding with a pivotal point where cultural divergence occurred in North China during the Upper Paleolithic. Certain adaptive strategies of hunter-gatherers underwent a dramatic transformation, shifting from highly mobile livelihoods to the mobility reduction, marking the emergence of a new adaptive mode. Compared to the abundant materials in North China, the archaeological materials in the middle and lower reaches of the Yangtze River are relatively limited. However, these limited materials evident significant changes in human adaptive strategies during the Upper Paleolithic from the Early Paleolithic, that is, the shift from a plant-dependent foraging economy to an emphasis on animal hunting. In this case, the transformation from typical foraging strategies to collector strategies could be seen, moving from stable and singular adaptive strategies to flexible and diversified ones. These changes in adaptive strategies in the middle and lower reaches of the Yangtze River have contributed to the accumulation of resource endowments that paved the way for the origin of agriculture.

Notes

1. MSA, the Stone Age of Sub-Saharan Africa, is divided into three phases. The middle phase ranges from approximately 150,000 years ago to 30,000 years ago, which corresponds to the Middle Paleolithic in Eurasia. The late phase extends from 30,000 years ago to the historical period. In this region, the advent of agriculture occurred relatively late, and a typical Neolithic period is lacking.
2. It is necessary to clarify the concept of hunting by distinguishing between opportunistic hunting and effective hunting. The former, also known as chance hunting or random encounters, relies on luck and has low reliability. The latter refers to controlled hunting, where hunters can actively decide whether to kill or abandon the prey. Based on this criterion, it can be argued that humans were able to hunt effectively only from the late phase of the Upper Paleolithic.
3. Regarding the dating of the Xujiayao site, in 2004, Japanese scholar Tsuneto Nagatomo and others announced at the International

Symposium on the Centenary of Mr. Pei Wenzhong's Birth and the 75th Anniversary of the Discovery of the First Skull of Peking Man in Beijing that the latest dating of Xujiayao is approximately 660,000 years with an error margin of ± 150,000 years.
4. There are numerous terms used to refer to the microblade technology, such as microblade, microliths, microlithic assemblage, microblade assemblage, microlithic technology, microblade technology, and bladelets. In international academic literature, these terms often have specific geographic connotations. Microliths and microlithic are commonly used in West Asia and South Asia, while bladelets are frequently employed in North Africa. Microblade, on the other hand, is mainly used in East Asia and Northeast Asia. To make a clear distinction, the concept of "microblade" seems more appropriate. The terms "microblade technology" and "microblade assemblage" are both valid. The concept of microblade technology is purely technical and not restricted to specific time periods or regions. Any archaeological remains that utilize microblade technology fall within the scope of study. In contrast, when we refer to microblade assemblage, it must be accompanied by temporal and spatial limitations. For example, the Late Paleolithic microblade assemblagein Shanxi, microblade assemblage in Siberia, or microblade assemblage in the northern steppe region during the Neolithic period. In other words, the concept of microblade assemblage is based on the combined characteristics of archaeological remains, and it is not contradictory to the concept of microblade assemblage.
5. Information obtained through communication with Professor Chen Quanjia.
6. Personal communication with Mr. Wang Youping.
7. The difference between wedge-shaped microcores and boat-shaped micro cores lies in the ratio of the platform width to the height of the core. The ratio for wedge-shaped microcores is less than 1:2, while for boat-shaped microcores, it is greater than this value. A typical boat-shaped microcores should have a ratio greater than 1.
8. The solid line encompasses the main distribution area of microblade technique during the Late Paleolithic, while the dashed line indicates that the distribution of microblade technique in North China during the Late Paleolithic can be roughly divided into two

regions. The arrow shows the process of microblade technique disappearing from south to north.

9. The different curves in the figure represent the impact of different mobility and resource energy levels. For example, a high-mobility, hunting-based livelihood can achieve the highest energy return, while a low-mobility, gathering-based livelihood has a lower energy return.
10. Personal communication with Dachangzhengshan.
11. Method 1: The technology of producing microblades from wedge-shaped microblade cores;

    Method 2: Derived from prismatic core technology.

    (Figure cited from T. Kobayashi, Microblade industries in the Japanese Archipelago. *Arctic Anthropology* 7 (2): 38–58, 1970).
12. 1. Yuanyang Pond in Yongchang, Gansu Province; 2&3. Xinglongwa, Aohan Banner, Inner Mongolia; 4. Australia; 5. Siberia.
13. The areas surrounded by dots indicate the areas most likely to rely on hunting as a primary means of livelihood.
14. Each polygonal block represents a resource patch. The left figure represents a period of abundant resources, with shorter distances between resource patches; the right figure represents a period of resource scarcity, with the distance between patches expanding, and the distance hunter-gatherers need to travel to obtain the same amount of resources also increasing accordingly.
15. The area shown by the arrow is the forest-steppe region where the microblade process originated. The two point lines represent the southern boundary of the microblade process distribution and the southern boundary of the mammoth-woolly rhino fauna.
16. In order to speculate on the age, there is currently no good data for absolute dating.
17. It is currently not known whether it belongs to the Late Paleolithic. The division of early and late periods here is based on the lithic style and geological age of the Li River Basin.
18. Only 20 square meters of the upper layer of Jigongshan have been preserved, and most of it may have been destroyed by soil excavation, so the materials obtained may have significant deviations.
19. The naming of stone tools often has a certain subjectivity, and different researchers may give different names to the same type of artifact.

20. The use of stone balls may vary in different regions. The largest number of stone balls石球 found in the Xujiayao site in North China may have been used for hunting, such as throwing with a sling. In the middle and lower reaches of the Yangtze River, it may be different, and the use of composite tools is usually considered a characteristic of the late Paleolithic.
21. The Xujiacheng Site in Gansu Province, which exhibits characteristics of the Late Paleolithic, can be dated as early as 43,000 years ago (Layer 4B). However, the data from the deeper Layer 4C indicates a later date of approximately 34,000 years ago.
22. The core idea of evolutionary stable strategy is that any strategy adopted by the majority of individuals in a population that is superior to other strategies is an evolutionarily stable strategy. Once it is formed, any strategy adopted by an individual deviating from the evolutionarily stable strategy cannot compare with it. From an evolutionary perspective, after a major environmental change occurs, there may be a short period of evolutionary instability or even fluctuations within the population. However, once an evolutionary stable strategy is established, it will stabilize, and any behavior deviating from the evolutionary stable strategy will be punished and eliminated by natural selection.

References

A. P. Derev'anko, *The Palaeolithic of Siberia: New Discoveries and Interpretations.* University of Illinois Press, Urbana, 1998.

A. V. Tabarev, Paleolithic wedge-shaped microcores and experiments with pocket devices. *Lithic Technology*, 22 (2): 139–149, 1997.

An Zhimin, and Wu Ruzuo, *The Stone Age Remains in the Shayuan Area in Dali, Shaanxi. Acta Archaeologica Sinica*, 1957, 1.

An Zhimin, Excavation of Cave Deposits from the Paleolithic Age at Xiaonanhai Site, Anyang, Henan. Acta Archaeologica Sinica, 1965, 1.

An Zhimin, *The Mesolithic Remains in Hailar - On the Origin and Tradition of Microliths. Acta Archaeologica Sinica*, 1978, 3.

An Zhimin, *Trial Excavation of the Xiaonanhai Cave Deposit in Anyang, Henan Acta Archaeologica Sinica*, 1965, 1.

Anhui Provincial Institute of Cultural Relics and Archaeology, Jilin University Border Archaeology Research Center: *Brief Report on the Excavation of the*

Hualongdong Paleolithic Site in Dongzhi County, Anhui. Archaeology, 2012, 4.

B. Hayden et al., Evaluating lithic strategies and design criteria. In G. H. Odell ed., *Stone Tools: Theoretical Insights into Human Prehistory*, pp. 9–45. Plenum, New York, 1996.

B. P. Winterhalder et al., Risk-sensitive adaptive tactics: models and evidence from subsistence studies in biology and anthropology. *Journal of Archaeological Research*, 7: 301–348, 1999.

C. S. Henshiwood et al., An early bone tool industry from the Middle Stone Age at Blombos Cave, South Africa: implications for the origins of modern human behaviour, symbolism, and language. *Journal of Human Evolution*, 41: 631–678, 2002.

Chen Hui, and Chen Shengqian, *Experimental Research on Chopping Tools at the Yuzui No. 2 Paleolithic Site in Yun County, Hubei. Acta Anthropologica Sinica*, 2012, 31 (01).

Chen Shengqian et al., *Excavation Report of the Yuzui No. 2 Paleolithic Site in Yun County, Hubei. Acta Anthropologica Sinica*, forthcoming.

Chen Tiemei et al., *Uranium-series dating of mammalian fossils from the Xujiayao Site. Acta Anthropologica Sinica*, 1982, 2 (2).

Chen Tiemei et al., Reliability study of uranium-series dating for fossil age determination and uranium-series chronology of major paleolithic sites in North China. *Acta Anthropologica Sinica*, 1984, 3 (03).

Chen Tiemei et al., Accelerator mass spectrometry redetermination and discussion on the age of the Zhoukoudian Shandingdong Site. *Acta Anthropologica Sinica*, 1989, 8 (3).

Chen Tiemei et al., The second batch of accelerator mass spectrometry 14C age data and discussion on the Zhoukoujian Shandingdong Site. *Acta Anthropologica Sinica*, 1992, 11 (02).

Chen Zheying, *Paleolithic Artifacts of Tashui River in Lingchuan*. Wenwu Jikan, 1989, 2.

Chinese Academy of Social Sciences Institute of Archaeology Inner Mongolia Work Team: Brief Report on the Excavation of the Xinglongwa sedentism in Aohan Banner, Inner Mongolia. *Archaeology*, 1985, 10.

Chinese Academy of Social Sciences Institute of Archaeology Inner Mongolia Work Team: Brief Report on the Excavation of the Xinglongwa sedentism in Aohan Banner, Inner Mongolia in 1992. *Archaeology*, 1997, 1.

Chun Chen and Xiangqian Wang, The upper paleolithic microblade industries in North China and their relationships with Northeast Asia and North America. *Arctic Anthropology*, 26(2): 127–156, 1989.

Chun Tang and Pei Gai, The upper paleolithic cultural traditions in North China. In F. Wendorf and A. D. Close eds., *Advances in World Archaeology*, Vol. 5, pp. 339–364. Academic Press, New York, 1986.

Chun Tang and Pei Gai, Upper paleolithic cultural traditions in North China. In F. Wendorf and A. D. Close eds., *Advances in World Archaeology*, Vol. 5, pp. 339–364. Academic Press, New York, 1986.

D. A. Brooks et al., Dating and context of three Middle Stone Age sites with bone points in the Upper Semliki Valley, Zaire. *Science*, 268: 548–553, 1995.

D. B. Madsen et al., Dating Shuidonggou and the Upper Paleolithic blade industry in North China. *Antiquity*, 75: 706–716, 2001.

D. E. Young et al., Low-range theory and lithic technology: Exploring the cognitive approach. In R. Bonnichsen and D. G. Steele eds., *Method and Theory for Investigating the Peopling of the Americas*, pp. 209–237. Center for the Study of the First Americans, Oregon State University, Corvallis, 1994.

D. N. Grimstead, Ethnographic and modeled costs of long-distance, big-game hunting. *American Antiquity*, 75: 61–80, 2010.

Dong Guangrong et al., *New insights into the age of the Late Pleistocene Salawusu Formation. Chinese Science Bulletin*, 1998, 43 (17).

Du Shuisheng, Classification of wedge-shaped microcore types and the origin of microliths. *Acta Anthropologica Sinica*, 2004, 23 (Supplement).

F. D'errico et al., Archaeological evidence for the emergence of language, symbolism, and music—an alternative multidisciplinary perspective. *Journal of World Prehistory*, 17: 1–70, 2003.

F. D'errico, The invisible frontier. A multiple species model for the origin of behavioral modernity. *Evolutionary Anthropology*, 12: 188–202, 2003.

Fang Yingsan et al., Burial environment and age of paleolithic sites in Xinyu, Jiangxi. *Acta Anthropologica Sinica*, 2003, 22 (02).

Feng Jianping, Late paleolithic sites in Shiligang, Li County, Hunan. In *Proceedings of the International Symposium on Mesolithic Culture*, Guangdong People's Publishing House, 1999.

G. A. Clark, Through a glass darkly: conceptual issues in modern human origins research. In G. A. Clark and C. M. Willermet eds., *Conceptual Issues in Modern Human Origins Research*, pp. 60–76. Aldine de Gruyter, New York, 1997.

G. Clark, *World Prehistory in New Perspective, an Illustrated Third Edition.* Cambridge University Press, Cambridge, 1977.

G. G. Politis, Moving to produce: Nukak mobility and settlement patterns in Amazonia. *World Archaeology*, 27: 492–511, 1996.

G. H. Odell, Investigating correlates of sedentism and Domestication in prehistoric North America. *American Antiquity*, 63: 553–571, 1998.

G. M. Feinman and T. D. Price eds., Fully modern humans. In *Archaeology at the Millennium*, pp. 109–136. Kluwer Academic/Plenum, New York, 2001.

Gai Pei, and Wei Qi, Discovery of the late paleolithic site at Hutouliang. *Journal of Vertebrate Paleontology and Paleoanthropology*, 1977, 15 (3–4).

Gansu Provincial Museum Cultural Relics Work Team, Wuwei Regional Cultural Relics Survey Team: Excavation of the Neolithic Cemetery at Yuanyangchi, Yongchang. *Archaeology*, 1974, 5.

Gansu Provincial Museum: The Liujiacha Paleolithic Site in Huan County, Gansu Province. *Acta Archaeologica Sinica*, 1982, 1.

Gao Xing et al., New dating and discussion of related issues at Shuidonggou. *Acta Anthropologica Sinica*, 2002, 21 (03).

Gao Xing, Exploration of the Middle Paleolithic in China. *Acta Anthropologica Sinica*, 1999, 18 (1).

Ge Zhigong, and Lin Yipu, *Mesolithic Microliths from Daxianzhuang* Southeast *Culture*, 1985, 1.

Ge Zhigong, and Lin Yipu, Microliths from the Middle Stone Age at Daxianzhuang. *Southeast Culture*, 1985, 1.

Ge Zhigong, Discovery and Significance of Ten Thousand-Year-Old Pottery Fragments in Shenxiandong, Lishui. *Southeast Culture*, 1990, 5.

Gezidong Excavation Team: Excavation Report of the Gezidong Paleolithic Site in Liaoning. *Journal of Vertebrate Paleontology and Paleoanthropology*, 1975, 13 (02).

H. Leith, Modeling the primary productivity of the world. *Nature and Resources*, 8 (2): 5–10, 1972.

H. Thieme, Lower Paleolithic hunting weapons from Germany. *Nature*, 385: 769–771, 1997.

Hebei Provincial Institute of Cultural Relics and Archaeology: Late paleolithic microlithic site at Jijitan. Wenwuchunqiu. *Wnwuchunqiu*, 1993 (2), 1993.

Hebei Provincial Institute of Cultural Relics, et al.: Excavation brief of the Mengjiaquan Paleolithic Site in Yutian County, Hebei. *Wenwu Chunqiu*, 1991, 1.

Hebei Provincial Institute of Cultural Relics, et al.: Microlithic Site at Tingsijian, Changli, Hebei. *Wenwuchunqiu*, 1992 Supplement.

Hebei Provincial Institute of Cultural Relics: The Upper Paleolithic Microlithic Site at Jijitan. *Wenwuchunqiu*, 1993, (02).

Hou Yamei, *Stone Industry of the Donggutuo Site in the Nihewan Basin*, Ph.D. dissertation, Institute of Vertebrate Paleontology and Paleoanthropology, Chinese Academy of Sciences, 2000.

Huang Wanbo et al., Excavation Report of the Zhangnaodong Paleolithic Site in Fang County, Hubei. *Acta Anthropologica Sinica*, 1987, 6 (01).

Huang Weiwen et al., Stratigraphy/chronology and ecological environment of Fossil Homo sapiens in Ordos. *Acta Anthropologica Sinica*, 2004, 23, Supplement.

Huang Weiwen, et al., Paleolithic of Ang'angxi, Heilongjiang. *Acta Anthropologica Sinica*, 1984, 3 (03).

Huang Weiwen, *Late Paleolithic Cultures in China*, found in *Ancient Humans in China*, Science Press, 1989.

Hubei Provincial Museum, Danjiangkou City Museum: Investigation Report of the Paleolithic Site on the Back Slope of Shigushan in Danjiangkou City. *Jianghan Archaeology*, 1987, 4.

J. B. Sollberger and L. W. Patterson, Prismatic blade replication. *American Antiquity*, 41: 517–531, 1976.

J. D. Speth and E. Tchernov, Neanderthal hunting and meat-processing in the Near East: evidence from Kebara Cave (Israel). In C. B. Stanfrod and H. T. Bunn eds., *Meat-Eating and Human Evolution*, pp. 52–72. Oxford University Press, Oxford, 2001.

J. F. Hoffecker and S. A. Elias, Environment and archeology in Beringia. *Evolutionary Anthropology*, 12: 34–49, 2003.

J. F. Hoffecker, *A Prehistory of the North: Human Settlement of the Higher Latitudes*. Rutgers University Press, New Brunswick, 2005.

J. F. Hoffecker, Innovation and technological knowledge in the Upper Paleolithic of northern Eurasia. *Evolutionary Anthropology*, 14: 186–198, 2005.

J. J. Flanniken, The Paleolithic Dyuktai pressure blade technique of Siberia. *Arctic Anthropology*, 24 (2): 117–132, 1987.

J. Mulvaney and J. Kamminga, *Prehistory of Australia*. Smithsonian Institution Press, Washington, 1999.

Ji Duxue et al., The impact of the deterioration of the last glacial maximum environment on paleolithic cultures in Northern China. *Acta Anthropologica Sinica*, 2005, 24 (3).

Jia Lanpo and Wei Qi, The Xujiayao Paleolithic Cultural Site in Yanggao County. *Acta Archaeologica Sinica*, 1976, 2.

Jia Lanpo, et al., New materials from the Shuidonggou Paleolithic Site. *Journal of Vertebrate Paleontology and Paleoanthropology*, 1964, 8 (01).

Jia Lanpo, et al., Excavation report of the Shiyu Paleolithic Site in Shanxi. *Acta Archaeologica Sinica*, 1972, 1.

Jia Lanpo, et al., The excavation report of the Xujiahe Paleolithic Cultural Site in 1976. *Journal of Vertebrate Paleontology and Paleoanthropology*, 1979, 17 (03).

L. G. Straus, The Upper Paleolithic of Europe: an overview. *Evolutionary Anthropology*, 4: 4–16, 1995.

L. H. Kelly, Hafting and retooling: effects on the archaeological record. *American Antiquity*, 47: 798–809, 1982.

L. R. Binford, *Constructing Frames of Reference: An Analytical Method for Archaeological Theory Building Using Hunter-gatherer and Environmental Data Sets*. University of California Press, Berkeley, 2001.

L. R. Binford, *Constructing Frames of Reference: An Analytical Method for Archaeological Theory Building Using Ethnographic and Environmental Data Sets*. University of California Press, Berkeley, 2001.
L. R. Binford, *In Pursuit of the Past*, translated by Chen Shengqian, Shanghai Sanlian Bookstore Co.Ltd., 2009.
L. R. Binford, Were there elephant hunters at Torralba? In M. H. Nitecki and D. V. Nitecki eds., *The Evolution of Human Hunting*, pp. 47–105. Plenum, New York, 1987.
L. R. Binford, Willow smoke and dog's tails: Hunter-gatherer sedentism systems and archaeological site formation. *American Antiquity*, 45: 4–20, 1980.
Li Chaorong, et al., Excavation Brief of the Wangfujing Dongfang Square Paleolithic Site in Beijing. *Archaeology*, 2000, 9.
Li Chaorong, et al., New progress in Paleolithic Archaeology in Beijing Area. *Acta Anthropologica Sinica*, 1998, 17 (02).
Li Feng et al., Excavation report of the Xujiacheng Paleolithic Site in Gansu Province. *Acta Anthropologica Sinica*, 2012, 31 (03).
Li Xingguo et al., The chronology of the Hetao people and the Salawusu culture. *Proceedings of the First National Radiocarbon Conference*, Science Press, 1984.
Li Yanxian, *The Classification of Late Paleolithic Cultures in China*, *Acta Anthropologica Sinica*,1993, 13 (03).
Lin Shenglong, A comparative study of technological patterns in old stone age cultures in China and the West. *Acta Anthropologica Sinica*, 1996, 15 (01).
Linyi Cultural Relics Management Committee, Tancheng County Library: Heilongtan Microlithic Site in Tancheng, Shandong. *Archaeology*, 1986, 6.
Liu Deyin and Wang Youping, Preliminary report on the excavation of the Jigongshan Site. *Acta Anthropologica Sinica*, 2001, 20 (02).
Liu Yuan, New discoveries of paleolithic artifacts in Xigou, Quwo, Shanxi. *Acta Anthropologica Sinica*, 1986, 5 (04).
M. C. Nelson, The study of technological organization. In M. B. Schiffer ed., *Archaeological Method and Theory*, Vol. 3, pp. 57–100. University of Arizona Press, Tucson, 1991.
M. C. Stine et al., Paleolithic population growth pulses evidenced by small animal exploitation. *Science*, 283: 190–194, 1999.
M. C. Stiner et al., Paleolithic population growth pulses evidenced by small animal exploitation. *Science*, 283: 190–194, 1999.
M. C. Stiner, Carnivory, coevolution, and the geographic spread of the Genus Homo. *Journal of Archaeological Research*, 10: 1–63, 2002.
M. G. Winkler and P. K. Wang, The late-quaternary vegetation and climate of China. In H. E. Wright eds., *Global Climates since the Last Glacial Maximum*, pp. 221–261. University of Minnesota Press, Minneapolis, 1993.
M. H. Nitecki, The idea of human hunting. In M. H. Nitecki and D. V. Nitecki eds., *The Evolution of Human Hunting*, pp. 1–9. Plenum, New York, 1987.

M. Shott, Technological organization and sedentism mobility: An ethnographic examination. *Journal of Anthropological Research*, 42: 15–51, 1986.

Ningxia Institute of Cultural Relics and Archaeology: *Shuidonggou - Excavation Report of 1980*, Science Press, 2003.

O. Bar-Yosef, The Upper Paleolithic Revolution. *Annual Review of Anthropology*, 31: 363–393, 2002.

O. Bar-Yosef, Eat what is there: hunting and gathering in the world of Neanderthals and their Neighbours. *International Journal of Osteoarchaeology*, 14: 333–342, 2004.

P. Bleed, The optimal design of hunting weapons: maintainability or reliability. *American Antiquity*, 51: 737–747, 1986.

P. Mellars, The impossible coincidence. A single-species model for the origins of modern human behavior in Europe. *Evolutionary Anthropology*, 14: 12–27, 2005.

P. Savolainen et al., Genetic evidence for an East Asian origin for domesticated dogs. *Science*, 298: 1610–1613, 2002.

R. G. Elston and P. J. Brantingham, Microblade technology in Northern Asia: a riskminimizing strategy of the Late Paleolithic and Early Holocene. In R. G. Elston and *Think Small: Global Perspectives on Microlithization*, pp. 103–116. Archaeological Papers of the American Anthropological Association Number 12. Arlington, American Anthropological Association, 2002.

R. G. Elston et al., New dates for the north China Mesolithic. *Antiquity*, 71: 985–993, 1997.

R. G. Klein, The archaeology of modern human origins. *Evolutionary Anthropology*, 1: 5–14, 1992.

R. L. Kelly, Hunter-gatherer mobility strategies. *Journal of Anthropological Research*, 39: 277–306, 1983.

R. L. Kelly, *The Foraging Spectrum: Diversity in Hunter-Gatherer Lifeways*. Smithsonian Institution Press, Washington, 1995.

R. L. Kelly, The three sides of a biface. *American Antiquity*, 53: 717–734, 1988.

R. S. MacNeish, Lithic typology. In *Origins of Rice Agriculture: The Preliminary Report of the Sino-American Jiangxi (PRC) Project SAJOR*. Publications in Anthropology No. 13, El Paso Centennial Museum, The University of Texas at El Paso, 1995.

S. C. Gerlach et al., Blood protein residue on lithic artifacts from two archaeological sites in De Long Mountains, Northwest Alaska. *Arctic*, 49(1): 1–10, 1996.

S. E. Churchill, Hand morphology, manipulation, and tool use in Neanderthals and early modern humans of the Near East. *PNAS*, 98: 2953–2955, 2001.

S. Kuhn, A formal approach to the design and assembly of mobile toolkits. *American Antiquity*, 59: 426–442, 1994.

S. L. Kuhn and R. G. Elston, Introduction: thinking small globally. In R. G. Elston and S. L. Kuhn eds., *Think Small: Global Perspectives on Microlithization*, pp. 1–8. Archaeological Papers of the American Anthropological Association Number 12. Arlington, American Anthropological Association, 2002.

S. McBrearty and A. S. Brooks, The revolution that wasn't: A new interpretation of the origin of modern human behavior. *Journal of Human Evolution*, 39: 453–563, 2000.

Shenyang Municipal Cultural Relics Management Office, Shenyang Imperial Palace Museum: Second Excavation Report of the Shenyang Xinle Site. *Acta Archaeologica Sinica*, 1985, 2.

Shizitan Archaeological Team: Excavation report of site no. 9 at Shizitan, Jixian County, Shanxi. *Archaeology*, 2010, 10.

Shizitan Archaeological Team: Site S14 at Shizitan, Jixian County, Shanxi, in the Paleolithic. *Archaeology*, 2002, 4.

Sun Ruyong, *Principles of Animal Ecology* (3rd edition), Beijing Normal University Publishing Group, 2001, 258.

T. G. Schurr and S. T. Sherry, Mitochondrial DNA and Y chromosome diversity and the peopling of the Americas: evolutionary and demographic evidence. *American Journal of Human Biology*, 16: 420–439, 2004.

T. G. Schurr, The peopling of the New World: perspectives from molecular anthropology. *Annual Review of Anthropology*, 33: 551–583, 2004.

T. Goebel, Pleistocene human colonization of Siberia and peopling of the Americas: an ecological approach. *Evolutionary Anthropology*, 8: 208–227, 1999.

T. Goebel, The "Microblade Adaptation" and recolonization of Siberia during the Late Upper Pleistocene. In R. G. Elston and S. L. Kuhn eds., *Think Small: Global Perspectives on Microlithization*, pp. 117–132. Archaeological Papers of the American Anthropological Association Number 12. Arlington, American Anthropological Association, 2002.

T. Kobayoshi, Microblade industries in the Japanese Archipelago. *Arctic Anthropology*, 7 (2): 38–58, 1970.

T. S. Uinuk-ool et al., Ancestry and kinships of native Siberian populations: the HLA evidence. *Evolutionary Anthropology*, 12: 231–245, 2003.

Tan Yuanhui, *Investigation Report of the North Slope Paleolithic Site in Huzhuashan. Hunan Cultural Relics Collection*, 1999.

W. A. Niewoehner, Behavioral inferences from the Skhul/Qafzeh early modern human hand remains. *PNAS*, 98: 2979–2984, 2001.

W. C. Pei. The upper cave fauna of Choukoutien. Pal. *Sin. New Series* C. (10): 1–100, 1940.

Wang Enlin, New materials from the microlithic site at Tingsijian, Changli, Hebei. *Acta Anthropologica Sinica*, 1997, 16 (01).

Wang Jian, et al., *Investigation and Excavation Report of the Dingcun Paleolithic Site Gropus Wenwu Jikan*, 1994, 3.

Wang Jian, et al., The Xiachuan Culture. *Acta Archaeologica Sinica*, 1978, 3.

Wang Linghong, *New Advances in the Chronology of Ancient Humans in China*. In *Ancient Humans in China*, Science Press, 1989.

Wang Shejiang, et al., Study on the stone artifacts excavated from Longyadong in Hualong, Luonan, 1995. *Acta Anthropologica Sinica*, 2004, 23 (02).

Wang Xiaokun et al., Excavation report of the Jinsitai Cave Site in Inner Mongolia. *Acta Anthropologica Sinica*, 2010, 29 (01).

WangJian, The Xiachuan Culture. *Acta Archaeologica Sinica*, 1978, 3.

Wei Lanying et al., Preliminary study on the climate and environmental changes during the last deglaciation in Beijing area. *Quaternary Sciences*, 1997, 2.

Wu Jia'an, On the relationship between the Xiachuan culture and Xiaonanhai culture. *Archaeology and Cultural Relics*, 1984, 1.

Wu Maolin, The human fossils excavated from Xujiayao Site in 1977. *Journal of Vertebrate Paleontology and Paleoanthropology*, 1980, 18 (03).

Wu Xianzhu, Excavation report of the Xiniudong Paleolithic Site in Shennongjia. *Acta Anthropologica Sinica*, 1998, 17 (02).

Xi'an Banpo Museum, Dali County Cultural Museum: *Archaeological Survey Report of the Shayuan Area in Dali, Shaanxi. Prehistoric Research*, Inaugural Issue, 1983.

Xiaokongshan Joint Excavation Team: *1987 Excavation Report of the Xiaokongshan Paleolithic Site in Nanzhao, Henan Huaxia Archaeology*, 1988, 4.

Xie Fei and Yu Shufeng, *Research on late paleolithic stone tools at Xibaimaying Site, Yangyuan, Hebei*, *Wenwuchunqiu*, 1989, 3.

Xie Fei, An outline of the transition between neolithic and paleolithic cultures in the Circum-Bohai Sea Region. In *Review and Prospect of Chinese Archaeology across the Century*, Science Press, 2000.

Xu Xinmin et al., *Significant achievements in the excavation of the qiliting paleolithic site in Changxing, Zhejiang*, Chinese Cultural Relics News, July 26, 2006.

Yin Gongming, and Huang Weiwen, Optically Stimulated Luminescence Dating of the Fanjiagouwan at the Salawusu Site. *Acta Anthropologica Sinica*, 2004, 23, Supplement.

Yuan Jiarong, *Jigongdang Paleolithic Site in Li County*, *Chinese Archaeology Yearbook 1989*, Cultural Relies Press, 1990.

Yuan Jiarong, Paleolithic archaeology in the Middle Yangtze River Region. In *Review of the Century of Chinese Archaeology: Paleolithic Archaeology Volume*, Science Press, 2004.

Yuan Jiarong, Regional types and status of the paleolithic culture in Hunan. In *Proceedings of the Prehistoric Culture and the Second Asian Civilization Symposium in the Middle Yangtze River Region*, Yue Lu Press, 1996.

Yuan Sixun et al., *Chronology and cultural study of the Shizitan Site in Jixian, Shanxi, Archaeology*, 1998, 6.

Yuan Sixun et al., Uranium-series dating of the Hetao people and Salawusu culture. *Acta Anthropologica Sinica*, 1983, 2.

Yuan Sixun, Accelerator mass spectrometry dating of Xinglong decorated antler and the Shiyu site samples. *Acta Anthropologica Sinica*, 1993, 12 (01).

Zhang Juzhong and Li Zhanyang, Excavation report on the Microlithic Site at Dagang, Wuyang, Henan. *Acta Anthropologica Sinica*, 1996, 15 (02).

Zhang Senshui, *Old stone age culture in China*, Tianjin Science Technology Publishing House, 1987.

Zhang Senshui, Regional progression and cultural exchange of old stone age industries in Northern China. *Acta Anthropologica Sinica*, 1990, 9 (04).

Zhang Songlin and Liu Yanfeng: Excavation report of the Zhijidong Paleolithic Site. *Acta Anthropologica Sinica*, 2003, 22 (01).

Zhang Wenjun, *Excavation report on the Shizitan Paleolithic Site in Jixian, Shanxi, Archaeology and Cultural Relics*, 1990, 1.

Zhang Zhenhong, et al., Excavation brief of the Xiaogushan Site in Haicheng, Liaoning. *Acta Anthropologica Sinica*, 1985, 4 (01).

Zhao Jingfang and Huo Baoqiang, *Important discoveries in the late paleolithic site at Shizitan, Jixian County, Shanxi, Chinese Cultural Relics News*, Nov. 30, 2007.

Zhou Guoxing, *Stone age remains from Lingjing Site, Xuchang, Henan, Archaeology*, 1974, 2.

CHAPTER 5

Initial Environmental Conditions for the Emergence of Chinese Agriculture

If there is not sufficient depth,
water will not float large ships.
—*Zhuangzi*

5.1 Introduction

Undoubtedly, environmental conditions and their variability are essential for the livelihoods of hunter-gatherers who rely on natural resources. Successful adaptation depends on whether humans are able to accurately judge the sources of natural food resources and know how to effectively find and get access to them. Environmental conditions and their variability determine the distribution characteristics of food resources in time and space. In this case, the hunter-gatherer cultural system had to adjust the adaptive mechanisms based on the changing resource conditions. During the transition from the hunter-gatherer cultural system to the agricultural one, environmental conditions change was an important factor that must be taken into account.

However, this does not mean that changes in environmental conditions directly led to the origins of agriculture. Environmental determinists view cultural systems as an organism, which could change automatically without explanation. Cultural systems could completely ignore the

S. Chen, *Prehistoric Modernization*,
https://doi.org/10.1007/978-981-97-9423-2_5

changes in internal mechanisms, and see the changes in external environmental conditions as the most fundamental agent. However, this idea seems insufficient. For example, even though some external elements, such as temperature, sunlight, and precipitation, play significant roles in plant growth, however, we cannot ignore the growth mechanism of plants. Archaeologists, who study the cultures of ancient humans and the mechanisms of cultural changes, are not geologists. For the hunter-gatherer cultural system, the natural environment and its changes are the initial conditions that precede changes within the system. Archaeologists are concerned with the impact of changes in external conditions on cultural systems, in particular the mechanisms of cultural response and the processes by which the environment influences culture.

As aforementioned in the previous section, we have already discussed the influence of environmental conditions on the cultural adaptive mechanism of hunter-gatherers in a theoretical perspective. Now, we need to turn to the origins of Chinese agriculture at the concrete level. How did the natural environment, the initial condition, change before and after the origins of Chinese agriculture? I will discuss this question on three levels: firstly, the best habitat for hunter-gatherers; secondly, the ecotones, which were crucial for hunter-gatherer adaptation; thirdly, the environmental changes in hunter-gatherer habitats before and after the origins of Chinese agriculture. The former two factors will discuss the regional characteristics of hunter-gatherer cultural adaptations from a theoretical level, helping us to understand the places where the earliest agriculture took place.

5.2 Best Habitats for the Hunter-Gatherers

In archaeological research, the most common mistake made is to take today's perceptions of the environment to be those of the ancients. Researchers who live in the Industrial Age regard prosperous places in present days as places of environmental superiority in ancient times, consuming that the hunter-gatherers during the Stone Age thought so as well. These researchers apply the perceptions in the Age of Agriculture to judge the ones in the hunter-gatherer society. In the Industrial Age, resource allocations were determined by the markets, thus, the places with convenient transportation tended to become concentrated. Marine transport was the least costly, so modern developed cities were mostly located on the coast. In the Agricultural Age, on the other hand, the most attractive places were the areas with favourable climates. The most fertile arable lands were located in places with flat terrain, fecund soil, and accessible irrigation conditions, such as Central China, where the centres of civilisation were formed.

For hunter-gatherers in the Stone Age, the least portable items were water, stone, and fuel (as was the case for ethnographic hunter-gatherers), which were important factors in choosing habitats. Besides, the other important factors for hunter-gatherers to choose shelter were to protect themselves from wind, rain, and cold. If the hunter-gatherers had made use of natural caves and rock shelters, they could have saved the efforts in residences. After that comes proximity to food resources. A combination of the conditions aforementioned would lead to the habitat choice, with water being the foremost factor considered. Except for desert and Gobi areas, which lack surface runoff, hunter-gatherer habitats must be associated with rivers.

Rivers originate in areas of high ground, often with high-mountain valleys or plateaus upstream where terrains are steep and rivers are well-powered with huge gravels. Full of steep terrains and rapid rivers, upstreams of mountain valleys and plateaus are not suitable for human activities and, therefore, not conducive to the mobile livelihood of hunter-gatherers. While, in the lower reaches, the rivers are slow-moving due to sediment accumulation and numerous waterways. Without ships or vessels, it would be difficult for hunter-gatherers to move around. At the same time, a still-water environment breeds mosquitoes. Due to the slow-moving water, hunter-gatherers lacked clean drinking water unless they drilled wells. In terms of a river basin, the middle reaches of the river were the best place for hunter-gatherers to survive.[1] With wide valleys and basins providing wide space for activities, small tributaries of the river with spring outcrops provided safer drinking water, and gravels of the right size provided convenient raw material for stone tools (Fig. 5.1).

Within China, the suitable zone mainly refers to the tributary basins of great rivers (such as the Yangtze River and the Yellow River), such as the Han River and Xiang River, etc. The great rivers are the confluence of these tributaries and are further downstream of them. Compared with the downstream areas of the tributaries, the middle reaches of great rivers are often densely covered with lakes, thus, it is less suitable for hunter-gatherers to survive. It is reasonable to assume that the tributary basins of China's great rivers are the best habitat for prehistoric hunter-gatherers.

Fig. 5.1 Optimal raw material zone

The natural situation is more attractive to human livelihood and more conducive to human survival, thus, certainly more densely populated.

As hunter-gatherers expanded from their best habitat to the surrounding areas, moving upstream on the river took them into marginal environments that lacked the wide land and irrigation needed to develop agriculture. Hunter-gatherers expanded downstream of the river with wide land and convenient irrigation conditions, which were conducive to agricultural development. Also, humans had to overcome the disadvantages of frequent flooding, heavy clay soils, and weeds. Hunter-gatherers needed greater labour input, better organisation, as well as more sophisticated and advanced techniques.

How did the hunter-gatherers' best habitats influence the origin of agriculture? According to this theory, we can assume that the tributary basins of the great rivers were the heartland for hunter-gatherer existence before the origin of agriculture. These basins, being the most favoured by hunter-gatherers, were the most likely to reach the population limit for hunter-gatherers. Merely based on best habitats theory, agriculture initially emerged in tributary basins, broad valleys, and foothills. Amid further population increase, humans expanded their habitats upstream and downstream. The lower reaches, with wide and flat terrain, fertile land, and convenient irrigation conditions, were conducive to agricultural development. Meanwhile, the disadvantages in the lower reaches, such as flooding, lack of clean water, and mosquito breeding, were gradually overcome amid cultural development. That is to say, we should see a process whereby the agricultural population expands downstream of the river; and correspondingly, the hunter-gatherers and agricultural groups expand upstream of the river, where the marginal habitat conditions pose a serious challenge to human survival: for hunter-gatherers, upstream of rivers is not easy for mobility; for agricultural development, upstream of rivers lacks arable land and irrigation water.

5.3 Ecotones

In Chapter 1, the concept of ecotone has been introduced, indicating a transition area between two biological communities. Compared with a single biological community, an Ecotone possesses more diverse resources, but it is more sensitive to environmental changes. As for typical ecotones, they are junctions of forest and grasslands, seas, and lands. The most macroscopic ecotone is the temperate zone, between the tropical and frigid zones, with North China and the lower reaches of the Yangtze River included. The average winter temperature is normally below zero centigrade along the middle and lower reaches of the Yangtze River because of the influence of Eurasian continentality and colder areas of the same latitudes. Low winter temperatures can effectively control some infectious diseases. At the same time, species in temperate regions are less diverse than those in the tropics. However, in temperate regions, large groups or communities of species tend to be formed, such as bison herds in North American grasslands and wild wheat grasslands in West Asia. Unlike the frigid zone environment, this resource distribution in temperate regions is beneficial for humans to use and control, as it is not too homogeneous and barren to take advantage of hunter-gatherers who rely on walking.

Besides the factor of latitude, altitude also plays a crucial role in the distribution of the human population (Beall 2001). On the one hand, altitude affects the temperature, that is, for every 100 metres of rise, the temperature decreases by 0.6 °C. Therefore, altitude can also effectively control certain infectious diseases; on the other hand, altitude controls the distribution of natural resources, that is, flora and fauna generally distributed vertically along altitudes. For example, at a certain latitude, the forest has to give way to grassland, creating an ecotone zone between grasslands and forests. Similar ecotones could be found in the boundary zones between the mountains and plains, such as the edges of basins and the terraces of river valleys.

The best habitats for hunter-gatherers lie on the edges of basins and along the terraces of river valleys where there are often junctions of forest and grasslands (steppes or marshy grassland). In Northeastern China, river valleys form meadows due to waterlogging; while the mountain slopes, especially the shaded ones, are wooded (the sunny side is also mostly grassy in the Great Khingan region due to high evaporation). Similarly, in the Loess Plateau region, forests can grow within the slopes because of the thin loess and shallow groundwater. On the contrary, grasslands are formed on the deep Loess Plateau as the deep groundwater makes it difficult for tree roots to reach (Zhang and An 1994; Zhu 1994; Liu et al. 1994). The edge of the basin and terrace of river valleys, serving ecotone between forest and grasslands, provide hunter-gatherers with diverse resources. The ecotones and the excellent raw materials, water, and terrains constitute multiple advantages for forming the best habitats.

While the above illustration is from a microscopic perspective. The most macroscopic ecotone is the forest steppe zone that runs through China from the northeast to the southwest, also known as the Heihe-Tengchong Line that divides the area of China into two parts with contrasting population densities, dense in the east and sparse in the west. This long line stretches from northeast to southwest China and spans different altitudes of terrain and different climate zones. However, based on materials from different historical archaeological phases, cultural similarities could be found in different zones (Tong 1986). This ecotone attracted hunter-gatherers, especially those with preferences for hunting. Prehistoric materials, including Paleolithic archaeological materials, prove this point. For example, lithic technology in the Mousterian Industry is found across a wide area from the Jinsitai site in eastern Inner Mongolia to the Dahe Site in Fuyuan, Yunnan Province (Wang et al. 2010; Ji et al. 2006; Zhang 2005). The eastern part of this zone, that is, north and south of the Yan Mountains, is the area where the ancient Great Wall was distributed and where agricultural and nomadic people fought over. It is also an area of repeated fluctuations as far as hunter-gatherers and early agriculturalists are concerned (see Chapter 7). The changes are remarkable.

There are two other typical ecotones in China. One is the transition zone of Qinling-Funiu Shan-Huai River, dividing the temperate deciduous forest of North China adjoins to the evergreen broad-leaved forest of South China, as well as the traditional Northern dry farming from the Southern rice farming, and the cultural–geographical division between the north and the south. The other ecotone is the Nanling Mountains transition zone, which divides subtropical regions and tropical ones. These two ecotones as well as the north and south of the Yan Mountains region separate eastern China into four main ecological zones: Mongolian grassland, temperate Central China, subtropical lower and middle Yangtze, and tropical South China.

It is specifically necessary to emphasise that the changes of these ecotones, bound with mountains, fluctuate back and forth, with an increase in temperature, especially at the end of the Last Glacial Maximum (LGM). These ecotones and four major ecological areas divided by them moved from south to north. How did this change affect prehistoric hunter-gatherers? The amount of research related to the environment is sufficient, but only a few of them mention the effects on humans, and there is a particular lack of research on the cultural ecology of hunter-gatherers.

Ecotones, as a preferred habitat, attracted hunter-gatherers for their diverse resources. Living in ecotones, hunter-gatherers can develop more resilient livelihood strategies to utilise diverse and changing resources. However, because ecotones are sensible to environmental changes, hunter-gatherers in these ecotones gain access to resource diversity at a corresponding opportunity cost. Within these ecotones, the adaptation risk is relatively high. The higher risks also motivate hunter-gatherers to utilise wider and more resilient adaptation strategies to cope with unpredictable changes, such as livelihood strategies with richer diet, collector strategies with mobility, and livelihood methods depending on various resource conditions, etc. Thus, the overall migration and repeated fluctuations of ecotones allow archaeologists to see more variations in cultural adaptation strategies. The origins of agriculture and cultural adaptation strategies are intertwined. Agriculture inevitably originates in adaptive strategies that are more sensitive to environmental change and poses the significance of exploring ecology within agricultural origin studies.

5.4 To New Niches

Since the LGM, because of multiple factors influenced (e.g. Milankovitch cycles and the rapid change in ventilation of the deep North Atlantic), certain phenomena occurred: the rise of temperature, the end of ice ages, the melt of glaciers, the rise of sea level, the faunal extinction, the increased seasonality, stable climate fluctuations, etc. Evidently, not all the variables directly influenced the habitats of hunter-gatherers, only several variables directly played crucial roles in the subsistence change of hunter-gatherers at the Pleistocene-Holocene boundary. Here I summarise the variables under the headings of the rise of sea level, faunal extinction, seasonality, and climate stability.

5.4.1 *The Rise of Sea Level*

An apparent change in the habitats of hunter-gatherers came from the rise of sea levels after the LGM. The rise of sea level inundated former land and narrowed the areas of hunter-gatherer habitats. With abundant materials from boreholes and internationally cooperative work, now we not only know the depth of the lowest sea level in the LGM but also know the gross rise rate of the sea level rise. With these data, we can calculate the land area that was lost with the rise of sea level, and remark upon the influence it had on the living space of hunter-gatherers at that time. Generally speaking, the effect of rising sea levels is associated with the geographical features of a region. In general, the wider the continental shelves, the more severe influence would be expected. For example, African and South American continents have narrow shelves so that the rise of sea level would not even slightly change the habitats of hunter-gatherers in these regions. It is noteworthy that Europe and North America had shelves as wide as East Asia, but these two continents were covered by colossal ice caps in the north during LGM, resulting in uninhabitable conditions for hunter-gatherers. New niches came into being when ice caps melted after LGM, compensating for the continental shelves inundated by the rise of sea level. The Asian situation was different, in that no colossal ice cap existed inland, even in the LGM, so the loss of continental shelves to the sea was simply a deficit of living space. This impact brought by the rise of sea level was less in the inland regions. Instead, the melting glaciers and the warm and wet climate expanded hunter-gatherers' habitats. On the contrary, the impact was greater in zones that are closer to the coast.

In the LGM, the whole Yellow Sea and Bohai Sea became land with a total area of more than 457,000 km^2, while half of the East China Sea also became land equal to about 350,000 km^2. The coastline of the East China Sea moved eastward to a maximum distance of more than 1000 km (Xie 1996). Including the exposure area of the South China Sea, the total area of shelves exposed reaches 1,600,000 km^2, one-third of the shelves in the present (see Table 5.1 and Fig. 5.2). East China was connected with the Korean Peninsula and the Japanese archipelago by land, and the vast exposed and had the same plant cover as the fluvial plain of North China (Zhao and Yu 1996), after a process of desertification on the continental shelves during the period of 24,000–18,000 BP.

The process of rising sea levels was rapid. From 15,000 to 10,000 BP, there was an annual rise of sea level of about 24 mm in height, and 100 metres in distance forward for the coastline of East China (Huang 1998). However, the rate of sea level rise was not even. For instance, the sea level fell again, probably in response to the return of the cold climate of the Younger Dryas Event (Li et al. 2010). The continuous rise of sea level reached the highest level in the Holocene Climatic Optimum at about 7000 BP, marking a height of 4–5 metres above the present sea level (see Table 5.2).

On the basis of the data, we can then evaluate the influence on Eastern China from the rise of sea level. All terrestrial animals and hunter-gatherers had to migrate westward along the same latitudes when the exposed shelves were inundated by sea transgression. Then they could stay in habitats similar to the one being inundated. Including three main river systems: Yellow, Yongding, and Huai, the North China Plain covers an area of 387,000 km^2, which is less than the exposed land

Table 5.1 The loss of sea to land during the LGM (Xie et al. 1996) (in square kilometres)

Sea	*Present Area*	*Area in the LGM*	*Loss to Land*	*Inundation Percentage (%)*
South China Sea	3,500,000	2,800,00	700,000	20
East China Sea	770,000	350,000	420,000	55
Yellow Sea	380,000	0	380,000	100
Bohai Sea	77,000	0	77,000	100
Total	4,727,000	3,150,000	1,577,000	33

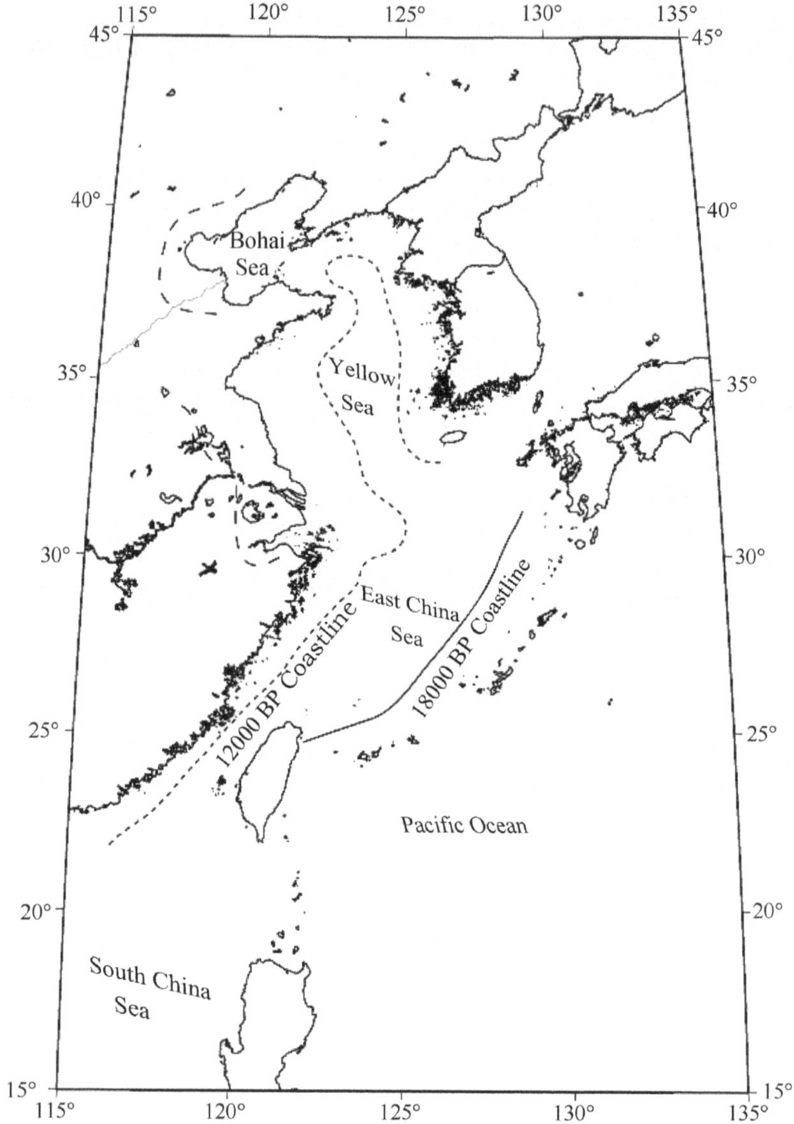

Fig. 5.2 Changes of coastline 18,000–6000 years ago

Table 5.2 The sea level rise from the LGM to the Holocene (Huang 1998)

Time (BP)	*Relative Height of Sea Level (m)*
8,000 ~ 7,000	+ 4 ~ + 5
9,000	?
10,000	−20 ~ −30
11,050 ~ 10,900	−33
11,600	−26
12,400 ± 200	−35 ~ −50
13,000	−70
13,500	−83
14,400 ± 750	−115
14,780 ± 500	−154
18,000	−185

area in the LGM. This means that the population density of hunter-gatherers would be easily doubled, even if there were no population growth around 10,000 BP. Although the retreating population moved inland in waves, they would have stopped in front of the mountains, where they would not move any more westwards because the mountains represent different habitats. The hunter-gatherer groups were assembled as sand was deposited on the beach, with the high density of populations increasing in the hillside zone. Being close to the best habitats, this area has been considered an ecotone that has attracted hunter-gatherers. It is unavoidable that such an accumulation of population in the ecotone would lead to an early build-up of population pressure.

Similarly, sea transgression inundated the 350,000 km^2 continental shelf of the East China Sea. This compares with the present area of fluvial plains along the middle and lower Yangtze Valley, which is less than 200,000 km^2 including lakes. With the decrease of sea levels in the LGM, the channel of the Yangtze River cut into the plain owing to the lower river mouth; therefore, few lakes were formed on either side. After the LGM the headward channel aggradation, resulting from sea level rise, could have reached as much as 1750 km above the present estuary, and have reached a height of 40 m above the present sea level along the Yangtze River channel and formed lakes. Evidently, most of the lakes along the Yangtze River are very young, only about 2000–7000 years old (see Table 5.3). The lake basins are generally not closed or are closed only by natural levees of the Yangtze River. The lake levels are controlled by the water phases of the Yangtze River. Therefore, in most cases, the

Table 5.3 Age of the formation of the major lakes located along the Yangtze River (Fang 1991, pp. 527–536)

Lake	*Elevation (m)*	*Age (BP)*
Tai Hu	−1	2300
Penli Ze	−4	>4000
Poyang Hu	6	3000
Runmeng Ze	9	7000
Hong Hu	19.5	350
Dongting Hu	12	2500

lakes on both banks are almost completely in a dried-out condition during the seasonal low water phase. Thus, at the boundary of the Pleistocene-Holocene, there are still no huge lakes in the middle and lower Yangtze River except those expanded by seasonal floods (Fang 1991). Hunter-gatherers who retreated to this region lost habitat that includes not only the exposed shelves but also areas inundated by seasonal flooding. Of course, the formation of lakes also created new habitats and access to certain aquatic resources.

In short, the impact on hunter-gatherer habitats in East China brought by the rise of sea level is obvious, especially those that reside in North China and the Middle and Lower reaches of the Yangtze River. In contrast, due to the total area of shelves exposed in the South China Sea being relatively low, with the predominantly mountainous coastal terrain, the hunter-gatherers are not highly affected by the rise of sea levels. However, an accumulation of retaining population gathers in the ecotone and best habitats, forming a new population centre that would lead to population pressure.

5.4.2 *Palaeomonsoon*

Modern climatological research divides the Asian monsoon system into three relatively independent subsystems: the Indian monsoon, the East Asian monsoon, and the Plateau monsoon. The Asian monsoon system developed from Cenozoic (Ruddiman and Kutzbach 1991) has experienced various fluctuations with the Earth's orbital variables, the configuration of sea and land, and the uplift of the Tibetan Plateau (An 1990). In southeast China, the East Asian monsoon has the dominant influence on climate and environment, making it significantly different from the desert-dominated landscapes of the same latitude (around 30 degrees of northern latitude), where the monsoon brings abundant precipitation and creates the subtropical evergreen broad-leaved forest landscape that is now the main area of rice agriculture in China (An 2000). However, the monsoon also brings abrupt and concentrated rains that erode soils and lead to floods, which pose a challenge for regional adaptation.

Precipitation associated with the East Asian monsoon comes from the interaction between the north-moving moist summer monsoon air and a northern mass of cooler air. The rain belt moves northward with the advancing summer monsoon front. The first pause occurs in the first second week of May in the coastal areas of South China, the second pause in the third and fourth weeks of June in areas between the Yangtze and Yellow rivers, and the third in mid-July in the northern part of North China and Northeast China. If the pause time of the rain belt is abnormally prolonged, devastating floods would form; while if the pause time were too short or arrived too late, droughts would occur. This happens more frequently in North China. In comparison, the Indian monsoon mostly influences the areas of Southwest China, with relatively fewer influences on East China.

The Holocene Climate Optimum in China is regarded as a precipitation maximum (moisture or moist period) and is viewed as the time of maximum postglacial warmth since LGM (Winker and Wang 1993). The monsoon precipitation and effective moisture (related to precipitation, evaporation, and temperature) are effectively controlling biological productivity, especially in arid, semi-arid, and semi-humid areas that are more sensitive to the variations of the monsoon precipitation. Based on climatic data, the maximum precipitation (effective precipitation) occurred ca. 12,000–10,000 years ago in Northeast China, ca. 10,000–7000 years ago in North China, and ca. 8000–7000 years ago in the middle and lower valleys of the Yangtze River. In Southwest China, the earliest appearance of a precipitation peak from the LGM is ca. 12,000 BP under the influence of the southwest monsoon (An et al. 2000). This abundance of precipitation is a favourable condition for the origin and development of food production. Precipitation/effective moisture gets later in the further south. The enhancement of the East Asian summer monsoon continuously moves northward until ca.6000 BP reaches the maximally northern limit, contributing to a warm and humid summer climate (see Fig. 5.3).

The monsoon is a regular climatic phenomenon, but the climatic materials show high-frequency variability on time scales of 1000-year, or even

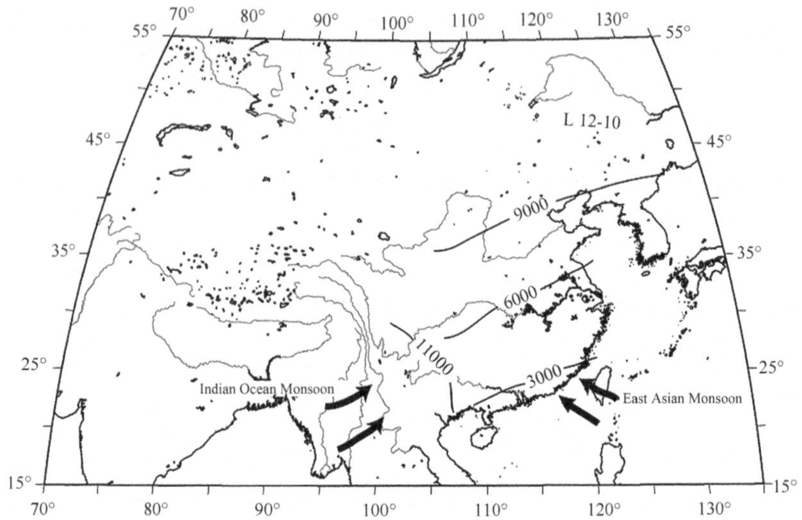

Fig. 5.3 Changes in monsoon from 11,000 to 3000 years ago

shorter, and instability of the East Asian paleomonsoon system (An 2000). The long-period change and annual fluctuation of the paleomonsoon, to prehistoric hunter-gatherers, are reflected in the changes in vegetation, and the migration of the fauna that feeds on the vegetation. Other effects are seasonal, especially the seasonal floods happening in the middle and lower reaches of the Yangtze River that would inundate habitats and thus lead to the inadequate of seasonal resources.

In a word, the warm and moist East Asian Monsoon brings many favourable conditions, which increase the biological productivity of the area and brings a climate pattern of precipitation and heat in the corresponding time period forms that are very favourable for the plant's growth. It would not have been possible for wild rice, which usually lives in tropical environments, to grow in the middle and lower reaches of the Yangtze River, and the same goes for millet-pig agriculture in north China due to drought. Notably, the wide variability of the paleomonsoon strengthens the seasonality and instability of hunter-gatherers' adaptations in such habitats.

5.4.3 Vegetation

As an initial condition that must be taken into account, the association between Vegetation turnover and the domestication of crops cannot be underestimated. Synthesised palynological research has produced a substantial frame of vegetative changes since the LGM. However, some regional features were different from the general tendency of vegetative changes. For instance, there is the Sichuan Basin, where the climate is warmer than the areas of the same latitude because of its relatively closed environment, and where features were changed less in the LGM. Southwest China, including the Sichuan Basin, is interspersed with mountains, plateaus, and basins, the vertical distribution of vegetation overwhelms the latitudinal tendency. In comparison, the vegetative changes of North China and the middle and lower Yangtze Valley are dominated by latitudinal movements.

Evidence based on palynological records sees replacements of vegetative types, especially in North China (Sun and Chen 1991). At 18,000 BP of the LGM, most of China was cold and dry. *Bucella frigida*, a cold-water foraminifera, shifted 5 degrees southward in latitude while spruce and fir grew at least 1200 m lower than at present. Frozen steppes covered almost all of North China, and the southern limit of continuous permafrost in the east moved 11 degrees to the south (Sun and Chen 1991). Because of the decrease in sea level and expansion of the land area, the Continental Climate is strengthened. From the Fenzhuang profile in the west of Beijing, Artemisia dominates the pollen assemblages (ca. 90%) with very little tree pollen during the period from 17,000 to 13,000 BP; thus indicating tundra vegetation. Then between 13,000 and 11,000 BP, there was an increase in conifer pollen assemblages and a few broad-leaved deciduous, similar to the vegetation growing above 1800 m above sea level in the Beijing area today. This represents a cold and wet climate with a temperature of 6–9 °C lower than today (Wei et al. 1997). Notably, there is another Artemisia dominance from 11,400 to 11,000 BP, indicating a short-term climatic deterioration, comparable to the "Younger Dryas Event" in Europe. Except that, the palynological records suggest that the herbaceous pollen percentage in the North China plains is higher than in the mountainous in West China.

Glacial conditions were over in most parts of China by 9000 BP. Temperatures were similar at present, with more moisture at the same time, but North China was still colder than today. The pollen data suggest that the vegetative change was abrupt and remarkable from 12,000 to 9,000 BP. The transition from Betual and Abies/Picea forests to warmth-loving deciduous forests in only fewer than 500 years. Similarly, evergreen broad-leaved forests also showed a rapid replacement of the earlier vegetation at this time.

Such a quick transformation of vegetation inevitably challenged the adaptations of hunter-gatherers in terms of quantity and types of food resources. In particular, this challenge was the most obvious in the boundary zones of vegetation, that is, ecotones. Hunter-gatherers probably had to form a lifeway as "niche chasers" to gain habitually used resources and began to adapt the new resources. Under proper interventions, vegetative replacement provides the opportunities to intensify some resources that have the greatest potential to increase production, that is, the predecessors of crops. In this period, the availability of starch-protein staples was significantly enhanced in the Fertile Crescent zone of the Middle East, attracting hunter-gatherers to utilise them (Hillman 1996, pp. 159–203). Similar changes can also be found in North China and the middle and lower Yangtze Valley. Wild rice (*Oryza rufipogon*) usually lives in tropical environments, rarely seen in subtropical zones. Its northern limit coincides with the line of 4 °C, the average winter temperature of January. With the warming process from the LGM, the distribution of wild rice (*Oryza rufipogon*) continuously expanded northward and provided the chance for people to make use of it. A similar situation synchronously appeared in North China, where the predecessors of millets occupied the frozen steppe of the LGM with the vegetative assemblages of deciduous forest. In short, the vegetative replacements in both North China and the middle and lower Yangtze Valley provided a crucial initial condition for the origins of agriculture in China.

5.4.4 *Faunal Variability*

Similar to vegetation variability, faunal variability determines the potential domestication of animals. Faunal variability influences the adaptations of hunter-gatherers in ways such as distribution, migration, and extinction. So far, studies of fauna have focused on some of the most representative species. From the Late Pleistocene, the dominant assemblage of North China is the Elaphus-Ultima (red deer and hyena) fauna, which is mainly represented by some extinct species such as Bubalus wanjocki, Bos primigenius, and Megaloceros ordosianus. In Northeast China, the woolly rhino-mammoth fauna is widely distributed; also, the mammoth fauna has been found in the Yellow Sea, about 100 km south of the Liaodong Peninsula with a latitude of 38.5 degrees (Dai 1984). Mammoth is also found in Tongwei (103°E, 35°N), Gansu province of Northwest China, possibly a local sub-species of mammoth (Liu and Li 1984). During the LGM, mammoth immigrated and penetrated the northern part of North China. Comparatively, woolly rhinos were not regarded as rigorous a tundra animal as mammoths since it was vastly distributed in almost all of North China. With the southward movement of woolly rhinos-mammoth fauna, the Elaphus-ultima fauna of North China also moved southward. Then, in the middle and lower Yangtze Valley, there was discovered a mixed fauna, including species from both the north and the south. From the Early Pleistocene, the fauna of the south was dominated by a very old faunal assemblage, that is, Ailuropoda-Stegodon fauna. At the boundary of the Pleistocene-Holocene, the Ailuropoda-Stegodon fauna was gradually replaced by modern fauna without an evident hiatus.

In Chapter 4, when discussing the origins of the microblade technique, I mentioned the significance of the southward-penetrating woolly-mammoth fauna, which attracted those big-game hunters to North China. In this case, cultural contact was achieved between the hunter-gatherers of North China and those of the northerly regions. With the exchanges of techniques, the microlithic technique has been formed. Another boundary of faunas, Elaphus-Ultima in the north and the Ailuropoda-Stegodon in the south also had the same effect on the adaptations of hunter-gatherers. Currently, the data and evidence from the Upper Paleolithic in the Lower and Middle Yangtze Valley are insufficient. The existing evidence is especially on a coarse chronological scale. All we can detect is that the miniaturisation of lithic technology occurred here at this time, dominated by the flake industry, with the predominance of flint, quartzite, and flakes that was more consistent with the small-size flake industry that has been prevalent in North China.[2] This change might be related to the southward movement of the fauna of North China during the Last Ice Age.

The extinction of fauna at the end of the Pleistocene is a very conspicuous and well-discussed issue (Martin and Wright 1967). The mass extinction of large mammals in North America and Australia is attributed to the direct effect of human predation (Martin and Steadman 1999). The hypothesis is challenged by some researchers, in that it cannot explain why extinctions stopped 10,000 years ago instead of other time periods

(Alroy 2001, pp. 1893–1896). Among the total 40 species of herbivores (the main target of human hunting) found in the Upper Paleolithic sites of North China, 14 species went extinct, that is, a 35% extinction rate. However, when examining the list of all mammals that lived in the Pleistocene, extinction was merely a normal phenomenon: the earlier the fauna; the more extinct species in it. From the archaeological records, these herbivores were commonly found in the Early Paleolithic sites but dramatically decreased in the Upper Paleolithic sites, especially in North China. Is this the result of overkill by hunter-gatherers? In fact, in the Early or Middle Neolithic sites, the remains of wild animals are still considered in the faunal assemblages. It implies that hunting is still a very important part of the subsistence of Neolithic hunters. Is this an abnormal phenomenon? To some extent, hunting can explain the extinction of large mammals by overkilling, but this does not necessarily need to be at the very end of the Pleistocene, since it could occur earlier such as in the Early Upper Paleolithic. The abrupt and rapid environmental change after the LGM might have changed the habitats of mammals in both size and quantities as they have to adopt a new diet (Pushkina 2006). It is implied that hunting can be only seen as the last straw.

Overall, before the origin of Chinese agriculture, the massive extinction of fauna was a fact of life. It is merely a question of how soon and why. The challenges faced by hunter-gatherers were obvious, the days of hunting large herbivores for subsistence were gone, and thus, the pressure on livelihoods had increased as a result.

5.4.5 *The Younger Dryas Event*

As a short and abrupt climatic episode, the Younger Dryas Event is argued to be the main impetus to food production in the Middle East (Bar-Yosef and Meadow, 1995, pp. 39–94). Although the Younger Dryas Event has been proven to take place globally, the environmental responses were diversified in different regions. For example, South America became wetter when the Younger Dryas Event occurred while in contrast the climate episode generally led to a colder and drier environment elsewhere (Rodbell 2000, pp. 285–286). The ice cores of Greenland present date records of high-resolution, showing that the middle-latitude climate index changes were earlier than the more northerly regions (Tylor et al. 1997). The ice cores of Greenland also show that the Younger Dryas Event occurred a few decades ago and that it took about 1500 years (from ca.11,700 to ca.10,200 BP) to reach the typical Holocene Climate (Grootes et al. 1993). Moreover, a nearly annually resolved, multiparameter record from the cores of Summit, Greenland, indicates that most of the transition occurred in a series of steps with durations of only about 5 years. Therefore, the beginning date of the Younger Dryas Event could be traced at 11,645 BP in higher latitudes and 11,660 BP outside of the Arctic (Tylor et al. 1997). This episode terminated as rapidly as it was initiated, in a 50-year period, so the environmental changes that it led to were abrupt and intense.

Take the Beijing area as an example, where almost all incipient Neolithic sites of North China. The lacustrine data suggests two short transitions towards a cold and dry climate, ca.11,600 ~ 11,300 and ca.10,950 ~ 10,480 BP respectively. It was relatively wetter and warmer between these two periods (Wei et al. 1997). This coincides with the ice record, displaying regional diversity in the in-between amelioration and then the re-deterioration. The Younger Dryas Event was widespread in China, from Northern to Southern China, reaching the middle and lower Yangtze Valleys, Nanling, Southwestern regions, and even the tropical areas, with diversified extents and environmental responses in different regions (Xiao et al. 1998; Wang and Liu 2001; Zhang et al. 2011). Peat deposits in the northern Yellow Sea region indicate a cold and wet rather than a dry and cold climate (Li et al. 2010).[3] The desert and the lake deposits in the Loess transition zone showed a pattern of dry, wet, and cold from morning to night (Zhou et al. 1996). The cooling rate in Southwestern China was slower than in North China because of the Tibetan Plateau and local topography. The Nanling area showed a cold and wet climate, which differs from the cold and dry trend in North China, while in the Qinghai-Tibet region, the effects are amplified (Shen et al. 1996).

The joint zone between Northwest and North China also witnessed the Younger Dryas Event as a period of cooler and dryer conditions marked by widespread aeolian deposition. It also suggests that climatic variability immediately, during and immediately after the Younger Dryas Event interval, is associated with rapid technological elaboration as well as innovation in the production and use of chipped stone tools, and perhaps, ground stone tools (Madsen et al. 1996). The abrupt climate fluctuation brought about by the Younger Dryas Event led to the instability of food resources, increasing the pressure on hunter-gatherers in regard to their substances. Stability became a scarce resource and the highest priority for hunter-gatherers, which was the advantage of agriculture. The cultural adaptation systems of hunter-gatherers adjusted to the push and pull of subsistence pressures and the attractiveness of food production, with small-scale adjustments involving technological innovations and large-scale adjustments involving changes in overall food access.

5.4.6 *Seasonality and Stability*

Seasonality is an integrated variable that comes from the effects of a series of environmental factors, expressed in the length of the growing season, the wet and dry, and the hot and cold of the year. For hunter-gatherers, seasonality can be measured in terms of the temporal and spatial diversity of food resources. The more distinct the seasonality, the more pronounced the differences in the spatial and temporal distribution of resources; the weaker the seasonality, the more evenly distributed the resources.

An environment like tropical forests or polar areas has only one season (merely summer or winter); so its seasonality is the lowest, although tropical forests have the highest diversity of species and resources, there is little spatial and temporal variation in resources for hunter-gatherers, as is the case in the polar regions. Thus, for the hunter-gatherers who lived in these regions, a toolset was sufficient to cope with subsistence issues throughout the year. The temperate zone mainly has the strongest seasonality because there are four distinct seasons and certain resources are only found at certain times of the year; the resources are highly patchy in this zone, thus, certain resources can only be found in certain areas. Resource diversity in the temperate zone is less than in tropical rainforest areas, but the quantities of every species in the temperate zone could be very large, especially in the grassland areas. Thus, the hunter-gatherers living in this region needed to widely adopt tool sets to exploit the seasonal resources.

The further increase of seasonality in North China and Middle and Lower Yangtze Valley, where the four seasons were already distinct, was seen between 12,000 and 8000 BP in terms of the difference in summer–winter solar radiation relative to the present (Kutzbach and Webb 1993, pp. 5–11). Hunter-gatherers also relied on storage to survive between winter and spring, and the increased seasonality has also raised the demands for storage.

Interestingly, the increased seasonality of the climate was accompanied by much greater stability. Based on Richardson et al., climate stability, high concentrations of carbon dioxide in the air, and increased precipitation constitute initial conditions for the origins of agriculture. They also argue that the highly unstable climate of the Pleistocene hindered the possibility of intensified plant use, which evidence was found in the ethnographic material instead of among Pleistocene hunter-gatherers (Richerson et al. 2001). This view overemphasises the importance of environmental conditions, even though the evidence of plant intensification is actually found in the Late Upper Paleolithic, at least in North China. However, Richardson et al. can hardly exclude that the hunter-gatherers, who supposedly had the slowest cultural evolution, were not influenced by the surrounding agricultural groups.

5.5 Conclusion

At the end of the Late Pleistocene, the environmental changes were undoubtedly dramatic and fluctuating, differing considerably from those of the Holocene in terms of sea–land area, paleomonsoonal patterns, and, especially, changes in plant and fauna. Basic climatic patterns, such as temperature and precipitation showed latitudinal and longitudinal shifts of several degrees upwards, shifting around 5 degrees in latitudes at the boundary of the warm temperate zone. During the Terminal Pleistocene, especially around the Younger Dryas Event, there was a further recurrence of climatic and environmental changes. These climatic changes could lead to a crisis of survival for hunter-gatherers. However, the term "crisis" is itself dialectical, with new opportunities alongside pressures. Hunter-gatherers accumulated in the best habitats and the hillside zone, enjoying the abundance of precipitation brought by ancient monsoons, the new and intensified species that replaced old plants and animals, and the stable climatic mechanism, which provided important environmental conditions for the origin of agriculture. The advantages and disadvantages of environmental change were evident and constituted a double impetus.

The environmental association is multifaceted: the end of the turbulent Pleistocene climate marks constraints on agricultural development; the patterns of cultural adaptation influenced by different environments constitute the resources for agricultural development. In addition, resilient adaptive strategies and broader patterns of resource utilisation are fundamental conditions in the evolution of the adaptation approach to agriculture in different ecotones. In mid-latitude regions, the need for resilient adaptive strategies was further enhanced due to the increased seasonality. The accumulation of populations within optimal habitats provided the demographic base needed for agricultural development. At the same time, environmental responses have diversified since the LGM according to the differences in regional conditions, with North China and the middle and lower Yangtze Valley being the most affected by sea-level rise and the latitudinal migration of plants and animals. On the contrary, the tropical, mountainous, and basin regions are much less affected.

It is easy to generate environmental determinism from the environmental association, yet an overemphasis on the role of environmental conditions and their variation would neglect the exploration of adaptive cultural mechanisms by hunter-gatherers and the role of humans. In this case, the environment, supposedly being regarded as a supreme being, determines human destiny, seeing the environment as a religion rather than a scientific study. As a necessary but not a sufficient condition for the origin of agriculture, environmental variation and change will not necessarily lead to the agricultural origin. This makes environmental determinism biased.

Notes

1. It should be noted that the term "middle reaches of the river" used here is not a geographical division in the strict sense. Instead, it refers to the size of gravel and the degree of terrain openness. The middle reaches are considered to be the region that falls within a moderate range in terms of suitability for the production of stone tools and the activities of hunter-gatherers. While there may be some similarities to geographical divisions, it is not an exact correspondence.
2. In recent years, archaeological surveys and excavations conducted in conjunction with the South-to-North Water Diversion Project have revealed significant changes in the technological aspect of the Late Paleolithic in the middle and upper reaches of the Han River. In the early period, the predominant tool industry was characterised by gravels and choppers, while in the later period, the main industry was characterised by quartzite and chert blade technology, with clear distinctions between the two phases. Unfortunately, there is currently a lack of reliable absolute dating data. It can only be speculated that the quartzite and chert blade industry belongs to the late Paleolithic phase. Sites containing such technologies are very

common along both banks of the Han River. I have conducted surveys, excavations, and research on the Paleolithic in this region.
3. New research shows that the sea level in the Yellow Sea was still 50 meters lower than the present during the Younger Dryas Event.

References

An Zhisheng et al., Asynchronous Holocene optimum of the East Asian monsoon. *Quaternary Science Reviews*, 19: 743–762, 2000.

An Zhisheng et al., *Preliminary study on ancient environmental changes in China in the past 20,000 years, in loess, quaternary geology, and Global Change*, Science Press, 1990.

An Zhisheng, The history and variability of the East Asian paleomonsoon climate. *Quaternary Science Reviews*, 19: 171–187, 2000.

C. M. Beall, Adaptations to altitude: a current assessment. *Annual of Review of Anthropology*, 30: 423–456, 2001.

D. Pushkina, Dynamics of the mammalian fauna in Southern Siberia during the Late Plaeolithic. *Vertebrata PalAsiatica*, 2006, 44 (03).

D. B. Madsen et al., settlement patterns reflected in assemblages from the Pleistocene/Holocene transition of North Central China. *Journal of Archaeological Science*, 23: 217–231, 1996.

D. T. Rodbell, The Younger Dryas, cold, cold everywhere? *Science*, 290: 285–286, 2000.

Dai Guohua, Ancient geographical evidence of cultural exchange between China and Japan in the Late Paleolithic. *Prehistory*, 1984, 1.

Fang Jinqi, Influence of sea level rise on the middle and lower reaches of the Yangtze River since 12, 100 BP. *Quaternary Science Reviews*, 10: 527–536, 1991.

G. Hillman, Late Pleistocene changes in wild plant-foods available to hunter-gatherers of the northern Fertile Crescent: possible preludes to cereal cultivation. In D. R. Harris ed., *The Origins and Spread of Agriculture and Pastoralism in Eurasia*, pp. 159–203. Smithsonian Institution Press, Washington, DC, 1996.

Huang Chunchang, *Environmental Changes*. Science Press, 1998.

J. Alroy, A multispecies overkill simulation of the end of pleistocene megafaunal mass extinction. *Science*, 292: 1893–1896, 2001.

J. E. Kutzbach and T. Webb, Conceptual basis for understanding late-quaternary climates. In H. E. Wright eds., *Global Climates since the Last Glacial Maximum*, pp. 5–11. University of Minnesota, Minneapolis, 1993.

Ji Xueping et al., A batch of Mohe culture stone artifacts unearthed from the Fuyuan River, Yunnan. China Cultural Relics News, Aug. 18, 2006.
K. C. Tylor et al., The Holocene Younger Dryas transition recorded at Summit, Greenland. *Science,* 278: 825–827, 1997.
Li Tiegang et al., The Younger Dryas impact hypothesis and the Formation of the North Yellow Sea Peat Layer. *Earth Science Frontiers*, 2010, 17 (01).
Liu Dongsheng, Li Xingguo, The time of mammoth survival in China and its significance in distribution. In *Proceedings of the Second National Radiocarbon Conference*. Geology Press, Beijing, 1984.
Liu Dongsheng et al., Prehistoric vegetation landscape of the Loess Plateau—forest or grassland? *Acta Geographica Sinica*, 1994, 3–4.
M. G. Winkler and P. K. Wang, The late-quaternary vegetation and climate of China. In H. E. Wright et al. eds., *Global Climates since the Last Glacial Maximum*, pp. 221–261. University of Minnesota, Minneapolis, 1993.
Modified according to Xie Chuanli et al.: The paleogeographic outline of the last glacial maximum in China and its climatic effects. *Quaternary Sciences*, 1996, 1.
O. Bar-Yosef and R. H. Meadow, The origins of agriculture in the Near East. In T. D. Price and A. B. Gebauer eds., *Last Hunters-First Farmers*, pp. 39–94. School of American Research Press, Santa Fe, 1995.
P. J. Richerson et al., Was agriculture impossible during the Pleistocene but mandatory during the Holocene? A climate change hypothesis. *American Antiquity*, 66: 387–411, 2001.
P. M. Grootes et al., Comparison of oxygen isotope records from the GISP2 and GRIP Greenland ice cores. *Nature*, 366: 552–554, 1993.
P. S. Martin and D. W. Steadman, Prehistoric extinctions on islands and continents. In R. D. E. MacPhee ed., *Extinctions in Near Time: Causes, Contexts, and Consequences*, pp. 17–55. Kluwer Academic/Plenum, New York, 1999.
P. S. Martin and H. E. Wright Jr. eds., *Prehistoric Extinction: The Search for a Cause*. Yale University Press, New Haven, CT, 1967.
Shen Yongping et al., Climate and environment of the Younger Dryas impact hypothesis on the Qinghai-Tibetan Plateau. *Journal of Glaciology and Geocryology*, 1996, 18 (03).
Sun Xiangjun and Chen Yinshuo, Palynological records of the last 11,000 years in China. *Quaternary Science Reviews*, 10: 537–544, 1991.
Tong Enzheng, A discussion on the cultural transmission belt from Northeast to Southwest in China. In *Collections of Cultural Relics and Archaeology*, Edited by the Editorial Department of Cultural Relics Publishing House, 1986.
Wang Wenyuan and Liu Jialin, Record of the Younger Dryas impact hypothesis in the Holocene Sediments of Lake Guangyan, a Maar Lake in the Tropics. *Scientia Geographica Sinica*, 2001, 21 (01).

Wang Xiaokun et al.: Brief report on the Excavation of the Jinsitai Cave Site in Inner Mongolia. *Acta Anthropologica Sinica*, 2010, 29 (01).

Wei Lanying et al., Preliminary study on the climate and environmental changes during the last deglaciation in the Beijing area. *Quaternary Sciences*, 1997, 2.

W. F. Ruddiman and J. E. Kutzbach, Forcing of late Cenozoic Northern Hemisphere climate by Plateau uplift in southern Asia and the America West. *Journal of Geophysical Research*, 94: 18409–18427, 1991.

Xiao Jiayi et al., Palynological evidence of the Younger Dryas impact hypothesis in the Eastern Nanling Mountains. *Acta Botanica Sinica*, 1998, 40 (11).

Xie Chuanli et al., The paleogeographic outline of the last glacial maximum in China and its climatic effects. *Quaternary Sciences*, 1996, 1.

Zhang Lirong et al., Record of the Younger Dryas impact hypothesis in the Desheng Lake Sediments in the Subei Basin. *Geological Science and Technology Information*, Wu Han, 2011, 30 (01).

Zhang Senshui, The Mohe Industry in China. In *Prehistoric Research 2004*. Sanqin Publishing House, 2005.

Zhang Xinbao and An Zhisheng, The relationship between forest and loess thickness in the Loess Plateau Region. *Bulletin of Soil and Water Conservation*, 1994, 14 (6).

Zhao Songling and Yu Hongjun, Formation of desertification environment on the Huangbo Sea Shelf at the end of the Late Pleistocene. *Quaternary Sciences*, 1996, 1.

Zhou Weijian et al., High-resolution peat records of the Younger Dryas Period in the Desert-Loess Transition Zone: an example of East Asian monsoon climate variability. *Science in China (Series D)*, 1996, 26 (02).

Zhu Zhicheng, Basic characteristics of forest and grassland in the Loess Plateau. *Scientia Geographica Sinica*, 1994, 14 (02).

CHAPTER 6

Basic Modes for the Origins of Agriculture in China

Therefore, the phenomena of the Dharma are vast in heaven and earth, and the adaptability is vast in the four seasons.

——*The Book of Changes*

6.1 Introduction

The previous five chapters have theoretically discussed the happening and unhappening of agriculture, and predicted the areas in China that might have seen the emergence of agriculture from the perspective of ecological modelling and suggested the initial conditions of the origins of agriculture in China from the perspectives of both environmental change and the Upper Paleolithic adaptation. This chapter will discuss the formation, process, and characteristics of Chinese agricultural centres through the perspective of archaeological material by analysing, generalising, and comparing.

Although research on the origins of Chinese agriculture starts relatively late, mainly centred on archaeological materials, the relevant studies are adequate due to the significance and multiple disciplines involved. However, from the perspectives of the cultural adaptation mechanisms of hunter-gatherers and the trace from the Paleolithic archaeology, theoretical studies have been lacking. The preceding chapters are mainly intended to fill these gaps. Moving into the Neolithic context, the main

S. Chen, *Prehistoric Modernization*,
https://doi.org/10.1007/978-981-97-9423-2_6

achievements of the existing research lie in the acquisition of certain archaeological materials and the evidence for the occurrence of domestication in paleoethnobotany and zooarchaeology. However, the research related to the processes of agricultural formation, that is, how agriculture developed from a simple state to eventually become a complex agricultural ecosystem in relation to characteristics of species and environmental conditions. Besides, as an order of parameter influencing cultural systems, how agriculture and other social factors interplay should be further detected. Although this chapter cannot contribute to new material or find new evidence of domestication, we might gain some new insights into the race to "characterise the Earliest Neolithic Cultures" at many sites through systematically integrating and examining existing research materials based on a cultural-ecological perspective.

I will start with the most conclusive material on the Neolithic cultural features and trace them back to more ancient and ambiguous times, linking the Neolithic to the Paleolithic. In this context, the Neolithic cultural features are defined as pottery vessels, ground stone tools, settled sedentisms, domesticated species, etc. Undoubtedly, these features did not appear at the same time. Moreover, there is considerable controversy in defining these features (which is indicative of the fact that these features had their developmental formations). However, overall, they are all inextricably linked to the Neolithic. To describe this period, I am applying the concept adopted by most Chinese scholars: Paleolithic-to-Neolithic transition, which corresponds to an absolute age of ca. 15,000–8500 BP. It is not an accurate time range, but only a rough estimation. The earlier the cultural features of the Earliest Neolithic, the more ancient its lower limit; likewise, as we learn more about the final phase of the Paleolithic, its upper limit fluctuates. After clarifying the time sequence of prehistoric cultures, I will discuss the processes that shaped agriculture in both North China and the middle and lower Yangtze Valley, exploring the interrelationships among humans, plants and animals, and environmental conditions. Followed by comparison with other agricultural origin centres in the world, the basic patterns of Chinese agricultural origins will be summarised.

6.2 Essence of "Neolithic Revolution" from the View of the Early Neolithic Culture

The beginning of the Neolithic is open because its meaning changes continuously with new discoveries. In recent years, there have been frequent and significant discoveries in the PNT. However, notably, after nearly a century of archaeological work, a relatively complete system of the Neolithic chronological framework has formed, particularly in North China and the Middle and Lower Yangtze Valley areas. Even though the Neolithic archaeological discoveries changed rapidly, the newly discovered evidence serves to enrich the established system instead of subverting it. In other words, the cornerstones of the Early Neolithic have already been laid, forming the basis and standpoint for us to discuss the origins of agriculture. We can trace the cultural history of agricultural emergence from the known to the unknown.

The classification of the Neolithic has undergone many changes, while the classification into the Early, Middle, and Late phases has been well-established, with only the beginning of each phase still differing in date. In this chapter, the classification scheme adopted is to separate out the "Paleolithic-to-Neolithic transition" (ca. 15,000–8900 BP) (Yan 1987, 1989).[1] Without rich and adequate materials, the cultural features are clearly transitional in nature, representing a key phase of cultural changes. Between ca. 8500 BP and ca. 7000 BP (ca. 5500~ 7000 BC), the Early Neolithic was the period of the establishment of the Neolithic and the initial formation of primitive agriculture. Representing the heyday of its cultural identity, the Middle Neolithic indicates a period when primitive agriculture was fully established. The Upper Neolithic marks ancient society began to become more complex and moved towards the origins of civilisation. This classification considers the existing archaeological evidence, as these published archaeological materials have always referred to this phase of archaeological culture as the Early Neolithic, which is impossible to change. In this case, it would be better to separate out the Paleolithic-to-Neolithic transition from the existing system. At the same time, this classification also takes the developmental phases of prehistoric agriculture into account, distinguishing the distinctive properties of the Early Neolithic from the Paleolithic-to-Neolithic transition. Thus, this classification possesses the significance of a cultural phase development rather than purely being a subjective cultural chronology.

In contrast to the sporadic finds from the PNT, the materials of the Early Neolithic show a character of "emergence", as the Early Neolithic culture seems to have abruptly flourished with large-scale sedentisms, patches of house sites, complete village structures, and storing pit, moat, tomb, etc. There was a dramatic increase in terms of the number of sites, the thickness of the cultural accumulation, and the richness of the relics. Undoubtedly, the "emergence" needs to be explained rather than simply being seen as a "natural" cultural development. How does the "emergence" phenomenon relate to the development of the PNT? Why did it happen? What factors were involved? We need to examine the Early Neolithic archaeological materials to answer these questions.

There have been several studies related to the temporal and spatial frameworks of Neolithic cultures (Su 1999; Zhao 2008; CA, CASS 2010). With the previous discoveries of Early Neolithic cultures so far, I divide the Early Neolithic cultures of North China and the middle and lower Yangtze Valley into eight cultural regions. The chronological intervals of each culture see Table 6.1.

6.2.1 North China

The Early Neolithic cultural pattern in North China is diverse: from north to south there is Xinglongwa culture in Western Liaoning (Suo and Guo 2004); in the northerly part of the North China Plain there is Cishan culture, with the Peiligang culture in the middle and the Jiahu type of the Peiligang culture in the south[2]; and the Xiaoshankou Phase I in the Huai River basin. From east to west, there is the Houli culture in the Shandong region, with the Peiligang culture in the middle, the Laoguantai culture in the Guanzhong region to the west, possibly including the Lijiacun type, and the Dadiwan culture in the Longdong region to the west, possibly to include the lower Beishouling. The transitional phase from the Paleolithic-to-Neolithic period to the Early Neolithic marks the cultural style change from vagueness to distinction. North China abruptly emerged numerous locally distinctive Early Neolithic archaeological cultures, and this explosive cultural growth marked a striking phenomenon.

The upsurge in the number of sites is an obvious phenomenon, and it is particularly noteworthy that the changes in the distribution of sites are more compelling. Compared with the site distributions of the Late Upper Paleolithic (LUP) and Paleolithic-to-Neolithic transition (PNT),

Table 6.1 The early neolithic sequences and chronology in North China and the Middle and Lower Yangtze Valley

Cultural Regions	*Sequence*	*Date (BP)*	*Calibrated Date (Cal BC)*	*Date Range (BC)*
Zhong-Yuan	Cishan	7355 ± 100 (ZK0439)~7060 ± 100 (BK78029)	6100–5960~5820–5630	ca. 6100~5600
	Peligang	7455 ± 110 (ZK0754)~6855 ± 110(ZK0747)	6230–5589~5640–5480	ca. 6200~5500
Guanzhong	Laoguantai	7150 ± 90 (BK80025)~6465 ± 120 (ZK0159)	5960–5720~5340–5083	ca. 6000~5000
Shandong	Houli	7675 ± 90 (BK91037)~6966 ± 70 (BK91036)	6384–6179~5680–5582	ca. 6300~5600
	Beixin*	6725 ± 200 (ZK0632)~5645 ± 140 (ZK0640)	5630–5243~4470–4167	ca. 5600~4100
Long-Dong	Dadiwan	7150 ± 90 (ZK2138) ~6245 ± 90 (ZK1267)	5960–5720~5203–4864	ca. 5900~5000
Liaoxi	Xinglongwa*	6925 ± 95(BK80025) ~5660 ± 170 (ZK1389)	5712–5530~4510–4159	ca. 6000~5300
Two Lakes	Pengtoushan	8200 ± 200 (BK87050)~7745 ± 90 (BK89017)	7195–6548~6424–6219	ca. 7200~6200
	Lower Zaoshi*	7210 ± 110 (BK87046)~6405 ± 90 (OXA2218)	5961–5713~5256–5050	ca. 5960~5050

(continued)

Table 6.1 (continued)

Cultural Regions	*Sequence*	*Date (BP)*	*Calibrated Date (Cal BC)*	*Date Range (BC)*
	Chengbeixi*	6970 ± 127 (ZK2645)~6800 ± 80 (BK84028)	5740–5540~5619–5477	ca. 5740~5400
Lower of the Yangtze Valley	Shangshan	9610 ± 160(BA02236)~8180 ± 35 (BA06137)	9220–8790~7190–7070	ca. 8000~6500
	Kuahuqiao	7595 ± 242(HL91002)~6180 ± 90 (BK200170)	6700–6050~5480–5230	ca. 6000–5400

*Only partly in the Early Neolithic; # the Xinglongwa culture has obvious transitional properties, spanning the north and south of the Yan Mountains, located at the intersection of North China, Northeast China, and the Mongolian steppe region. This will be further discussed in Chapter 7, and The Shangshan culture is dated to the Paleolithic-to-Neolithic transition, but the cultural landscape belongs to the Early Neolithic and its dating will be discussed later. BK/BA: Department of Archaeology/College of Arts and Sciences, Peking (Beijing) University; ZK: Institute of Archaeology, Chinese Academy of Social Sciences; OXA: Department of Archaeology, Oxford University; HL: The Second Research Institute of the State Oceanic Administration; BP: from 1950 AD; Half-life: 5568 years for Upper Hill and Kuahuqiao sites, 5730 years for the rest the Early Neolithic sites are mainly located on the border between mountains and the North China Plain, closer to the plain areas. Different from the LUP, sites in the PNT move towards the fluvial plains, mainly located on the margins of basins and in the terraces of river valleys. The sites of the Early Neolithic culture take another step forward into the fluvial plains, although the scales are not as widespread as the fully diffused Neolithic culture in the Middle Neolithic. Take the Houli culture as an example, around ten discovered sites are relatively limited distributed, being found only in Changqing and Wande to the south of Jinan and in the area around Zhangqiu, Zibo, and Qingzhou to the east which are located in the hilly terrain of the Mount Taishan and Mount Lushan. The Peiligang culture sites are more widely distributed with over a hundred discovered, mainly located in Henan province along both sides of the Beijing-Guangzhou line, that is, in the transition zone from the Taihang Mountains, the mountains of Western Henan and the Dabie Mountains to the plains of North China (Zheng et al. 2005; CA, CASS 2010).[3] Compared to the later phase of the Hongshan culture, the Xinglongwa culture located in the mountains of western Liaoning is at a higher altitude and further away from the valley plains.[4] On the other hand, the sites in Shanxi possessed rich discoveries in the LUP, however, the situation was the opposite in the Early Neolithic (Zhang 2002). This sharp contrast indicates that early agricultural producers chose the hills and fluvial plains of North China rather than the Loess Plateau. The uncommon scarcity of Early Neolithic in Shanxi cannot simply be attributed to the inadequate archaeological fieldwork, since the most abundant Middle and Upper Paleolithic sites in China have been found here. Similarly, the burial environment is not the reason for the scarcity of Early Neolithic in the Loess Plateau, where several Early Neolithic sequences have been identified in Longdong and Guanzhong, such as Dadiwan Phase I, Lijiacun, Laoguantai, and Baijia. A reasonable explanation is that the environment of Shanxi in the Early Neolithic was not the best for early plant cultivation, although adequate microblade technology remains have been found in this region, including Xiaguan, Xueguan, Dingcun, Shizitan, etc. To be more cautious, the Neolithic in Shanxi merely developed slower compared to other areas (except for Southwest China).

In 1986, Xingbang Shi noticed the distribution of the pre-Yangshao culture (Early Neolithic culture) sites and summarised four types. The first type is the basin site which is located on the hill slope, slightly away from the river and high in topography. This type of site generally only has deposits of a single archaeological culture. The second is the hillside site, where a stream commonly rises nearby, and gentle slopes extend to conjoin with the fluvial plains. The third is the terrace site by a river, especially in the triangle zone by the intersection of two rivers. The last one is the riverside site on the plain, similar to the villages nowadays. Shi argues that there is a common tendency among these sites, with the first two types generally having earlier radiocarbon dates than the latter two types (Shi 1986). This Early Neolithic culture site distribution might show that people gradually move towards the plains which are evidently more favourable for gaining plain lands (beneficial for transportation and harvest), irrigation, and fertilising farmlands. In fact, the soil contains higher clay materials with viscousness making the lush plant grow, thus, making the cultivation activities in plains more difficult than those on the hillsides. The expansions into plains reflect the development of agricultural production. Thus, according to the geographical distribution of sites, the extent of agricultural production can be roughly determined. This judgement is based on the simple premise that the best arable land is on the plains and agricultural producers have to live nearby. In other words, with most of the Early Neolithic sites located on the slopes, it implies that the size of plant cultivation was fairly limited or at least not dominated in the subsistence. The beginnings of agriculture changed the relationship between humans and the environment: the criteria by which humans chose the environment changed, and humans changed the environment at the same time. The emergence of new features reflected the creation of a new way of life and production.

The beginnings of agriculture also dramatically changed the human body, including the morphology of the skeleton, the biochemical composition, and genetic selection (Pinhasi 2011). Chinese archaeological research applies paleo-diet analysis, which includes trace element analysis and stable isotope analysis of human and animal bones. Cai Lianzhen and Qiu Shihua ever used measurement 13C to study the food composition of prehistoric people of North China, and found that the food of the early Yangshao culture (ca. 5000~3000 BC) came from 4C plants, to which millets belong in a proportion of 48%, with the ratio increasing to 67% in Taosi culture (ca. 3000~2000 BC) (Cai and Qiu 1984). The possible proportion of 4C plant food as a maximum of 30% at about 7000 BC if the dependence on this type of plant increases linearly, that is, increases one per cent every 100 years. This hypothesis is feasible because the evolution from Early Neolithic to middle Neolithic is gradual and transitional, no marked jump is seen from the integrated cultural features. The paleo-diet analysis has become popular in recent years. The carbon isotope analysis of human bone from the Xiaojingshan site of the Houli culture and animal bones from the Yuezhuang site shows that the 4C plant (millets) only accounts for 25% (Hu et al. 2008), which roughly conforms to the previous speculation. The analysis of stable isotopes in the Dayuewan site (δ13C and δ15N) implicates two distinct phases of the domestication and utilisation of plants and animals. The first phase (5900~5200 BC) was a slightly less intensive phase of agriculture, during which people grew and stocked millets, not only for their own consumption but also to feed their dogs.[5] The second agricultural phase (ca. 3900 BC) was more intensive, during which not only broomcorn but millets were sowed, besides, people fed pigs and dogs. In the Dadiwan site, the Early Neolithic agriculture was low-intensity dry farming, which developed more advanced primitive agriculture brought by the Yangshao culture from the Central Plains (Barton et al. 2009). Isotopic analyses of other Early Neolithic sites such as Xinglongwa (Zhang 2003) and human bone material detection from Jiahu (Hu et al. 2007) also indicate that humans applied a mixed economy through utilising crops alongside gathering, hunting, or fishing. Isotopic analysis of human and animal bones from more recent Neolithic sites such as Jiangzhai and Xipo shows that millets were not only the main food for humans but also feed for domesticated animals such as pigs, dogs, and chickens (Pechenkina et al. 2005).

The paleo-diet analysis serves as direct evidence that humans consumed crops. Besides, more importantly, the analysis possesses a culturally significant implication, showing that agricultural development was a gradual process. Based on the 25% dependency ratio of crops from the Houli culture, we could speculate on an even older date that humans would have begun to consume millets (the dependency ratio would be higher) 10,000 years ago if the dependency rate of crops was constant. Clearly, the older the age, the more problematic this speculation becomes. People in the LUP might occasionally consume the wild ancestors of millets (broad-spectrum food), while only beginning a sustained period of intensifying exploitation during the PNT, which marks the beginning of the domestication of the species. The cultural mechanisms and related motivations/agents shaped by the utilisation predated the beginning of domestication. Biologists are concerned with the domestication of species; archaeologists are concerned with the cultural processes that led humans to the species' domestication. Based on the paleo-diet analysis, even though the incipient agriculture was somehow insignificant, however, it constantly developed and influenced humans. Eventually, an archeological system and agricultural society formed, constituting the foundation for agricultural civilisation.

Plant and animal remains represent the most direct evidence of agricultural emergence. Through microfossils (e.g. phytoliths) and macrofossils (e.g. seeds of plants), phytoarchaeology has been studied for evidence of domestication. Recent studies have further validated the previous viewpoint. Based on phytoliths and molecular biology analysis evaluated by Lv Houyuan et al., millets might have been the earliest domesticated crop (Lu et al. 2009) as a small amount of millets that dated to ac. 8300~6700 BC was found at an ancient pit at the Cishan site. The study of plant microfossils can also be used to reveal the regional sequence of plant domestication. Studies of plant remains from 26 sites in the Yiluo River basin and 10 AMS dating data indicate the dominant domesticated crop in this area was millet during the Early Neolithic, followed by broomcorn. Rice appeared during the Upper Neolithic (ac. 3000 BC) and then wheat appeared during the Shang Dynasty (Lee et al. 2007).

There is also some strong evidence detected from the archaeological sites, particularly those that are relatively well preserved. The Cishan site was excavated in three excavation areas totally 2579 square metres,[6] including eighty pits that contain deposits of millets with a depth generally from 0.3 to 2 metres and ten pits have deposits of more than 2 metres (Hebei Provincial Cultural Relics Management Office, Handan Municipal Cultural Relics Storage Institute 1981). Parts of grains are still visible in excavation and are identified as millets, pecans, hazelnuts, small-leafed parka, etc. With abundant stone tools (Table 6.2), the differences between the upper and lower cultural layers are noteworthy. The increased proportion of ground stone tools and the marked decrease of chipped stone tools in the upper culture reflect the greater emphasis on grinding techniques. Merely taking stone axes into consideration, 53% are found in the first cultural layer and 48% in the second, showing little difference between the early and late Cishan culture periods, with slightly more in the early period. If taking stone shovels as a digging tool, the proportion is 10.8% in the first cultural layer and 8.4% in the second cultural layer, marking an even smaller difference. However, the difference in harvesting and grain processing tools is distinctive: no stone sickles have been found in the first cultural layer, but six in the second cultural layer; four pieces each of grinding slabs and handstones in the first cultural layer are in disrepair, while the second cultural layer reaches 52 and 50 pieces respectively. The selective difference implies that more harvesting and grain processing activities could happen in the later culture phase. Otherwise, it is not possible to explain the small difference in the proportions of stone shovels and axes, but the huge difference in stone sickles, grinding slabs, and handstones. The differences between the Early and Late cultural layers at the Cishan site suggest that there was a clear primitive agriculture development during this phase of the Early Neolithic.

In the study of the Cishan site, the interesting process of abandonment has been overlooked for a long time. Why is the site so abundant in food remains and cultural relics? In particular, many intact assemblages of relics were placed in pits, with 45 toolkits of grinding slabs, handstones, and groups of pottery pots and holders in the second cultural layer of the three excavation areas (Hebei Provincial Cultural Relics Management Office, Handan Municipal Cultural Relics Storage Institute 1981). These usable items were obviously intentionally placed (as Schaeffer suggests, de

Table 6.2 Comparison of the stone tools excavated from the two cultural layers at the Cishan site

	Second Cultural Layer (Upper Cultural Layer)	*First Cultural Layer (Lower Cultural Layer)*
Number	687 items	193 items
Tool-making Techniques	21.8% of chipped stone tools, 65.4% of ground stone tools, 8.8% of both	34.2% of chipped stone tools, 57% of ground stone tools, 12.8% of both
Chipped stone tools	144 items, including 10 shovels, 107 axes, 3 knives, 24 whetstones	56 items, including 2 shovels, 48 axes, 4 smashers, 2 scrapers
Grinding and chipping	94 items, including 18 shovels, 56 axes, 7 chisels, 2 adzes, 5 lids, 6 projectiles	17 pieces including 2 shovels, 10 axes, 5 chisels
Ground stone tool	443 items, including 30 shovels, 167 axes, 6 sickles, 8 adzes, 13 chisels, 105 balls, 10 hammers, 52 grinding slabs, 50 handstones, 2 grinders	109 items, including 17 shovels, 45 axes, 3 hammers, 2 adzes, 27 balls, 4 pieces of grinding slabs, 4 handstones, 4 projectiles
Relative Remains	282 pits, including 18 grain piles, 2 seed piles, 128 bone tools, and 321 pottery tools	186 pits, including 62 grain piles, 343 pieces of bone, antler, ivory, and mussels, and 156 pottery tools

facto refuse [Schiffer 1987]), rather than damaged and unusable waste. This suggests that the ancient inhabitants would come back as considered, thus, only leaving the place temporarily. Besides, the grain cellars provide hard evidence for storage. Ancient people would not have left grain in a site if they had intended to abandon it. At the Cishan site, two house sites and a large number of pits, some of which might be merely house sites without doorways or hearths based on the planar graph, have been found. The same situation was also excavated at the Jiahu site in the same era, where some house sites were initially classified as pits but it was found incorrect when publishing (Henan provincial Institute of Cultural Heritage and Archaeology 1999). The easy confusion between house sites and pits reflects the fact that the cultural sedentism capacity of Early Neolithic humans was not strong, as people may have left the sedentism from time to time to forage for certain seasonal resources. Very complete pottery and tools can often be found in Early Neolithic sites rather than Late Neolithic sites. If we analyse the contrast via a perspective of the process of abandonment, more recent sites are usually planned for slow abandonment over a long period; whereas the abandonment of Early Neolithic sites is usually rapid, that is, people did not intend to discard the sites but would return the previous residents as planned, but for some unexpected reasons stopped their original plans.[7] Thus, these sites could be well preserved, allowing archaeologists to see a more complete picture of the prehistoric culture.

The early and late cultural layers at the Cishan site illustrate the development of early agricultural society. In the lower cultural layer of the Cishan site, chipped stone tools, ground stone tool, and stone tools with both technologies constitute 34.2, 57, and 12.8% respectively. In comparison, the proportions of the upper cultural layer correspond to 21.8, 65.4, and 8.8%, reflecting a continuous decrease in chipped tools but a predominant position in polished tools. The benefit of polished tools is their durability, which is an indicator of mobility declination, as no group is willing to put effort on tools that are disposable after usage. All polished tools are manufactured for repeated use since edge and surface polishing is a kind of redundancy design with a limited increase in efficiency. However, a wholly polished tool has higher durability because of the lower risk of breaking. The materials discovered from the Cishan site show that the prehistoric inhabitants remained somewhat mobile and less settled than the Upper Neolithic inhabitants; on the other hand, we can see that their mobility was decreasing and their sedentism capacity was increasing further.

6.2.2 *The Middle and the Lower Yangtze Valley*

Compared to North China, the features of Neolithic cultures in the middle and the lower Yangtze Valley are relatively vaguer and more complicated. Broadly defined, there are two sub-regions: the middle and the lower Yangtze Valley, sometimes named around Dongting Lake and around Tai Lake cultural areas. Some scholars incorporated Boyang Lake Plain with Dongting Lake Plain and named them the "Two-lake sub-region" (Yan 1989). Considering Boyang Lake Plain combines with Pearl River Delta and forms the axle of Neolithic cultures of Southeast China (Su 1999). In recent years, with the discovery of the Shangshan culture in the lower Yangtze Valley, the cultural pattern of the Early Neolithic has become much clearer. The existing materials show that the Dongting Lake and Tai Lake cultural areas formed two sub-regions in the middle and lower Yangtze Valley, constituting the basis of the later regional civilisation centres.[8] The intermediate Boyang Lake Plain is more akin to a transitional zone between the middle and lower reaches of the Yangtze Valley and the Lingnan and southeast coastal areas.

At present, Dongting Lake region possesses the richest Early Neolithic cultural remains in the middle and lower of the Yangtze Valley, with some typical sequences: Pengtoushan (ca. 7000~5000 BC), lower Zaoshi (ca. 5500~5000 BC), and Chengbeixi (ca. 6500~5000 BC).[9] Because of the abundant cultural remains, including rice remains, Pengtoushan culture, typically represented by the sites of Pengtoushan and Bashidang, has been well-dated and studied. For the sites of Pengtoushan culture, there have been 29 radiocarbon dates from three laboratories including the Research Laboratory of Archaeology and the History of Art in Oxford, England. Thirteen dates come from the common technology, 14 from AMS, and the remaining two with the Mossbauer method. Chronologists notice that once the coarse old carbon from Southern soil or silt is mixed with clay, it could lead to earlier dating of pottery (Chen 1994), thus, the carbon components should be excluded from pottery. Most of the AMS dates use rice husk and straw as the materials for measurement. While synthesised measurement analyses make the date of the Pengtoushan culture reliable.

The Pengtoushan culture site is mainly distributed in the Li Yang Plain at the lower reaches of Li Shui, between the Wuling Mountains and the Dongting Lake basin. It is a transitional zone with the lake area to the east and the mountains to the northwest (Hunan Institute of Cultural Relics and Archaeology, Lixian Cultural Relics Management Office 1990, 1989). Interestingly, so far ten sites of Pengtoushan culture have been found on the low hillside north of Li Shui. This pattern of site distribution might correlate with the southward shifting of the river channel of Li Shui because of the Coriolis Force, so that the south of the river would suffer more erosion (Gu 1991). To prevent flooding, the ancients chose to live on the northern riverbank, reflecting the proximity to water and indicating the presence of seasonal flooding during the Pengtoushan culture. Compared to the Pengtoushan culture, the site distribution expanded to the marginal area of Dongting Lake during the period of the lower Zaoshi culture. It is noteworthy that Dongting Lake did not come into being until 2500 BP (Fang 1991) and possibly even later. Before the formation of Dongting Lake, small ponds, seasonal floods, and fluvial plains constituted the main landscape. Spore pollen analysis at the Pengtoushan site shows a lack of aquatic plants, mostly fir and wet fern spores, and a forest-grassland environment dominated by warm coniferous forests. Entering the Middle Neolithic, the representative sequence—Daxi culture, entirely expanded its sites in the lake area. In retrospect, the typical Upper Paleolithic sites, including Yan'er dong, Shiligang, Zhuma, etc., reveal the tendency of prehistoric human sites: a gradual transition from the wide valleys of the middle of the river and the foothills of the lower reaches to the fluvial plain, that is, from the best habitats for hunter-gatherers to the best habitats for agricultural producers. The changes in habitat centres are undoubtedly a gradual process, and at the same time reflect a change in the main ways of human substance.

The most important discovery of the Pengtoushan culture is the rice crop remains, which are found at the Bashidang site with 15,000 grains of rice and paddy (9800 grains in total in physical terms) (Hunan Provincial Institute of Cultural Relics and Archaeology 2006), as well as a large number of animal bones, bamboo and bone tools, and woven fabrics. The cultivation of rice during the Pengtoushan culture has already reached a consensus, but there is not a congruent viewpoint on whether the rice planted in Pengtoushan is domesticated or not (Zhang and Pei 1997; Zhang et al. 2003). It is still unknown how much-planted rice contributed to the food composition at that time either. The remains of rice cultivation at the Pengtoushan site are found within the pottery and in the red-burned clay, but no physical evidence of rice cultivation has been found in the cultural deposits (Hunan Provincial Institute of Cultural Relics and Archaeology 2006). The rich physical evidence found at the Bashidang site suggests that the ancient people lived nearby water and that their substance included hunting, fishing, and gathering rhizomes, gorgonians, lotus roots, and lotus seeds. These plants are fairly easy to domesticate, and have high yields, in that water caltrop can be dried or fried for storage, while lotus root is available for 11 out of 12 months in the year. During the Pengtoushan culture period, rice cultivation and plant collection were equally important. If the degree of crop utilisation at this time was similar to that in North China, then rice farming was a fairly limited part of livelihoods, occupying no more than a quarter, so it is reasonable to explain that no significant remains of rice cultivation have been found at the Pengtoushan site.

There are only two categories of faunal remains found at the Pengtoushan site: five fragments of crushed cow teeth and two small fragments of bird bones. There are more faunal remains discovered from the Bashidang site, totalling over a thousand. The identifiable specimens are mammals, birds, fish, and other aquatic animals. Mammals include cattle (19), pigs (only 3), deer, water deer, and black rats. The skeletons of the animals are broken, so it is unknown if pigs and cattle are domesticated species. People continuously increased the use of the resources of meiofauna where they could be caught close to the site of residence, including birds and aquatic animals, indicating the decline of the hunting economy in their substance and the increase in sedentism. These two aspects are interrelated: sedentism meant that large animals could not be caught as often, and thus, the decline of hunting activities required greater reliance on cultivation or animal husbandry. In the sequence of the lower Zaoshi culture, hunting did not disappear, but the same emphasis was placed on the usage of aquatic resources. A large number of animal bones, mostly teeth and neck bones, were found at the site of Zaoshi. These represent an assemblage that includes spotted deer, buffalo, goat, pig, barking deer, porcupine, turtle, and mouse. The teeth of deer, buffalo, and pig are dominant at the site, but it is still unknown if they are domesticated species (Hunan Provincial Museum 1986). At another site of lower Zaoshi culture, Hujiawuchang site, it is reported that the earlier the layers, the richer the faunal remains. The bones in the remains are smashed, which phenomenon is possibly related to the full use of food resources.[10] The identified assemblage in the Hujiawuchang site is similar to that in Zaoshi but has more aquatic animals. The abundance of pig, buffalo, and goat remains at the Hujiawuchang site lead to the speculation that these animals are possibly domesticated (Hunan Provincial Institute of Cultural Relics and Archaeology 1993). Clearly, from the available evidence, there are no signs of domestication until the Middle Neolithic. Thus, even if animal domestication existed during the Pengtoushan culture, the proportion would be very low.

The tool assemblage may also reflect a shift in adaptation. The items excavated at the Pengtoushan site could be divided into two categories: tools and debitages (1406 items) and ornaments (62 items). Among the tools and debitages, there are 344 flint artefacts, approximately a quarter of the total; and 25 polished tools, which account for only 2% of the total in this category. While gravel tools make up the rest. The polished tools include axes (11), adzes (7), shovels (2), chisels (1), pebbles, and grinding slabs. The flint artefacts are small, with 2 to 5 centimetres in maximum diameter, which includes 58 stone tools, predominantly by scrapers, augers, and burins. The amount of pebble tools is 81, dominated by smashers and scrapers. The amount of stone ornaments is 47 pieces with a fine polish. The situation at the Bashidang site is similar, with 2070 pieces of stone tools for the first category, 23% of flint artefacts, and still less than 2% of ground stone tools; with 167 ornaments for the second category (Hunan Provincial Museum 1986). The similar situation in the two sites identifies that the grinding techniques have already been presented during the Pengtoushan culture, with tools such as axes, adzes, and chisels being used for woodworking, which can be seen as evidenced by the wooden components excavated from the site. Due to the fragility of flint, the stone tools and scrap made from it were very small, thus greatly expanding the total statistical sample and making the proportion of polished tools of daily activities more than merely 2%. As polished tools are used for daily routines such as cutting and smashing, the lifespan of the polishing tool is relatively short. In contrast, durable polished tools (with a resharpenable edge) played a far greater role in real life than its proportion excavated. Undeniably, in the Pengtoushan culture, stone tool grinding techniques were initially applied to the production of ornaments, as nearly all stone ornaments, instead of stone tools production, applied such technique. By the time of the lower Zaoshi culture, at the Hujiawochang site, chipped stone tools occupied 89.77% of the assemblages, with the proportion of polished tools increasing significantly compared to the Pengtoushan culture. Its assemblage of chipped stone tools is the same as that of the Pengtoushan culture, including flint microlithics and gravel artefacts, as well as a flake made by the ridged-hammer bipolar flaking technology similar to that from the Guizhou area (Hunan Provincial Institute of Cultural Relics and Archaeology 1996). The proportion of ground stone tool increased to 30% by the late Lower Zaoshi period (Yin 1996). However, the assemblage of chipped stone tools from both the Pengtoushan and the lower Zaoshi cultures is consistent with the characteristics of the Upper Paleolithic in the region. As a new cultural element, the emergence of grinding techniques particularly applied in felling and woodworking tools, together with mortise and tenon construction, building materials such as planks, and the discovery of a variety of dwelling site types, reflecting the emergence of settled life at this time.

Four types of dwelling sites were revealed at the Bashidang site: pit-dwelling, semi-crypt, dry-pen, and high-platform. The high-platform structure has pristine pad soil with no inclusions, indicating that the pad soil might have been deliberately carried from around the Bashidang site (Hunan Provincial Institute of Cultural Relics and Archaeology 1996). An enclosing wall and moat at the site of Baishidang imply the ancient input into the construction of the sedentism. Encompassing an area of about 16,500 m^2, the remains of the wall have a height of 0.5~1 metre and a base width of 6 metres. The moat has been cleared of silt at least three times during the usage period. The reasonable functions of the moats might defend the floods and to drain water. Compared with the sophisticated urban sites of the Middle and Late Neolithic, for example, the nearby urban site of the Chengtoushan of the Qiujailing culture, the Bashidang site shows no planning layout with a simple structure of walls and small size. In this case, the Bashidang site is more likely a cofferdam in a location that is needed to prevent flooding. The construction of a cofferdam would have required a large amount of labour input, which cannot be spent on a site that was only used for temporary residence. Thus, inputs in the residential construction also reflect the presence of sedentism during the Pengtoushan culture.

As mentioned above, the main characteristic of the Early Neolithic is the "emergence" nature. Early Neolithic archaeological cultures began to distribute in patches in certain areas, forming what archaeological research defines as "archaeological cultures": artefact assemblages (especially pottery assemblages) with a clear spatial and temporal identity. Applying this criterion to measure archaeological materials from the lower Yangtze Valley, among the Shangshan culture sites, only Pujiang Shangshan and Xiaohuangshan sites in Chengzhou have the same type of remains. More than 70 km apart, both Shangshan and Xiaohuangshan sites are located in the upper valleys of two rivers that converge towards Hangzhou Bay.[11] Thus, we argue that more material may be found in the future. The Shangshan culture of 14C (9600~8100 BC) is similar but with more mature cultural landscapes compared to cultures in the middle and lower Yangtze Valley during the PNT, such as Yuchanyan in Dao County, Xianrendong in Wannian County and Diaotonghuan sites. However, the archaeological community is sceptical about the date of 14 C (Hebei Provincial Cultural Relics Management Office, Handan Municipal Cultural Relics Storage Institute 1981). Five out of the six published radiocarbon dating for the Shangshan site, the oldest data are all from charcoal-tempered pottery (Zhejiang Institute of Cultural Relics and Archaeology, Pujiang Museum 2007). It is questionable whether the radiocarbon dating materials were mixed with the coarse old carbon. Only one data is over 9000 years old, exceeding the other dates by as much as 1200 years. Excluding this abrupt data, the date of Shangshan culture could be from 8250 to 7310 BP. With the intervention of coarse old carbon, there would be about 1000 years earlier than the actual date for the Pengtoushan site. Excluding this fact, the date of Shangshan culture and Pengtoushan culture should not be far away.

The standpoint to identify Shangshan culture as an Early Neolithic culture similar to the Pengtoushan culture lies in the cultural landscapes of Shangshan culture. The ceramic assemblages in the Shangshan culture are mainly represented by the cup with a large open-mouth, double-ear jars, large flat-bottomed plate, and hollowed plate with ring foot. Most of the pottery is made of charcoal,[12] with a small amount of sand. During the late Shangshan culture, the number of sand-tempered pottery gradually increased. The ceramic body is normally thick with more than 2 centimetres; the firing temperature is low (about 800 °C) with uneven fire, resulting in a seemingly red coating appearing on the surface of the vessel; the fractured surface of the ceramic body is commonly laminated, with obvious grains of rice husk visible in the body. According to the preliminary statistics, 85% are flat-bottomed potteries, with additional ring foot, circular base, and serrated or creamed feet. The pottery surfaces are mostly plain, with occasional rope and stamped decorations. In addition, polygonal-edged potteries have been found (Zhejiang Province Institute of Cultural Relics and Archaeology, Pujiang Museum 2007). Although pottery production techniques are primitive, their function and form have been differentiated, such as cooking vessels, water jar, storage jar, and so on. Besides, a variety of vessel types has emerged in the same category of pottery, such as circular-based vessels, ring-footed vessels, flat-bottomed vessels, etc. artefact modelling has begun to be shaped with necks, lugs, and ring-feet, whereas Wannianxianrendong and Yuchanyan in the Dao County only have jars with a circular base and iron pots(*fu*) (Yan 2006), and only a few pottery sherds have been unearthed from the other sites. Thus, there is no specific evidence to prove that the Neolithic culture in the lower Yangtze Valley originated earlier or developed faster, in this case, the Shangshan culture is tentatively regarded as the Early Neolithic culture.

Different from sites discovered during the Paleolithic-to-Neolithic transition, the scale of the Shangshan site (covers an area of over 20,000 square metres) and the Xiaohuangshan site (covers an area of over 50,000 square metres) are much larger than sites in the earlier phase such as Ma'anshan in Yangyuan County, Hebei, Donghulin, and Zhuannian in Beijing, Xianrendong in Wannian County amid the middle Yangtze Valley and Yuchanyan in Dao County. The architectural remains at the Shangshan sites in the form of amphiprostyle foundations and moats are found. The house F1 found in the southern area under the 4th floor has clear structural units,[13] with a total length of 14 metres, a total width of approximately 6 metres, a floor area of nearly 90 m^2, and an arrangement in a northwest–southeast orientation. There are several pillar holes left after the decay of the wooden columns in the amphiprostyle foundation (Sheng et al. 2006). The pillar holes are in three rows, all with upright walls, circular bases, and small stones at the bottom. Besides, these pillar holes have a diameter from 27 to 50 centimetres and a depth from 70 to 90 centimetres, with small stone pieces at the bottom occasionally. Some excavators argue this is a Stilt House that is akin to that of the Hemudu site because the scale of the construction and the construction techniques are not found at any of the sites during the PNT.

Three types of ash pits were found at the Shangshan site (based on items): the first type of ash pit mostly contains more than one complete or near-complete pottery vessels, which might be associated with rituals from excavators' view; the second type contains pottery sherds, stone tools, charcoal debitages, bone debitages, etc., with steep, straight walls, a flat bottom, and regular shape, generally more than 70 cm in depth, which excavators regard as storage pits; As for the third type, the items are irregularly distributed and irregular shaped, but some pits are rich in excavated materials. Represented by layers 5 and 6 of Area B and associated with relic units, the first phase of the Xiaohuangshan site revealed a large number of possible storage pits, which are mostly square or circle in shape with straight and regular walls and flat bottoms with a diameter (side length) of around 1 metre. For example, BH12 has a diameter of 1.9 metres and a depth of about 1 metre, making the volume of nearly 3 cubic metres. The remains of a relatively regular distribution of pillar pit sites were found around it, and it is assumed that the storage pit might have been covered with canopy constructions. Part of the pits have rectangular upper openings and rounded lower foundations, with grinding slabs at the bottom; some other pits have small bottom openings resembling pockets; some other pits have steps, possibly facilitating storage and getting access to resources. As aforementioned, Early Neolithic cultural sites normally possess stockpiled facilities, and particularly well preserved and still usable artefatcs. This remaining feature is related to the lack of sedentism ability in the Early Neolithic, as prehistoric agricultural producers needed to periodically leave their sedentisms and exploit certain seasonal resources. Before leaving sedentisms, prehistory agricultural producers stockpiled artefacts, tools, and even food (such as the Cishan site) that were not easily portable, for use upon return. At the Shangshan site, all three types of pits are possible for storage; the situation is the same at the Xiaohuangshan site.

The Shangshan culture is represented by a relatively primitive lithic assemblage, including grinding slabs, handstones, spheroid, perforated tools, and a few polished adzes and chisels. The grinding slabs have large and concave curved grinding surface, and raw basis; the handstones are mostly in the form of chunks with multiple abrasions and a high proportion of fragments; natural pebbles with flat bases are also considered as a kind of grinding tool. Many specimens in the spheroid are highly rounded. Only a few tools were used for grinding and processing, such as adzes and chisels, as well as perforators, whetstones, and gravel tools, which were perforated by the two-sided faceting method. From the existing materials, it seems that small tiny flint tools, which are commonly discovered in the middle Yangtze Valley, do not exist in the Shangshan culture. The stone tools are mostly made from aleurolite, a material that is unsuitable for making choppers based on experimental studies (Chen and Chen 2012). The lithic assemblage at the Shangshan site is mainly used for food processing, including grinding, smashing, and a few fine woodworking (represented by adzes and chisels). However, stone tools are rarely used for hunting-related activities, such as cutting and scraping. In contrast, the Pengtoushan culture lacks stone grinding tools, instead, wooden tools are mainly used (e.g. wooden grinding slabs were found at the Bashidang site). On the one hand, different processing tools reflect different cultural habits; on the other hand, it might be related to the differences in resource utilisation. Generally speaking, the strength of the processing tool is consistent with the strength of the artefact being processed. For example, in the American Southwest, the grinding slabs used to process corn are very heavy, closely related to the hardness of the corn husk. In comparison, the grinding slabs of the Cishan culture used to process millets are much lighter and thinner. If people in the Cishan culture processed rice, a wooden pestle or a mortar would have been sufficient. The heavy grinding tools in the Shangshan culture indicate that people might have made greater use of nuts. Based on the ethnographic analogy, perforated stones could be used as weighing stones during dibbing in agriculture. In South China, tools, mostly made of organic materials including bamboo, wood, and bone, might have been used; for example, bone ploughs were used at the Hemudu Site; the wooden ploughs and shovels were used at the Bashidang site and possibly the Shangshan culture. The characteristics of lithic assemblage and those of the middle Yangtze Valley share more in common instead of divergences.

In the later phase of the Early Neolithic, most of the stone tools were ground in the Kuahuqiao culture (6200 to 5000 BC), with the dominance of adzes, along with axes, chisels, and arrowheads. Bone tool assemblages include equipment of farming, fishing, and hunting, and household utensils, such as bone ploughs, bone awls, arrowheads, needles, and daggers. During this period, potteries became thinner and more refined, seeing a significant improvement in craftsmanship compared to the Shangshan culture. However, the firing temperature remains at 750 to 850 °C, with little differences between the two cultures (Zhejiang Provincial Institute of Cultural Relics and Archaeology, Xiaoshan Museum 2004). Some wild accessible fruits, such as rhizomes, walnuts, walnuts, plums, apricots, pine cones, gorgonians, southern sour jujubes, and, notably, a pit full of acorns, were unearthed at the Kuahuqiao site (Bryant 2003; Piperno and Stothert 2003). These accessible fruits were also found at the Hemudu site (Zhejiang Provincial Institute of Cultural Relics and Archaeology 2003). Acorns have hard shells, thus, they are resistant to stockpiling. However, containing tannins, acorns have to be processed before accessible. Besides, filtering and precipitating are essential after filtering. Similar processes are required to process gravy into gravy powder, and lotus roots into lotus root powder, etc. These activities are associated with grinding tools and potteries, such as the flat-bottomed basin with an open-mouth that is commonly seen in the Shangshan culture (Fig. 6.1). However, this basin is not suitable for cooking, and on which no soot has been found by researchers, leading to the speculation of "stone-burning" cooking method but there is no burned stone unearthed from the site. This basin would have been most suitable for filtering sediment, and the intention of the polygonal-rim design might be associated with repeated filtering and fluid pouring. The discovery of various nuts at the Kuahuqiao site indirectly demonstrates that these resources might also have been used in the Shangshan culture, and the specific shapes of the various tools support this hypothesis. At the Kuahuqiao site, wooden mortars and grinding slabs, bone ploughs, wooden shovels, thousands of rice and grain, and possibly the bones from the earliest domestic pigs in the southern region were found. This evidence is similar to that found at the Bashidang site which belongs to the late phase of the Pengtoushan culture. In this case, it is reasonable to infer that the proportion of rice consumed in people's diets increased significantly during this period.

The early parts of this chapter have elaborated developmental sequences of the Neolithic of North China and the middle and lower of the Yangtze Valley. What does the developmental sequence of early cultures tell us? What does it tell us about the origins of agriculture? Exploring the unknown from the known could help us to build some solid standpoints, rather than conjectures. According to the cultural materials of the two regions from the Early Neolithic cultures, the main grain crops did not occupy a high proportion of people's diet until the Neolithic was fully established (or, based on some classification criteria, the Middle Neolithic). The paleo-diet analysis shows that the main grain crops only make up a quarter of people's diets. Undoubtedly, this figure was even lower in earlier times. Correspondingly, the main human habitats gradually expanded from valleys and basins to foothills, terraces, fluvial plains, and waterways. Similar to the lithic assemblage in the Upper Paleolithic, a

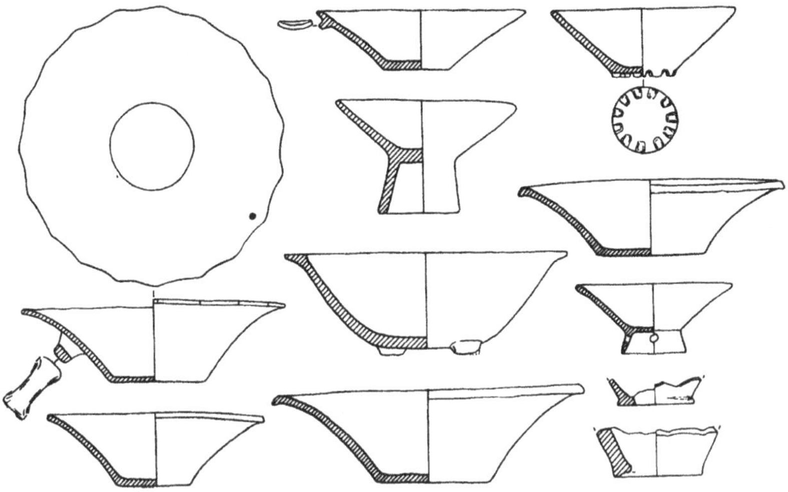

Fig. 6.1 Different types of open flat pans in the Shangshan culture

large number of chipped stone tools remain and the proportion of ground stone tool is gradually increasing in those of the Early Neolithic, particularly in the middle and lower Yangtze Valley. It is perhaps because of this apparent transitional nature that modern scholars oppose Childe's term "Neolithic Revolution", which means that agriculture occurred gradually and developed slowly. Thus, the "Neolithic Revolution" has gradually been replaced by the term "Neolithic transition" or "transition to agriculture".

While emphasising the gradual and transitional nature of agriculture, we also need to emphasise the "emergent" nature of the Early Neolithic culture. It is true that agriculture developed gradually, but also periodically with a rapid rate of development between them. Just like a person grows up day by day, there are several periods of rapid growth in between. The Early Neolithic is such a period: the archaeological cultural sequences sprang up; the number and scale of sites increased rapidly and their structures became more complex (including house sites, moats, dykes, storage pits, tombs, etc.); a complete pottery system has formed; and ground stone tools began to play an irreplaceable role. These characters all marked significant differences between the UP and the PNT (see in the next section). Because of its obvious differences, it is difficult to interpret the

Early Neolithic culture as an entirely subjective archaeological determination that the archaeological material supports the existence of an "Early Neolithic" phase with artefactive validity.

In the Early Neolithic cultures, another notable phenomenon is the extraordinary emphasis on storage facilities, a large number of which could be seen in the Cishan and Shangshan cultures. Why was such an emphasis given to storage? Humans had been living in temperate and frigid zones for tens of thousands of years, in this case, storage would not have been considered a mere product of the Neolithic. For hunter-gatherers in the temperate and frigid zones, storages were a necessity. In a state of nature, hunter-gatherers stocked their food in nature and took it as they went (hunting or gathering). However, not all foods are equally conducive to "storage" in their natural state, but animals are exceptional. Being alive is the best storage, better than any other type of frozen facility. In contrast, plant foods, such as fresh fruits, are different, as they would go rotten if not being consumed in time. Similarly, nuts are not easy to store in their natural state because animals would compete for them. In addition, nuts are prone to go rotten if not dried, thus, collecting and processing are essential before storage. Plants such as wild millets and rice have a large range of distribution and can be harvested in large quantities once a year, making them ideal for stockpiling and easy to control compared to nut trees. Nevertheless, hunter-gatherers may be able to make use of animal resources, a natural "stockpile". In winter, plant food is scarce, but it is a good time to hunt, since the falling leaves make the landscape open and fatten animals. But if the food resources of animals are scarce in winter, then plant food has to be stocked by humans. The widespread faunal extinction at the end of the Pleistocene and the extremely broken animal bones found at the site (which show full exploitation) suggest that animals serving as food resources had already become scarce during the Paleolithic-to-Neolithic transition. The development of storage is also linked to sedentism (or reduced mobility) because extensive storage is a post-sedentism adaptation. After the sedentism of groups, access to food would be largely reduced, thus, storage becomes necessary. In this case, the development of reserves is also indicative of an increased capacity for sedentism.

However, a full sedentism had not yet been achieved in the Early Neolithic. Studies related to the abandonment process in the Early Neolithic suggest that the remains were mostly retained in their original locations, such as intact artefacts and even food. This phenomenon reflects the original intention to return to the sedentism, that is, the ancient people would also periodically leave the sedentism to exploit certain seasonal resources. The lithic assemblage from the numerous storage facilities in Cishan and Shangshan sites reflects people's sedentism and periodic leave. During this period, some of the lithic assemblage also retain tools that are suitable for mobility and hunting, typically the microblade technology products excavated from the Xinglongwa culture and the flint stone artefacts from the Pengtoushan culture. The lithic assemblage suggest that hunting and mobility were considerably important for people to maintain substance at that time but agriculture has not fully and rapidly replaced other means of livelihood. It is unrealistic to assume that the beginning of the Neolithic must have marked a complete livelihoods replacement of the Paleolithic. Besides, it is equally wrong to take the remnants as a sign to indicate the beginning of the Neolithic; or even questioning the importance of agriculture in ancient people's livelihoods, as the view ignores the importance of agriculture as an "order parameter" for social conditionality and development. However, the earliest agriculture might have been insignificant in terms of subsistence, but it was able to induce changes in other cultural and social factors, forming the "emergence" cultural character of the Early Neolithic.

6.3 Exploring the Paleolithic-Neolithic transition

What exactly were the changes from the Paleolithic to the Neolithic in China? How did they happen? And why did they happen? Many scholars have engaged in exploring these unanswered questions. Undoubtedly, prehistoric culture landscapes have dramatically changed from the Paleolithic to the Neolithic, however, did these conversions occur suddenly or gradually? When did the typical features of the Neolithic emerge? Did the typical features appear simultaneously or chronologically? Did all cultures of the Upper Paleolithic move towards the Neolithic? If not, where was the divergent point? What significance do the typical features of the Neolithic period have in prehistoric cultural systems? Why did the typical features of the Neolithic appear? These related questions have fascinated archaeologists and scholars who are interested in the cultural changes of prehistoric mankind.

6.3.1 Why Is the Transition Period Called "Paleolithic-to-Neolithic Transition" rather than the "Mesolithic"?

In the previous part, I have argued strongly for maintaining the original chronological range of the "Early Neolithic", which conforms to published material. Thus, I have separately defined the period between the pre-Early Neolithic and the Upper Paleolithic as the "Paleolithic-to-Neolithic transition". Normally, this transition period is referred to as the "Mesolithic". The validity of "Mesolithic" has been very much discussed by scholars in both China and abroad (Chen 1995, 2000; Chen 1990; Price 1987, pp. 225–306). The views could be categorised as follows: some scholars argue that the Mesolithic was a widespread transitional phase between the Paleolithic and the Neolithic periods; thus, there is no need to elaborate on this range of ages; while, others view that the Mesolithic refers to a particular type of adaptation, with the predominance of aquatic resources utilisation. As a territorial discipline, archaeology requires evidence when a new concept is proposed. However, even African prehistoric archaeologists did not accept the most widely adopted divisions of the Paleolithic and Neolithic. Under the colonial and post-colonial contexts, non-Western scholars are forced or actively pursue conformity with the dominant culture (i.e. Euro-American), which is one of the key features in academia, while was the same in the archaeological research but only constitutes a small part. The concept of the Mesolithic originated in Europe, more specifically, in the Western European region. During the formation of the Mesolithic concept, long-term archaeological practices (archaeological discovery and research) have given this concept a specific meaning. Through the interpretations from different research paradigms, Mesolithic has evolved from a concept of cultural evolutionary phases in the cultural-historical archaeology paradigm to a specific mode of adaptation within the paradigm of process archaeology. The processual archaeology paradigm challenged the cultural contexts of colonialism and postcolonialism, as well as the connotations of scientism (the idea that the universal laws of cultural development exist, and everything can be as precise as natural science).

After reflecting on the ideological backgrounds of the three different paradigms, I tend to avoid the concept of "Mesolithic" in the study of Chinese agricultural origins. Not only because of the criticism of post-process archaeology but also due to the significance of the concept as a specific adaptation based on considerations of the historical and regional specificity of the study. The definition of material specificity in the Mesolithic context is often overlooked by European archaeologists. Compared to the core area of Western Asia, Western Europe has a long transitional history from the Paleolithic to the Neolithic. The unique geographical features of Western Europe construct the basis of the concept of Mesolithic. From this perspective, the Mesolithic also contains regionalism of materials, even though the emphasis on the universal property of the Mesolithic cultural evolution is highlighted.

Denying the applicability of the Mesolithic in the question of the origins of Chinese agricultural does not mean that the concept of "Mesolithic" is useless. The Mesolithic has an irreplaceable role in the study of the prehistoric changes in the prehistoric cultures of southern and northeastern China (see Chapters 7 and 8). The Paleolithic-to-Neolithic transition in Southern and Northeastern China lasted for a long time, and the adaptation patterns are very comparable to those of the Mesolithic in Western Europe, in that both of these two cultures emphasise aquatic resource exploitation. Thus, we can learn from the long research history and rich research experiences and outcomes from the Western Europe. In short, a critical assimilation of Western archaeology is the basic attitude of Chinese archaeology.

6.3.2 *Date and Materials: Why Did the Paleolithic-to-Neolithic Transition in China Begin 15,000 Years Ago?*

The end of the Paleolithic-to-Neolithic transition phase is the beginning of the Early Neolithic, and the emergence of a range of Neolithic cultures proves the beginning of the New Age. In contrast, the beginning of the Paleolithic-to-Neolithic transition is not clear. Based on the available archaeological material and paleoenvironment changes, the incipient date of the Paleolithic-to-Neolithic transition was set at ca. 15,000 BP, when the LGM is over with the rise of sea level and the migration of plants and animals. In terms of archaeological materials, 15,000 BP marks a vital phase when the microblade technique was at its peak as almost all sites in North China were dominated by this process (Zhang 1991); however, at the same time, it marks the decline of the microblade technique in North China. Besides, DNA research proved that 15,000 BP also represents the date of dog domestication somewhere in East Asia (Savolainen et al. 2002). In other words, domestication, the process of adapting wild plants and animals for human use, has been mastered by human beings. Then humans just need to apply the domestication method to other species. At the same time, as one of the basic elements of agriculture, the domestication of species has occurred. 15,000 BP is also the time when other typical features of the Neolithic period emerge, such as pottery,[14] especially ceramic vessels. Within China, the earliest pottery was found at the Miaoyan site in Guilin; both dating data are at ca. 15,000 BP; while, the 14C of the pottery from the same layer at the Wannian Xianrendong site is even earlier than 15,000 BP, but this figure is viewed as inaccurate by scholars as it was early. Three dating data of the pottery sherd excavated from the Yuchangyan site Data are also around 15,000 BP (Yuan 2000).

In spite of pottery, grinding slabs, handstones, and adze-shaped tools for repairing and sharpening edges have been mainly found in North China. An adze-shaped tool is excavated from the Mengjiaquan site in Yutian, Hebei Province. The adze-shaped tool is polished on one side with a smooth surface but subtle longitudinal scuff marks. Accordingly the Mengjiaquan site is dated to 24,000 to 17,000 BP (measured date) (Xie 2000). A grinding stone shovel and a grinding stone are excavated at the Longwangshan site in Yichuan, Shanxi Province. The excavators claim that the absolute date of the incipient site should be ca. 20,000 to 15,000 BP (CA, CASS, Shaanxi Provincial Institute of Archaeology 2007). However, no dating data has been available from 14 C, so the dating issue is still suspicious for the moment. Site at Locality S9 of Shizitan in Shanxi Province is excavated "in a Layer of 0.5 metres in thickness with two pieces each of used grinding slabs and handstones, and one piece each of used grinding stone and pigment, distributed within a 2 metres diameter position. With a slightly earlier dating date, Locality S1 at the Shizitan site (20,000–10,000 BP) also discovers two grinding slabs." In 2002, the dating date of 14C for the S9 is only 8340 ± 130 BP (the selected specimen is a S9DH 01 charcoal chunk from the site with fire remains) (Shizitan Archaeological Team 2010), which is not ancient. Adze-shaped tools and grinding slabs are also excavated from the Xiachuan site, dating to between 24,000 BP and 16,000 BP (Wang et al. 1978). The dating of the Xiachuan site is still controversial. In general, the date of the emergence of groundstone tools is not well-established, and is more likely to predate the emergence of pottery. A date of ca. 15,000 BP would be a medium estimate.

The Paleolithic-to-Neolithic transition has been the subject of archeological study only for the last twenty years because of the lack of relative materials on it, or, if they have not been recognised. Later, the outline of this transition period has become clearer with the increasing number of new discoveries, especially archaeologists could directly understand the ages and characteristics of plant remains and pottery with the usage of paleoethnobotany and AMS dating. In recent years, a few important sites have been discovered, such as Lijiagou (School of Archaeology and Museology, Peking University, Zhengzhou Municipal Institute of Cultural Relics and Archaeology 2011; Zhengzhou Municipal Institute of Cultural Relics and Archaeology, School of Archaeology and Museology, Peking University 2011), Shizitan (Shizitan Archaeological Team 2010; Zhao & Huo 2007), Bianbiandong (Sun 2007), etc.; along with some previous materials were collected and published, for instance, the Nanzhuangtou site (Baoding Cultural Heritage Management Office et al. 1992; Hebei Provincial Institute of Cultural Relics, et al. 2010), leading to a significant increase in the total amount of materials (Table 6.3). Notably, as the Paleolithic-to-Neolithic transition has become an important issue in academia, together with advanced research methodologies have been utilised, the level of sophistication of newly discovered materials has increased considerably compared to the previous studies, which provides essential conditions for an in-depth analysis of the characters of this period. The materials discovered in the Upper Paleolithic and Early Neolithic periods are also increasing, notably, some sites have revealed a continuous cultural sequence of three epochs, such as Dadiwan (Zhang et al. 2010). However, compared to the Early Neolithic and the Upper Paleolithic, the archaeological items excavated in the Paleolithic-to-Neolithic transition are still sparse and fragmentary. The limited material remains one of the main constraints on research.

Take a comparison of Table 6.3 and Table 6.4; a number of Neolithic features have been identified in archaeological sites located in North China. Some other well-dated sites are not typical during the Paleolithic-to-Neolithic transition because of their age and level of sophistication. For example, the earliest pottery, grinding slabs, and grinding balls from the Hanshui basin were reportedly found in the upper cultural layer (layers 2 and 3) at the Kengnan site in Xichuan, Henan Province (Song et al. 2011). Through analysing the characteristics of pottery sherd, excavators suggest that the date of pottery excavated at the Kengnan site is in the Paleolithic-to-Neolithic transition, which might be slightly later than that at the Lijiagou site. Notwithstanding, due to the lack of 14C dating evidence, the specific period cannot be determined yet. However, there are many sites that possess certain features of the Neolithic that

Table 6.3 Typical sites in the paleolithic-to-neolithic transition

Date (BP)	*Period*	*Typical sites*			
		Middle and Lower Yangtze Valley		*North China*	
ca. 15,000–85,00	PNT	Yuchanyan, Xianrendong, Diaotonghuan	Zengpiyan, Miaoyan, Dayan	Zhuannian, Donghulin, Nanzhuangtou, Yujiagou, Ma'anshan, Bianbiandong, Lijiagou, Dadiwan	Shizitan, Xiachuan, Xueguan
Before 15,000	LUP	Lower layer of Bashidang, Shiligang, Yan'erdong, etc		Shizitan, Xiachuan, Tingsijian, Lingjing, Dagang, etc	

are beyond the typical sites during the Paleolithic-to-Neolithic transition, providing significance for us to explore the different origins of the different features and their cultural adaptation.

I have listed the Guangxi sites of Zengpiyan, Miaoyan, and Dayan in the middle and lower Yangtze Valley. Geographically, the sites listed above belong to the Basin of the Pearl River, Southwest China. However, there is no major geographical barrier between the Xiangjiang River basin and the Li River basin, which belong to a tributary of the Yangtze River and the Pearl River respectively. Thus, this geographical feature enabled people in the Qin Dynasty to dig Lingqu and bridge the Xiangjiang River basin and the Li River basin. The Yuchanyan site in Dao County, which belongs to the middle and lower Yangtze Valley, is almost on the same latitude as the Guangxi sites, with a linear distance of only 120 kilometres. Thus, there are no major differences in physical settings between these sites. From 15,000 to 8500 BP, the incipient temperatures of the Paleolithic-to-Neolithic transition were lower than they are at present. By 12,000 BP, the climate in this area was still the temperate forest one, which was equivalent to that of North China ca. 9000 BP (which was very similar to the present climate). Thus, these sites will be taken into consideration in the study of the agricultural origins in the Middle and Lower Yangtze Valley, as well as in Southern China. Notably, the prehistoric cultural development of the Pearl River basin diverged from that of the Middle and Lower of the Yangtze Valley after the Holocene. In this

Table 6.4 New features of the paleolithic-to-neolithic transition

		North China							
		Nanchuntou	*Zhuannian*	*Donghulin*	*Lijiagou*	*Bianbiandong*	*Kengnan*	*Yujiagou*	*Ma'anshan*
Utensils	Pottery vessels	√	√	√	√	√	√	√	√
	grinding slabs	√	√	√	√	√	√		
	Stone handstones	√	√	√		√			
	Adze-shaped tools				√			√	√
	Stone vessels		√						
	Ground stone tools		√	√	√				
	Diversity		√	√	√	√	√		√
Site Structures	Hearth	√		√	?	√			√
	Diverse remains	√	√	√	√	√			√
	architectural remains	?		?	?	?			
	graves			√		?			
	Central site	√		√	√	√			√
	Ash moats	√		√					
	Ash pit	√				√			

(continued)

Table 6.4 (continued)

		North China							
		Nanchuntou	*Zhuannian*	*Donghulin*	*Lijiagou*	*Bianbiandong*	*Kengnan*	*Yujiagou*	*Ma'anshan*
Animals and plants remains	millet			?					
	broomcorn			?					
	rice								
	dog	√							
	wild boar	√							
	Undomesticated chicken								

		North China						
		Hutouliang	*Mengjiaquan*	*Dianjitan*	*Xueguan*	*Shizitan*	*Xiachuan*	*Longwangchan*
Utensils	Pottery vessels	?		×				
	grinding slabs					√		√
	Stone handstones					√		√
	Adze-shaped tools	√		√			√	
	Stone vessels							
	Ground stone tools		√		?			√
	Diversity		√			√	√	√
Site Structures	Hearth							?
	Diverse remains							√

		North China						
		Hutouliang	*Mengjiaquan*	*Dianjitan*	*Xueguan*	*Shizitan*	*Xiachuan*	*Longwangchan*
	architectural remains							
	graves							
	Central site							
	Ash moats							
	Ash pit							
Animals and plants remains	millet							
	broomcorn							
	rice							
	dog							
	wild boar							
	Undomesticated chicken							

		*Middle and Lower Yangtze Valley**				
		Yuchanyan	*Huadang*	*Zhujiagang*	*Zhuma*	*Xianrendong*
Utensils	Pottery vessels	√	√	√		√
	grinding slabs					
	Stone handstones					
	Adze-shaped tools					
	Stone vessels					
	Ground stone tools					
	Diversity		√	√		
Site Structures	Hearth					
	Diverse remains					
	architectural remains					
	graves					

(continued)

Table 6.4 (continued)

		*Middle and Lower Yangtze Valley**				
		Yuchanyan	*Huadang*	*Zhujiagang*	*Zhuma*	*Xianrendong*
	Central site					
	Ash moats					
	Ash pit					
Animals and plants remains	millet					
	broomcorn					
	rice	√			√	√
	dog					
	wild boar					
	Undomesticated chicken					

		*Middle and Lower Yangtze Valley**			
		Shenxiandong	*Zengpiyan*	*Miaoyan*	*Dayan*
Utensils	Pottery vessels	√	√	√	√
	grinding slabs				
	Stone handstones				
	Adze-shaped tools		√		
	Stone vessels		√		
	Ground stone tools				
	Diversity		√		
Site Structures	Hearth		?		
	Diverse remains		√		
	architectural remains				
	graves		√		
	Central site		?		
	Ash moats				

		*Middle and Lower Yangtze Valley**			
		Shenxiandong	*Zengpiyan*	*Miaoyan*	*Dayan*
	Ash pit				
Animals and plants remains	millet				
	broomcorn				
	rice				
	dog		?		
	wild boar	√	√	√	
	Undomesticated chicken		√		

Note *The Middle and Lower Yangtze Valley include several sites in northern Guangxi Province for comparison; the site of Shizitan is represented by Locality of 9 and 29; √ means yes; × means no; ? indicates that relative remains might exist as they have been found at the site, but cannot be ensured. The site of Yuchanyan includes Sanjiaoyan; the Xianrendong site in Wannian County includes the nearby Diaotonghuan site case, the Guidongbei site and other typical ones will be examined in the context of the Southern area's prehistoric cultural changes.

6.3.3 *Generalisation and Analysis of New Features: What Exactly Changed During the Paleolithic-to-Neolithic Transition?*

(1) The choosing of features

A common issue in current studies is purely from the morphological and spatiotemporal characteristics of archaeological evidence, instead of being based on theoretical studies of hunter-gatherers; either, scholars ground on the theoretical studies of hunter-gatherers to discuss the cultural adaptations (that is, agriculture is a cultural adaptation, rather than the sum of artefactual characteristics). However, the standpoint often depends on the common sense of contemporary life, applying the knowledge of contemporary society to the Paleolithic hunter-gatherers. Undoubtedly, this research approach is problematic. Therefore, it is impossible to understand the mechanisms of human cultural adaptation in the Paleolithic without the study of hunter-gatherers. The archaeological material itself cannot speak nor automatically reveal its meanings. The generalisation of spatiotemporal and morphological features cannot directly reveal the behaviours of ancient people. In this case, the process of logical reasoning is essential from static material to dynamic behavior. However, any logical reasoning requires certain theoretical frameworks. The cultural and behavioural ecology of hunter-gatherers would provide such a theoretical framework.

The Paleolithic-to-Neolithic transition represents the transition from hunter-gatherer to agricultural producer. The features needed to be generalised and analysed would be based on this period. For any artefact, there are an infinite number of its features. As archaeologists, we can only choose features that are relevant to the main motif of our research. Therefore, the characteristic selection should reflect the distinctions between hunter-gatherers and agricultural producers, possessing the significance of cultural adaptation. On the other hand, the features must be visible in the archaeological evidence. The former is primary. If an archaeological researcher is unaware of the cultural significance of the features, then the researcher is merely assembling material without knowing the implications. Thus, before summarising and analysing materials, we must be clear about the meaning of the elements we are studying.

The greatest difference between hunter-gatherers and agricultural producers is that the former are mobile hunters and the latter are food producers.[15] The most striking difference between these two groups lies in mobility (this has been elaborated on in Chapter 1). Hunter-gatherers and agricultural producers use different types of tools, techniques, resources, and ecological spaces, determining at least three different aspects of the archaeological materials: (1) the artefact assemblage, including the new artefacts usage, such as pottery and other new tools (e.g. tools for grinding grains, ground stone tool, etc.). In addition, certain features of the tool assemblage would be emphasised; (2) the existence of different site structural forms. This difference is not only in the site distributions but also in the internal structures. Mobile hunter-gatherers and sedentary food producers differ in the requirements of site selections and usages; (Sherrat 1996) the differences in the resources utilisation: hunter-gatherers tend to use a greater variety of resources but agricultural producers use domesticated species.

This study focuses on the Paleolithic-to-Neolithic transition, that is, from the life of hunter-gatherers to agricultural producers. Therefore, it must contain features of two cultural adaptations. In other words, during the transitional period, some typical features of the Paleolithic declined but those of the Neolithic grew. As for species utilisation, the undergoing domesticated species would be seen during this period. However, it is crucial to notice the emergence of new elements in spite of the transitional features.

(2) Summary of the New features

Table 6.4 summarises the new features exhibited by the different sites based on three aspects as follows. It is easy to see that some so-called new elements are not only found in typical Neolithic transitional sites but also in some Upper Paleolithic sites. This phenomenon reflects that the incipient Paleolithic-to-Neolithic transition is more complicated than we understood, and our research related to certain atypical sites needs to be strengthened. It would also be significant if we could not determine that there are no new discoveries, that is, there is already a differentiation between different sites even in the same area during the Paleolithic-to-Neolithic transition. The emergence of new features is like a spark that gradually spreads and then develops into a prairie fire.

Correspondingly, the new features appear in different periods with different degrees of change, reflecting the modes of cultural adaptation. For example, compared to the emergence of certain new techniques, major changes in the structure of a site would be more profound and its appearance necessarily later than those less complicated factors. On the other hand, the time taken for changes in different features to manifest themselves in archaeological material varies. Taking domesticated species as an example, the act of domestication might happen very early, however, the recognisable domesticated features must have appeared after the act of domestication. Time length is related to the domesticated intensity because conscious and sustained intensive utilisation might result in the species' domestication in a shorter time period; while, unconscious, non-sustained and less intensive utilisation would take a longer time for the species to show domestication characteristics. Thus, the different domesticated intensities would lead to the fact that features presented at the same time are not necessarily simultaneous, and features presented at different times might be simultaneous. Thus, specific features need to be analysed case-by-case. On the contrary, merely starting from the chronological data in an archaeological sequence will not work.

It is noteworthy that there are differences as well as commonalities of new features between North China and the Middle and Lower Yangtze Valley. For example, polished tools appear very early in North China while late in the Middle and Lower Yangtze Valley (based on the existing materials). Besides, the sites found in the Middle and Lower Yangtze Valley are cave sites, which are different from those in North China. On the other hand, in both areas, the considerable synchronicity could be seen in the emergence of pottery, domesticated species, etc. Among the differences and commonalities, some differences do not necessarily reflect differences. For example, typical tools in North China, such as the grinding slabs and grinding handstones, have not been found in the Middle and Lower Yangtze Valley. This absence cannot reflect that there have been no grain processing tools in the Middle and Lower Yangtze Valley. In the prehistoric period of this area, people processed rice by using wooden grinding slab and mortars, which are difficult to preserve (unless the sites are situated in a swampy environment, as in the case of Bashidang, Kuahaiqiao, and the Hemudu sites could preserve a large amount of organic matter) compared to grinding slabs. Similarly, some commonalities do not necessarily reflect the same adaptations. For example, pottery in North China might have been used to cook wild or domesticated grains; while conditions in the Middle and Lower Yangtze Valley might have been different, with the possibility of using pottery to cook plant roots, aquatic artefacts, and rice. That is to say, the presence of pottery is not necessarily used for cooking domesticated grains.

(3) Analysis of the artefact and Lithic Assemblages

The analysis of artefact assemblages can be broadly divided into three levels: firstly, the emergence of pottery; secondly, other new types of artefacts; and thirdly, changes in the characteristics of chipped stone tools artefact assemblages. These three levels of characteristics still reflect mobility, the most central variable of the cultural adaptation in prehistoric human groups. Ancient people had to go to nature for food seeking during a time when food could not be produced. Without mobility, people could not go for food hunting. However, only a very few environmental zones in nature (such as the Northwestern European region passed by North Atlantic Current, the Pacific coast passed by Humboldt Current, and the Japanese archipelago passed by Kuroshio Current) could get access to abundant food resources, allowing hunter-gatherers to obtain necessities for life without long-distance mobility. The reduction and even loss of mobility forced hunter-gatherers to transfer to agricultural producers.

The emergence of pottery can be seen as a "thumb rule" for judging the dramatic reduction in mobility, as pottery was prone to damage during transportation; besides, the production process of making pottery is expensive, because pottery is not a ready-to-use utensil, requiring processes such as clay handling, kneading, firing, etc. Even though pottery systematic features had been constructed in the Early Neolithic period, it was still common to see the signs of repairing holes in broken pottery, implicating how much the ancient people cherished pottery. Pottery, or pottery vessels, were rarely produced for groups with high mobility.[16] In Chinese archaeological research, the presence of pottery is also seen as the most important feature of the Paleolithic-to-Neolithic transition. The typical sites in both North China and the Middle and Lower Yangtze Valley have been noticed for the excavation of pottery. Although the appearance of pottery does not necessarily link to agriculture, for example, agriculture in the Japanese archipelago was not established until the Han Dynasty in China, instead of the emergence of pottery dated back to 12,800 BP in the Jomon period of Japan (Aikens 1995). Early pottery from 10,000 BP has been discovered in the Russian Far East where there was no agriculture (Vandiver et al. 1989). Likewise, pottery appeared in the Late Mesolithic in Europe where there was no agriculture either. Even though we have illustrated exceptions, the presence of pottery is an appropriate indicator of mobility reduction. The lifeway of sedentism occurred in both the Jomon period and the European Mesolithic Pottery period.

The loss of mobility will inevitably lead to cultural adaptation among hunter-gatherers, except for aquatic resource utilisers who could have been settled without agriculture. In areas where aquatic resources are not available, the emergence of pottery is possibly associated with plant utilisation. Except for food roasting, hunting groups could process meat without pottery, as they could use animal stomachs to make soup and boil water, just like Oroqen hunters (Zhao 1990), or they could use the skins of animals. Nevertheless, when making use of seeds of wild millets and rice, or nuts like acorns, pottery becomes indispensable. During the food processing procedures, pottery is irreplaceable as it could be repeatedly used for cooking and filtering sediments (for example, the flat-bottomed jars with open-mouth of the Shangshan culture). There are numerous explanations for the origins of pottery. Prudence Rice has systematically summarised the existing theories (Rice 1999), which either from functional theory or from Hayden's "feasting model theory" that views the pottery originality is used to display (Hayden 1998). All these theories acclaim that the origins of pottery are linked to the intensification of resource utilisation. There are diverse reasons for the intensification of resource utilisation, ranging from socio-economic reasons (differentiation of social inequality, competition for higher social status, and prestige within groups) to environmental change, overpopulation, etc. Based on the analysis of European materials, pottery was originally designed for display and then moved on to utility. In the Chinese context, there are no pottery artefacts in the Upper Paleolithic. If pottery was initially for display function, burials would have been accompanied by the pottery to realise this purpose. However, according to the complete burial items discovered at the Donghulin site during the Paleolithic-to-Neolithic transition, burials were accompanied merely by ornaments, small grinding stone axes, etc., without pottery. Besides, burial accompanied by pottery was a custom that began in the Early Neolithic. In this case, at least based on the evidence from North China, the origin of pottery is linked to utility and then turns to non-utility aspects.

Interestingly, no pottery was unearthed at the Shizitan site, nor at the Xiachuan site where grinding slabs and handstones were found. The paucity of Early Neolithic culture in Shanxi Province has already been summarised in the previous section. It seems that the development path in Shanxi Province is different from that of the piedmont zone in North China (including the terraces, hills, and fluvial plain of the Yan, Taihang, and Yuxi mountains) during the Paleolithic-to-Neolithic transition. However, we cannot draw a conclusion merely based on the presence or absence of pottery, but we will further discuss the existence of this divergence in the Paleolithic-to-Neolithic transition in North China by examining liths, site structure, faunal and plant remains.

Except for pottery vessels, stone vessels have been excavated at the Zhuannian site (Yu 1998). A similar find has been discovered at the Baiyinchanghan site of the Xinglongwa culture in the Early Neolithic (Inner Mongolia Autonomous Region Institute of Cultural Relics and Archaeology 2004). Stone vessels are sturdier than pottery vessels, requiring more laborious inputs. Excavated at the Zhuannian site, the stone vessels, which are made of less hard material, are larger than earthen bowls used for pigment grinding. The development of containers facilitates starchy food processing. In addition, the sturdy property of stone vessels is easy to transport, which is consistent with a means of livelihood in the Paleolithic-to-Neolithic transition where intensifying the utilisation of plants while maintaining a certain degree of mobility. Requiring investing intensive labours, intensification of utilisation means the processing of unavailable food resources into accessible ones through patient collection, drying, grinding, filtration, and cooking. Intensification of utilisation is often linked to broad-spectrum food utilisation, which means consuming something that was rarely consumed before. Even though hunter-gatherers know many accessible food resource types, only a relatively limited range of them could be used. When hunter-gatherers encountered a food crisis, they could use these food resources as "lifesavers" in agricultural society. On a broad-spectrum basis, intensification means the usage of food resources that are available in large quantities and easily accessible, but require a great deal of processing. Paradoxically, intensification and the broad spectrum are contradictory: more labour spent on food processing means less time spent on food searching, and this would lead to the reduction of mobility. It seems that the food is available in a large range; however, except for the resources being intensified utilisation, the proportion of food resources has fallen dramatically, especially large granivores could be hunted. For example, once a deer could be caught every few days, but now it would take every few months to catch one. Even though deer are still included in food resources, their practical significance has been greatly reduced. The development of pottery and stone vessels indicates a reduction in mobility, while, the deeper reason for the lower mobility is related to the intensification of resource utilisation.

During the Paleolithic-to-Neolithic transition, pottery grew from nothing to a gradually abundant quantity and category, until it formed a system of Early Neolithic pottery assemblages. We could see the gradual development of the culture from the rare changes of sequences, containing three successive cultural layers dating from 10,500 to 8600 BP, at the Lijiagou site in Xinmi, Henan Province. The lower layer is dominated by the remains of a microblade technology, with partially polished stone adzes and plain sand-tempered pottery flakes; the middle layer contains coarse sand-tempered pottery and grinding slabs with generally embossed ornaments; and the upper one is the Peiligang cultural layer. The charcoal sample from the lower layer is dated from 10,500 BP to 10,300 BP; the middle layer is dated from 10,000 BP to 10,300 BP. In the lower layer, two pieces of pottery sherds were found. In the middle layer, the number of pottery sherds discovered increased significantly to over 200 after two excavations. The texture of pottery sherds is fairly hard, indicating a higher firing temperature. There are several straight-sided cylindrical pottery with different pottery rims. The vast majority of the pottery sherds are decorated with motifs, including interrupted rope-like motifs, a combination of rope-like and interrupted rope-like motifs, and incised floral motifs. The richness of the motifs might often be associated with female pottery making[17] (Chen and Chen 2012), at the same time, suggesting that males were less involved in plant collection, exploitation, and utilisation, as men might engage in predominantly hunting activities.

The widespread distribution of pottery in the Middle and Lower Yangtze Valley and the adjacent Lingnan area marks a striking feature in the Paleolithic-to-Neolithic transition, especially under the circumstance of the absence of other Neolithic features, making pottery almost the only symbol. Secondly, based on the existing materials, the firing temperatures of the early unearthed pottery are very low, for example, the firing temperature of the pottery pieces excavated at the Zengpiyan site does not exceed 250 °C (CA, CASS 2003), which is much lower than those of 750 ~ 850 °C at the Shangshan culture (which proves that the Shangshan culture does not belong to the Neolithic). Similarly, the loose pottery pieces explored from the Yuchanyan site seem to be made of clay dough and thus easy to break (Yuan 2000). Another feature is the very early date of the pottery from both areas, which makes many archaeologists hard to accept. 34 data of pottery pieces discovered at the Xianrendong and Diaotonghuan sites in Wannian were measured through AMS method by Chinese and American archeologists, with several dating dates from 19,780 ± 360 BP to 15,050 ± 60 BP; and from the upper layers of the Xianrendong site (3B1), the latest dating data could be dated to 12,430 ± 80 BP. The pottery sherds from Yuchanyan and Miaoyan sites also date to ca. 15,000 BP; and the mollusk shells at the Liyuzui site in Dalong County, Liuzhou date to over 20,000 BP. It is uncertain whether the age of pottery in the Middle and Lower Yangtze Valley and adjacent Lingnan area is actually earlier than that in North China. Even though pottery is a significant indicator, it is not the only index. Thus, even though the answer is yes, we cannot view plant intensification in the Middle and Lower Yangtze Valley and Lingnan area predating North China merely based on the pottery.

Plant grinding tools are directly associated with the intensification of plant resource utilisation, including grinding slabs and handstones. Undoubtedly, compared to eating plants directly, grinding and processing plants is more complicated, time-consuming, and labour-intensive, as it requires labourers to make tools and grind plants for further operation. Thus, unless hunter-gatherers had to, there was no reason for them to choose grinding and processing plants, the more costly method. The emergence of grinding slabs and handstones represents one of the signs of intensification. Among all the sites of the Paleolithic-to-Neolithic transition, these plant grinding tools are only found in North China, especially at the Donghulin site where the grinding slabs are generally oval in surface; there are two types of handstones, one with a rounded square profile and the other with a rounded profile, notably, the latter one is more delicate. Grinding slabs and handstones have been found at the sites of Nanzhuangtou, Zhuannian, Bianbiandong, Lijiagou (no handstones are mentioned in existing materials), Kengnan,[18] Xiachuan, Shizitan, and Longwangchan in the west of the Taihang Mountains. Marking a distinction from grinding slabs in the Neolithic (Wang et al. 1978), three grinding slabs in the shape of a disc due to multiple rotational grinding in the middle have been discovered at the Xiachuan site. The grinding slabs are more akin to tools for processing pigments and nuts. grinding slabs and handstones were excavated from layer 4 of Locality S9 at the Shizitan site, however, layer 5 was missing. Grinding slabs, pigment chunks, mussel shells, and ostrich egg shells perforated ornaments were found in cultural layer 4 at Locality S29, next to the fire remains." grinding slabs have been found in strata at the Shizitan site 20,000 BP, and researchers have extracted plant starch remains from the surfaces of grinding slabs" (Anonymous 2011). The result remains unknown, but this has already indicated that grinding slabs might have been used to process starchy plants. If the dating and strata of the Shizitan site are correct, the intensification of plant resource utilisation would begin in the Late Upper Paleolithic and as antique as the earliest pottery from the Middle and Lower Yangtze Valley and Lingnan area.

As aforementioned, intensification is associated with mobility reduction, which means hunter-gatherers settle in one place for a long time and find it worth investing in shelter building and durable tools. On the contrary, short-term stays are not worthwhile for such investments, as immediate disposal of shelter and tools would waste labour invested. Ground stone tools are more durable than chipped stone tools, but the former is more labour-intensive. For example, durable tools, such as grinding slabs and handstones, would be used repeatedly, thus being called "site furniture". At the same time, in order to build a relatively permanent dwelling, firm building materials are needed, such as large logs and stones. Thus, more durable felling and excavating tools are necessary under this circumstance. Until now, only a few grinding tools have been found, such as a small partially polished stone axe and adze at the Donghulin site; a small number of grinding stone axes at the Zhuannian site; a grinding stone shovel at the Longwangshan site, with the top two sides ground and into a curved blade; adze-shaped wares with some of the belly surface after polishing at the Mengjiaquan site; similar stone axes, with the possibility of being chipped semi-finished product, were found at the Xueguan site, consistent with stone axes and adzes in the Neolithic period (Wang et al. 1983). Relatively speaking, the amount of adze-shaped tools is slightly larger, being discovered at sites of Yujiagou, Ma'anshan, Hutouliang, Jijitan, Xichuan (where seven adze-shaped tools were found) in Shanxi Province, and Zengpiyan in South China. According to the recent micro-trace analysis and experimental studies, the main function of the adze-shaped tools revealed at the Hutouliang site were designed for wood processing (Zhang et al. 2010). Being found at the Nanzhuangtou site, the felling and processing of wood could be used to make handles of composite tools, such as the hand-chiselled wooden stick (handle) segments. On the other hand, by cutting and processing trees, ancient people could construct long-lasting inhabited buildings. In addition to felling tools, digging tools are also notable. At the Yuchangyan site, the long and flat hoe-shaped tools explored are suitable for digging. Besides, the bone shovel with smoothly worn edges were also formed during the digging process (Yuan 2000). These tools are undoubtedly linked to mobility reduction.

The obvious landmark artefacts include pottery, food processing tools (including grinding slabs and handstones), felling and digging tools. Besides these tools, the chipped stone tools assemblage is often overlooked. Generally, without the emergence of pottery, sites belonging to the Paleolithic-to-Neolithic transition would be considered the Upper Paleolithic ones. However, regardless of the existence of pottery, the main artefact assemblage is chipped stone tools during the Paleolithic-to-Neolithic transition. Unfortunately, chipped stone tools are often regarded as a remnant of the Paleolithic, thus, the research values are generally overlooked. However, the chipped stone tools assemblage could reflect the changes in ancient people's mobility. Thus, from a different perspective, the analysis of chipped lithic assemblage could provide us with a more comprehensive picture of changes in human adaptation. It is noteworthy that the state of the raw material of chipped stone tools could have an essential influence on the assemblage characteristics, sometimes exceeding the influence of mobility on the assemblage. Thus, when selecting artefatcs for comparison, archaeologists should choose materials with similar raw material supply states. Unfortunately, there are only a few materials related to chipped lithic assemblage available among the published comprehensive reports. Thus, we could only extract useful information from the limited material available. From basic principles, measurement methods, and desired outcomes, Table 6.5 summarises a number of research avenues into the level of mobility (sedentism) in relation to chipped stone tools, examining the possibility of stone cores analysis, repair methods, waste proportions, raw material conditions, and tool characteristics. Then, through comparing these data with archaeological excavations, we could detect the mode of cultural adaptations that reflected in our materials. Here I select Locality S9 at the Shizitan site and the Lijiagou site as representatives of the Eastern and Western

Taihang Mountains respectively. The reports related to these two sites are relatively enriched, with more than two cultural layers of the Paleolithic-to-Neolithic transition, providing very rare analytical material for intra- and inter-site comparisons.

The tool characteristics required by hunter-gatherers are different from agricultural producers, as the former possess higher mobility when collecting food. Mobile foraging faces greater uncertainty about resources. At the same time, the hunter-gatherers need to carry tools, which should be more widely adapted and lighter in this case. A more versatile tool, being identical to a "Swiss Army knife", with a wider range of adaptations would be made under standardised retouch and repair, such as the production of bifaces and the use of the microblade technology. In order to reduce the risk of foraging, it is predictable that the hunter-gatherers were likely to use tools made from the best quality raw materials and repair them carefully. In contrast, the less mobile agricultural producers needed durable tools because they were sedentary in one place for a long time. At the same time, agricultural producers did not encounter the same risk of foraging as the hunter-gatherers, thus, the former were more likely to apply techniques of ready-to-use when making stone tools, since the discarded flakes remained could be used the next time as they did not need to deliberately repair and processing of the stone cores or tools. Besides, the raw materials were relatively abundant, so there was no need for fine processing of the chipped stone tools. Furthermore, sedentism makes standardised and multi-purpose techniques less valuable, for example, we rarely use a Swiss Army knife at home nowadays because there are more specialised and larger tools by hand. Diversity in mobility allows us to make predictions about the characteristics of stone tool techniques and possible artefact assemblages. Although it is unclear whether agricultural production begins in the Paleolithic-to-Neolithic transition, however, an analysis of the mobility levels allows us to speculate on the changes that occurred in human cultural adaptation during this phase.

There are three cultural layers at the Lijiagou site: the upper, the middle, and the lower. The upper layer belongs to the Peiligang culture period, which will not be elaborated here. In the Middle Cultural Layer (reported as Early Neolithic remains), the lithic assemblage is characterised by the report saying that "the number of microblade core is limited and dominated by those with wide surfaces; microblade are numbered significantly less than that in the Early Neolithic phase. The

Table 6.5 Effect of sedentism on the chipped lithic assemblage

	Predictions	*Measurement methods*	*Expected results*
Stone cores	Relatively speaking, more informal tools would be produced for the sedentism, therefore, more stone cores to be produced, then more flakes; more formal tools and standardised artefacts would be produced by mobile groups	Ratio of standardised artefacts to stone cores (except fine stone cores)	Proportion progressively declines, with fewer standardised artefacts
Repairs	Mobile groups would spend more time for repairing stone tools in order to complete them; repairs of surfaces would be more time-consuming than edges	Ratio of surface repairs to edge repairs	The ratio progressively declines with fewer surface repairs. That is, the repairs would be more haste
Debitage	More bificial wares repairing flakes represent formal tool production; flakes with repairing platforms also mean more time was spent on tool production	Ratio of repair stone pieces for two-sided ware, ratio of repair stone surface pieces to other stone pieces	The ratio progressively declines, less two-sided ware
Raw materials	The shorter residence times, the smaller changes in raw materials; vice versa. Short-term sedentisms are more likely to lead to exotic raw materials and semi-finished artefacts with great qualities. Oversized specimens could be more likely produced locally	Proportion of exotic raw materials; degree of diversity of raw material; proportion of specimens without natural stone surfaces; presence or absence of oversized specimens	More local raw materials, abundant types of raw materials, but less raw materials in great qualities, fewer oversized specimens, low percentage of natural stone artefacts
Tools	Low-mobility groups tend to use ready-to-use tools and durable tools; high-mobility groups tend to use standard and formal tools	Ratio of two types of tools; presence or absence of durable tools	Durable tools are available, at the same time, the proportion of informal tools increases

figure of common stone tools is more numerous, but which are mostly for convenience. Those common stone tools are mainly made from quartz sandstone and quartz, with the technology of direct hammer percussion, thus, the application of smashing techniques could be seen. The tools are mainly side scrapers and choppers, but in a variety of forms" (Peking University School of Archaeology and Museology, Zhengzhou Municipal Institute of Cultural Relics and Archaeology 2001). In contrast, the lithic assemblage in the lower culture layer is also made of "high-quality flint, which is not normally found locally. Thus, it might be the result of long-distance collection and transportation. Although there are only a few flints and quartz artefacts found in the middle layer, deliberately polished artefacts can rarely be discovered. The statistic of convenient stone artefatcs made from sandstone or quartz sandstone is high. Based on the remains of grinding slabs found at this phase, some of the flat sandstone chunks were used as raw or semi-finished materials for processing stones" (Zhengzhou Municipal Institute of Cultural Relics and Archaeology, Peking University School of Archaeology and Museology 2011). Grounded in raw material analysis, the ancient population was less mobile in the Middle Culture Period than in the Lower Culture Period. During this period, the reduction in human mobility is clearly reflected in the following aspects: with few exotic raw materials, local people tended to use local materials; convenient tools were widely used; a great variety of artefact forms could be seen; and standardised artefatcs (the microblade technology artefacts are typical of standardised artefatcs) were simplified, etc.

Here, we will analyse the mobility of Locality S9 at the Shizitan site through the chipped lithic assemblage. The Shizitan site contains four cultural layers: the items excavated from the layers 1 and 3 are few, thus, lack values for comparative studies. However, the items detected from the layers of 4 and 5 are relatively enriched, with 1119 and 439 stone artefacts discovered respectively. Based on the previous hypothesis, the settlers (or less mobile groups) would have made more stone tools at the site, while the mobile groups might have carried semi-finished or finished artefacts. In this case, the ratio of stone cores and tools should occupy a lower proportion among settlers, as more of the various tools produced by stone cores would have been left at the site. Counting the proportion of stone cores (cores and microcores) and tools made from them (flakes, microblade, broken chunks, broken flakes, used flakes), the figure is approximately 0.032 in Layer 4, with a lower figure in Layer 5–0.024. Debitage was excluded from the statistics considering they are likely to be more associated with repairs. If debitages are considered artefacts of edge repair, flakes and broken flakes are artefacts of surface repair, then the ratio in Layer 4 would be 2.64 (the sum of flakes plus broken flakes divided by the number of fragments), and 1.74 for Layer 5. That is to say, there might have been more surface repair in Layer 4. This hypothesis is consistent with the results of comparing the ratio of stone cores to their artefacts. Compared to the 4 and 5 cultural layers in Locality S9 at the Shizitan site, the mobility is higher in the late period, even though stone grinding slabs and grinding slabs are excavated in Layer 4. The existing materials cannot support a clear lower mobility in the two cultural layers at Locality S9 in the Shizitan site. On the contrary, mobility might be higher.

Possibly, the Lijiagou and Shizitan sites possess the chronological difference, that is, Locality S9 at the Shizitan site is older. However, the available dating date material from 14C does not support this explanation, as the fire remains from Locality S9, Layer 3 is dated to just over 8300 BP, and the date in Layers 4 and 5 still remains unknown. Supported by materials in the Early Neolithic, another possibility is a geographical difference. Until now, explorations of the Early Neolithic culture in Shanxi are quite limited, while the existing material suggests that the Neolithic process in Shanxi is slightly later than that in the east area of the Taihang Mountains. The differences between the Shizitan and Lijiagou sites further support the view that hunter-gatherers in Shanxi during the Paleolithic-to-Neolithic transition might have retained high mobility, which opposed the prerequisite for the agricultural origin, that is, the reduction of mobility.

(4) Structural Analysis of the Sites

The mobility reduction is not equivalent to sedentism, but reflects a reduction of people's activity range and frequency of relocation, that is, people's mobility is confined to a smaller area. At the same time, the sum of sites is reduced, and there might be some centralised camps with a residence period near a one-year period (more than one year can be called sedentism). Based on the archaeological materials, firstly, people frequently visit one area and leave many similar remains; furthermore, the emergence of central sites, which contain a variety of remains, especially relics and remains, could reflect long-term sedentism. Examples of relics, including architectural structures, pits, and hearths, were discovered (Bar-Yosef and Meadow 1995); the remains include items that reflect mobility reduction as discussed before. A gradual decrease in mobility could reflect in the process from a repeated visitation in a certain area to the formation of a central location. If a hunter-gatherer group regularly prepares to return to the central sedentism, they would consider leaving behind durable tools and facilities that can be used upon return. In other cases, ethnographic record a division of labour: hunter-gatherer groups left women, children, and the elderly in the central sedentism to collect food, intensify certain resource utilisation, and hunt small animals in the vicinity of the sedentism; at the same time, adult males went out to hunt. This practice strikes a balance between mobility and sedentism, until mobile hunting eventually has become unsustainable and full sedentism has been achieved. Through analysing the structural features of the site, a further vision of changes in prehistoric human mobility could be understood.

During the Paleolithic-to-Neolithic transition, the Lijiagou site revealed active surfaces of the two cultural layers: the lower part of that revealed an oval-shaped stone circle formed by stone artefacts and hand-carried stones, with an approximately 3.5 m height from east to west and a 2.5 m width from north to south. To the east of the stone circle are mainly animal skeletal remains, mostly fragmented limb bones and antlers of large herbivores. The middle cultural layer thickens considerably, where an area of the stone assemblage is found with a length of approximately 3 m east to west and a width of approximately 2 m wide from north to south. The centre of the remains consists of grinding slabs and stone anvils, alongside a number of flattened stones, interspersed with a high proportion of burned stone fragments, pottery shreds, and animal bone pieces. The shape of the stones is mostly flattened, predominated by sandstone and quartz sandstone, which could come from primary rock formations in the vicinity of the site. The artefatcs and durable tools (such as grinding slabs and stones transported) presented in sites reflect a greater investment in the construction of the sedentism, mirroring a proportional duration of the sedentism. Edge-ground stone adze and a large number of pottery sherds are discovered at the Lijiagou site, demonstrating the characteristics of central sedentisms.

In contrast, the Shizitan site is still in the phase of repeatedly using one single site. Locality S29 at the Shizitan site has eight cultural layers, totalling 232 fire remains are found ranging from top to bottom: 4

remains (culture Layer 1, 0.3 metre thick); 75 remains (culture Layer 2, 1 metre thick); 17 remains (culture Layer 3, 0.5 metre thick); 94 remains (culture Layer 4, 1.7 metres thick); 54 remains (culture Layer 5, 1 metre thick); 38 remains (culture Layer 6, 0.8 metre thick); 3 remains (culture Layer 7, 3.5 metre thick); and no evidence was detected in the cultural Layer 8, although charcoal flecks are scattered within. The dense fire remains indicate that ancient people repeatedly used this site, but this place has not been formed as a centralised sedentism at this time.

The most representative Paleolithic transition sites for centralised sedentism are the Nanzhuangtou and Donghulin sites (Fig. 6.2). Both of these sites excavate a very diverse range of archaeological remains, such as tombs, pits, ash moats, and even architectural remains. At the Nanzhuangtou site, pits, moats, and hearths are discovered. In 1992, a report stated that three antlers were excavated from H1, one of which was relatively intact and placed in a vertical orientation, seemingly intentionally buried, with traces of artificial cutting on the shank. In 1997, two hearths were excavated, as well as wood stripes with artificially carved grooves (14C dating to 9875 ± 160 BP), which seem to be the remains of architectures. Based on the HD3 measures, more than 10 hearths were found at the Donghulin site, approximately 80 cm × 60 cm with a depth of 30 cm with a large amount of black ash presented and a relatively large number of gravel chunks with fire marks and animal bones (dominated by deer) revealed in the central area of the hearth. The upper gravel chunks are placed in an untidy way, while the lower ones are stacked in a roughly circular pattern with a neat appearance, which must have been intentional (Zhao 2003; Peking University School of Archaeology and Museology, et al. 2006). Complete burials were found at the Donghulin site, all of which were vertical pits, with two types of burial style: supine with straight limbs and supine with flexed limbs, accompanied by small polished stone axes and ornaments. The characteristics of such remains often appear in Neolithic sites, differing from typical characteristics of scattered and disordered remains in the Paleolithic sites.

The characteristics of the Ma'anshan site in the Nihewan Basin appear to lie between the two categories of Nanzhuangtou and Donghulin sites. Possessing the excavation area of 50 m^2, the Ma'anshan site discovers 34 traces of fire remains in Layer 3, including firing place, hearths, and fire piles.[19] On the one hand, the Ma'anshan site is characterised by repeated visitation, similar to the Shizitan site; on the other hand, the stoves are dug with a surface of oval-shape, a depth of 15 to 18 cm, a wall and

Fig. 6.2 Landscape in Donghulin Sites

base of red-burned clay of 1.5 to 3 cm thick. AMS analysis gives it a reliable date of 13,080 ± 120 BP (Mei and Wang 2001), which has the characteristics of a Neolithic stove (Yan 2000). Its lithic assemblage is similar: nearly 20,000 lithic artefatcs of various types are excavated, made from predominantly quartzite, with a certain proportion of flint, agate and siliceous mudstone, and occasional tuff, hornfels, and limestone. The raw materials are mainly from local sources (quartzite[20]), but also from exotic sources such as agate. The lithic assemblage includes nearly 300 wedge-shaped cores from blanks to the final phase of flaking, as well as a large number of microblades and crest blades, in addition to side scrapers, spearheads, carving tools, concave notched scrapers, end scrapers, points, and adze–shaped tools. All of the tools mentioned are in a stable type and beautiful shape. Generally, standardised tools echo a higher degree of mobility, while the complete sequence of lithic production with adze-shaped tools reflects the opposite trend. It seems that some people in the ancient group remained highly mobile as a means of livelihood (such as male hunters), while the others stayed in residence for longer than their ancients.

The same situation happens at the Bianbiandong site. Clear active surfaces are found on the surfaces of layer 4, Layer 3A, and Layer 2. There are several burnt earth surfaces exploited on the surface of Layer 4, with ashes piled up nearby and interspersed with many fragmented animal bones with signs of fire. Next to a burnt surface, a stone grinding slab and two handstones are found. There is a layer of sandy soil about 3 cm thick on the surface of Layer 3A, grey-brown in colour with a flattened and thick uniform, which apparently has been deliberately tidied. In addition to the burnt surface, there are four pits (no opening levels are reported) found, with the shape of a sub-circular pot-bottom. Having been explored and tapped, human skulls with limb bones were unearthed from one of these pits, thus, it could be a tomb based on the description by local villagers. Another ash pit was found with a burnt earth surface on the slope of the pit wall. A pile of ashes is on the surface of the pit, thus, it is possibly a post-abandonment cooking pit (Sun et al. 2007). The structure of the site alone has all the characteristics of a Neolithic site, such as a man-made dwelling surface, stoves, ash pits, burials (still in uncertainty), grinding slabs, and handstones, but on the other hand, the site is situated in a natural cave, which is low and not far from the river, but cannot be seen as an ideal shelter as it faces north. Compared to this site, the ancient people preferred a less suitable cave, reflecting the temporary character of it.

Paleolithichunter-gatherers and Neolithic agricultural producers have different cultural and ecological characteristics when choosing dwellings. Agricultural producers needed to be close to arable land, while hunter-gatherers needed to be close to animal migration routes; agricultural producers built permanent sedentisms, while hunter-gatherers preferred to use natural caves temporarily. Sites in the Paleolithic-to-Neolithic transition hold a distinctly transitional character, with structures and relics similar to those of Neolithic sites on the one hand, and the location of Paleolithic sites on the other. The geographical location of a site can mirror the extent of agricultural development. Through the investigation of the geographical location of sites in Western Asia, archaeologists have confirmed that agriculture has not emerged in the Natufian culture period (Bar-Yosef and Meadow 1995, pp. 39–94), but in the phase of the intensification of natural resource utilisation. Through examining the geographical distributions of the Neolithic transition sites, most of the sites still retain the distribution features of Paleolithic sites, such as the Bianbiandong site in Yiyuan, the Yuchanyan site in the Dao County, the xianrendong site, the Diaotonghuan site, and the Lingnan Cave sites. All the sites mentioned draw on natural caves and significantly differ from the Early Neolithic site distributions. Even though the sites are located in the wilderness, the distributions are in different locations compared to those of later Neolithic cultures. Taking the Xiachuan site as an example, the excavators noted: "We have worked at the Xiachuan site for a long term but have never seen a mixture of ground stone tool and pottery sherds from the Neolithic and later periods, either in the stratigraphy or on the surface. However, in the area around Zhongcun and the terraces of the Qinhe Valley which is a dozen and dozens miles away from Xiachuan culture respectively, there are Neolithic remains, but Xiachuan cultural traces are excluded. Thus, it seems that Neolithic man did not settle in Xiachuan" (Wang et al. 1978).

From the geographical perspective, the distribution of sites in the Paleolithic-to-Neolithic transition seems closer to the piedmont zone than that of the Upper Paleolithic, such as typical Donghulin and Zhuannian sites; Lijiagou sites in the mountain-hill area; and Nanzhuangtou site in the terraces of the mountain plain. The remains of high platforms architecture discovered at the Linlizhuma site (Chu 1999), in the Middle and Lower of the Yangtze River, although relatively few in number, reflect a developmental trend that is closely associated with the distribution characteristics of sites from Early Neolithic cultures, that is, resources utilisation alters the distribution of the sites. The gradual geographical site displacement strongly indicates a progressive nature between the Paleolithic and the Neolithic.

(5) Faunal and Plant Remains

As an essential part of the archaeological material, plant and fauna remains could reflect not only the mobility of prehistoric hunter-gatherers but also directly the basic form of incipient agriculture. Thus, without discussing plant and fauna remains in the Paleolithic-to-Neolithic transition, it is analogous to disconnecting the Industrial Revolution and commodities. However, before discussing plant and fauna remains, we also ensure the authenticity of archaeological materials. As a real being, the existence of archaeological material is not without a doubt (even if the existence is being questioned, it is still in the paradigm of archaeology). Archaeologists are mainly concerned about whether the archaeological materials could adequately reflect the ancient people's way of life, that is, whether the substantial patterns of ancient people could be revealed from the plant and fauna remains. The preservation of plant and fauna remains builds upon several conditions, especially natural conditions, such as soil pH, the exposure span of plant and fauna remains on the surface, and the state of groundwater activity, etc. Therefore, without distinguishing different preservation conditions, the result would be unconvincing. Thus, when choosing artefatcs for comparison, it would be ideal for archaeologists to compare different layers of the same site, instead of different sites in the same region (for example, sites that all belong to the Loess Plateau in North China), and even better than between sites in different regions (such as comparing different sites in North China with those in South China). Besides, the function of sites, especially Paleolithic sites, should be scrutinised by archaeologists. The sites with different functions feature distinctive characteristics of plant and fauna remains, such as the lithic workshop site and the slaughterhouse site. An abundant fauna remains would be left after a slaughter happened on a certain herd. Most sites in the Paleolithic-to-Neolithic transition have the property of central sedentism and repeated visitation, in this case, the extreme situations aforestated are unlikely to happen. When analysing the sites of the Paleolithic-to-Neolithic transition, the mobility changes of prehistoric hunter-gatherers should be particularly noted since different mobility patterns would directly affect the preservation of faunal and plant remains. As for highly mobile groups, it would be impractical for them to transport all hunted animals back to a central sedentism, instead, they would consume the animals immediately. Even for relatively sedentary groups, hunting activities are less likely to happen in the central sedentism and nearby areas. It would be very uneconomical to transport a bison of over 500 kg for a long distance, thus, only a few meaty bones could actually be brought to the central sedentism. Therefore, the highly mobile groups left more bones of megafauna at sites, especially with less fleshy ones, which would be consumed after hunting. As for the less mobile groups, people would be more likely to bring the whole small animals and fleshy parts of large animals to the central sedentism. Finally, the faunal and plant remains on the site are related to the people's substantial way. At the sites in the Paleolithic-to-Neolithic transition, we hope to find evidence of domestication, broad-spectrum adaptations/revolution, and intensification of resource utilisation. In short, we wish to derive from the plant and fauna remains to find changes in mobility and the utilisation of plants and animal resources.

In hunter-gatherer adaptations, the utilisation of aquatic resources could reflect an adaptive intensification of agriculture, such as the utilisation of shellfish, fish, birds, amphibians, and mammals. The aquatic resources utilisation could effectively relieve the livelihood pressures because the substance does not merely depends on terrestrial resources, instead, could support exceeding the population density threshold of dependence on terrestrial resources (see Chapter 3). The utilisation of aquatic resources often demands for more sophisticated technologies, ranging from oars, bows and arrows to a wide range of fishing tools. The utilisation of aquatic resources has been occurring since the Upper Paleolithic, as harpoons have been discovered at the Xiaogushan site. The possibility and intensity of the aquatic resources utilisation should be considered in this case. At the end of the Late Pleistocene, the sea levels had not yet risen to the present height. At the same time, lakes and floodplains had not formed in the Middle and Lower Yangtze Valley and North China. Thus, aquatic resources were not universally available. Even if the conditions for the utilisation of aquatic resources exist, they cannot be comparable to the sustained supply capacity of marine resources. The utilisation of aquatic resources has left a rich legacy at the Zengpiyan site: the weights of the snail shell and carbonated plant remains are not significantly different in Phases I and II, but show disparities in Phases IV and V (CA, CASS et al. 2003). In accordance with the contrasts of different layers located at the same site, it is clear that humans enhanced the utilisation of aquatic resources at a later phase. Clearly, this is associated with the warmer and wetter climate of the Holocene, resulting in an adaptation pressure relief and an incentive of labour-intensive agricultural production reduction. Even in North China, where aquatic resources were not abundant, evidence of aquatic resources utilisation still exists. For example, at the site of Donghulin, there are abundant and various skeletal shells of molluscs, including conches, mussels, and snails, with the largest shells reaching over 20 cm in diameter (Zhao et al. 2003, Peking University school of Archaeology and myseology, et al. 2006). Animal specimens unearthed from the Nanzhuangtou site in Xushui include mussels, fish, turtles, and birds.

Except for the Nanzhuangtou site (which was partially destroyed), the faunal remains at the sites in North China are not abundant. Regularly, the faunal remains feature with bone pieces, which cannot entirely blame on the natural conditions. The excavators and scholars universally argue that the fragmented bones are related to the utilisation of humans, since the sites in the Upper Paleolithic (such as Shiyu and Xiaonanhai) preserve abounding fossils of animal bones. That is, animal resources are rarer than before, making hunting more difficult, thus, the opportunities to hunt megafauna decrease. Hunting, as an essential part of substance in the Early Neolithic, assuredly became a fundamental trend, although it might have nuances from site to site. From the lower cultural layer at the Lijiagou site, a large number of animal remains were discovered, including large herbivores such as horses, cattle, deer of different sizes, omnivorous pigs, carnivores, rodents, and birds. Calculating based on the minimum number of individuals, the proportion of large herbivores, such as cattle, horses, and large deer, occupy more than 50%. Most of the bone remains are heavily fragmented, with light signs of weathering and abrasion on the surfaces of the specimens. Researchers assume that this phenomenon might be associated with bone cracking or processing launched by humans. From the Middle Culture layer (Peking University School of Archaeology and Museology, Zhengzhou Municipal Institute of Cultural Relics and Archaeology 2011), the bones of smaller animals, especially deer, are dominated within. Only a few bone fragments of larger herbivores could be discovered in the Middle Culture layer (Zhengzhou Municipal Institute of Cultural Relics and Archaeology, Peking University School of Archaeology and Museology 2011), from bovine, equine, sheep, pigs, and carnivores. The changing characteristics explored from the two cultural layers during the Paleolithic-to-Neolithic transition are consistent with the trend aforementioned, where megafauna hunting is gradually shrinking.

Amid the decline of megafauna numbers, people have to hunt smaller animals, even some very nimble ones, such as rabbits and small carnivores. Nearly 695 animal skeletal specimens were excavated at the Locality S9 of the Shizitan site, from where 72.2% specimens were burnt bones with a length of 1 to 2 cm. Ratiocinating from teeth and joints of the samples, the preliminary identifications show that these animals include rodents, lagomorphs, carnivores, even-toed ungulates (antelopes), and birds (including ostriches). The faunal assemblage in Locality S9 at the Shizitan site shows an expansion of the hunting range, but the large herbivores are excluded. Among the herbivores, deer are generally the preferred hunting species during the Paleolithic-to-Neolithic transition in North China. At the Nanzhuangtou site in Xushui, 1068 vertebrate skeletons were recovered, which are all from Layer 5 and Lower Layer 5. The faunal assemblage is mostly mammals, with deer being the main species, including 12 species of pheasant, bird, rat, rabbit, dog, plum deer, small deer, and buffalo (Yuan and Li 2010). Based on the preliminary identifications, animal bones excavated at the Hudonglin site are distinguished as deer being predominant, alongside pigs, badgers, and other animals. Similar discoveries are detected at the Xianrendong in Wannian County, which is located in the Middle and Lower Yangtze Valley, indicating that deer are also the main hunting target (Redding 1995). Amid the expansion of forests at the time, deer are usually regarded as woodland edge species. Typically in temperate regions, grassland animals are often herd-like, that is, not rich in species but incredibly abundant in each type, analogically to the case that over 20 million bison lived before large-scale commercial hunting on the North American continent (Alroy 1999, pp. 105–143). The sporopollenin analysis at the Donghulin site shows a remarkable increase in pollen of woody plants during the Early Holocene (10,000–8200 BP), up to 55%, including heat-tolerant pine, fir, spruce, hemlock (found individually and now living in the subtropics), quercus and hoodia. Different from the late Pleistocene, the herbaceous plants see a remarkable decrease during the Paleolithic-to-Neolithic transition, suggesting an intense temperature rise. The average annual temperatures are probably similar to or slightly warmer than the present, making the dominant vegetation species as temperate broadleaf and mixed forests during the Donghulin culture period (Xia et al. 2011). The expansion of forests may result in the decline of large herbivores. Surprisingly, a lack of large herbivores happened at the site of Shizitan, which lacks forest coverage. In this case, we could attribute this situation to overhunting by humans.

In contrast, abundant animal bones are found in sites located in the Lingnan region (where pottery sherds are found) and the Yuchangyan site. Twenty-eight species of mammals are found at the Yuchangyan site, among which 27 have been identified, dominated by deer and small carnivores (at least 11 species); the number of bird skeletons is also very high, accounting for 30% of all animal skeletons; 18 species are associated with the aquatic environment (occupying 67% among birds), including shellfish, fish, and turtles. With 26 mussels and 7 clam species identified, the most abundant of the aquatic resources are mussels, generally living in freshwater lakes, rivers, and riverside ponds. The faunal assemblage at the Yuchangyan site presents a characteristic of aquatic resource utilisation. Faunal remains discovered at the Zengpiyan site are even richer, with well-preserved 81,233 shellfish and 27,211 animal bones revealed in 2001. To be more specific, there are 47 species of shellfish, 20 species of birds, 37 species of mammals, each one species of crab, fish, and reptile, totalling 108 (Yuan 2000). The plentiful faunal remains assemblage at the Zengpiyan site features a distinctly broad-spectrum adaptation. Seventeen species of mammals, fifteen species of shellfish, and zero fish or birds were discovered at the Miaoyan site, which is in the same region as the Zengpiyan site (Zhang et al. 1999). The third phase of the Dayan site remains is also dominated by snail shell assemblage (Fu et al. 2002). Such a shell-dominated assemblage is also found at sites during the same period, such as Bailiandong in Liuzhou, Liyuzui, and Huangyandong in Fengkai; after entering the Holocene, such findings became more common and could be found from the site of Dingshishan, Zengpiyan, the late Dayan, and the Baozitou group of sites. Utilisation of shellfish signifies an essential cultural adaptation in the Lingnan area, starting from the beginning of the Paleolithic-to-Neolithic transition and lasting throughout the whole Neolithic.

Compared to other sites where fewer shellfish are utilised in the Lower and Middle Yangtze Valley, abundant faunal remains are found at the Xianrendong in Wannian County. The excavation materials are not identified in the early phases, but found out that nearly 82% specimens were deer and zero shellfish among the 1672 animal bones in the 1993 excavation (Redding 1995). At the well-preserved Kuahuqiao site of the Neolithic, over 5000 animal bones are excavated but none shellfish remains were found (Zhejiang Provincial Institute of Cultural Relics and Archaeology, Xiaoshan Museum 2004). In the later period, the Hemudu site is abundant in faunal remains, but only three species of shellfish are recovered (Zhejiang Provincial Institute of Cultural Relics and Archaeology 2003). Among all the sites in the Middle Yangtze Valley, the Bashidang site possesses the richest animal bone remains, with only a few extremely fragmented and unidentifiable shellfish specimens. The difference in the shellfish utilisation represents one of the main indicators of distincting sites in the Middle and Lower Yangtze Valley and Lingnan area, as it could reflect the disparity in the abundance of aquatic resources and the intensification of aquatic resource utilisation. Shellfish gathering is suitable for everyone, since it does not require skills and complex tools needed in hunting and fishing activities. Through shellfish collection, ancient people could obtain protein. This nutrition could compensate for the lack of hunting resources, thus, reducing the need for animal domestication. Aquatic resource dependence (not purely utilisation) is a product during the Paleolithic-to-Neolithic transition, at the same time, being a hallmark of the intensification of resource utilisation.

During the Paleolithic-to-Neolithic transition, the number of large herbivores declined, particularly grassland animals; simultaneously, ancient people started to utilise and rely on aquatic resources. Besides, domestication had already begun, as a canid animal was found at the Nanzhuangtou site. The canid animal has a dentary row of 79.4 mm in its left lower jaw, which is smaller than that of wolves (averaging 90 mm). In this case, archaeologists assume this canid animal as a dog, which was domesticated, consistent with the mtDNA evidence (Yuan and Li 2010). Junfeng pang et al. assume that dogs have been domesticated in the South Yangtze River ca. 16,300 BP (Pang et al. 2009). Canine specimens have also been found at the Xianrendong in Wannian County. Pig bones are found at the Nanzhuangtou site, but there is no evidence to show their domestication (Yuan and Li 2010). Besides, based on the Age Structure studies, abundant pig bones at the Zengpiyan site might belong to boars (CA, CASS et al. 2003). Chickens are found in the upper layer of the Xianrendong, with possibilities of being domesticated and potentially dated to the Middle to Late Neolithic. Based on existing evidence, the main domesticated animals (such as pigs, chickens, and cattle) appear after plant domestication, as the former is fed by wastes generated from plant processing, such as hulls and stalks. In this case, a large number of wild animal remains could be seen at the sites of the Early Neolithic. At the same time, evidence of the incapacity of humans to establish stable sedentisms has been found. Thus, animal hunting was always an indispensable means of obtaining meat and skins for substances at that time.

The main resources of plant remains discovery and research come from the sites of Yuchanyan, Zengpiyan, and Donghulin. Through sifting and rinsing the deposits of the Yuchangyan site, a large number of plant kernels are collected. Archaeologists initially sorted 40 species of plant kernels, with 17 kinds identified. Among them, the seed of hackberry is the most abundant. In Burial 4 and its fill at the Donghulin site, dozens of celtis koraiensis (similar species) are found. In the midst of archaeological sites, the earliest discoveries of hackberry seeds are discovered at the Peking Man Site at Zhoukoudian, moreover, the seeds are scattered at sites in the Neolithic, such as Xinglongwa, Bancun, Banpo, Longgangsi, etc. Referring to the Indian custom of eating the pulp from the hackberry tree then grinding the kernels into powder as a seasoning and eating it with meat or corn (Hao et al. 2008), Shougang Hao et al. speculate that the grinding slabs excavated at the Donghulin site might have been used to grind the kernels of hackberry trees (Peking University School of Archaeology and Museology, et al. 2006). The seeds of hackberry trees are straightly accessible, while the kernels are only used as a seasoning, instead of resources for intensive utilisation. The artefatcs of intensive utilisation require considerable labour processing procedures. Typical examples include acorns, wild corn, rice, and wheat, all of which are not editable directly but require tedious processing.

Eleven grains of rice are excavated from the Yuchanyan site. Based on research, these grains of rice characterise a mixed feature of wild, indica, and japonica rice, still situated in the differentiation phase (Zhang and Yuan 1998). Archaeologists have applied the rice phytoliths method to explore the origin of the intensification of rice utilisation at the Xianrendong in Wannian County and Diaotonghuan site. According to Zhijun Zhao's study of rice phytoliths at the Diaotonghuan site, 14 soil samples are collected from layers B to O (from top to bottom). Except for layer F, rice phytoliths are abundant above layer G (that is, layers B, C, D, and E), notably, nearly 100 rice phytoliths are found in less than 10 grams of soil sample in layer G. Layer G is dated to 17,040 ± 270 BP. The Layer F, which has no rice phytoliths detected, is dated to 15,180 ± 90 BP. The study shows that rice phytoliths recovered from Layer G belong to wild rice, while those in Layer E and D are equally likely to be wild and cultivated. In Layer C, which is much later in date, the proportion of cultivated rice is greater than wild rice. In this case, it is possible to date cultivated rice to layer E, which is estimated at 11,900 BP (Zhao 2000). Why is there no rice uncovered in Layer F? If the Younger Dryas Event occurs at that time (the Younger Dryas is at 13,900–12,800 BP, which is a speculation considering that the 14 C dates of both Xianrendong and Diaotonghuan sites are earlier than actual dates), the distribution of wild rice would have moved south amid the decline of local temperatures, thus, the wild rice could not have been utilised. Based on the different rice remains excavated from the Diaotonghuan and Yuchanyan sites, the rice cultivation might be ca. 12,000 BP, while the utilisation of rice by humans might have begun much earlier.

Systematic collection and analysis of plant remains at the Zengpiyan site have been carried out: archeologists collected charred plant remains through the floating collection, weighed the charred artefatcs, counted the number of plant seeds, and compared the weight of root tubers. The comparative studies show important changes between Phase III and V at the Zengpiyan site, showing that the root tubers the later the weightier. There are no charred rice grains or wild rice remains found at the Zengpiyan site. In addition, the sampling analysis of plant phytoliths does not consist of rice, indicating that Zengpiyan people are less likely to engage in rice agriculture (CA, CASS et al. 2003). Instead, depending on root tubers (possibly both gathering and cultivation) and the utilisation of aquatic resources, Zengpiyan people formed a mode of "low level of food production" created by Bruce D Smith, which allowed them to maintain the substance without solely relying on species domestication and become fearless of local resources depletion even in the case of mobility decline. In the Neolithic, the domestication of species remains an irrelevant role of the subsistence approach for a long time. Moreover, it is easy to cultivate asexual reproduction of root tubers, which are not easy to spot during the domesticated process. In this sense, the Zengpiyan site might represent another type of food production (see Chapter 8), seeing a different path compared to the development of rice-based agriculture such as the sites of Yuchanyan, Xianrendong, and Diaotonghuan.

(6) Development of Culture Branches

The development of prehistoric cultures is a process of constant differentiation and integration, creation and extinction. Culture serves as a means of environmental adaptation, also a development path of human beings themselves (including societies and individuals). Differentiation and integration represent channels of solving practical problems; either, marks certain elements of the emergence or decline of a new culture. The essential issue should not lay on the macro-discipline of cultural development but realise the spatial and temporal conditions in cultural development. Thus, we can recognise the structural differences in cultural development beyond appearances, then the internal mechanisms of cultural evolution could be understood.

Current Chinese archaeology has noted two distinct developmental patterns between North and South China during the Paleolithic-to-Neolithic transition. In North China at this phase, the popularity of composite tools produced by the microblade technology indicates the existence of a "Mesolithic" period in which microblade technology predated the origins of agriculture. In southern China, although ancient people used small tiny flint tools, however, the microblade technology was not applied; instead, the traditional gravel industry of the Upper Paleolithic was continued during this time, with the appearances of ground bone awls, perforated mussels, and burned non-container clay chunks (CA, CASS 2010). However, the existing views do not take the middle and lower Yangtze Valley into consideration, nor does it take note of the cultural differentiation within regions.

In fact, the transitional modes go far beyond the two patterns in North and Southeast China during the Paleolithic-to-Neolithic transition. In addition to the divisions within the core region of agricultural origins, there are also several patterns such as North and South of the Yan Mountains, Northeast China, Southwest China, and Qinghai-Tibet. Undoubtedly, the routes of agricultural development are different in the core region of agricultural origins, that is, North China and the Middle and Lower Yangtze Valley. Meanwhile, the important divisions within the two regions should be noted. In these two regions, the agricultural modes seem to be divergent; however, there is no substantial difference, as they eventually entered the Neolithic almost simultaneously. However, substantial differences occur in several regions within North China, as well as Lower Yangtze Valley and areas adjacent to Lingnan. These regions chose separate developmental paths at key phases of the Paleolithic-to-Neolithic transition, constructing the basis of a rich and varied local culture in the Neolithic.

North China In the previous sections, I have already analysed the main modes of human adaptation in North China during the early phase of LUP. Based on the existing evidence, at least four modes of human adaptation could be detected: the Shuidonggou mode, the Shiyu mode, the Shandingdong—Dongfang Square—Xiaonanhai mode, and the Xiaogushan mode. Representing a period of cultural differentiation, the early phase of LUP in North China witnessed successful human mobility worldwide and further differentiated social development. Gradually defined reciprocal relationships lead to the strengthening of group territoriality, thus forming cultures with local characteristics. During the LGM, there was another trend towards the integration of Paleolithic cultures in North China, with the prevalence of the microblade industry spread through Northeast Asia and even parts of North America. As a period of cultural differentiation, the Paleolithic-to-Neolithic transition indicates an almost simultaneous intensification of resource utilisation, with certain iconic Neolithic features emerging in the following sites: the Locality 12 at the Shuidonggou site, Gezishan site, Shizitan site, Xiachuan site, Xueguan site, Donghulin site, Zhuannian site, and Lijiagou site; however, in the Early Neolithic, a flourish of Neolithic culture appears in the foothills and plain areas of North China, while there is no strong archaeological evidence to prove the emergencies in other areas.

Using Shizitan and Lijiagou sites as examples, the cultural divisions between the areas in the west and east of the Taihang Mountains in North China have been analysed above. Compared to the eastern area of the Taihang Mountains, the entrance of the Neolithic of the western area was fairly late. Instead, until the Middle Neolithic, the local Neolithic culture in the western area started to flourish. In the further west area, the microblade technology remained popular. After the cultural sequence of Xiaogushan culture, no material from the Paleolithic-to-Neolithic transition has been found in the Southern part of the Northeastern region (which is still geographically part of North China). However, the Xinglongwa culture, the Lower Xinle culture, and possibly some earlier cultures (such as the Xiaohexi culture[21]) show that the transformations in these regions are almost as rapidly as in central China, but a more mixed and diversified economy is prevalent. The differentiation during the Paleolithic-to-Neolithic transition in North China is linked to the LUP, but a uniform pattern of the Paleolithic-to-Neolithic transition in North China does not exist as a variety of different forms exist within this region. The incipient agricultural origin started in the foothills and plains to the east of the Taihang Mountains, in the mountains of Western Henan, in the Yan Mountains, and on the edge of the Shandong Mountains; while hunting and gathering, as a "Low-Level Food Production", still existed in marginal areas until the maturity of primitive agriculture. After the spread of agriculture to the east of the Taihang Mountains, the local Neolithic culture was established and developed.

In North China, some geographically transitional features could be shown in the Neolithic during the process of agriculture origin: a gradual increase in the proportion of new features from west to east, reaching a maximum in the foothill areas. Humans began to rapidly expand their living space once primitive agriculture developed on a relatively wide scale, forming the flourishing culture in the Early Neolithic (Fig. 6.3). The establishment of agriculture means the formation of a new niche, which is different from a cultural-ecological relationship with hunter-gatherers. The hunter-gatherers do not rely on the lifeway of sedentism in the areas of the plain, thus, the significance is more important than animals and plants domestication. The centre of cultural adaptation gradually shifted from the best habitats for hunter-gatherers (broad valleys in the middle rivers, marginal areas of basins, foothills, etc.) to the best zone for agricultural producers, that is, the plains. The Paleolithic-to-Neolithic transition could reflect this transitional property, that is, humans settle in the foothills, between the mountains and plains, gaining resources from the well-established hunter-gatherer mode, on the other hand, trying a new mode for substance.

The Lower and Middle Yangtze Valley During the LUP, a clear divergence occurred in the Lower and Middle Yangtze Valley: the gravel industry remained in the west, and the microlith began to occur in the east with the emergence of flint assemblages. Unfortunately, the chronological framework has not been fully established, thus, the exact date of the LUP onset in the Lower and Middle Yangtze Valley still remains unknown, neither the date of the tiny flint tool. In the previous chapter, the LUP in the Lower and Middle Yangtze Valley is categorised into the early and late phases in terms of Layer and cultural appearances, however, with speculation.

In discussing the situation of the Lower and Middle Yangtze Valley, the situations in the Lingnan region are also taken into account, as the

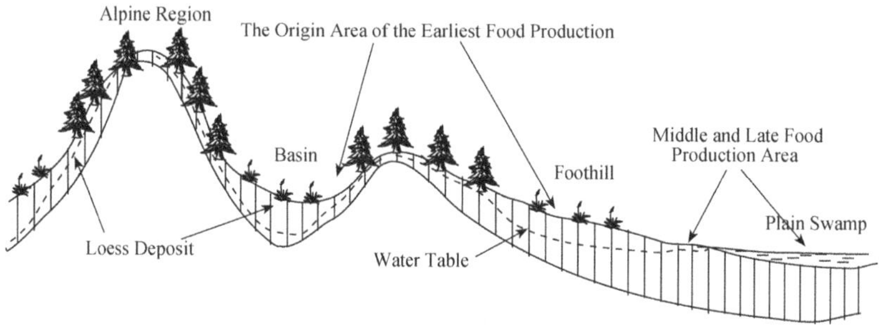

Fig. 6.3 Agricultural origin zone in North China

Paleolithic-to-Neolithic transition almost began simultaneously in both areas. At the same time, between the end of LGM and the beginning of the Holocene, the climate of the Lingnan region was warming up, but not as warm and humid as that in the Holocene. Overall, its climatic environment was closer to that of the Lower and Middle Yangtze Valley. In this case, the Lower and Middle Yangtze Valley and the Lingnan region will be put together in our discussion. At this time, some typical features of the Neolithic began to appear at certain typical sites, including the Miaoyan site, the Dayan (Phase III) site, the Zengpiyan site, and the Dingsishan (Phase I) site. The earlier ones, such as the Dayan (Phase II) site where burned clay chunks are discovered and Bailiandong (Phase II) the edge-ground tools and perforated gravels. Besides, abundant tiny flint tools are found in other sites, signifying this transitional pattern for ca. 16,000 BP to 11,000 BP (Guangxi Liuzhou Bailiandong Science Museum 2009). Similar discoveries are excavated at the Dushizai site in Yangchun, the Huangyandong site in Fengkai, and the Niulandong site in Yingde. The synchronous development of the Lingnan region and the Lower and Middle Yangtze Valley represents the major character of the entire southern region during the Neolithic. This can draw an analogy to the predominance of microblade technology in North China.

Later, the subsequent developments witnessed a crucial divergence between the two regions: the Lower and Middle Yangtze Valley started to develop rice farming, while the Lingnan region formed the "Low-Level Food Production" with a preferred utilisation of root tuber cultivation and aquatic resources until the matured rice farming in this region expanded into the Lingnan region until the Middle and Lower Neolithic. In ca. 11,000 BP, the incipient Neolithic culture emerged in the Lower and Middle Yangtze Valley, at the same time, the adaptive method of the Paleolithic-to-Neolithic transition remained in the Lingnan region, representing an overt divergence between these two regions.

Neither North China nor the Lower and Middle Yangtze Valley entered the Neolithic in a standardised or unified pattern. The differentiations, divergences, innovations, and decay of the prehistoric culture intertwined together, composing a rich and varied pattern of changes during the Paleolithic-to-Neolithic transition. We cannot isolatedly understand any single culture, instead, we need to consider the transitional phase by taking the cultural development in both the Upper Paleolithic and Early Neolithic into consideration. Besides, the mechanisms of cultural adaptation of hunter-gatherers and the general patterns of cultural system changes should be understood, building the basis of the connection of fragmentary archaeological evidence, thus, we are able to obtain a complete understanding of this transitional phase.

6.3.4 *Explanation of Changes: What Do These Changes Mean? Why Did They Happen?*

During the Paleolithic-to-Neolithic transition, the changes mainly took place in two elements: firstly, the emergence of new cultural elements, which is regarded as typical characteristics of the Neolithic, appeared or sprouted during the Paleolithic-to-Neolithic transition; secondly, the advent of cultural differentiation, which chiefly happened in the Paleolithic-to-Neolithic transition. However, not all areas with new features appeared entering the Neolithic at the same time. It is essential to realise these two aspects. If the new features are ignored or overlooked, then the critical changes happening in the cultural system and the "order parameters" happening in the critical state within the cultural system would be disregarded. Besides, if we ignore the cultural differentiation, it would be hard to explain the differences in archaeological material, in this case, the comprehension of the agricultural origins would become impossible either.

From the Paleolithicto the Neolithic, the changes from hunter-gatherers to agricultural producers witnessed the steady loss of human mobility. We could divide this process into four phases: firstly, the construction of a forager strategy alongside stable central camps; secondly, the repeat utilisation of one area before gradually forming a number of central areas; thirdly, the formation of central sedentisms with initial Neolithic characteristics; lastly, the formation of Neolithic sedentisms. During the Paleolithic-to-Neolithic transition, the representative remains are the central areas and the central sedentisms, the former being represented by Shizitan site, and the latter by sites such as Donghulin, Zhuannian, Nanzhuangtou, Lijiagou, etc. Possessing ample and diverse lithic assemblages, remains, and relics, the Neolithic central sedentism sites show significant differences from the Paleolithicones. The loss of mobility does not only reflect on the site structures but also on the artefactual remains, for instance, fragile and non-mobile pottery, durable but labour-intensive polished tools, grinding slabs and handstones that are often used as "furniture". The loss of mobility has a radical influence on hunter-gatherers, just as conventional agricultural production gives way to commerce and industry, leading to a qualitative transformation in the whole cultural system.

The loss of mobility or the beginning of sedentism[22] determines that the hunter-gatherer culture has to apply a new cultural adaptive way to cope with the resource shortages that arise. “New cultural adaptation” is not a fixed concept, instead, it is an ongoing process that includes a range of adaptation strategies with the essence of utilisation intensification, that is, gaining the largest possible number of resources within a limited area. Eventually, a new way of accessing resources, that is, food production, or primitive agriculture (often referred to as agriculture), has been developed on the basis of these strategies. From the existing archaeological evidence, the intensification strategies can be included in the following four aspects.

Firstly, the repeated utilisation of one area. In fact, as a cultural trend that began in the Upper Paleolithic, the repeated utilisation of one area is particularly well-represented in North China, especially at the sites in Shizitan and the Nihewan Basin. The frequency of the utilisation still remains unknown, however, the hunter-gatherer mode of excessive resource acquisition within one area would lead to the exhaustion and deterioration of the resource environment. The essence of the practice is the same as that of contemporary agricultural issues. Draining the pond to get all the fish will make sustainability impossible. Once the prehistoric hunter-gatherers encountered resource pressure, their first response was to increase mobility and obtain resources as much as possible. However, this cannot be regarded as a stable adaptive approach. Amid the decreasing quality of the environmental resources, mobility enhancement would be hard to sustain.

Secondly, a broad spectrum of animal resource utilisation. During the Paleolithic-to-Neolithic transition, the number of large herbivore bones declined in the sites; on the other hand, animal bones are very fragmented due to the intensification of fauna resource utilisation. The decline in the herbivore amount is related to the ongoing warming process after LGM, when grasslands are replaced by forests; however, large herbivore remains are lacking at the Shizitan site on the western side of the Taihang Mountains. In this case, we could blame the phenomenon on human overhunting. At sites with well-preserved animal bones in South China, the preys include small carnivores and agile animals, which are small and difficult to hunt, such as rabbits. Besides, the discoveries also cover birds, amphibians, and various aquatic species. Compared to the Upper Paleolithic, the Paleolithic-to-Neolithic transition witnesses a significant expansion of the range of animal resource utilisation. This utilisation range is almost equal to North China or South China (including the Middle and Lower Yangtze Valley and the Lingnan region) in the Upper Paleolithic.

Thirdly, the intensification of certain plant resource utilisation. Based on Binford's study on nearly 400 contemporary hunter-gatherer groups, intensifying plant resource utilisation becomes inevitable when hunting becomes unsustainable (Binford 2001). Clearly, grinding slabs and handstones compose evidence of the plant resource-use intensification. Although it is unclear what exact plants ancient people processed, at least we could know that it is the plant they handled. Notably, under the circumstance of mobility decline, durable tools are developed to proceed plants. That is to say, intensification of plant resource-use promoted the occurrence of agriculture. The appearance of pottery (vessels) indicates a reduction in mobility and the intensification of plant resource utilisation. However, pottery is not necessarily used to solely process plants, it could also be used to process aquatic resources or other resources. Nevertheless, from the functional and symbolic perspectives, the emergence of pottery represents a product of intensification.

Fourthly, the utilisation of certain aquatic resources. Aquatic resources and the terrestrial ones show different temporal patterns. Besides, the former is much more abundant (as the aquatic resources are mobile and replenished, thus, could be used repeatedly in the same place. However, terrestrial resources do not possess this character). Moreover, aquatic resource utilisation allows everyone (even elders and children who cannot participate in hunting) to join in protein acquisition activities (shell-fish collection). When hunter-gatherers lose mobility, aquatic resource utilisation could support a sedentary life without agricultural production. Unlike agricultural production, aquatic resources utilisation does not require sedentism as humans could move along rivers and coastlines. The loss of mobility leads to aquatic resource utilisation, not the other way around. Similar to agricultural production, aquatic resource utilisation, which serves as an important means of intensification, relieves pressure on terrestrial resources. During the Paleolithic-to-Neolithic transition, evidence of extensive aquatic resource utilisation could be seen in sites in the Lingnan region, followed by the Middle and Lower Yangtze Valley, and even in North China, where aquatic resources were relatively limited. With certain conditions, aquatic resources were utilised in certain places, such as the Nanzhuangtou site. Thus, aquatic resource utilisation plays an essential part of substance in the Lingnan region, while a limited one in North China.

Then, we will discuss the reasons for mobility loss, especially why this happened during the Paleolithic-to-Neolithic transition instead of earlier phases. In the early phase of LUP, before the emergence of agriculture origins, the mobility of hunter-gatherers increased instead of decreased, as if in the darkness before dawn. All of these questions need to be explained.

Let's start with the beginning of the LUP because people at that time (anatomically modern humans) were already physically the same as us and mastered basic techniques. In this case, they could theoretically develop agriculture at least 30,000 BP but failed, why? What prevents humans from agricultural production? Some scholars attribute the agricultural failure to the unstable climatic conditions of the Pleistocene (Richerson et al. 2001, pp. 387–411). However, this hypothesis is difficult to test, as there is no archaeological evidence to prove that humans ever tried and failed agriculture production. Based on archaeological materials, the formation of distinctive regional cultures with unique styles of artefacts becomes a cultural trend in the LUP, marking the intensification of regional ideas. The concept of territoriality equals to an exclusive way of resource utilisation, that is, an area that serves a certain group cannot be unconditionally used by other groups. Martin Butzer explains territoriality based on restricted beneficent relations, that is, groups cannot bear unlimited beneficence but have to limit the scope of beneficent relations amid the population increases, hence, a territorial culture is formed (Wobst 1976, pp. 49–58). Butzer notices the importance of population increase. The LUP saw a large-scale human immigration, occupying all continents except Antarctica. If there were still enough space to contain immigrates, it would not have been possible to develop a high population density that makes the universal beneficent relations untenable. At the latest of 15,000 BP when humans spread to the American Continent, almost all areas that could be exploited by hunter-gatherers had already been used. This constitutes an important background for the agricultural origins.

In addition to the population increase, the unfeasibility of population migration further established a higher population density. A higher population density is associated with the development of social complexity, increased competition for resources, the failure of universal beneficence, and the strength of territorial perceptions. In this case, to obtain sufficient resources, people could only move more frequently within their own sphere. This idea has been supported by the archaeological evidence from the LUP (such as the microblade technology from North China and the tiny flint tools from South China). After entering the Paleolithic-to-Neolithic transition, this trend became more obvious. The raise of population density and the development of social complexity leads to the strength of territorial perceptions and intensified competition within groups. At the same time, social complexity affects different groups but also makes the competition within groups fiercer. Thus, some scholars argue that the purpose of pottery emergence is for display and gaining prestige (Arnold 1996). Besides, scholars even argue that agricultural production is developed for winning in social competitions instead of maintaining substance. The acceleration of population growth and increase of population density constitute one of the initial conditions for the agricultural origins, that is to say, excessive population spreads to new habitats becomes unrealistic, or what Binford calls "packed" (Binford 2001), which also serves as an impetus to push forward the agricultural origin.

Population is one of the initial conditions, but population growth will not necessarily lead to an agricultural origin. We have already discussed in detail about the structural differences in the resource endowment of hunter-gatherer culture. The existing archaeological evidence indicates that the occurrence of agricultural origins happens in North China and the Middle and lower Yangtze Valley. In Chapter 4, I have already analysed the structure of resource endowment in the LUP, especially the shifts in adaptive strategies seen in the Middle and lower Yangtze Valley where hunter-gatherers shifted their strategies from a predominantly forager ones to a predominantly collector ones, and from a stable single adaptive one to flexible multiple adaptive ones. In the Paleolithic-to-Neolithic transition, the accumulation of resource endowment structures and their impact became more obvious. Closely linked to the development of the LUP, several Neolithic cultural elements are excavated from North China instead of surrounding areas. Owning new culture elements, North China accommodates both the LUP hunting and gathering economies, with a slight preference for hunting or gathering. It was the areas with a preference for gathering that formed the earliest agriculture rather than areas dependent on hunting and aquatic resources.

As an initial condition, paleoenvironmental change produces a hard constraint on cultural development and plays a major impact on the cultural synchronisation and differentiation during the Paleolithic-to-Neolithic transition. Arriving of the LGM promoted the emergence of the microblade technology in North China. With the cease of the LGM, temperatures continue to rise, postglacial vegetation is replaced by new types, and fauna is simultaneously affected. The resource-use intensification in cultural adaptation begins to emerge, as has been confirmed by a series of archaeological finds. Interestingly, the paleoenvironmental changes also could be seen in the Lingnan region (and possibly including Southwest China) simultaneously and even earlier. Based on Fig. 6.4, temperate forest vegetation just covers south of the Middle and Lower Yangtze Valley ca. 12,000 BP; synchronously, subtropical vegetation covers the Lingnan region. Both the Middle and Lower Yangtze Valley and Lingnan region need to reserve resources (Chapter 3), as the shorter growing seasons cannot provide enough food supply for the whole year. More interestingly, as the tropical climate zone recovered after Holocene, the Lingnan region did not move forward to develop grain agriculture, instead, a "Low-Level Food Production" combining the utilisation of root tubers and aquatic resources was developed in this area (Chapter 8). Around the onset of the Holocene, North China also saw a divergence: agricultural production developed in the foothills and plains; while hunting and gathering continued to develop in marginalised areas, or a mixed economy was adopted.

In the process of differentiation, the influence of the absolute constraint environment is very obvious. The development of primitive agriculture is far more difficult in the tropics than in the temperate zones, the reasons are as follows: the tropical soils are barren as they contain low organic matter; the lush growth of plants that compete with grains and severe pests; plant diseases and insect pests in this area. At the same time, primitive agriculture is replaced by the plant resources that are available for collection and the abundant aquatic resource conditions in tropical areas. Hard constraints and new opportunities intertwined together, leading to the divergent development paths of the Lingnan region and the Middle and Lower Yangtze Valley. Within North China, there is a large difference in water and heat conditions between the west and east of the Taihang Mountains. Agriculture in the west of the Taihang Mountains was either not established or unstable by the Early Neolithic, and the importance of agriculture in substance was far less important than that of the North China Plain. Similarly, with the warmer temperatures after LGM, these areas became an intersection of forest and grassland

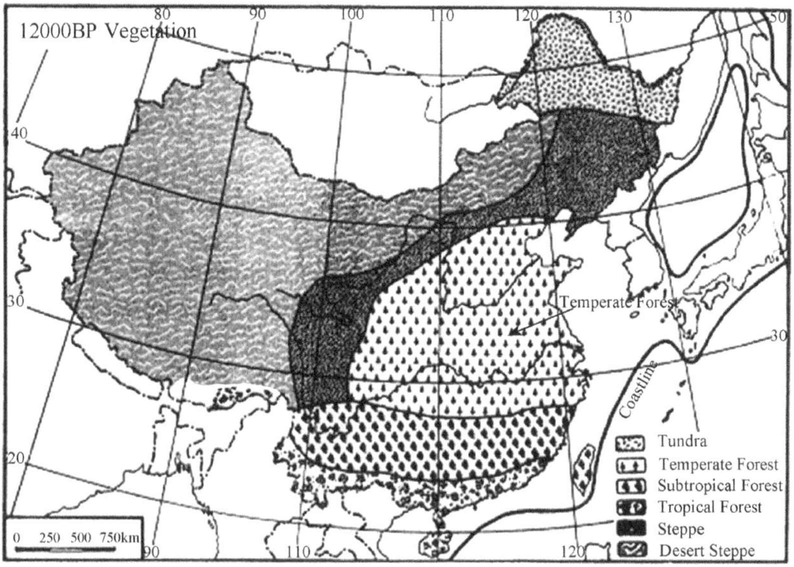

Fig. 6.4 Vegetation from 12,000 years ago

again, possessing natural favourable conditions for hunting. In this case, hunting continued to serve a major part in people's substance. Based on the same principles, the Lingnan region diverges from the plain regions in North China.

Primitive agriculture probably began as a buffer to fight against the substantial risks of the time, especially in seasons of resource scarcity. Substance risks not only include the shortages of resources, but also uncertainty about them. Radically, agricultural production aims at crops that are easy to reserve. The reserve facilities in the Early Neolithic conform to this property. Gathering and wild resource-use intensification is a necessary process that must be undergone before agriculture is formed, being reflected in the Paleolithic-to-Neolithic transition. This can be traced to the environment of that time and the complication of cultural development, which includes the advancement of technology and the complexity of social structure. Primitive agriculture is only one form of intensification, not the only means. The cultural differentiations in the

Paleolithic-to-Neolithic transition witnessed distinctively developmental trajectories in different regions.

6.4 Summary: A Revolutionary Transition

During the Paleolithic-to-Neolithic transition, the changes occurring in the core area of China's agricultural origins are revolutionary in nature and transitional in process, lasting at least 5000 years (from the emergence of new features to cultural differentiation in the Holocene). In this case, I adopt the term "revolutionary transition" in this context. Even the Industrial Revolution, which took place only for decades, is now considered a transition rather than a revolution. The issue of time scale is involved under this circumstance. Archaeologists are good at analysing issues on a long time scale. Undoubtedly, 5000 years is not a short period, however, it is a relatively short fragment when being put in the hunter-gatherer history that lasts millions of years. On the other hand, we need to understand the profound influences brought by cultural development. Agriculture comprehensively changes hunter-gatherer cultural systems in all directions, including the structure of human society and the cultural-ecological relations of humankind. These profound changes are unique in human history. Therefore, it cannot be dismissed as revolutionary. Although the archaeological material available in China is still quite limited in terms of both quantity and refinement, it is sufficient to confirm the existence of the "Paleolithic-to-Neolithic transition", its basic characteristics, and cultural processes. Probably we cannot understand the social transition from hunter-gatherer to agricultural societies in the same way that we understand the one from agricultural to industrial societies. However, at least it is clear that some regions are fast, some slow, and some barely affected during the transitional period. Also, different cultural processes are determined by the factors of environment, population, the accumulation of cultural development, etc.

6.5 A Comparative Perspective: Basic Features of the Origins of Agriculture in China

Why do we need to apply a comparative perspective? Chinese archaeological researchers rarely consider the origins of Chinese agriculture within a world framework due to a lack of academic environment and educational background. Chinese archaeologists, who are concerned with the origins of Chinese agriculture, would find it difficult to conduct relative research as a consequence of insufficient information or relevant academic background. However, the cultural mechanism of agricultural origins cannot be understood if we merely look at Chinese agricultural origins in isolation. As a result, studies of the origins of Chinese agriculture could be only launched in a world context by scholars outside of China. However, the situation has changed in recent years, but the research carried out by Chinese scholars is still limited, requiring collaboration with scholars outside of China. In addition, scholars from both China and outside of China normally just focus on the origins of Chinese agriculture, while largely ignoring cultural changes in other locations that coincided with the origins of Chinese agriculture. In this case, it seems that only the agricultural origins, especially the origins of grain agriculture, are worth studying; while the rest of the cultural changes are less advanced, conservative, insignificant, and therefore unworthy of exploring. This idea is biased and unreasonable. Analogically, it is unreasonable to detect the origins of modern society by merely focusing on the Industrial Revolution that happened in a few major countries, such as Britain and France.

In light of the whole world, China is not the only place where agriculture originated; in the context of China, agriculture origin is not the only cultural change that occurred at that time. A comparative perspective acts as a fundamental way of understanding the unity and diversity of cultural development, helping us to explore the mechanisms of cultural changes and occurrence conditions and fulfilling the archaeologist's purposes of understanding and interpreting archaeological material. Thus, a comparative perspective is essential to understanding the basic features of China's agricultural origins. Without a comprehensive understanding of the pattern of the world's agricultural origins, a comprehension of the characteristics of Chinese agricultural origins would become impractical. Undoubtedly, comparisons are important, but artefatcs being contrasted are equally essential. Based on the order of archaeological research, the spatial and temporal framework should be considered the first artefact being examined, that is, what kind of dates and areas of occurrence lead to the environmental conditions. In other words, what kind of climate change arised? What kind of special natural conditions appeared in the area where agriculture took place? What kind of influences were brought about by climate change? Then, we need to consider the species' domestication. What kind of species were domesticated? What features did these domesticated species have? What's the process of domestication? Finally, we need to scrutinise the depth of cultural change, with the combination of domesticated species and the society of the time. What kind of changes took place in society? What kind of structures did they form? And what is the historical significance? By comparing these aspects, we could have a better understanding of the process and significance of Chinese agricultural origins.

6.5.1 *Major Centres of Agricultural Origins in the World*

The materials about the origins of agriculture in the world are strikingly abundant. However, due to the limited space, I would only describe it in a macro and rough way. Current archaeological research has identified that there are at least seven regions where plants or animals being domesticated independently: West Asia, North China, the Lower and Middle Yangtze Valley, the Sachel zone in Africa, Southeast North America, the Mexican Plateau in Central America, and the Altiplano Plateau in South America. Besides, there are many more places that claimed independent domestication events, such as the New Guinea Highlands, the Indus Valley, the western deserts in Egypt, the Ethiopian Highlands, and the tropical jungles in South America, etc. It is predicted that more areas might be added to the domestication sequence in the future.

Actually, different research perspectives should be paid attention to, as there are two main groups of scholars who currently study the origins of agriculture. Biologists constitute the first group, highlighting the origins and biological mechanisms of domesticated species. For example, in the nineteenth century, Darwin studied the evolution mechanisms through the artificial domestication of plants and animals. The Swiss botanist A. de Candolle published *Origin of Cultivated Plants* (1882), which set out the main criteria for determining the crop origins. Later, the Soviet scholar Vavilov identified "eight centres of crop origin" and three sub-centres (Vavilov 1982). On the basis of theories proposed by Vavilov, the American geneticist Jack Harlan recognised the distinctions between origin centres and dispersal areas. Then Harlan proposed the theory of "seed dispersal", which classifies different types of domesticated species based on their spatial and temporal distribution (Harlan 1971). Next, Hawkes went further, realising the differences between centres of crop plants and agricultural origins, thus, defining diversity regions as the regions that spread crops from the core centre to form a rich variety. In addition, Hawkes applied the concept of "small centres" to describe places where only a few species originated (Hawkes 1983). Chinese scholars have identified five major centres of crop domestication: China, the Near East,[23] Central and South America, South Asia, and Africa (Li and Dong 2006).

Archaeologists compose the second group, primarily concerning cultural changes, that is, the beginning of domestication, and the role of domestication in prehistoric culture and human history. As aforestated, the archeologist's perspectives are as follows: from the aspect of time scale, the incipient and development of agricultural origins, as well as its eventual historical impacts could be examined; from the mindset of cultural systems, investigating the depth of the impact of agricultural origins could be realised. Crop cultivation, as substance complementation for hunter-gatherer groups, profoundly transferred hunter-gatherer groups into farmers and pastoralists. The cultural impacts, which are scrutinised by archaeologists in the study of the origins of agriculture, are historically crucial, especially the influence on the process of social sophistication or the origins of civilisation. Many regions are abundant with domesticated species, but only several of them established "civilisation" (state organisation, writing characters, cities as signposts) on the basis of domesticated species. The four cradles of civilisation include West Asia, China, Central America, and the Andes in South America. The research proposed by archaeologists and biologists is complementary and corroborative with different highlights. In actual research, the perspectives of archaeologists and biologists are often intertwined, thus, making it difficult to distinguish from each other. In modern research, which is dominated by science, the archaeological perspective is prone to be overlooked and marginalised. The archaeologist's perspective combines the social, cultural, and historical (with an emphasis on cultural origins and development in a spatial and temporal framework) elements. Conflating the two perspectives would lead to the neglect of the role of human culture. Thus, archaeologists would become insignificant, because they would merely become excavators and providers of analytical material.

The perspectives for understanding the origins of agriculture are diverse. In addition to the perspective of disciplinary distinctions, the study of the level of agricultural development also provides an important perspective in this field. After recognising the diversified origins of world agriculture, archaeologists need to further understand its structural features, where each perspective would contribute to a more detailed and profound understanding of the agriculture origins, similar to a Computed Tomography Scan. From hunter-gatherer to agricultural establishment is a continuous sequence with intermittent evolutionary features (Chapter 1). In terms of the origins of world agriculture, Smith distinguishes several developmental phases from low to high levels regarding to food production, that is, from limited/extensive to intensive agriculture. Such classification boosts our understanding of diverse forms of "agriculture" in the archaeological material. In the Chinese context, the areas where intensive agriculture developed initially include North China and the Middle and Lower Yangtze Valley. The Liaoxi region, located in the north of the Yan Mountains, showed a fluctuating mixed economy. The Lingnan region grew another type of low-level food production. In addition, the agricultural patterns were different in Northeast and Southwest China. Compared to the regions illustrated, the Tibetan Plateau evolved a stable adaptive strategy much later. Worldwide, intensive agriculture was foremost established in West Asia and China; then in Egypt, the Indus Valley, and the European region; and finally in Africa and Southeast Asia. Intensive agriculture was formed much later in Central America and the Andes in the New World, but it eventually developed independently. Similar to the Lingnan region of China, low-level food production persisted for a relatively long time in southeast North America, Africa, and Southeast Asia.

To a large extent, the differences between low and high levels of food production are the ones between tropical and temperate agriculture. For example, although both Central America and the Andes belong to tropical areas in latitude, the environment is similar to temperate regions because they belong to highland environments with vertical distribution characteristics. A similar situation existed in the Ethiopian highlands and New Guinea, where situated on a highland and claimed evidence of domestication. In the tropical lowlands, there are strong hard constraints to intensive agriculture: barren soil conditions, plant diseases and insect pests, intensive labour of clearing jungle and weeds, the limited lifespan due to tropical diseases, and negative effects brought by agriculture (such as the pollution of water sources, the destruction of vegetation, and soil erosion) severely constrain the development of agriculture. The current archaeological evidence seems to be able to support the existence of agriculture in tropical areas. For instance, evidence of human intervention, dense sedentisms, and population density has been found in the Amazon jungle, indicating the existence of agriculture (Heckenberger et al. 2003; Stokstad 2003; Mann 2000, 2002). Analogously, researchers argue that intensive plant utilisation and cultivation began tens of thousands of years ago in New Guinea, followed by mounding cultivation, ditched cultivation, domestication of taros (*Colocasia esculenta*), bananas, and possibly sugarcanes (Denham et al. 2003; Newman 2003). Through analysing spore powder and phytolith, evidence of domestication has also been found in Ecuador (Bryant 2003; Piperno and Stothert 2003), however, where other types of archaeological evidence have not been found. Notably, in the tropics, the suitable conditions for agriculture and the origin of agriculture are different. In the tropics, the utilisation of domesticated species complemented hunter-gatherer-based substances, marking a low-level of food production because the changes in social structure are limited.

Worldwidely, domestication can be broadly divided into the following categories in terms of species: grain farming, root and tuber crop farming, animal domestication, that is, animal husbandry, cultivation of fruit and trees to supplement substances, and other types of domestication to supplement substances. In the process of low-level of food production, domesticated artefatcs are used as condiments, other non-food artefatcs (for example, gourd shells are used as religious ceremonial vessels, and gourds are used as buoys for fishing nets), or limited food supplements (for example, sunflowers and amaranth in North America). Hayden notices non-food functions of domesticated artefatcs, thus, he emphasises the socio-economic aspects of the origin of agriculture. Major artefatcs of the domesticated fruit trees include bananas and other food supplements. Only root and tuber crop farming and grain farming reached the level of intensive agriculture. The former mainly distributed in the Andean region where potatoes were domesticated and corn/maize cultivation was introduced here afterwards, constituting the basis of South American civilisation.[24] In summary, grain agriculture is the groundwork of almost all civilisations, as grains are annual plants that are accomplished by sexual reproduction. These properties of grain are easier for people to select improved breeds. Compared to root and tuber crops that are attained by asexual reproduction, grains are more stable, thus, facilitating their spread. In addition, it is easier to initiate grain cultivation on a large scale and seed storage. Notably, the advantage of storage has in turn contributed to the development of social complexity (Testart 1982). Thus, the differences in natural species distributed among various geographical regions somehow determined subsequent civilisations. According to the statistics from Jared Diamond, the largest number of large-seeded wild crops is allocated in West Asia, with 33 of the 56 large-seeded wild crops in the world occurring within (32 species occur in the Mediterranean region). While, only 2 species existed in South America, and 5 in Central America where the amount of large-seeded wild crops was far fewer than that in West Asia, as well as its adjacent areas (Diamond 2000).

Regardless of the condition of the species, the utilisation of domesticated artefatcs is crucial. Based on less than six large-seeded plants occurring in East Asia, two major centres of agricultural origin are formed here. In East Asia, domestication is not only the conscious cultivation, selection, and utilisation of species but also a revolutionary change in cultural systems, for example, sedentism, pottery, polished tools, architecture, etc. Thus, unlike a hunter-gatherer society, a different structure of cultural systems is formed, that is, an agrarian society. Similar to China, West Asia underwent changes in cultural systems before the real formation of agriculture. In other words, the change in cultural systems simultaneously promoted domestication. Different from this conscious and systematic domestication, there is the other type of domestication named "symbiotic domestication", which could be seen from the domestication of pumpkins in Ecuador and it happened in southeastern North America. Symbiotic domestication is associated with long-term utilisations by people. Domesticated artefatcs have only occupied a limited proportion of human substance, thus, they lack further systematic development. Likely, cattle were once found at the site of Nabta in the Western desert of Egypt (Wendorf and Schild 1998), indicating the cattle domestication was achieved under human invention since the species could not be able to survive dependently in desert oases where they possessed insufficient surface water. However, domestication did not continue, thus, the subsequent history was not influenced by it. In this case, domestication, which shapes agriculture, is a cultural ecosystem, including a whole set of cultural concepts, tool systems, social organisation, etc. However, in areas with a low-level of food production, domestication merely indicates the domestication of species, with specific species only utilised for humans without affecting the structure of the cultural system in a comprehensive way.

Notably, domestication is a concept that can spread and be inherited rapidly. Therefore, there are centres and radiating areas of domestication, as well as primary and secondary development of domestication. Biologists indicate that the centre of domestication is the sum of the entire historical period in which the idea of domestication is applied to further develop the domestication of species. Instead, the centre of domestication is not where domestication first occurred or where the idea of domestication arose and profoundly affected cultural systems. Throughout the Old World, the domestication process in Egypt, Europe, and the Indus Valley is associated with the spread of agriculture in West Asia. In West Africa and East Africa, the agricultural process is more recent, suggesting that the onset of agriculture is closely linked to ancient Egypt. The process of civilisation in East Asia is correlated with the spread of farming in China. In short, there were only two authentic centres of agriculture and civilisation in the Old World, that is, West Asia and China. In this case, these two areas constitute the best examples when analysing the issue aforestated in a comparative perspective.

6.5.2 *The Origins of Agriculture in West Asia and China*

Most people view contemporary West Asia as not suitable for developing agriculture, due to the arid climate and barren soil. While certain areas are prosperous because of the abundant oil resources. Undoubtedly, this stereotypical impression is a misconception. During the Paleolithic-to-Neolithic transition, West Asia was unique in human history. The most complete and systematic archaeological materials are preserved in West Asia. Until now, more than 200 hunter-gatherer sites, to 20,000–7000 BP (corrected chronology), have been found in West Asia. Besides, over 800 14C dating data in the Neolithic have been obtained from nearly 100 sites in this region, attracting scholars and researchers from over a dozen countries and discipline backgrounds who participated in the research. The richness of the theoretical background and the application of multi-disciplinary approaches are advanced worldwide, which is well worthy of Chinese archaeological research to refer to. Initially, American geologist Pumpelly suggests that the drought events resulted in the origin of agriculture in West Asia. Later, on the basis of Pumpelly's hypothesis, Childe proposes the Oasis Theory and refines Pumpelly's assumptions. In the 1950s, Braidwood proposed the "Fertile Crescent Theory", regarding agriculture as a result of human cultural progress rather than climate change. The reason that agriculture initially occurred in West Asia could be simply attributed to the existence of domesticated plants and animals. Binford proposes the Marginal Zone Theory, emphasising that the adaptive pressure leads to the origins of agriculture (Chapter 1). Thanks to the abundant archaeological material and pioneering theoretical research,

West Asia has become a classic and typical region for the study of the origins of agriculture.

Western Asia is not a separate geographical unit as it can be divided into at least three geographical parts: the Anatolian Plateau, Mesopotamia Plain, and the Mediterranean Climate Zone. Most of the regions between the Tigris and Euphrates have a climate of tropical desert that is terribly dry, which results in the infeasibility of cultivation without irrigation. Similarly, inland of the Anatolian Plateau has a continental semi-arid climate with less than 250 millimetres of annual precipitation. The Mediterranean climate zone covers the coastal areas of Anatolia, the Levant (the eastern shore of the Mediterranean), and the northern part of the Mesopotamia Plain. The "Fertile Crescent" region proposed by Braidwood is also the area encompassed by the 300 millimetres precipitation line that divides diverse plants. This precipitation line runs along the arc of the mountains, from the Levant highland and the Toulouse Mountains to Zagros Mountains. Diamond summarises East Asia as a place with the most abundant resources that profoundly influenced the origins of agriculture (Diamond 2000).

As for the climatic character in the Mediterranean climate zone, the winters are mild and wet, while the long-lasting summers are hot and dry, with the agricultural growth pattern of "in one year, withers and thrives once each". In this area, a large number of annual grains and legumes with large seeds lie dormant in the dry season and germinate in the rainy season without wasting energy on woody parts and fibrous stalks. The wild ancestors of these crops are already lush and productive, especially in the fertile areas where they generate plentiful wild crops and can harvest over 130 pounds of seeds per acre (this figure is almost equal to the early agricultural one). According to the wild wheat harvest experiment in Turkey initiated by Harlan, collecting one kilogram of seeds purely by hand equals to harvesting 50 calories of food energy by costing one calorie of labour; while, with the help of sickles, a family of four could harvest wheat that is enough for the whole year consumption within three weeks. Wild wheat has great nutritional value (Harlan 1992), with a protein content of 8–14%, which is higher than rice and corn/maize (which lack two important amino acids needed by humans).

The Fertile Crescent has the largest Mediterranean climate zone in the world, with a variety of climate changes and wild plants, as 33 out of 56 of the world's most valuable species of poaceae could be found here. In terms of the size of seeds, barley and emmer wheat are ranked first and thirteenth. Bar-Yosef et al. compare 23 species of seeds that are still growing in the Jordan Valley, selecting the largest and the most delicious poaceae. Then they found that barley and emmer wheat were the best: emmer wheat has the largest seeds and barley the second largest. From the domestication perspective, barley is the easiest one to be domesticated, and emmer wheat is easy to collect and dehull. Ancient hunter-gatherers selected plants based on the criterion of seed size, tastiness, and yield (Bar-Yosef and Kislev 1989).

In the Fertile Crescent, there is a high proportion of monoecious self-pollinated plants and occasional cross-pollination ones. The former makes domestication easy to achieve, while the latter helps people to select new types of plants, such as bread wheat. Einkorn wheat, emmer wheat, and barley are all self-pollinated. The domestication mechanism of annual plants is easy, allowing people to continuously observe and improve the process. In this case, humans just need to control one or two genes, improving characteristics (such as grain-falling properties), thus making domestication easy to achieve. The domestication experiment carried out by Hillman has proved that domestication could be achieved in two hundred years with artificial selection (Hillman and Davies 1990). In comparison, due to the long growing period and multiple bitter-taste genes, oaks are hard to domesticate. The ancestor of American corn/maize has a different seed and flower structure from general corn/maize. In a wild state, the ancestor of American corn/maize is less productive than wheat, as the seeds of the former are small with a hard shell covered. Thus, it would take a long time to select and evolve from the wild species to the present one. In addition, these annual plants grow fast and thus can be harvested in a few months and stored after drying.

Another advantage of the Fertile Crescent is reflected in its high variation of height and topography over short distances. This property indicates that a large variety of plants are easy to collect and transplant since the harvesting season can be staggered. In the Fertile Crescent, there are also concentration areas for large wild mammals that are particularly suited to domestication, including goats, sheep, pigs, and cattle. As herd animals, goat and sheep groups are easily controlled after managing the head goal. The eight founder crops include emmer wheat, einkorn wheat, barley, lentil, pea, chickpea, bitter vetch, and flax. Except for flax and barley, which have wild ancestors in other areas, the rest of the species are only available and domesticated in the Fertile Crescent. Because of the existence of suitable plants and animals, this area could quickly establish an effective and balanced environmental association and form a complete cultural ecosystem: three types of crops can be used as a source of carbohydrates; four types of legumes, four species of domestic animals (goats, sheep, cattle, and pigs), wheat, fibre, and flax (40% oil content) as rich supplement of protein; finally, the animals can also be used for milk production, shearing, and servitude. In this case, the basic needs of the farmers could be met through crops and livestock. In this case, domesticated crops can provide foragers for animals, creating a complementary economy (when the amount of domesticated animals does not exceed the amount of forage available from the crop).

From the other aspect of geography, the other types of resources in a hunter-gatherer society are fewer, including aquatic resources and root and tuber plants. In hunting activities, the main prey is merely gazelle, suggesting the possibility of over-exploiting/overkilling.[25] On the basis of wild grain gathering and processing, sedentism villages have already occurred in the Fertile Crescent before the incipient food production. In this case, it is easier for hunter-gatherers in the region to transfer to agricultural producers more quickly than those in China. In comparison, sedentism in Central America was achieved until 1500 BC. In Central America, the available domesticated animals were merely turkeys and dogs, and available domesticated crop corn/maize, hindering the process of sedentism.

Advantageous conditions in the domesticated plant and animal resources constitute an important precondition for the origins of agriculture in Western Asia. However, the population and culture also influence the origins of agriculture. Andrew Sharrat proposes a "population bottleneck" theory (Sherrat 1996). Surrounded by oceans and deserts, West Asia is situated at the crossroads of three continents, Asia, Europe, and Africa. Before the occurrences of seafaring boats, human mobility must pass through this region, thus, a population bottleneck is created with higher population density and pressure. At the same time, water constraints result in a highly patchy resource distribution. Hence, people concentrate on nearby water resources. At the beginning of LGM, population growth started and concentrated in places where environmental conditions were more ideal (Belfer and Morris 2011). There is a positive correlation between population and the origin of agriculture: population does not necessarily lead to agriculture, but agriculture requires a certain population density. Except for the population aspect, West Asia, as the crossroads of three continents, also serves as a crossroad of prehistoric culture and the first area in Eurasia to accept the culture of modern man as it was the first area reached by modern humans out of Africa. Thus, abundant cultural resources could be available in West Asia, forming an advantageous condition for agricultural development.

Evidence of the intensification of plant resource use at the end of LGM is found in West Asia. At the Ohallo II site, which is 23,000 BP, over 100,000 charred plant remains are found, as well as grinding slabs covered with starch residues (Kislev et al. 1992). After the end of LGM, the increment of temperature resulted in the expansion of forests and grasslands. In this case, humans had more abundant resources to utilise and more opportunities for further resource exploitation, thus, new natural habitats were established. However, being regarded as a result of climate change, this new natural habitat is different from those created by agriculture. The Geometric Kebaran culture, dating from 14,500 to 13,000 BP, expanded into the former desert area at the time, which has become grassland by now. The occurrences of grinding slabs and stone bowls suggest the existence of plant processing practices. Generally speaking, the popular geometric microblade are associated with the utilisation of aquatic resources or small-sized animals (Binford 1968) whereas in this case, they are more likely to be associated with plant harvest and processing.

As a representative during the Paleolithic-to-Neolithic transition, the Natufian culture is characterised by complex hunter-gatherer property and the earliest rural societies are formed within. Some sites of the Natufian culture are located in caves, which is similar to those in the Paleolithic, however, some sites already show a larger size with obvious architectural structures. Taking the Ain Mallaha site of Natufian culture as an example, it is located near a spring (Ain means spring) and covers an area of over 2000 square metres, with 200 to 300 people living here. A large number of circular buildings are scattered on this site, indicating a dense construction. Besides, caries have been found in the human skeletal material, implying the intake of carbohydrates; grinding slabs, handstones, grinding stone bowls, plates, and mortars are found at the site as well. Wild plant collection, processing, and storage techniques have already been developed based on the occurrences of flint scythes, baskets, stone mortars, grinding slabs, stoves for roasting grain, and cellars for grain storage (Perrot 1966, pp. 39–94).[26] The excavations explored in the Natufian culture are very comparable to those from sites such as Donghulin and Zhuannian. In the burials of Natufian, necklaces made of shells are found (such as the El-Wad Cave and the Jiedi site), which are akin to those at the

Donghulin site. Besides, stone vessels and grinding slabs are discovered at the sites of Natufian and the Zhuannian. The main difference reflects on the sedentism buildings and pottery. At the sites of the Natufian culture, sedentism buildings are found but with no pottery, whereas the situation is reversed at sites in North China during the Paleolithic-to-Neolithic transition (Table 6.6).

Bar-Yosef considers the people under Natufian culture to be sedentary hunter-gatherers (Zhang et al. 1999). Also, he suggests that hunter-gatherers might apply mobile strategies of both foragers and gatherers to maximise adaptability. That is, certain individuals within certain groups, such as males, maintained high mobility. In terms of sedentism, Bar-Yosef implies that biological evidence could include living beings (such as house mice and sparrows), and seasonal flora and fauna, such as winter and summer utilised species coexisting at the same site. The archaeological evidence mainly comprises storage pits, house sites, and durable tools (such as stone grinding tones and ground stone tool). Based on the study related to the discard process, the Natufian people do not adapt to a sedentary life. The discarding pattern of artefacts is more akin to that of mobile hunter-gatherers who lack maintenance of the site and recycling of artefatcs, which are typical characteristics of sedentary life. A large number of food debris, tool-making debitages, de facto refuse, ornaments, and human bones that are ritually significant are found in these sites. The evidence shows different discarded features of that in the Neolithic (Smith and Edwards 2004). From the Pre-Pottery Neolithic A, the discarded mode has changed, with the occurrences of secondary refuse.[27] By the Pre-Pottery Neolithic B, more secondary refuse are discovered, showing the character of sedentism.

Which statement is more reasonable? Perhaps we should highlight the definition of "sedentism". Regarding the new artefact features and site structures, the Natufian culture is different from LUP. However, the difference does not mean that sedentism (living in one place consistently for more than a year) was able to be achieved, instead, the lower mobility might be more reasonable. According to Bar-Yosef, the evidence of sedentism might be found at some "semi-sedentary" sites. The discarded mode is judged on the basis of the eventual discarding outcome. Compared to the features of abandonment in the Neolithic, it is definitive evidence. Thus, the situations in both West Asia and North China are similar. Some sites, such as Donghulin, Zhuannian, and Nantouzhuang, represent a substance where mobility is significantly reduced. The males

Table 6.6 Comparison of characteristics of the PNT in West Asia and North China

Characteristics	*West Asia*	*China*
Chronology	ca. 20,000 to 10,000 BP, Post-Paleolithic Early Phase: Kebaran culture Middle Phase: Geometric Kebaran culture Late Phase: Natufian culture (Paleolithic-to-Neolithic transition)	ca. 22,000 to 15,000 BP, LUP ca. 15,000 to 8000 BP, Paleolithic-to-Neolithic transition, similar to West Asia
Cultural conditions	Fertile Crescent, extending from the Levant region to Mesopotamia, Mediterranean climate, semi-arid	Piedmont zone in North China, has a temperate climate, semi-arid, and similar natural conditions to West Asia
Artefacts	Apart from pottery, tool technology is not divergent, with more developed stone vessels and larger amounts of grinding tools	Pottery has been discovered, but the findings are relatively limited at present
Ground stone tool	Found	Found
Stone vessels	Found	Found
Pottery	Not found	Found
Ground stone tool	Small amount	Small amount
Chipped stone tools	Containing microliths	Containing microblade technology products
Site structure	The scale and complexity of the sites in Western Asia are impressive, with sedentism villages appeared where hundreds of people could live assembly	The site structures are distinguished from those in the LUP. However, the scale and complexity are not as good as that in West Asia
Sedentism	Over a thousand square metres in size with a complex structure	The area is relatively small
Building construction	Clear and complex	No evidence is shown
Storage facilities	Found and complicated	Rarely seen
Burial	Found	Found

(continued)

Table 6.6 (continued)

Characteristics	*West Asia*	*China*
Flora and fauna remain	Good conditions for domesticated species, with wild species of all eight categories of domesticated species distributed here, allow for a more complete cultural ecosystem	The wild ancestors of millet and broomcorn are distributed here with relatively less productivity
Potentially domesticated species	Goats, sheep, cattle, and pigs	Dogs, pigs, and chickens
Main preys	Antelopes	Deer
The possible intensification of plant resource utilisation	Wheat and legumes	Millet

within groups would continue to hunt and maintain mobility, which is evidenced by the occurrence of microblade technology artefacts and the common sense that animals cannot be perpetually hunted in the vicinity of the sedentism. In this case, periodic or occasional migrations become necessary. In North China, by the Early Neolithic, the sedentism was not fully established, which could be reflected at the Cishan site. In addition, the ability of sedentism is even worse in sites situated in marginal areas, such as Xinglongwa culture. Sedentism is a gradual and transitional process. Even though sedentism was not fully achieved in the Paleolithic-to-Neolithic transition, the significant decrease in mobility enables the achievement of semi-sedentism. Until the Early Neolithic, although sedentism was achieved, the sedentism capacity was weak because the time of continuous sedentism in one place remained relatively short. In terms of wild resource supplement, West Asia had a much more abundant supply, such as adequate wild wheat. Thus, the mobility of hunter-gatherers in West Asia might have been lower than that in North China during the Paleolithic-to-Neolithic transition. From the archaeological material, sites with clear sedentism structures and a large number of dense constructions appeared in West Asia instead of North China, further proving the decline of mobility.

The transition from the Natufian culture to the Neolithic one is influenced by The Younger Dryas Event, which is also considered a triggering factor in the origin of agriculture (Bar-Yosef and Meadow 1995). The Younger Dryas Event occurs at 10 to 20-year intervals, accompanied by a falling temperature, stronger seasonality, and drier summers. Thus, the Younger Dryas Event poses a more serious challenge for hunter-gatherers in the Natufian culture for adaptations. Amidst the depletion of animal hunting resources, the Natufian people are obligated to increase reliance on plant resources. During this period, the wild grains are mainly distributed from the Levant, upper Mesopotamia, and the southern edge of the Anatolian Plateau. The Holocene observes a temperature rise and expansions of the range of forest and grassland, which provide more favourable conditions available for agricultural development. At the site of Abu Hurairah, 80% of the faunal remains are identified as antelopes, 10% as goats and sheep, and the rest as wild asses, bison, wild boar, and deer in 9000 BC; goats and sheep accounts for 80% in 6500 BC. In the LUP, more than 160 species of plants are detected from food remains, while the figure decreased to 8 to 9 in the late Pre-Pottery Neolithic (Hillman et al. 1989).

The earliest cultivation of wild grains occurs in the northern Levant at the end of the Younger Dryas Event (9700/9500 BC), and later in the south. There is an obvious expansion in the earliest farming sedentism sites, with 8 to 10 times larger than the largest ones in the early Natufian culture. Typical sites, such as Jericho, have already been capable of constructing public works, for example, stone tower construction. However, early agricultural cultivators also hunt animals and collect wild plants. In West Asia, the domestication of animals was slightly later than the domestication of plants. The evidence shows that the domestication of goats might begin in the Zagros Mountains at ca. 8000 BP (Zeder and Hesse 2000). However, further research exhibits that the domestications of animals and plants occur almost simultaneously but in different regions. Besides, domestication might occur multiple times (Zeder 2011). The eight founder crops and founder livestock in West Asia form a balanced structure, constructing a complete cultural ecosystem. By 6500 BC, the occurrence of pottery made storage and cooking more convenient. By 6000 BC, the full establishment of the Neolithic in both West Asia and North China was almost simultaneous. However, during this time, agriculture in West Asia began to expand, seeing a different pattern from that in North China. In West Asia, the cultural ecosystem not only involves domesticated species but also the corresponding technological assemblages and social organisations. This agricultural package in West Asia initially spread into the west of Asia Minor Peninsula, then into

Europe, later spread southwestwards into Egypt, as well as spread northwards, eastwards, and southeast-awards. Due to the unclear archaeological evidence, the earlier cultural origins in the Early Neolithic of the various cultural zones in North China are unknown. A large-scale cultural spread occurs in North China in the Middle Neolithic, which dates later than in the Middle and Lower Yangtze Valley. Agricultural development in North China and the Middle and Lower Yangtze Valley are accompanied by the migration of human sedentisms from the foothills to the plains, which is similar to the early agriculture spread from the east part of the Southwest Asia region, the Mesopotamia plain, but with a very limited area of plains to spread in the eastern Mediterranean. For both regions, irrigation or flood control become the impetus of the complexity of social organisations (Wittfogel 1957).

During the agricultural process in West Asia, the origin of animal husbandry is much more advanced than that in China. As aforesaid, the domestications of the eight founder crops and four founder livestock could form a complementary relationship and a more complete economic structure. The domesticated poaceae in West Asia also serves as fodder for herbivores; at the same time, crops on farmlands attract herbivores. This process is beneficial for both hunting and animal husbandry, facilitating domestication. Simultaneously, livestock could eat the stalks of the crops, thus, a virtuous production cycle is created. However, as the scale of animal husbandry increases, issues arise. Firstly, crops cannot provide enough forage for livestocks. Secondly, labourers are required to look after herds. The more difficult issue to deal with is the irreconcilable contradictions between animal grazing and agricultural cultivation. Lastly, grazing (especially goats) would destroy grassland, leading to soil erosion, ecological degradation, and increased social competition.

The development in North China is very different from that of West Asia. The earliest domesticated animals in North China are dogs, pigs, and chickens; while sheep and cattle are domesticated much later. Based on Isotopic analysis, dogs, pigs, and chickens could survive by relying on grain wastes such as chaff and stalks (Barton et al. 2009). The early domesticated plants, such as corn and millet, could be harvested by pinching the earheads. Thus, most of the crop stalks and leaves are left in the fields, serving as nourishments for wild herbivores. In this case, the food competitions between humans and animals were not competitive. The Middle and Lower Yangtze Valley is full of rivers and lakes, thus, having abundant aquatic resources that could be utilised by humans for supplementing protein. This geographical character shapes people's residence near water. In West Asia, animal husbandry (raising animals such as cattle, goats, and sheep) requires large areas of grasslands, contradicting the cultivation of crops and eventually forming divergences. That is, animal husbandry occupies the grassland areas in the mountainous regions, exchanging spaces with agricultural cultivators. In this case, a symbiotic but competitive relationship between the industries has been established. Eventually, animal husbandry became dominant in West Asia. To prevent the establishment of a self-sufficient economy through pig raising by agricultural producers, later, this region made pig raising a taboo (Harris 1988).[28] Therefore, animal husbandry started to dominate the cultural development in West Asia. The situation was different in China. Agriculture became mature and then spread to the surrounding areas in North China and the Middle and Lower Yangtze Valley. The competition between animal husbandry and farming penetrates throughout the whole of Chinese history, however, the agrarian society maintains the domination.

The rapid development of sedentism is an important feature of the agricultural origins in West Asia. For a long time, sedentism has been regarded as a prerequisite for the origin of agriculture in West Asia. However, domestication occurs 2000 years after sedentism. In comparison, after several thousand years of the incipient of the earliest domestication happened in the New World region, the sedentism is finally realised (Chu 1999). Certainly, the earlier sedentism also occurs in the New World region, however, it might be established by marine mammal hunters alongside the Peru coast as they utilise the aquatic resources. Sedentary hunter-gatherers once inhabited the Nile River Valley in Egypt during the Paleolithic-to-Neolithic transition. Similarly, the sedentism might be associated with the utilisation of aquatic resources. The aquatic resources are inadequate in West Asia, thus, establishing a sedentism here is exceptional. We have previously discussed the sedentism in Natufian culture from the perspective of an abandonment process. If we choose the criteria of mobility reduction instead of sedentism to compare the mobility in the incipient agricultural origins among West Asia, the New Continent area, and China, West Asia is undoubtedly the least mobile, while the New World region is the most mobile and China ranked the second. Two reasons possibly could explain this: firstly, due to the natural conditions and demographic history of West Asia, the population is prone to assemble and reach the situation of "packed"; secondly, the predecessor of the "agricultural packaging" in West Asia includes several wild crops that are abundant and productive, thus, the intensification of the resource utilisation is easy to launch. Besides, these two reasons also determine the severe competition for natural resources in West Asia, not only between human groups for better-watered habitats and areas with more abundant wild grains and animal resources but also between humans and animals for wild grains; in the later phase, the competition happens between animal husbandry and crop cultivation. In this case, the competition for territory control became more intense. From this perspective, the public construction in West Asia, such as settlements, architectures, rituals, and other functioning artefatcs fully represents the intensification of regional ownership, which in turn boosts the development of the settlements. Probably because of the acceleration of social complexity, West Asia and Egypt became the first two countries to establish "civilisation".

Except for the differences in mobility, which is a key variable in hunter-gatherer cultural systems, another key difference in the origins of agriculture is a lack of pottery for thousands of years from the Neolithic. It is hard to explain this phenomenon by merely saying it is "not needed". Instead of making bread, the earliest wheat was ground into flour for pastry or porridge, which required pottery. Bread wheat is a result of selection in later phases. Ground stone tool containers are often discovered at sites of the Natufian culture and the Pre-Pottery Neolithic, such as cups, plates, and jars, but none are burned. At that time, the technology to fire pottery was already available. For example, before the onset of pottery vessels, typical Neolithic sites have already discovered pottery statues, lutes, pottery wheels, "tokens", pottery beads, and other ornaments, most of which were unburned or poorly burned. There are two reasons to explain the phenomenon: firstly, the origin of pottery is not for practical purposes, but more related to rituals, displays, etc.; secondly, the origin of pottery is associated with social changes, as pottery vessel makers are real settlers (Hoopes 1995, pp. 39–53), while the complete sedentism was not achieved by Neolithic inhabitants in earlier phases. In West Asia, pottery serves as an icon for social changes, as the occurrence of pottery might link to practical functions and intra-social competitions.

6.5.3 The Inherent Differences in the Origins of Chinese Agriculture

Examining the origins of Chinese agriculture from a comparative perspective, we need to realise the existence of the zones of cultural ecology (Chapter 3). Secondly, we need to compare the discrepancies between the zones of agricultural origins and the surrounding areas. Thirdly, the inherent variations within the zones should be compared. Finally, comparing the differences among surrounding areas. We have already distinguished the major zones of cultural ecology within China, thus, I will not illustrate again. The development of each region's culture is discussed in detail in the previous section; now I will illustrate the diversity of the origins of Chinese agriculture and clarify some misconceptions based on the comparisons of various regions.

With its vast territory and diverse environmental conditions, considering China as one geographical unit is inappropriate. The origins of Chinese agriculture can be divided into a central region, a transition zone, and a zone of influence. The central region indicates North China and the Middle and Lower Yangtze Valley. Within the central zone, there is the earliest germination area and development area. In North China, the earliest agriculture emerges in the foothills and terraces of the Taihang Mountains, the Yan Mountains, the Luxi Mountains, and the Qilu Mountains, before continuously expanding into the plains; the same pattern of development could be observed in the Middle and Lower Yangtze Valley, that is, the agriculture spread from the middle basins and wide valleys of the tributaries to the lower reaches and finally arrive at the floodplain of the Yangtze River.[29] As for the transition zone, the situation is more complicated. The Lingnan region, as a transitional region from temperate to tropical, is the earliest one being affected by LGM environmental changes. However, after the rise of temperatures, the climate in the Lingnan region transfers from a temperate to a tropical one. Interestingly, the agricultural origins in the Lingnan region also progress quickly initially and then slowly, establishing the Low-Level Food Production that relies on root tubers and aquatic resources (Chapter 8). The northern and southern regions of the Yan Mountains are the eastern part of China's natural transition zone from the northeast to the southwest. With the unstable climate in this region, human cultural adaptations fluctuate between farming and hunter-gatherer substances, establishing the dry farming and nomadic economy (Chapter 7). In short, the origins of Chinese agriculture go beyond the origins of millets or rice farming but include root and tuber agriculture, which is more similar to horticulture and nomadic economy. Within North China, the agricultural patterns between the plains and marginal areas are different: the former relies on millet crops; while the latter counts on broomcorn that has shorter growing periods and greater tolerance to low temperatures and drought. Thus, in order to adapt to the marginal environment, the dry farming economy has been developed, such as diversified planting to lower risks, and the economics of hunting increased, and divided livestock groups.

The biggest debate concerning centres of agricultural origins in China lies on the number of central areas, one, two, or multiple? The existence of multiple centres marks a deconstruction of centres of origin. Different forms of agriculture described above are all linked with the domestication of different species. The root tubers in Lingnan are domesticated independently; the origin of rice indica and japonica originate in different regions; the millets in North China and corn in South China have separate origins; with the origins of nomadic economy, the polycentricity of the origins of Chinese agriculture is formed. The final form of polycentricity equals the origins of Chinese agriculture to zonal adaptations, or popular concept spread. The polycentric theory is appropriate for developing diverse agricultural forms. However, when tracing the origins of agriculture, the existing material only supports two centres: North China and the Middle and Lower Yangtze Valley. Besides this view, there is the monocentric model (Zhengzhou Municipal Institute of Cultural Relics and Archaeology, Peking University School of Archaeology and Museology 2011), which implies that the Middle and Lower Yangtze Valley are earlier and the origins of agriculture in North China. Agricultural origins in North China are merely a product of the expansion of agriculture in the Middle and Lower Yangtze Valley, because the utilisation of rice cultivation in south China has a history of more than 20,000 years (such as the Diaotonghuan site). Besides, the pottery history is earlier. The trace of rice cultivation, which is over 10,000 BP, in the Early Neolithic is found (such as the Shangshan site). In contrast, there is no sign of agricultural domestication at the Shizitan site. However, the Peiligang culture of the Early Neolithic in North China, especially the Jiahu site shows clear evidence that rice cultivation has already started under the influence of South China. Besides, the cultural exchange between North and South China might have started much earlier. We cannot deny that the 14C dating data for south China is possibly earlier. However, from a perspective of cultural development during the Paleolithic-to-Neolithic transition (Chapter 5), North China had already possessed a series of new features and a resource endowment for agricultural development that was even more complete than that in the Middle and Lower Yangtze Valley. Also, as the comparative artefact, the Shizitan site is not situated in the central zone of the origin of agriculture. As for the expansion of rice cultivation to North China during the Jiahu culture and later phases, it is related to the expansion of the utilisation of the lake and marsh areas at the onset of the Holocene Warm Period, and there is no evidence to show that the earliest domesticated crop in North China was rice. The differences in cultural processes between North China and the Middle and Lower Yangtze Valley can be traced back to the Early Paleolithic, when gravel industries are dominated in the south and blade industries in the north; during the Late Upper Paleolithic when cultures are widely integrated, the cultural changes occurred also differed in these two regions. The predominant microblade technology in North China and Northeast Asia did not extend to the Middle and Lower Yangtze Valley. As a period of cultural differentiation in the Paleolithic-to-Neolithic transition, the bicentric view is more convincing than the monocentric theory.

Beyond the paradigm of archaeology, most people hold the popular pattern to associate the origins of Chinese agriculture with the Loess, thus equalling Chinese civilisation to a Loess civilisation. In this case, the stereotype that the Loess Plateau is the birthplace of Chinese culture is formed. Based on the existing archaeological material, there is no evidence to show that the Loess Plateau breeds Chinese agricultural origin. Evidence of subsistence intensification since the LUP is found from the sites of Shizitan, Xiachuan, Xueguan, etc. which are located in the Loess Plateau. However, they differ in excavations and site structures compared to sites in the eastern Taihang Mountains, such as Donghulin, Zhuannian, Nanzhuangtou, Ma'anshan, and Lijiagou. The earliest agriculture occurred in the terraces (foothills), since the loess is thinner with shallow groundwater, better vegetation, and various available resources. With the property of looser and greater fertility, the loess provides great conditions for the early agricultural development. However, after the Early Neolithic, agriculture expanded to the plains, where the proportion of clay content of the soil was higher, with flourished vegetation, seasonal floods, and wet soil that contained lots of plant roots, making cultivation hard to proceed. In the case of the Western region in the

Liaohe River, ancient people gradually started to make use of river valleys in the mid to late Neolithic period, while in earlier times more of the hillsides. The origin of agriculture in the Middle and Lower Yangtze Valley is not linked to loess, considering the looseness of the soil is not a necessary condition for the origin and development of prehistoric agriculture. The notion of "Chinese civilisation was a Loess civilisation" is shaped in modern Chinese society under colonialism, being reproduced by historical scholars and thus influencing all social statuses. This idea is disapproved by contemporary archaeological studies.

The current research supports the application of a complex and diverse adaptive approach in prehistory. Beyond the central areas and transition zones in the origins of agriculture, there are a number of unique adaptations. Northeast China sees a high degree of substance, mixing with agriculture, animal husbandry, hunting, gathering, and utilisation of aquatic resources (Chapter 9). Due to different local resource conditions, the substance approach varies. In the Songnen Plain and other areas where rivers and lakes are widely spread, utilisation of aquatic resources dominates the substance approach. Agriculture is prevailing in some areas. Generally speaking, the Paleolithic-to-Neolithic transition is long-lasting in Northeast China. In this case, to some extent, the term "Mesolithic" applied in the European region is also applicable to Northeast China. In Southwest China, the cultural patterns are more complex (Chapter 10). By virtue of the hunter-gatherer subsistence mode, Southwest China entered the Neolithic relatively late. The division of geographical units resulted in difficulties in cultural exchange. In the case of the Tibetan Plateau, the successful adaptation of the Tibetan hinterland is a product of the emergence of agriculture (Chapter 11), which seems a bit radical. Hunter-gatherers were unable to establish effective adaptations on the Tibetan plateau, resulting in many so-called "Paleolithic" materials being actually from the Holocene. In vast Northwest China, where the arrival of the Neolithic is still rather obscure. The establishment of successful adaptations is not suitable for hunter-gatherers, as the established adaptations required nomadic pastoralism in the desert, Gobi, and alpine zones. Amid the arrival of the Neolithic, space that is available for human exploitation has greatly expanded. The hinterland in the Tibetan Plateau and Northwest China symbolise the result of population spread in the Agricultural Age. Besides, the coastal region, which is normally overlooked, is small but rests on a boundless ocean that embraces abundant marine resources. The coastal region is a paradise for hunter-gatherers and is the most solid base of resistance to the spread of agriculture. Once the coastal region adopts agriculture, it would become the vanguard of its spread to marine areas. In short, the cultural adaptations within China were rich and varied over 10,000 years ago, and the origins of agriculture were just one way of adaptation. However, after the maturation of agriculture, the coastal region began to spread, displace, and integrate indigenous cultures from different regions with its greater efficiency of resource access, more complicated social structure, and ability to support higher population density. This process is quite similar to the rise and expansion of the modern Industrial Revolution, Only the transition took a longer period and the scale of social conflicts was moderate. In any case, the origins of agriculture start a new chapter in human history.

6.5.4 Basic Features of the Origins of Chinese Agriculture: Unification and Differentiation

Before the origins of agriculture, history was created by hunter-gatherers, thus, the study of hunter-gatherers should be necessarily involved in the research on mechanisms of agricultural origins. The hunter-gatherers' mechanism of cultural adaptation features both unification and differentiation. In Chapter 1, we have already discussed the long-term trends of cultural adaptation for hunter-gatherers and the significance of mobility, which is the key variable in the hunter-gatherer cultural system. The millions of years of hunter-gatherer history is not a simple repetitive process but is characterised by a cumulative development. As an integral part of human culture, hunter-gatherer history is accompanied by the continuous development of technology, culture, and the complexity of social organisation. Based on this, the perpetuated trend of pursuing adaptive strategy to maximise energy throughout human social history. At the same time, the pursuit of adaptive strategy means that human culture seeks to optimise fit and minimise risk. Braidwood views agriculture as a product of human cultural progress, which is valid. However, many archaeologists assume that Braidwood's view is similar to Spencer's theory of social evolution. The development of cumulative culture creates the structure and resources required for the origins of agriculture, including a certain population size and density, which is a cumulative result of long-term development. Besides, an increased population also leads to a series of social changes, such as increasing territoriality, limited beneficence, and intra-group competitions. These aspects reflect the long-term adaptive trend of hunter-gatherers, that is, one aspect of what I call unification. There is another aspect that is not related to historical trends but is linked to the subsistence strategy of hunter-gatherers. The ancient and modern hunter-gatherers have to seek natural resources as long as they are not engaged in agricultural production. In this case, they are not likely to have a sedentary life. This is similar to the relationship between industrial societies and cities. Urban life does exist in agricultural societies, but cities represent the dominant form of sedentism that could only be supported by industrial societies. It can be analogised to the hunter-gatherer society. Under certain circumstances, sedentary societies could be formed by aquatic resource exploiters. However, large-scale sedentism is a residential form in agricultural societies (large-scale sedentism normally happens in agricultural societies). From the perspective of diachronic and synchronicity, a unification could be found in the cultural adaptation of hunter-gatherers, while in the former case in some trends and in the latter case in mobile foraging.

Based on the cultural development of the Upper Paleolithic of North China and the Middle and Lower Yangtze Valley, human habitations have spread throughout the world (because of population growth), and an advancement of stone tool technology with refining raw materials and unique style in regions (territoriality, a limited beneficence); during the Paleolithic-to-Neolithic transition, the occurrences of new techniques, such as pottery, ground stone tool, and food grinding tools (the intensification of plant resource utilisation is a strategy for energy maximum[30]). During this period, the development of ornaments (intra-group competitions), etc., reflect the cumulative growth in the long-term trends of hunter-gatherers. Different mobilities mirror regional differentiation in hunter-gatherer cultural systems, typical comparisons could be found between sites groups of Shizitan and Xiachuan, and sites of Donghulin and Nanzhuangtou. These origins of agriculture could only happen once the mobility is reduced to a minimum degree, which could explain the fact that the earliest zones of agricultural origin are located east of the Taihang Mountains. The long-term trend in the hunter-gatherer society is assembled "constant", representing a self-evident presence. While archaeologists are chiefly concerned with the differentiations among different regions, that is, they wish to explain the reasons for the agriculture origins in certain areas instead of others. However, the long-term trend does not equally develop in every region; when encountering the differentiations, trends are closely connected to the different environmental conditions.

Environmental diversity influences cultural adaptation twofold: on the other hand, the hard constraints, that is, extreme characteristics of certain environments hinder the cultural adaptations. For instance, aquatic resources are not available inland; in a dense forest environment, herbivorous fauna resources are not as rich as those in grasslands. Agricultural practices require several conditions, such as suitable temperatures, precipitation, soil, pest control, population, etc. As for the origins of agriculture, the existence of wild species is vital. Without suitable species for domestication, it would be pointless to master domestication techniques. In the Americas, agricultural systems are always fragile because of a lack of large animals that are suitable for domestication. In this case, agricultural producers had to hunt as well, thus making the establishment of settlements troublesome. Even though settlement could be built on the basis of agriculture, it would be fragile and unstable. Without large domesticated animals, the South American civilisations had to build based on human labour (alpaca consignment is very limited) because wheels (found on toys) cannot help local people to increase productivity. The situation is different in China where species do not pose a hard constraint in various zones of cultural ecology, as there are local species that are suitable for domestication in different parts of the country. In Northeast China, low temperatures and short growing seasons pose a challenge for agricultural development. The Tibetan Plateau is in an alpine climate, which is harsher than the polar climate, with low oxygen levels and a lack of marine mammals and reindeer herds that could be hunted and domesticated respectively in the polar climate. In Northwest China, water resources are scarce; thus, insufficient surface water would be hard for large herbivores to survive. Further, the hunter-gatherers would not choose this place for sedentism, let alone the origins of agriculture. However, environmental diversity also provides opportunities and certain advantageous resources that would form a soft constraint to the development of agriculture. For instance, the utilisation of aquatic resources becomes an alternative strategy for agriculture. Aquatic resources could be detected in a large range of areas, possessing the characteristics of high productivity, seasonality, complementary to terrestrial resources, stability, variety, and high quality (with high protein proportion, accessible to everyone, such as shellfish collection). In contrast to these advantages of aquatic resources, labour-intensive agriculture lost the selective advantage. People in Northeast China, the Lingnan region, and the coastal areas moved towards the utilisation of aquatic resources, relieving the adaptive pressure encountered by hunter-gatherers. Southwest China is the most suitable region for hunter-gatherers because of its highly diverse environment (Chapter 3). Cultures are flexible. Depending on different regional situations, humans have developed diverse adaptive strategies. This is not contradictory to the unification of cultural adaptations of hunter-gatherers, as it is just two sides of the same coin.

It is meaningless to compare the abstract significance of cultural unification and diversification. In the archaeological study of the origins of agriculture, the initial conditions of cultural adaptation onset should be analysed case-by-case. The initial conditions of agricultural origins include the natural logarithm, the changes in the natural logarithm, and the accumulation of cultural resources. Terrains with stepped features and a temperate monsoon climate are the key feature of the natural logarithm for the origin of agriculture in China, with four distinct seasons, simultaneous rain and heat. Correspondingly, the main rivers flow from west to east, leaving spacious room for hunter-gatherers shifting to agricultural producers. This is crucial for the germination, development, and eventual establishment of a well-established agricultural system in relation to the origin of Chinese agriculture. As for the Loess, it is not a sufficient condition for the origin and development of Chinese agriculture. Different cultural-ecological zones pose another aspect of the natural logarithm (base condition) and basic conditions for cultural diversity, being created by the different distributions of latitude and altitude, land and sea, and topographical divisions. The cultural adaptation of hunter-gatherers is highly resilient, but this does not mean that they do not have optimal habitats. Notably, we cannot equal the best habitat for hunter-gatherers to modern densely populated areas. The best habitats for hunter-gatherers are the natural transition zone from Northeast to Southwest China and the transition from the second to the first terrace of the Chinese terrain, where there are various and diverse resources for hunter-gatherers, such as accessible and abundant clean water, stone materials, and suitable fuel. With the origin and development of agriculture, humans gradually moved to plains. The plains in North China and the Middle and Lower Yangtze Valley, which are suitable for agriculture, are not the best habitat for hunter-gatherers. The situation is the same in the Qinghai-Tibet region, which is suitable for a nomadic economy instead of hunter-gatherers. Different modes of substance rely on different regions.

During the Pleistocene-Holocene transition, a considerable climatic and environmental change happened. LGM was followed by a temperature rise and the melting of glaciers. After the Pleistocene, several factors still influence the cultural adaptations of prehistoric hunter-gatherers, covering the rise of sea level, Paleomonsoon reconstruction, plant replacement, fauna extinction, the Younger Dryas Event, and increased seasonality, and stability of climate (Chapter 5). The changes in the climatic environment led to changes in natural logarithms. In the human cultural systems, these changes have both positive and negative effects, with varying degrees in different regions. The rise of sea level imposes the greatest impact on North China: due to land inundation, the area lost almost half of its human habit; even though the population remained the same, the population density would double as a result. The sea level rise also made a profound impact in the Middle and Lower Yangtze Valley. The Paleomonsoon, plant replacement, and stability of climate could all promote the onset of agriculture; while fauna extinctions, the Younger Dryas Event, and increased seasonality would benefit the origins of agriculture in the opposite direction, making crop cultivation a necessity. These elements all benefit the origins of agriculture in the Pleistocene-Holocene transition. The impact of climatic changes varied in different regions. The Paleomonsoon originally affected Southwest and Northeast China. At the same time, the strongest impact brought by fauna extinctions was reflected in North China, but with little impact in Southwest China.

The cultural shift process from hunter-gatherer to agricultural production is not a black box. Adding environmental conditions and proper changes would not necessarily lead to the origin of agriculture. During this process, natural species would lose certain abilities and be domesticated by human beings. Overall, the origin of agriculture is ultimately a cultural event. The biggest difference between hunter-gatherer cultural adaptation systems and agriculture lies in mobility, from which other changes are derived, including settlement patterns, tools, technology, substance, social structure, and ideology. From the Upper Paleolithic, the cultural landscapes in North China and the Middle and Lower Yangtze Valley changed dramatically, showing a transfer in hunter-gatherer adaptation strategies. During the Paleolithic-to-Neolithic transition, the appearances of certain Neolithic features symbolise an obvious reduction in population mobility. Such changes are widely seen across the zones of cultural ecology but to a markedly different degree, with Southwest China being the weakest. The Paleolithic-to-Neolithic transition witnessed a critical cultural differentiation. The earliest agriculture occurred in North China and the Middle and Lower Yangtze Valley, where the intensity of cultural differentiation is the strongest. The exceptions happened in the Lingnan region and Northeast China where the abundant aquatic resources led to a shift towards new hunter-gathering forms.

The centres of Chinese agricultural origins, as a part of the world's agricultural origins, provide us with a clear understanding of its characteristics through comparison with other regions in the world, especially West Asia. At the same time, the comparisons made within the centres of Chinese agricultural origins, and with neighbouring regions, further enrich the meaning of these features. In conclusion, I am summarising the characteristics of the agricultural origins in China into six aspects.

Firstly, China and West Asia could be considered the two earliest and most important independent centres of domestication in the world. Not only because of their early dates but also by their far-reaching influences. As a major centre of domestication, China has a long history of domestication and is home to nearly a quarter of the world's domesticated plants. The influence extends across East Asia, Southeast Asia, South Asia, and the Pacific. Possessing the characteristic of distinctly autochthonous, Chinese agricultural origins matured gradually from the Upper Paleolithic to the Paleolithic-to-Neolithic transition; while a surging explosion happened in the Early Neolithic. The entire development of Chinese agricultural origins largely synchronised with that in Western Asia. The origins of Chinese agriculture were initially temperate grain agriculture, which eventually reached a level of intensive agriculture, with a number of "low-level food production" formed in the surrounding transitional zones.

Secondly, the origins of Chinese agriculture consist of two relatively independent centres, which have different cultural backgrounds amidst the Paleolithic cultural contexts. Thus, the cultural landscapes are distinctively different in these two centres: in the LUP, North China is represented by the typical microblade industry; while the Middle and Lower Yangtze Valley is characterised by the conventional gravelly industry with an addition of a fine flint stone component. In terms of crop domestication, the main crops in North China are millets and broomcorn; while the Middle and Lower Yangtze Valley is grain. In the Early Neolithic, a crop exchange was seen in North China since rice cropping occurred in the north–south transition zone. During the

Paleolithic-to-Neolithic transition, the evidence of the intensification of resource utilisation occurred simultaneously in both regions. However, pottery appeared earlier in South China (based on the current 14C dating data); tools, such as grinding slabs, handstones, adze-shaped tools, etc., and later, pottery, were discovered in North China. There is no evidence to show that the Middle and Lower Yangtze Valley affected North China, nor is there evidence to the contrary. In this case, these two regions developed independently. Archaeological evidence in the Paleolithic-to-Neolithic transition shows that cultural change occurred more obviously and rapidly in North China, as several archaeological cultural areas have already emerged by the Early Neolithic; while, in South China, the level of development was somehow lower, with the only clear cultural landscapes occurred in Pengtoushan and Shangshan cultures. The spread of rice cultivation from South to North is mainly related to the warm and humid climate brought by the Holocene climatic optimum. Rice cultivation is beneficial for people in the Jianghuai region to utilise water network zones. Overall, in terms of agricultural origins, there is no evidence to show the Oneness of North China and the Middle and Lower Yangtze Valley. From the Early to Middle Paleolithic, the differentiations of cultural adaptations in these two regions existed. The continuous differential development during the two phases cannot support the view that these two regions belong to a single centre of agricultural origin.

Thirdly, there is no sufficient evidence to show that loess composes a sufficient condition for the origin of Chinese agriculture. Agriculture would still have taken place if it had been a temperate forest steppe soil instead of loess, as both the mechanism and natural logarithm in Northwest China do not rely on loess. The earliest agriculture in North China did not originate on the Loess Plateau, but in the foothills and terraces in the east of Taihang Mountains, south of the Yan Mountains, Qilu Mountains, and Yuxi Mountains. The perception of the so-called "Loess civilisation" is related to ancient Chinese history and the reality in modern China.

Fourthly, crop farming is the centre of early Chinese agriculture, and animal domestication serves for crop farming. The earliest domesticated animals include dogs, pigs, and chickens. Crop farming and animal domestication are complementary rather than contradictory, as the residues could be used in animal husbandry, which is unlike in West Asia. Compared to West Asia, pottery occurred earlier in areas of Chinese agricultural origins, but sedentisms later. This might be associated with the deficient Chinese "agricultural packaging" and the abundance of the one in West Asia. In the later phases, the development of prehistoric Chinese agriculture became more mature, with animal husbandry and crop agriculture complementing each other to form a better agroecosystem. Unlike in West Asia, animal husbandry and crop agriculture are contradictory, thus, a consequent division inevitably happened. Besides, animal husbandry had a clear destructive effect on agroecosystems in West Asia.

Fifthly, Chinese prehistoric agriculture spread from foothills to plains, indicating a clear geographical orientation. The large hinterland lays a foundation for agricultural development in China and provides sufficient material conditions for the Chinese prehistoric civilisation. Besides, the two agricultural centres in both North and South China could complement each other as a buffer space. In this case, the Central Plains civilisation could be partially preserved in South China after the collapse.

Sixthly, other forms of agricultural production and adaptations existed in the peripheral regions of China's agricultural origins, such as the development of dry farming and nomadic economies in north of the Yan Mountains (Chapter 7), root and tuber agriculture in the Lingnan region, complex hunter-gatherer societies based on the utilisation of aquatic resources in Northeast China, the colourful local styles in Southwest China, etc. The rich cultural diversity that developed in the surrounding areas, which are based on the centre of early Chinese agriculture, formed the foundation of the star-studded pattern of the origin of Chinese civilisation.

Notes

1. Mr. Yan Wenming proposed a chronological scheme in 1987, which placed the termination of the Early Neolithic period around 7000 years ago. In 1998, it was adjusted to 7000 BC, representing the transitional phase between the Neolithic and Paleolithics.
2. The Chengbeixi culture is mainly distributed in the southwestern part of Hubei Province and overlaps in time with the lower layer of the Zaozhi culture. It is a regional type of the Chengpengtoushan culture.
3. See page 9 for the distribution map of the Peiligang culture sites, which shows that they are mainly distributed in the piedmont

hills and plains. CA, CASS: *Chinese Archaeology: Neolithic Volume*. China Social Sciences Press, 2010. Figures 3–9 also supports this.

4. Xia Zhengkai and others have proposed a relationship between the elevation of archaeological sites and river incision. However, the tectonic movements and changes in sea level during the Holocene are extremely limited and insufficient to create significant river incision depths that would impact the distribution of archaeological sites over thousands of years. The distribution height of these sites is primarily associated with the development of agricultural production. In the later phases, people began utilising river valley terraces and floodplains, leading to a decrease in the elevation of the sites. This assertion is supported by traces of the use of digging tools found in stone tools and experimental archaeological research. The stone tools of the Xinglongwa culture, for example, are not suitable for excavating soils in river valley areas that are relatively sticky and contain gravel, but they are well-suited for excavating loess, a softer soil, on slopes. In comparison, the stone spades of the Zhaobaogou and Hongshan culture are suitable for overturning and excavating river valley soils. Detailed discussions can be found in Chapter 7.
5. It is more likely that dogs consumed human excrement rather than humans feeding dogs with millet, resulting in the presence of C4 plants in dog bones.
6. Area I and Area II are not more than 50 metres apart and the excavation area is nearly 1900 square metres, which can be considered as a single site. Two house foundations and 369 ash pits were discovered. Area III is approximately 200 metres away from Area I, with an excavation area of 473 square metres and the discovery of 68 ash pits.
7. This type of abandonment method was a common phenomenon in the early cultures of the Neolithic period. In the western Liaoning region, house foundations of the Xinglongwa culture often contain the most abundant remains, while sites of the Hongshan culture and its later archaeological cultures typically have fewer artefacts on the residential surfaces, except for those buildings that were abandoned due to sudden events such as fires or epidemics.
8. During my field survey in 1997, I found that this type of raw material could be seen exposed in the mountain gullies on the southern edge of the Nihewan Basin. Occasionally, high-quality flint can be found in these gullies.

9. Discovered during the first excavation in 1991, the types of fruits found at that time were not identified in a timely manner. Based on the findings from the second and third excavations, the excavators argued that they could possibly be fruits of Chinese chestnut (Castanea mollissima), cork oak (Quercus variabilis), and oak (Quercus spp.).
10. There is reason to argue that compared to the EUP (Early Upper Paleolithic) sites, the degree of fragmentation of animal bones significantly increases during the LUP (Late Upper Paleolithic), as is the case in the North China region.
11. Mr. Zhang Zhongpei argues that the age of the Xiaohuangshan section is definitely not earlier than 9000 years ago. Mr. Yan Wenming also argues that the Xiaohuangshan and Shangshan sections can only be placed together with the Pengtoushan culture and the Chengbeixi culture in the south, and the Peiligang culture and the Cishan culture in the north, and cannot be earlier; he even suggests that Xiaohuangshan and Shangshan belong to the Middle Neolithic, or the late phase of the early period (according to Mr. Yan's periodization scheme).
12. The age determined by humic acid dating of the pottery is 12,320 ± 120 years. The age determined by the pottery matrix is 14,810 ± 230 years, and the age determined by charcoal found in the same layer as the pottery is 14,490 ± 230 years. See Yuan (2000).
13. The reports by Sheng Danping et al. differ from the excavation report as they refer to the "lower three layers".
14. The earliest known pottery artefacts are sculpted figures made from fired clay, discovered in Europe and dating back approximately 27,000 years.
15. Agriculture also includes animal husbandry, and the lifeway of pastoralists is also nomadic, but it appeared slightly later than grain farming and developed based on it or as a differentiation from grain agriculture.
16. Pastoral nomadic societies during historical periods also produced pottery, which was influenced by surrounding agricultural societies. Additionally, these societies were not solely nomadic and also engaged in a small amount of farming. Nevertheless, their pottery was much simpler, and the design of the vessels also took into account the convenience of mobility.

17. Of course, the absence of decorations does not mean that it is not associated with female pottery making. It simply means that rich decorations are more likely to be related to female pottery making, just like how women often wear clothes with abundant decorations, but not all women do.
18. The mentioned grinding slab can be classified into oval, rectangular, and irregular shapes, but there are no accompanying illustrations.
19. The difference between hearth lies mainly in the degree of concentration and depth. Hearth sites are more concentrated and deeper, especially when intentional construction such as digging pits and stacking stones can be observed. This distinction needs to be considered in conjunction with the overall structure of the site and the characteristics of the remaining artefacts. Hearth sites are unlikely to be found in temporary residential sites.
20. Due to similarities in cultural style, some consider the Cishan culture and Peiligang culture to be part of the same culture.
21. It is possible that the Xiaohexi culture and Fuhe culture, some scholars argue, belong to the same period as the Xinglongwa culture. See Suo Xiufen and Guo Zhizhong (2004).
22. Even in the Early Neolithic period, sedentism may still have been unstable, so I often mention the "loss of mobility". In essence, it means the same thing, but it refers to the mobility of Hunter-Gatherers. The continuous loss of mobility also means that it becomes increasingly impossible to rely on natural resources for sustenance. The reason is simple: there are very few places where resources are abundant enough to support settled hunting and gathering.
23. This includes Western Asia, including Anatolia, the Mesopotamian region, the Arabian Peninsula, Palestine, and parts of northeastern Africa, such as Egypt and Sudan.
24. Currently, the earliest date for the domestication of maize has been significantly pushed back to around 3600 BC, rather than the 10,000 BC proposed by MacNeish. However, the domestication of gourds may have occurred as early as 8000 BC (the Guilá Naquitz site).
25. The age composition analysis of the remains found at the Abu Hureyra site reveals that newly born calves, one-year-old individuals, and 437-year-old adults were simultaneously hunted and

killed. In late April and early May, hunters were able to extensively hunt gazelles by driving them into enclosures or setting traps, leading to a rapid decline in the gazelle population.

26. Personal communication with Dr. Youssouf.
27. According to Shiffer's classification, primary refuse refers to items that have been discarded in their original usage area when they are depleted or near the end of their lifespan, while secondary refuse refers to items discarded outside of their usage area.
28. Marvin Harris has provided insightful discussions on this topic. Please refer to Harris (1988).
29. During the Spring and Autumn and Warring States periods, Chu, Wu, and Yue were the centers of civilisation in the middle and lower reaches of the Yangtze River. The middle reaches had the Shihaizi culture, while the lower reaches had the Liangzhu culture. The Dongting Lake area was connected to the Jianghan Plain, forming a large prehistoric agricultural activity zone. The lower reaches of the Yangtze River, centred around the Yangtze River Delta, connected to the north with the Subei Plain and to the south with the Qiantang River Basin. In comparison, the Poyang Lake area is relatively small, with the Dabieshan Mountains to the north, the Luoxiao Mountains to the west, the Wuyi Mountains to the east, and the Nanling Mountains to the south, with only a narrow and elongated plain along the river connecting it to the other two regions. So far, the cultural materials of the Early Neolithic period in the Poyang Lake area are not well understood, which reflects, to some extent, the intensity of human activities in this region during the Early Neolithic period. Mr. Yan Wenming and Mr. Su Bingqi have focused on different aspects of the cultural patterns in the middle and lower reaches of the Yangtze River.
30. Here, we are not discussing the efficiency of energy acquisition by ancient humans but rather the energy density per unit volume of food. Humans achieved this by intensively processing plant seeds.

References

A. Belfer-Cohen and A. N. Goring-Morris, Becoming farmers: the inside story. *Current Anthropology*, 52 supplement 4: 209–220, 2011.

A. Sherrat, Plat tectonics and imaginary prehistories: Structure and contingency in agricultural origins. In D. Harris ed., *The Origins and Spread of Agriculture and Pastoralism in Eurasia*, pp. 130–140. Smithsonian Institution Press, Washington DC, 1996.

A. Testart, Storage in hunter-gatherer societies. *Current Anthropology*, 23: 523–537, 1982.

A. M. T. Moore, The inception of potting in western Asia and its impact on economy and society. In W. K. Barnett and J. W. Hoopes eds., *The Emergence of Pottery: Technology and Innovation in Ancient Societies*, pp. 39–53. Smithsonian Institution Press, Washington DC, 1995.

Anonymous, *The Shizitan Paleolithic Site in Jixian*. Chinese Cultural Relics News, January 7, 2011.

An Zhimin, Peiligang, Cishan, and Yangshao. Archaeology, 1979, 4.

B. Hayden, Practical and prestige technologies: the evolution of material systems. *Journal of Archaeological Method and Theory*, 5: 1–53, 1998.

Bailiandong in Liuzhou, Guangxi Science Museum: "Liuzhou Bailiandong," Science Press, 2009.

Baoding Cultural Heritage Management Office, et al., *Trial Excavation Brief of Nanzhuangtou Site in Xushui, Hebei. Archaeology*, 1992, 11.

CA, CASS, Shaanxi Provincial Institute of Archaeology: *Yichuan County, Shaanxi: The Longwangchan Paleolithic Site. Archaeology*, 2007, 7.

CA, CASS, *Chinese Archaeology: Neolithic Archaeology Volume*. China Social Sciences Press, 2010.

CA, CASS, et al., *Guilin Zengpiyan*. Cultural Relics Press, 2003.

Cai Lianzhen and Qiu Shihua, *Carbon-13 Dating and Ancient Recipe Research. Archaeology*, 1984, 10.

C. C. Mann, Earthmovers of the Amazon, *Science*, 287: 786–789, 2000.

C. C. Mann, The real dirt on rainforest fertility. *Science*, 297: 920–923, 2002.

Chen Chun, On the Middle Neolithic. *Acta Anthropologica Sinica*, 1995, 12 (01).

Chen Chun, Research and Reflections on the Middle Neolithic. *Agricultural Archaeology*, 2000, 1.

Chen Hui and Chen Shengqian, Experimental study on the chopping tools of the Yu Zui No. 2 PaleolithicSite in Yun County, Hubei. *Acta Anthropologica Sinica*, 2012, 31 (01).

Chen Jiling and Chen Shengqian, Research on decorative art of Xinglongwa culture cylinder Jars. In *Frontier Archaeological Research*, Issue 11. Science Press, 2012.

Chen Tiemei, Pottery shreds from the Pengtoushan Site and accelerator mass spectrometry radiocarbon dating of the earliest rice remains in China. *Cultural Relics*, 1994, 3.

Chen Xingcan, Several issues regarding the middle neolithic. *Archaeology*, 1990, 2.

Chu Youxin, *Architectural Remains of the end of Paleolithic in Zhuma Village, Linyin County, Chinese Archaeology Yearbook 1997*, Cultural Relics Press, 1999.

C. M. Aikens, First in the world: The Jomon pottery of early Japan. In W. K. Barnett and J. W. Hoopes eds., *The Emergence of Pottery: Technology and Innovation in Ancient Socieities*, pp. 11–21. Smithsonian Institution Press, Washinton DC, 1995.

D. E. Arnold, The archaeology of complex Hunter-Gatherers. *Journal of Archaeological Method and Theory*, 3: 77–126, 1996.

D. R. Piperno and K. E. Stothert, Phytolith evidence for early Holocene Cucurbita domestication in southwest Ecuador. *Science*, 299: 1054–1057, 2003.

E. Stokstad, "Pristine" forest teemed with people. *Science*, 301: 1645–1646, 2003.

E. A. Pechenkina, et al., Reconstructing northern Chinese Neolithic subsistence practices by isotopic analysis. *Journal of Archaeological Science*, 32: 1176–1189, 2005.

F. Wendorf and R. Schild, Nabta Playa and its role in northeastern African prehistory. *Journal of Anthropological Archaeology*, 17: 97–123, 1998.

Fang Jinqi, Influence of sea level rise on the middle and lower reaches of the Yangtze River since 12, 100 BP. *Quaternary Science Reviews*, 10: 527–536, 1991.

Fu Xianguo, *Cave Sites of Stone Tool Age in Dayan, Lingshan County, Archaeology Yearbook 2001*, Cultural Relics Press 2002.

Lee, G., et al., Plants and people from the early neolithic to Shang periods in North China. *PNAS*, 104: 1087–1092, 2007.

G. C. Hillman and M. S. Davies, Measured domestication rates in wild wheats and barley under primitive cultivation, and their archaeological implications. *Journal of World Prehistory*, 4: 157–222, 1990.

G. C. Hillman, et al., Plant-food economy during the Epi-Paleolithic at Tell Abu Hureya, Syria: Dietary diversity, seasonality and modes of exploitation. In D. Harris and G. Hillman eds., *Foraging and Farming: The Evolution of Plant Exploitation*, pp. 240–268. Unwin Hyman, London, 1989.

Gu Haibin, Pollen Analysis and discussion on the paleoenvironment of the Pengtoushan Site in Lixian County, Hunan Province. *Cultural Relics*, 1990, 8.

Gu Haibin, Preliminary study on the evolution of quaternary paleoenvironment in the Dongting Lake Area and its impact on Human activities. In *Environmental Archaeology Research*, Series 1. Science Press, 1991.

Hao Shougang et al., Fruit stones from tomb no. 4 in Donghulin. *Acta Anthropologica Sinica*, 2008, 27 (03).

Harris, *The Cultural Materialism*, translated by Huang Qing, Huaxia Publishing House 1988.

Hebei Provincial Cultural Relics Management Office, Handan Municipal Cultural Relics Storage Institute: *The Cishan Site in Wu'an, Hebei Province Acta Archaeologica Sinica*, 1981, 3.

Hebei Provincial Institute of Cultural Relics, et al.: Excavation brief of Nanzhuangtou Site in Xushui in 1997. *Acta Archaeologica Sinica*, 2010, 2.

Henan provincial Institute of Cultural Heritage and Archaeology: *Wuyang Jiahu*. Science Press, 1999.

H. M. Wobst, Locational relationships in Paleolithic society. *Journal of Human Evolution*, 5: 49–58, 1976.

Hunan Institute of Cultural Relics and Archaeology, Lixian Cultural Relics Management Office: Investigation report on early neolithic sites in Lixian, Hunan Province. *Archaeology*, 1989, 10.

Hunan Institute of Cultural Relics and Archaeology, Lixian Cultural Relics Management Office: Excavation brief of the early neolithic site at Pengtoushan, Lixian, Hunan Province. *Cultural Relics*, 1990, 8.

Hunan Provincial Institute of Cultural Relics and Archaeology: Excavation report on the Pengtoushan Neolithic Site in Lixian County, Hunan Province. *Cultural Relics*, 1990, Issue 8.

Hunan Provincial Institute of Cultural Relics and Archaeology: The Hujiawu Site of the Neolithic Age in Linyin County, Hunan. *Acta Archaeologica Sinica*, 1993, 2.

Hunan Provincial Institute of Cultural Relics and Archaeology: Excavation report on the early neolithic site of Bashidang in Lixian County, Hunan Province. *Cultural Relics*, 1996, 12.

Hunan Provincial Institute of Cultural Relics and Archaeology: *Pengtoushan and Bashidang*. Science Press, 2006.

Hunan Provincial Museum: Neolithic remains in the Shimenzaozi Lower Culture, Shimen, Hunan. *Archaeology*, 1986, 1.

Hu Yaowu et al., Stable isotopic analysis of human bones from the Jiahu Site. *Science China (Series D)*, 37: 94–101, 2007.

Hu, Yaowu, et al., Stable isotope analysis of humans from the Xiaojingshan site: implications for understanding the origin of millet agriculture in China. *Journal of Archaeological Science*, 35: 2960–2965, 2008.

Inner Mongolia Autonomous Region Institute of Cultural Relics and Archaeology: *Baiyinchanghan: Excavation Report of Neolithic Sites*. Science Press, 2004.

J. Alroy, Putting North America's end-Pleistocene megafaunal extinction in context: Large-scale analyses of spatial patterns, extinction rates, and size distributions. In R. D. E. MacPhee ed., *Extinctions in Near Time: Causes, Contexts, and Consequences*, pp. 105–143. Kluwer Academic/Plenum Publishers, New York, 1999.

J. Perrot. Le gisement natoufien de Mallaha (Eynan), Israel. L'*Anthropologie*, 70: 437–483, 1966.

J. G. Hawkes, The Diversity of Crop Plants. Harvard University Press, Cambridge, 1983.

J.M. Diamond, *Guns, Germs and Steel: The Fates of Human Societies*, translated by Xie Yanguang, Shanghai Translation Publishing House, 2000.

J. R. Harlan, Agricultural origins, centers and noncenters. *Science*, 174: 468–474, 1971.

J. R. Harlan, *Plants and Man*, 2nd edition. American Society of Agronomy, Madison, 1992.

K. Newman, New Guinea: A cradle of agriculture. *Science*, 301: 1710–1714, 2003.

K. A. Wittfogel, *Oriental Despotism: A Comparative Study of Total Power*. Yale University Press, New Haven, 1957.

Barton, L., et al., Agricultural origins and the isotopic identity of domestication in northern China. *PNAS*, 106: 5523–5528, 2009.

Li Yu and Dong Yuchen, Introduction. In *Chinese Crops and Their Wild Relatives: Volume on Grain Crops*. China Agricultural Press, 2006.

L. R. Binford, Post-Pleistocene adaptation. In S. Binford and L. Binford eds, *New Perspectives in Archaeology*, pp. 313–341. Aldine, Chicago, 1968.

L. R. Binford, *Constructing Frames of Reference: An Analytical Method for Archaeological Theory Building Using Ethnographic and Environmental Data Sets*. University of California Press, Berkeley, 2001.

Lu Houyuan et al., Earliest domestication of common millet (*Panicum miliaceum*) in East Asia exteneded to 10, 000 years ago. *PNAS*, 106: 7367–7372, 2009.

M. A. Zeder and B. Hesse, The initial domestication of goats (*Capra hircus)* in the Zagros Mountains 10, 000 years ago. Science, 287: 2254–2257, 2000.

M. A. Zeder, The origins of agriculture in the Near East. *Current Anthropology*, 52 supplement 4: 221–235, 2011.

M. B. Schiffer, *Formation Processes of the Archaeological Record*. University of Utah Press, Salt Lake City, 1987.

M. E. Kislev, E. Nadel and I. Carmi, Epi-Paleolithic (19, 000 BP) cereal and fruit diet at Ohalo II, Sea of Galilee, Israel. *Review of Palaeobotany and Palynology*, 71: 161–166, 1992.

Mei Huijie and Wang Youping, *Upper PaleolithicSite at Ma'anshan, Yangyuan County*, in *Chinese Archaeology Yearbook 1999*. Cultural Relics Press, 2001.

M. J. Heckenberger et al., Amazonia 1492: Pristine forest or cultural parkland? *Science*, 301: 1710–1714, 2003.

N. I. Vavilov, *The World Center of Origin of Major Cultivated Plants*, translated by Dong Yuchen, China Agriculture Press, 1982.

O. Bar-Yosef and M. Kislev, Early farming communities in the Jordan valley. In D. Harris and G. Hillman eds., *Foraging and Farming: The Evolution of Plant Exploitation*, pp. 632–642. Unwin Hyman, London, 1989.

O. Bar-Yosef and R. H. Meadow, The origins of agriculture in the Near East. In T. D. Price and A. B. Gebauer eds., *Last Hunters First-Farmers*, pp. 39–94. School of American Research Press, Santa Fe, 1995.

P. Savolainen et al., Genetic evidence for an East Asian origin for domesticated dogs. *Science*, 298: 1610–1613, 2002.

Pang Junfeng et al., mtDNA data indicates a single origin for dogs south of Yangtze River, less than 16,300 years ago, from numerous wolves. *Molecular Biology and Evolution*, 26 (12), 2009.

P. B. Vandiver et al., The origins of ceramic technology at Dolní Vestoniçe, Czechoslovakia. *Science*, 246: 1002–1008, 1989.

Peking University School of Archaeology and Museology, et al., Prehistoric Sites in Donghulin, Mentougou District, Beijing. *Archaeology*, 2006, 7.

Peking University School of Archaeology and Museology, Zhengzhou Municipal Institute of Cultural Relics and Archaeology: Excavation Brief of Lijiagou Site in Xinmi City, Henan. *Archaeology*, 2001, 4.

P. J. Richerson et al., Was agriculture impossible during the Pleistocene but mandatory during the Holocene? A climate change hypothesis. *American Antiquity*, 66: 387–411, 2001.

P. M. Rice, On the origins of pottery. *Journal of Archaeological Method and Theory*, 6: 1–53, 1999.

R. Pinhasi and J. T. Stock eds., *Human Bioarchaeology of the Transition to Agriculture*. Wiley-Blackwell, Chichester, 2011.

R. W. Redding, Preliminary report on faunal remains recovered from the 1993 excavations. In *Origins of Rice Agriculture: The Preliminary Report of the SinoAmerican Jiangxi (PRC) Project SAJOR*. Publications in Anthropology No. 13, El Paso Centennial Museum, The University of Texas at El Paso, 1995.

School of Archaeology and Museology, Peking University, Zhengzhou Municipal Institute of Cultural Relics and Archaeology: Excavation brief of the Lijiagou Site in Xinmi City, Henan. *Archaeology*, 2011, 4.

Sheng Danping, Zheng Yunfei and Jiang Leping, Early Neolithic Site of Shangshan in Pujiang, Zhejiang. *Agricultural Archaeology*, 2006, 1.

Shi Xingbang, The discovery and significance of the Pre-Yangshao culture. In *Collected Papers on Chinese Archaeological Research: Commemorating the 50th Anniversary of Xia Nai's Birth*. Sanqin Publishing House, 1986.

Shizitan Archaeological Team: Excavation brief of site nine at Shizitan, Jixian County, Shanxi. *Archaeology*, Vol. 10, 2010.

Shizitan Archaeological Team: *S14 Site of the Shizitan PaleolithicSite in Jixian County, Shanxi*. *Archaeology*, 2002, 4.

Su Bingqi, *A New Exploration of the Origins of Chinese Civilization*. SDX Joint Publishing Company, 1999.

Suo Xiufen and Guo Zhizhong, The Xiaohexi culture relics of Baiyinchanghan Site. In *Frontiers of Frontier Archaeological Research*, 2004, 3.

T. D. Price, The Mesolithic of Western Europe. *Journal of World Prehistory*, 1: 225–306, 1987.

Shizitan Archaeological Team: *Excavation Brief of Site Nine at the Shizitan Site, Jixian County, Shanxi*. *Archaeology*, 2010, 10.

Song Guoding, Wang Tao and Jiang Hong'en, *Excavation of the Kengnan Site in Xichuan, Henan*. Chinese Cultural Relics News, Nov. 18, 2011.

Sun Bo et al., *Discovery of Early Neolithic Sites in Shandong*, China Cultural Relics News, Aug.15, 2007.

T. Hardy-Smith and P. C. Edwards, The garbage crisis in prehistory: Artifact discard patterns at the early Natufian site of Wadi Hammeh 27 and the origins of household refuse disposal strategies. *Journal of Anthropological Archaeology*, 23: 253–289, 2004.

T. P. Denham, et al., Origins of agriculture at Kuk Swamp in the highlands of New Guinea. *Science*, 301: 189–193, 2003.

V. M. Bryant, Invisible clues to New World plant cultivation. *Science*, 299: 180–181, 2003.

Wang Jian et al., The Xiachuan Culture—Investigation Report on Xiachuan Site in Shanxi. *Acta Archaeologica Sinica*, 1978, 3.

Wang Xiangqian, et al., Microliths in Xueguan, Pu County. *Acta Anthropologica Sinica*, 1983, 2 (02).

Xia Nai, Chinese archaeology in the past thirty years. *Archaeology*, 1979, 5.

Xia Zhengkai et al., Analysis of the living environment of Donghulin people in Beijing around 10,000 BP. *Science Bulletin*, 2011, 45 (34).

Xie Fei, An outline of the transition between the new and old stone age cultures in the Circum-Bohai Sea Region. In *Retrospect and Prospect of Chinese Archaeology across the Century*. Science Press, 2000.

Yan Wenming, New discoveries of early neolithic cultures in the Yellow River Basin. *Archaeology*, 1979, 1.

Yan Wenming, The unity and diversity of prehistoric cultures in China. In *Cultural Relics*, 1987, 3.
Yan Wenming, Investigation of settlement patterns in the Chinese Neolithic. In *Celebrating Su Bingqi's Fifty-Five Years in Archaeology: Collected Papers*. Cultural Relics Press, 1989.
Yan Wenming, The origins of Chinese Agriculture and Animal Husbandry. *Liaohai Cultural Relics Journal*, 1989, 2.
Yan Wenming, Three Questions on Neolithic Archaeology. In *The Origin of Agriculture and Civilization*. Science Press, 2000.
Mr. Yan Wenming's speech, see *Expert Discussion: Xiaohuangshan Site* in *Shengzhou, Zhejiang* Chinese Cultural Relics News, January 11, 2006.
Yin Jianshun, A preliminary analysis of the periodization and cultural attributes of the Zaoshi Lower Culture in the Dongting Lake Region, Hunan. In *Proceedings of the Prehistoric Culture and the Second Asian Civilization Symposium*, Yue Lu Press, 1996.
Yu Jincheng, Archaeological discoveries and research on the neolithic in Beijing. In *Collection of Travels*, Beijing Library Press, 1998.
Yuan Jiarong, Rice and pottery from Yuchanyan in Dao County, Hunan, over 10,000 years ago. In *The Origins of Rice Cultivation, Pottery, and Urbanization*. Cultural Relics Press, 2000.
Yuan Jing and Li Jun, *Research Report on Animal Remains from Nanzhuangtou Site in Xushui, Hebei. Acta Archaeologica Sinica*, 2010, 3.
Zhang Dongju et al., Archaeological records and the origin of Dryland agriculture at the Dadaiwan Site, Gansu, over the past 60,000 years. *Chinese Science Bulletin*, 2010, 55 (10).
Zhang Qingjie, A century review and prospect of archaeology in Shanxi. *Archaeology*. 2002, 4.
Zhang Senshui, A preliminary study on the classification of old stone tool industries in Northern China. *Wenwu Chunqiu*, 1991, 1.
Zhang Wenxu and Pei Anping, Research on the excavated rice grains from the Bashidang Site in Lixian County. *Cultural Relics*, 1997, 1.
Zhang Wenxu et al., Study on the rice husk double peaks and milk protuberance impression in pottery shreds from the Pengtoushan Site in Lixian County, Hunan Province. *Acta Agronomica Sinica*, 2003, 29 (02).
Zhang Wenxu and Yuan Jiarong, Preliminary Study on Ancient Cultivated Rice in Dao County, Hunan. *Acta Agronomica Sinica*, 1998, 24 (04).
Zhang Xiaoling, et al., Microtraces confirm the composite tools and their function thousands of years ago. *Chinese Science Bulletin*, 2010, 55 (03).
Zhang Xuelian, Elemental and isotopic analysis of ancient human bones to study food structure. *Acta Anthropologica Sinica*, 2003, 22 (01).

Zhang Zhenhong et al., Study on the Fauna of Miaoyan Site in Guilin. In *Collected Papers of the Symposium on Middle Stone Age Culture and Related Issues*. Guangdong People's Publishing House, 1999.

Zhao Chaohong, Yu Jincheng and Wang Tao, *Significant Discoveries from the Early Neolithic Site in Donghulin, Mentougou District, Beijing*, Chinese Cultural Relics News, May 9, 2003.

Zhao Fuxing, *The Oroqen Ethnic Group's Hunting and Gathering Culture*. Inner Mongolia People's Publishing House, 1990.

Zhao Hui, Research on the development spectrum of Chinese neolithic cultures. In *A Century Review of Chinese Archaeological Research: Neolithic Archaeology*. Science Press, 2008.

Zhao Jingfang and Huo Baoqiang, *Important Discoveries from the Upper PaleolithicSite at Shizitan, Jixian County, Shanxi*. Chinese Cultural Relics News, Nov. 30, 2007.

Zhao Zhijun, *Study on Rice Genus Phytoliths from the Diaotonghuan Site*. Chinese Cultural Relics News, July 5, 2000.

Zhejiang Institute of Cultural Relics and Archaeology, Pujiang Museum: "*Excavation Report on the Shangshan Site in Pujiang, Zhejiang*," Kaogu (Archaeology) 2007, 9.

Zhejiang Province Institute of Cultural Relics and Archaeology, Pujiang Museum: Excavation Report on the Shangshan Site in Pujiang, Zhejiang. *Archaeology*, 2007, 9.

Zhejiang Provincial Institute of Cultural Relics and Archaeology, Xiaoshan Museum: *The Kuahuqiao Site*. Cultural Relics Press, 2004.

Zhejiang Provincial Institute of Cultural Relics and Archaeology, *Hemudu: Archaeological Excavation Report of a Neolithic Site*. Cultural Relics Press, 2003.

Zheng Jiexiang, *Neolithic Culture and Xia Dynasty Civilization*. Jiangsu Education Publishing House, 2005.

Zhengzhou Municipal Institute of Cultural Relics and Archaeology, Peking University School of Archaeology and Museology: Major discoveries from the excavation of Lijiagou Site in Xinmi. *Cultural Relics of Central China*, 2011, 1.

Zhengzhou Municipal Institute of Cultural Relics and Archaeology, School of Archaeology and Museology, Peking University: Major discoveries from the excavation of Lijiagou Site in Xinmi. *Cultural Relics of Central China*, 2011, 2.

CHAPTER 7

The Crossroad of Emergence of Agriculture: The Yanshan Mountain-Great Wall Transitional Zone

Following the laws of nature brings the fortune and blessing

—*The Book of Changes*

7.1 Introduction

The Yan Mountain-Great Wall Transitional Zone is of great archaeological significance. From the perspective of physical geography, it is a transitional zone from the Mongolian Plateau to the North China Plain and the Northeast Plain. In terms of a river basin, it refers to the northern part of the Haihe River Basin, the lower part of the Luanhe River, that is, Yan Mountain-Great Wall southern zone, and the upper reaches of the Luanhe River, Daling River, Xiaoling River, and Liaohe River, that is, Yan Mountain-Great Wall southern zone. At present, the 400-mm annual precipitation line in China is located here which is also the transitional zone between northern arid and semi-arid regions and sub-humid regions. In terms of landform, Yan Mountain-Great Wall Transitional Zone is mainly composed of loess hills, loess terraces, and slopes, which are suitable for the development of early agriculture.[1] This region is the agricultural-pastoral boundary zone and their ecotone, and also the zone of ethnic integration, the main battlefield of nomadic and agricultural peoples in history. Further back in prehistoric times, especially in the Neolithic period, it was an independent cultural zone that was called Liaoxi (Gilman 1984) of the archaeological culture by Su Bingqi,

S. Chen, *Prehistoric Modernization*,
https://doi.org/10.1007/978-981-97-9423-2_7

which means a crucial part. This study emphasises that Yan Mountain-Great Wall Transitional Zone lies in an ecotone, which is the junction of temperate forest and grassland. The ecotone not only has a specific impact on plants and animals but also on prehistoric humans who relied on them for their livelihood. Moreover, it is worth emphasising that this ecotone fluctuates and migrates in the course of time, and the relationship between its changes and archaeological culture is one of the key parts of recent archaeological research (Gilman 1984). In this sense, Yan Mountain-Great Wall Transitional Zone can be called a cultural-ecological area, which ranges from the Late Paleolithic when the geographical characteristics of lithic assemblages are obviously strengthened with at least four types (Chen 2006d) in northern China, for example, to the pre-Qin times, during which Yan Mountain-Great Wall Transitional Zone began to exist as a distinct cultural-ecological area (it was not originally there), and eventually formed into three different economic forms.

7.2 Basic Issues

On the basis of current research, after nearly a century of fieldwork and research, the archaeological culture lineages of this area have been relatively clear, and the accumulation of archaeological materials is quite abundant, which does not need to be mentioned here. In addition to the study of culture lineages, the study of settlement patterns (Liu and Dong 1997; Hidenori 1997), economic forms (Shizuo 1997; Wang 2004), and even spiritual consciousness (Zhu 1990) is in the ascendant. However, the basis for such categorical research needs to be further strengthened, as archaeological materials are often fragmented or even distorted. In view of the particularity of archaeological materials, archaeologists cannot judge the life style and form of ancient human based on unilateral materials, and they must conduct multiple reasoning ways, just like how criminal investigators solve cases, putting the pieces of evidence together to reach a sound conclusion. The basic approaches of archaeological reasoning can be divided into macroarchaeology and microarchaeology. Macroarchaeology includes the application of deductive, inductive, and analogical reasoning, while microarchaeology includes the careful observation of archaeological remains, the utilisation of reference knowledge, the proposal of working hypothesis, and scientific analysis (Chen 2007). This study makes a combination of these reasoning approaches, such as deductive reasoning to build an ecotone cultural adaptation model, inductive reasoning to summarise numerous archaeological materials, and analogical reasoning to draw on the findings of hunter-gatherer ethnology. As the study does not focus on a certain site or a relic phenomenon, the application of microarchaeology is only partial, such as the investigation process of the abandoned site and the study of a few well-preserved relic units. In addition to the multi-way reasoning, we can also use two other methods: one is to take advantage of the unique advantage of archaeology, that is, to investigate the phenomenon to be studied over a long period of time. The second is to use comparative methods, which is the most fundamental to identify features. As for this study, what has to be done is to make a comparison between Yan Mountain-Great Wall Transitional Zone and its adjacent areas, the Central Plain area and the northeastern area. Obviously, the Central Plain evolved from hunter-gatherers to agriculture and then to civilisation during the period we studied, while the northeast kept up its hunter-gatherer adaptation (Yan 1993).

The reason why we choose "Adaptation Mode" as the start of the study is that it helps us to integrate all kinds of materials and evidence because of the lack and bias of unilateral archaeological materials, through multi-way reasoning to explain the living conditions of ancient human beings. As a unique environmental condition, the ecotone is closely related to the adaptation of prehistoric human beings. What is the relationship between the change of ecotone and the evolution of human adaption? A simple list of correlations between environmental climate change and archaeological cultural sequences is not sufficient and often gives the illusion that environmental changes dictate changes in human adaptation. In fact, prehistoric human adaptation was not cyclical like climate, but rather evolving, constantly forming new adaptation strategies. There is complex diversity even within the same cultural ecosystem, and not only do we need to understand these phenomena, but we need to explain these phenomena, how do they happen? Why is this so? Why do there exist differences between these areas?

7.3 General Model of Adaptation in the Ecotones

Ecotone, in short, is the intersection of two ecological zones (an ecological zone is an area with similar temperature, rainfall, and biological characteristics). Ecotone typically includes an estuary of a river (where fresh water meets sea water), a coastal zone (where land meets sea), and a zone where forests and grasslands intersect. Ecotone is generally characterised by biological types of two ecological zones, which have a mixed distribution, bringing richer biodiversity. Of course, not all ecotones mean increased diversity. For example, a gradual transition from temperate deciduous to coniferous forests does not necessarily mean greater diversity (King and Graham 1981). But even so, for organisms, living in ecotone means that they can easily move from one ecological zone to another. The concept of ecotone is not only of biological effect. In cultural ecology, the intersection zone of the two zones is also a kind of cultural "ecotone zone", which is usually the area where ideas and artefacts meet and is often more culturally creative (Sutton 2004). Yan Mountain-Great Wall Transitional Zone is a typical ecotone where temperate forests and grasslands intersect. Throughout the Neolithic period, it was a cultural "Cross zone" between the millet-dominated agricultural culture of Central Plain and the hunting and gathering culture of the northeast.

The advantage of ecotone for organisms is the diversity of resources, while its opportunity cost is the instability of resources. Ecotones are most sensitive to climate change, not only in the form of organic changes, such as changes in the composition of plants and animals, but also in inorganic parts, such as changes in soil. At the same time, changes are also reflected in the fluctuations of space and time. For example, in the Yan Mountain-Great Wall Transitional Zone, we can see the boundaries of the ecotone moving northward as grassland vegetation is replaced by temperate forests. In particular, it is worth noting that, such changes are very frequent compared to non-ecotone areas in the ecotone, especially in short-period climate events. In the case of Yan Mountain-Great Wall Transitional Zone where the annual precipitation rate is about 20%, continuous drought years cause the low stability or predictability of the resources (Shi 1989), which will undoubtedly affect human adaptation.

It is worth noting that the impact of resource instability in ecotone is different for people with different livelihoods and that humans can employ different adaptation strategies. For hunter-gatherers, their basic strategies for dealing with food risks and uncertainties include increasing liquidity and reserves, expanding diets (such as using aquatic resources), increasing exchange between groups (or relying on allies) (Sahlins 1965), and intensifying the use of certain resources (Chen 2004). Of course, these strategies can not be used at the same time, and the main adaptation strategies can differ from one situation to another, and there are differences in the complexity of the choice of these strategies. In other words, hunter-gatherers think about mobility first, and intensification last. Hunter-gatherers expand the range of their resource searches to increase mobility which is the simplest under the condition that there is enough room to expand, and there is a limit to what humans can cover in a day on foot. More reserves mean putting in more labour, and it can only be done during a certain season, especially in temperate regions (Soffer 1989). The use of aquatic resources requires a higher degree of technical complexity, such as fishing tools (boats, harpoons, nets, etc.) Reliance on allies comes at the cost of inferior status and increased social complexity. Finally, intensive use of some resources, such as toxic plants, requires a lot of labour to remove toxins, and the use of small seed plants needs heavy harvesting and processing. In addition, it should be noted that these strategies are contradictory. For example, intensive use of resources in a place means that liquidity must be reduced. Therefore, it can be predicted that these strategies should be used more frequently and to a greater extent in the ecotone than in the non-ecotone.

For people engaged in food production, a predictable risk-reduction strategy should begin with diversification of livelihoods, so as to avoid concentrating all labour investment in a single livelihood, such as agriculture whose failure can be devastating. For a food-producing economy, since production and consumption are both fixed, it is impossible to increase reserves unless the scale of production is expanded, which in turn requires additional labour; food production requires producers to pay attention to what they produce. Increased mobility does not increase food supply, as in the case of crops or livestock, but rather affects food production, which is itself an intensive use of certain resources, compared to hunter-gatherers, the only strategies food producers have for managing resource risks are diversification of livelihoods (that is, extending the continuity of the diet) and expansion of the social resources (that is, relying on the continuity of allies).

There are two strategies for expanding the social sources of resources. One is to exchange, to provide products that cannot be produced by the other through specialised production. For example, after the development of agriculture, some hunter-gatherer groups developed into specialised ones, who exchanged their prey, fur or honey with agricultural groups for necessities of life. Similarly, some food producers can develop into livestock groups and provide livestock to agricultural groups, establishing symbiotic relationship with them. The other way is to plunder, especially since the complexity of the social organisation makes the collective plunder possible, the profit is very obvious and attractive; in addition to the possibility of obtaining the necessary living resources, they also can obtain valuable labour without the upfront maintenance costs, which can be invested directly in the expansion of production; looting is not without costs, of course, but at the top of society, the benefits far outweigh the costs. Therefore, we predict that food producers in the ecotone will seek to scale up their production while relying on diverse livelihoods, and that, with available circumstances, their livelihoods could specialise to form a symbiotic relationship with a stable agricultural community or increase their earnings through war plunder, thereby avoiding the risk of resource instability.

Social change tends to become more complex over time. Societies with higher levels of specialisation, social stratification, and organisation are more effective at coping with adaptation risks, not only for its higher labour efficiency but also because they can reduce risks through the redistribution of resources within the society. Of course, this process is not without cost, and serious social stratification also leads to the intensification of social conflicts and increases the chances of warfare. Moreover, a society tends to constantly regulate social organisation through political and economic systems, and can also strengthen it through moral concepts, etiquette, and ideology. Thus, less risk of adaptation also facilitates the process of social complexity.

As archaeologists, we need to look for relevant archaeological evidence to prove the existence of these strategies and changes. Since the founding of the People's Republic of China, the archaeological focus has gradually developed from the unearthed artefacts to settlement patterns, animals and plants remains, and then the decorative theme of artefacts. While using stratigraphy and typology to establish the spatiotemporal framework of archaeological culture, Chinese archeological studies give more emphasis on the study of prehistoric economic forms, specifically the assemblage of tools in artefacts, especially in Paleolithic Archaeology. Archaeologists are limited in their knowledge at the time of excavating materials, and it is impossible to collect materials in accordance with the current standards, so now there will always be a lack of materials. Of course, this does not mean that research cannot be carried out. We may not have access to direct information, but it is feasible to obtain indirect information. For example, to judge mobility, when culture mobility is strong, artefacts can be seen in a wide range of consistency (Lin 1987); Judgements can also be based on whether the weight of objects, structure, and other characteristics adapt to the mobility of life (Chen 2008). Archaeologists are destined to interpret archaeological materials (Binford 2009), so they are responsible to find the necessary information from archaeological materials. In this book, we examine the evolution of human adaptation represented by prehistoric archaeological culture in terms of the characteristics and composition of artefacts, the settlement patterns, the remains of animals and plants, and the formation of symbolic systems.

7.4 Transformation of the Prehistoric Cultural Adaptations

Now the ecotone between temperate forests and grasslands in northern China is an ecologically transitional zone from the northeast to the southwest. It is closely related to the boundary where the monsoons extend inland because the monsoons bring precipitation. Yan Mountain-Great Wall Transitional Zone is in fact the eastern part of the ecotone, and its eastern part is the central and southern Inner Mongolia and the western part is located from Ordos to Longdong. In comparison, the eastern ecotone is wider, up to hundreds of kilometres, with east–west trend and north–south fluctuation. Its south boundary fluctuation is smaller, while the north boundary fluctuation is larger, the closer to the east, the more stable (Man 1996).

The distribution of archaeological cultures in this region is currently well understood. The cultural framework from the Late Paleolithic period to the pre-Qin era has been largely established, forming the foundation for research (Fig. 7.1). Of course, it must be admitted that archaeological culture is defined by archaeologists for the purpose of research, it is based on the spatial and temporal distribution characteristics of archaeological remains assemblage, as to whether it has an ethnicity significance is still uncertain. For example, an ethnicity in pre-Qin period probably had more than one way of living. Some might be engaged in agriculture, while others might focus on fishing and hunting. Therefore, it is problematic to simply equate archaeological culture with ethnicity. Archaeological culture is a highly generalised concept that includes all the remains left by ancient humans, including natural materials about adaptation. Here I must emphasise that archaeological culture is the research unit of

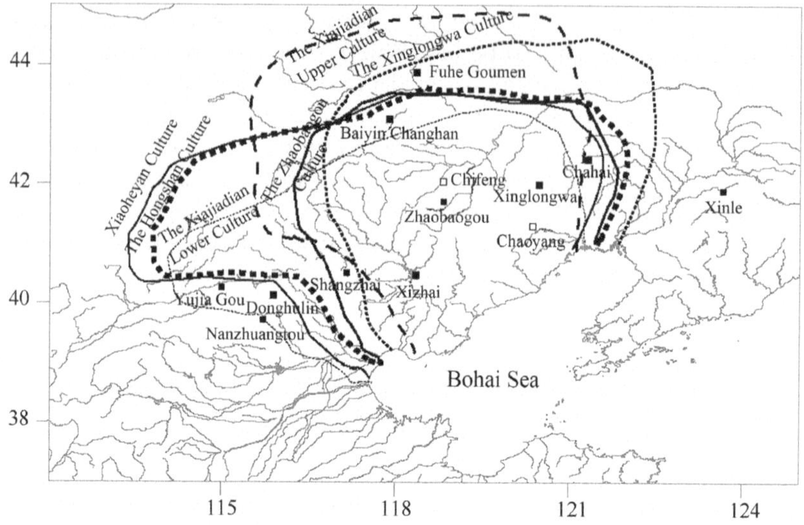

Fig. 7.1 Archaeological culture distribution map of the north and south of the Yanshan Great Wall

archaeologists. However, it is the comprehensive division of prehistoric people by archaeologists, which is not equivalent to our modern division by ethnicity, language, lineage, economic forms, etc. This concept corresponds to the population only for the small society that settled or relatively settled (Trigger 1969; Chen 2009). It is difficult to go beyond this scope, and the prehistoric cultures of Yan Mountain-Great Wall Transitional Zone that we are studying fall into this category and thus we have a necessary theoretical basis.

7.4.1 Adaptive Radiation in the Early Upper Paleolithic

In the early days of the Upper Paleolithic period, the culture of the whole northern China faces a substantial development. In addition to the progress of stone technology, there also appeared personal ornaments, bone antler tools, long-distance exchange items, and so on. At the same time, with the increase in population density, the range of social interaction shrank (Gilman 1984) and regional culture began to form. We can see at least four adaptation modes: Shuidonggou mode, Shiyu mode,

Shandingdong—Dongfang Square—Xiaonanhai mode, and Xiaogushan mode. The latter two models are related to Yan Mountain-Great Wall Transitional Zone. Shandingdong—Dongfang Square—Xiaonanhai mode is adapted to the temperate forest environment, and its stone assemblage is more diverse and irregular than that of the Shiyu assemblage which is mainly a hunter-gatherer livelihood form mainly for plant gathering. The harpoon in the tool combination of the Xiaogushan mode reflects an adaptation to the waterfront environment, which tends to be more stable, that is, it can live in an area for a longer time (Binford 2001). From the distribution of the four patterns, as the environment tends to get wetter, livelihoods ranged from animal-based hunting to diverse hunter-gatherers (more hunting in the Shiyu mode, and more gathering in the Shandingdong—Dongfang Square—Xiaonanhai mode, and to hunting and gathering using aquatic resources (Chen 2006d).

Yan Mountain-Great Wall Transitional Zone lie at the intersection of two adaptation modes. At present, the sites of the early stage of Late Upper Paleolithic in this area are mainly distributed in the southern foot of the Yan Mountains, represented by Shandingdong and Dongfang Square. Among them, the materials of Dongfang Square are well preserved, cultural relics can be pieced together, and leaves are preserved in the strata. Archaeological work has been extensive, with three-quarters of the specimens having three-dimensional coordinates (some are too small to be measured). Traces of fire were found, respectively two in the upper cultural layer and four in the lower cultural layer. Around them, dense stone artefacts and animal remains were found. The vertebrate fossils unearthed include primitive cattle, deer, Mongolian hare, ostrich, pheasant, and fish (Li 2000), reflecting a wide variety of resources from forest margins to grassland, desert, and aquatic. Tools accounted for 4% of the specimens (46 pieces in total), and the rest were flakes and debris, which was an industrial type dominated by flakes. Another 411 broken bones were identified as bone artefacts, of which 64 were identified as tools (Li 2000). The differentiation of the type of utensils is not very obvious. In addition to the round-headed scraper (or end scraper), there are no functionally stable and obvious tools in the Late Upper Paleolithic such as microblade core, microblade, backed knives, refined repair points, stone arrowheads, adzes, etc. It does not develop standardised long flakes like the Shuidonggou site, which reflects a more mobile way of living (Ningxia Institute of Cultural Relics and Archaeology 2003).

It is most similar to Shiyu mode, However, it lacks the relatively abundant and representative animal remains (Jia 1972) (mainly wild horses and wild donkeys, with at least 208 individuals represented by tooth statistics and more than 400 deciduous teeth of wild donkeys). Similarly, there are no specialised hunting and fishing tools like the Xiaogushan site, so this period of human adaptation is a gathering-based multiple hunter-gatherer economy.

The Early Upper Paleolithic sites found in the north of Yan Mountains include Sifangdong in Chengde, China (dated 27,880 ± 510 BP by 14C) (Institute of Vertebrate Paleontology and Paleoanthropology et al. 1992) and the Pigeon Cave (Gezidong) in Kazuo (Pigeon Cave Excavation Team 1975). The date is still controversial, the materials are few, and the working age is earlier, so we will not go into details here.

7.4.2 *The Development of Microblade Technology in the Late Upper Paleolithic—The Beginning of Intensification*

In the Late Upper Paleolithic of northern China, the lithic technique is characterised by the microblade technology, and it is no exception in Yan Mountain-Great Wall Transitional Zone. The main sites include Mengjiaquan in Yutian, Tingsijian in Changli, Donghuishan in Luan County, and Bajianfang in the west of Lingyuan. As a strategy to reduce livelihood risks (Elston and Brantingham 2002) the microblade technology represents a highly mobile livelihood. It can also be said to be the peak of a mobile hunter-gatherer livelihood, which was formed to adapt to resource changes around the LGM. According to the reconstruction of paleoenvironment, it should originate in North China, because the ecological conditions and cultural background of North China were most favourable for the production of microblade technology around the LGM (Chen 2008). Specifically, from the early to Late Upper Paleolithic, prehistoric humans in Yan Mountain-Great Wall Transitional Zone experienced an adaptation challenge. With the advent of the LGM, the distribution of resources that people used to adapt to become increasingly sparse, but for the new resources people have not adapted. Under hunter-gatherer conditions, the preferred strategy is to increase mobility and expand the range of foraging, thus requiring a portable set of tools capable of dealing with multiple situations or uncertainties. The microblade technology (MT) can meet this demand. The microblades produced by this technology are extremely lightweight and standardised products, which can be mounted on animal bone antlers and used as spears, knives, arrowheads, etc. The microblade core itself can also be used as a versatile tool, so the MT is a high-mobility technology. It is worth noting that most of the microblade cores unearthed from the Tingsijian (Hebei Institute of Cultural Relics Research et al. 1992; Wang 1997) and the Donghuishan (Tangshan Cultural Relics Management Office 1993) are the sizes of human fingernails, and the microblade scars are only two or three millimetres wide, which means that the microblades produced by them are difficult to be used as edge inserts, but may be used as barbs on harpoons, hooks, etc. If this is true, then we can assume that the diet was expanded at this point and that this instrument, which is studded with barbs of microblade, could stab smaller fish compared to the harpoon in Xiaogushan (Lü 1995). This exceptionally microblade core is not only found in this area but also in northern Jiangsu and southwestern Shandong. Obviously, in the Late Paleolithic period, the Yan Mountain-Great Wall Transitional Zone did not show the characteristics of ecotone. It used the same microblade technology as the entire northern part of China and the subtle difference in frequency and technology. It should be pointed out here that the ecotone was not located on the Yan Mountain-Great Wall Transitional Zone line around the LGM. According to Zhou Tingru et al.'s research on the melting fold and ice wedge of Hutou Liang, the local area of North China dropped 10–11° C during the LGM, which was larger than that in Europe and America at the same latitude (zhou et al. 1982). Wu Haibin et al. believed that during the LGM in the north, the desert expanded southward to 36° N and east to 125° E (Wu and Guo 2000). By comparing these results with the current climate, we can conclude that the ecotone is roughly in the Qinling-Huaihe River line. The average annual temperature in this area is 15°C, while that in the northern boundary of Yan Mountain-Great Wall Transitional Zone is about 5°C, which is roughly consistent with Zhou's prediction. It is particularly worth emphasising that the Qinling-Huaihe River line is also the southern boundary of the distribution of microblade technology in the eastern part of China, and the southern boundary of woolly rhinoceros-mammoth fauna during the LGM, and it is inferred that the microblade technology originated in the hinterland of North China due to the southward movement of this ecotone (Chen 2008). In short, Yan Mountain-Great Wall Transitional Zone was not a true ecotone before the end of the Last Glacial Maximum.

With the end of the Last Glacial Maximum, around 9000 years ago, the temperature rose back to almost the present level (Winkler and Wng 1993), and the ecotone in Yan Mountain-Great Wall Transitional Zone has gradually formed. The impact of this warming process on the ecological environment is undoubtedly enormous, including rapid climate fluctuations such as Younger Dryas Event (Grootes 1993; Tylor 1997), the reduction of human living space (Xie 1996), the replacement of plants (Sun and Chen 1991; Wang 1997), the extinction of animals (Martin 1999), and the enhancement of seasonality (Kutzbach 1993) caused by rising sea levels. At the same time, human beings had expanded to the Americas in the Late Upper Paleolithic, and all the continents suitable for human habitation had been occupied by human beings. Therefore, human beings could no longer reduce the population pressure by population diffusion. Furthermore, as mentioned before, the microblade technology represents the peak of mobile hunter-gatherer livelihoods. There is a limit to the mobility of hunter-gatherers, that is, without the help of livestock and transportation, the range of food intake that can be covered by daily walking is limited, and humans need new adaptive strategies to cope with the stress of survival. The internal and external pressures on cultural systems are not uniform everywhere. The stress is most evident in the Yan Mountain-Great Wall Transitional Zone, and human adaptation strategies are also clear.

In recent years, the discovery of the Donghulin and Neolithic sites in Beijing has confirmed a shift in human adaptation, in which humans have begun to make intensive use of some resources. Of course, fortification was not unique to hunter-gatherers at the end of the Late Pleistocene, and intensive use of small animals occurred in the Upper Paleolithic (Stiner 1999), the use of acorns by the California Indians of North and Central during the Holocene was intensified whenever there was a resource crisis (Basgall 1987), and the use of acorns required a considerable amount of time to wash out the toxic components, as with small animals, the ratio of human effort to profit has declined. The results vary depending on the cultural and environmental conditions of each region. North American Indians survived a temporary resource crisis through intensive use of resources, while West Asia eventually had to turn to food production. Pottery vessels and stone vessels unearthed at the Neolithic and Donghulin sites are signs of enhanced utilisation of resources. They were probably used for processing time-consuming plants, such as poisonous fruits and small seeds, and the grinding slab and handstones at the

Donghulin site further prove this point. Pottery, ground stone tools, burials, and the discovery of a variety of artefacts at the site (Zhao et al. 2003; 2006) are all signs of reduced mobility. Pottery was poorly suited to a life of mobility, and ground stone tools were a redundant design, and unless one lived in a place for a long time, it was not economical to put a great deal of labour into polishing them; The rich diversity of artefacts at the site suggests that humans lived there for a long time. The longer human beings live in a place, the more kinds of activities they engage in, and the more kinds of relics they leave behind (Binford 2009). Reduced mobility means that the scope of space for the utilisation of resources is reduced, so it is necessary to enhance the utilisation of certain resources. Of course, prehistoric man's coping strategies were flexible, and the microblade technology in the Neolithic and Donghulin sites suggests that there may have been a gender division here, with men hunting out, such as the farming Wahpeton Indian men who were still out hunting in the spring and autumn (Spector 1991). If the same was true of the men of the Neolithic and Donghulin sites, they still needed a tool set suitable for mobility.

Yan Mountain-Great Wall Transitional Zone is certainly not the only area where evidence of intensive use of resources has been found. Pottery and hearth have also been found at Yujiagou and Maanshan sites in the Nihewan Basin in the west (Yan 2000). This area can also be said to be within the ecotone. Intensive use is the first step in the origin of food production, but it cannot be said that it will necessarily lead to the origin of food production. Relevant conditions, such as appropriate species and initial cultural conditions, are closely related to it, which I have discussed in a special article (Chen 2006a; Chen 2006c) and will not be repeated.

7.4.3 *From Hunter-Gathering to Food Production*

Food production means that human beings obtain food through artificial means such as care, cultivation, and domestication, besides naturally growing food. It can include taking care of wild crops in the initial stage, horticulture, simple slash-and-burn agriculture, etc., sometimes indistinguishable from hunter-gatherers. Or it can be a budding form of agriculture. Agriculture, by contrast, refers to the full range of farming, planting, irrigation, harvesting, and other methods of production of human resources and mature means of livelihood. To avoid confusion, food production is used here to refer to early forms of agriculture.

After Yan Mountain-Great Wall Transitional Zone entered the Holocene, steppe unsuitable for hunter-gatherers during the LGM was replaced by forest-steppe ecotone, and the sandy land in the east completely disappeared during the Holocene optimum period (Wu and Guo 2000). The improvement of the environment can be said to provide a new ecological space for prehistoric people, and this once unsuitable place for human habitation can be used for human beings. What makes more sense is that when humans adopt the adaptive strategy of food production, they also start to build ecosystems to suit their needs, and thus create new ecological niches.

The transition from hunter-gathering to food production in the Yan Mountain-Great Wall Transitional Zone began in the south of the Yan Mountains, namely the places mentioned above such as the Neolithic and Donghulin sites and Nihewan Basin. However, with the expansion of new ecological niche and the formation of food production and livelihood, the north of the Yan Mountains seemed to be the central area, and the main body of Xinglongwa culture was also in the north of the Yanshan, while the south of Yan Mountains is the Dongzhai type, the Shangzhai culture site also belongs to the north of Yanshan system. From the north–south integration of Yanshan at the end of the Late Pleistocene to the North–South differentiation in the Holocene, Zhenjiangying Phase I, which was earlier than the rise of Xinglongwa culture, belonged to another system (Beijing Institute of Cultural Heritage 1999), which formed a system with Ciishan culture, Beifudi Class A remains, and Hougang culture. Of course, there is still a gap of nearly 2000 years between the Neolithic and Donghulin sites and the Xinglongwa culture, and there are still undetermined stages of cultural development, such as the Zhenjiangying Phase I and Xiaohexi Culture. The focus here is on Xinglongwa culture, exploring its adaptation transition from hunter-gatherer to food production.

According to Ren Shinan's summary, compared with the archaeological culture of the same period in North China, Xinglongwa culture has the most obvious settlement characteristics, the largest and most sophisticated buildings, and the significance of hearth is prominent. The decorative pattern of pottery is the most with only the sand-tempered pottery, and pottery types are not rich; stone tools are relatively simple, but still retain microblade technology, and cannot be seen in other areas. In the absence of exclusive graves, the system of using jade has been formed with far-reaching influence (Binford 2009). From the perspective of simple livelihood forms, the remains of cultivated plants of Xinglongwa culture have been found so far. In the 1000 soil samples collected at the first site of Xinglonggou, more than 10,000 seeds of plants were found, most of which were herbaceous plant seeds with small individuals, among which 1500 grains of charred millet and less than 100 grains of charred maize (Ren 1994). *Panicum miliaceum* is more tolerant to drought and cold conditions than *Setaria italica* (Li 1984), so it is understandable that the cultivation of broomcorn began in the Yan Mountain-Great Wall Transitional Zone. At the same time, it must be noted that cultivated plants only account for about one-fifth of plant seeds, which indicates that the gathering economy is still quite important at this time. Among the animal remains of Xinglongwa cultural site, deer and pigs (including wild boar and domestic pig) are the most common. Taking the Baiyinchanghan site as an example, the main hunting artefacts of human beings are red deer, zebra deer, and roe deer, and pig specimens occupy the fourth place. The measured length of the pig's third molar tooth is between that of wild boar and domestic pig, which is classified by researchers as wild boar (Inner Mongolia Autonomous Region Institute of Cultural Relics and Archaeology 2004). However, it also indicates that humans have begun to domesticate pigs. Judging from the complete skeleton of a male and a female pig buried with M118 in the Xinglongwa site tomb (Archaeological Research Team of the Institute of Archaeology, Chinese Academy of Social Sciences, Inner Mongolia 1997), pigs are more likely to be domesticated. It takes time for the morphologic evolution from boar to pig, and human intervention must be considered as the main aspect of the problem, otherwise domestication cannot be understood.

We know that the cultivation of crops and the domestication of animals in Xinglongwa culture is in its infancy, and it relies on hunter-gatherers at the same time. The problem is not the obvious diversity, but the differences between different sites. Does Xinglongwa culture have a common form of livelihood? Fish cellars and bone darts (Yang 1993), as well as a large number of stone net pendants (Hebei Provincial Institute of Cultural Relics Research et al. 1992) found in the first phase of Qianxi Dongzhai and Xizhai Phase I, indicate that fishing was very important in the life of these sites. These remains are not found in the sites of Baiyinchanghan and Chahai (Liaoning Provincial Institute of Cultural Relics and Archaeology 1994), which also indicates that the cultural livelihood of Xinglongwa is not a single mode, but a diversified livelihood according to the resource conditions of different regions.

According to Japanese scholar Okamura Hidenori, there was no formal agriculture in the Xinglongwa culture period. Grinding slabs and handstones were used to process nuts, and stone shovels were used only to break up topsoil and not to plough deeply. At this time, the environment was a forest area, which was not suitable for agriculture. He also cited North American ethnography materials to support the view that Xinglongwa culture residents were settled hunter-gatherers (Hidenori 1997). The dryland agriculture, mainly made of *Panicum miliaceum*, began with slope farming. In the Early Holocene, although the temperature and precipitation had increased, the slope itself had limited water conservation capacity, not the forest zone, but the grassland. The forest was distributed in the valley area with better water and heat conditions, so the slope cultivation did not need deep ploughing tools. Human beings, however, expanded lately farming into areas of valley forest where water and heat were better, requiring larger tools of cutting and digging. Such changes in farming methods can explain why cultural sites continuously decline into river valleys. From one perspective, it lies in the downcutting of rivers (Xia et al. 2000). In fact, the continuous length of Neolithic cultural sequence is relatively short in the context of geologic time, and the impact of downcutting of rivers on culture is limited. The more important reason should be the former. Humans have the ability and motivation to exploit the more inaccessible river valleys.

The means to determine the livelihood pattern is not only direct flora and fauna evidence, but also tool assemblages and some related evidence such as human bone 13C analysis, fields, and decorative themes (Liu 2002). Studies have been done to divide the stone tools and try to determine the economic form at that time (Liu and Dong 1996). But the functional division of the tool assemblage must be considered as a whole, rather than just stone tools. In addition, many tools in the Stone Age were multifunctional, especially the portable tools, such as the stone knives with bone handle. Therefore, the main functions of the tools must be considered first. Taking the remains of the Baiyinchanghan site phase II (i.e. Xinglongwa culture) as an example, we can classify the tools into excavation tools (including stone shovels, stone perforators, and hoe - shaped tools that may be weighted stones for digging rods), chopping tools (including stone axes, stone adzes and stone chisels, mainly used for processing wood), tools for processing plants (such as grinders, pestles, grinding slabs, handstones, small stone rods, mortars, cake-shaped tools, and smashers), hunting and fishing tools (e.g. projectiles, net pendants, spheroids), harvesting tools (e.g. stone knives, axe-shaped tools), and microliths. The microliths are all multi-purpose tools and are listed separately. According to the composition diagram of Baiyinchanghan stone tools (Fig. 7.2), the proportion of excavation tools is relatively high. Then there are the hoe-shaped tools from the Chahai site, which are numerous and broken. A complete hunter-gatherer economy would not have required these tool combinations. They did not have to invest a lot of labour to produce sharpened digging tools, since a very simple digging stick could quickly collect all the food resources around the site; if the distance from the site increased, it is very uneconomical to carry these large stone tools, so they should belong to the crop grower.

Although we say that the Xinglongwa culture adopted a diverse way of life, even if the Baiyinchanghan site found fewer hunting tools, it does not mean that the proportion of hunting is small. We need to think in terms of the abandonment process of the remains. Hunting tools are generally used outside the site. Taking the Baiyinchanghan site as an example, there are not many microblades found, but they are widely distributed on the nearby river terrace. In the graves of Xinglongwa site, hundreds of microblades are buried with them. It shows, in turn, that the reason why microblades are found less on the site is that most of them were not used or even made on the site, so it's not appropriate to show that hunting is not important for livelihoods because hunting tools are less found. It

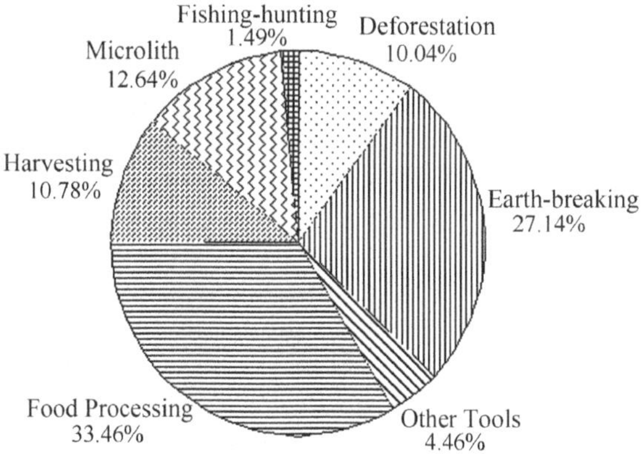

Fig. 7.2 The stone tool composition of the class B remains of the Baiyin Changhan phase II site

must also be combined with the investigation materials around the site and burial materials together.

The abandonment of sites of the Xinglongwa culture deserves attention. They have certain similarities, that is, the settlement pattern is complete and the remains are rich. The relatively well-preserved settlements include Xinglongwa, Baiyinchanghan, Chahai, Nantaizi (Inner Mongolia Autonomous Region Institute of Cultural Relics and Archaeology 1994), and Xinglonggou (Archaeological Research Team of the Institute of Archaeology, Chinese Academy of Social Sciences, and Aohanqi Museum 2000). Why does this happen? Why is this state of preservation not found in more recent sites? What kind of human behaviour did it reflect at that time? In the ruins of Xinglongwa culture, there are a certain number of utensils that can be used, especially the larger tools such as stone shovels and grinding slabs. For example, at the Baiyinchanghan site, many utensils remain in place on the housing surface, and the areas where people often move are cleaner such as the fine clay smear living surface. This means that the site was deliberately abandoned, but with the expectation that it would return at some point in the future, so that some useful artefacts were kept, without disrupting the basic distribution of relics. This abandoned feature of Xinglongwa cultural site also indicates that people's settlement ability was not strong enough at that time. While living on the site, they also needed to maintain a certain mobility, such as collective gathering and hunting in a season of the year. The microblade technology retained by Xinglongwa cultural site also supports this judgement. Baiyinchanghan site has preserved the house sites of Xinglongwa culture, Zhaobaogou culture, and Hongshan culture. Compared with the thickness of the braised soil of the hearth that of the Hongshan culture is much thicker (Inner Mongolia Autonomous Region Institute of Cultural Relics and Archaeology 2000). It is obvious that people used the site for a longer time during the Hongshan culture period.

The well-organised settlement structure of Xinglongwa culture has always been puzzling. It is different from the cultural settlements of the same period in the Central Plains. Why? As an ecotone, this area was a new ecological space for human beings at that time. The land was wide, and all that was lacking was sufficient labor; on the contrary, the land resources in Central China have decreased rather than increased due to the rise of sea level, the elevation of river erosion surface, and the increase of flooded and swamped areas. This very different resource situation is reflected in the settlement pattern, which is the emphasis on human resources in Yan Mountain-Great Wall Transitional Zone. The strict layout of the settlements and exaggerated scale of individual buildings (the area of individual buildings, in general, is 30–40 square metres, and large buildings are hundreds of square metres, far exceeding the architectural scale of prehistoric culture in Central Plains at the same time) reflect the sensitivity of the groups to the population size. Similar phenomena also occurred in the Neolithic in Europe (Hodder 1985). Settlement pattern itself has not only functional meaning but also symbolic significance. The strict structure shows people's attention to social belonging, and the exaggerated space shows people's pursuit. At this stage, we understand it as the competition of population size in different social units.

As I mentioned earlier, there was a lot of mobility in people's livelihoods during the Xinglongwa culture. Men were probably more mobile, and they might have had to go out hunting during certain seasons of the year like the Wapidon Indians, while women were responsible for other activities such as pottery making and household chores, which was supported by the pottery style and house layout of Xinglongwa culture. In general, if women presided over these affairs, this preference may have been expressed in the pottery decoration, home layout, and the number and variety of decorative items, as women preferred decoration layout. The pottery decoration style of Xinglongwa culture is rich, and the decoration is often all over the pottery. In the interior space, fine mud is applied to the living surface around the hearth. Compared with other cultures of the same period, its interior space is much more delicate. Decoration is divided into stone, jade, bone, and clam, especially jade is the most distinctive. On the other hand, other things being equal, a society that emphasises population size will inevitably value reproduction, and the status of women will be bound to rise. At the same time, a larger population also stimulates food production, as demand and supply (increased labor) expand in tandem.

The diversified livelihood forms of Xinglongwa culture are not only reflected in the tool combination, plant and animal remains, settlement patterns but also showed in the decorative structure. If we assume that all the decorative symbols of prehistoric times had some meanings, then different kinds of symbols would have different meanings. We may leave aside the content of the symbol for a moment and consider only its structure. The structure of a society is often similar on different levels, from livelihood, settlement, social structure, etiquette, and even ideology, which can be called the fractal structure of social existence. The decorative structure of the pottery of the Xinglongwa culture is generally divided into three sections, taking the remains of Baiyinchangkhan II Class B as an example, one section is generally a concave string pattern, the second and third sections have more variations, and the second section is particularly complex, indicating a combination of ornamentation that reflects transitions, contradictions, and harmonisation.

Since the Late Paleolithic, the phenomenon of cultural regionalisation has become more and more obvious, which is reflected in stone assemblage and technology. They may reflect the interaction circle of human groups, that is, the scope that humans can turn to when resources are poor (Gilman 1984). However, there is still a huge dispute about the relationship between stone assemblage and human groups (Bordes and D. Bordes 1970; Renfrew 1973). By the Early Neolithic, the scope of archaeological culture was quite clear. The significance of the assemblage of artefacts was not limited to stone tools but also extended to pottery, burials, dwelling sites, ornaments, etc., which could represent the clear interaction sphere of human groups. As mentioned above, the mobility of human beings is strong, and the consistency of the types of their artefacts is relatively high. It must be pointed out here that the commonality of the assemblage of artefacts is also closely related to the interaction within culture and the increase of social complexity. In Xinglongwa culture stage, we can see that a new and definite cultural region has emerged, which is much clearer than that in the Late Paleolithic. This means that people at this time had clear targets to turn to, and they also limited the number of people who could ask for help.

Summarising the basic features of human adaptation during the Xinglongwa culture, it is not difficult to find diverse strategies, starting with multiple livelihoods, where people also partially maintained mobility, developed reserves (e.g. cellars in dwelling sites) and expanded diets (e.g. fishing and gathering shellfish). At the same time, people put emphasis on population and developed well-defined cultural regions, showing an overall transitional nature of hunter-gatherer to food production, while employing both risk-sharing strategies.

7.4.4 *Comparison: Hunter-Gatherer Under the Influence of Food Production*

After the advent of the Neolithic in northern China, new cultural characteristics, represented by food production, settlement, pottery, and grinding techniques, began to affect the surrounding hunter-gatherers, and thus cultures of all shapes and sizes have developed around food producers. As far as Yan Mountain-Great Wall Transitional Zone is concerned, cultures such as Xinle Lower in Shenyang (Shenyang Municipal Cultural Relics Management Office, Shenyang Imperial Palace Museum 1985), Zujiashan I in Nong'an (Archaeology Teaching and Research Office, Jilin University 1989), and Xinkailiu in Mishan (Heilongjiang Provincial Cultural Relics and Archaeology Work Team 1979) emerged in the surrounding areas. The main characteristic of these cultures is that fishing and hunting tools are very developed. For example, no farming tools have been found at the Xinkailiu site, and there is no evidence of food production. However, they all have some elements of Neolithic culture, such as house structure, pottery, fixed graves, ground stone tools, and bone wares. The question is why did these new elements not emerge in the Upper Paleolithic period, but with adaptations to Neolithic food production? Which of these elements were part of the simultaneous independent development of hunter-gatherers? What are the influences of food producers? At present, it is known that pottery dating more than ten thousand years ago has been found in the Japanese islands and the Far East of Russia (Aikens and Higuchi 1982) which is basically at the same time as North China. At the same time, people in these areas expanded their diet, developed fishing and hunting livelihoods, and utilised aquatic resources to form a livelihood model that could support certain settlements (Binford 2001). The Indians of the northwest coast of North America even developed complex social forms using aquatic resources (Matson and Coupland 1995). In this regard, what we see is a synchronised development, that is, between the end of the Late Pleistocene and the Initial Holocene, the effects of environmental changes on hunter-gatherers were common, but not all hunter-gatherers chose food production. In some areas, hunter-gatherers developed a livelihood that supported a higher density of people than hunter-gatherers, namely hunting and fishing, according to the resource conditions of the area. Clearly, food production was not the only option for hunter-gatherers, nor was it a livelihood strategy they were striving for. However, they were undoubtedly influenced by food producers and chose to compensate for their livelihood by growing small amounts of crops or raising livestock. They may also have learned from food producers house building, tool grinding, pottery-making techniques, etc., so their cultural features were often similar to those of Neolithic food producers at the same time or earlier. For example, cylinder jars originating from the Yan Mountain-Great Wall Transitional Zone were very popular in Northeast China and formed cultural traditions (Yan 1993).

7.4.5 The Diversity of Mixed Livelihood Modes and the Self-development of Culture

After the Xinglongwa culture, several Neolithic cultures were developed in Yan Mountain-Great Wall Transitional Zone including Fuhe, Zhaobaogou, Shangzhai, Hongshan and Xiaoheyan cultures. What adaptation strategies have these cultures adopted? Are the policies the same? What are the characteristics of each? What kind of law of development does it reflect?

About 4000 years ago, the relatively warm climate of the Middle Holocene in Yan Mountain-Great Wall Transitional Zone ended (Wang 1996), although there were several small periods of climate stability and fluctuation (Shi et al. 1992). Compared with Xinglongwa culture, the cultural adaptation in this period still retained a diversified livelihood mode, mainly reflected in the combination of stone tools. In addition to tools suitable for cutting and farming, there were microblade technology products suitable for mobile hunting, which was obviously different from the Neolithic culture in the Central Plains at the same time. The animal remains at the site also support this point. This is distinctly different from the Neolithic culture of the same period in the Central Plains. This is supported by the faunal remains of the site, which are rich in animal skeletal remains from house sites F2 and F9 of the Zhaobaogou site, where the most identifiable animal specimens (NISPs) are pigs and deer-like animals (horse deer, spotted deer, and roe deer), and the third molar measurements of pigs from the Zhaobaogou site indicate that pigs were being domesticated (Institute of Archaeology, Chinese Academy of Social Sciences 1997) and that the main part of people's meat still came from hunting. It should be noted that there is no single adaptation pattern for a culture. The Qianxi Xizhai site of Zhabaogou culture is different from other sites with a large number of net pendants indicating that it was developed in fishing and hunting. Similarly, some sites of Hongshan culture (Zhu 1979), such as Daqintala and Erdaoliang (Zhang 1994; Inner Mongolia Autonomous Region Institute of Cultural Relics and Archaeology 1994) have unearthed very rich microblade technology products, which are different from other sites of Hongshan culture. Whether or not the remains are left from seasonal use of resources, it can be demonstrated that the livelihood composition was diverse.

What is particularly remarkable about these diverse livelihood modes at this stage is not their diversity, but a new phenomenon, namely the emergence of some almost specialised hunting economies. For example, in the Fuhe culture, the highly developed microblade technique in its tool assemblage and the abundant animal remains in the site indicate the existence of a specialised hunting livelihood. However, the Fuhe culture includes settlement sites, such as Goumen (Archaeology Work Team of the Institute of Archaeology, Chinese Academy of Sciences, Inner Mongolia 1964) and Jinguishan (Xu 1994), with hundreds of house sites and some cellars inside, reflecting that Fuhe culture has at least a certain degree of settlement capacity. However, hunting livelihoods are contradictory to settlement lifestyles, but this adaptation strategy can be justified when hunters exchange their prey with farmers for necessary supplies, at the cost of hunters becoming professional hunters. After the formation of food production livelihoods, a part of hunter-gatherers chose the path of specialisation and they constituted a symbiotic relationship with food producers (Chen 2006c). Similarly, no house sites have been found so far in the Xiaoheyan culture, but a large number of ash pits and graves, which are presumed to be similar to the present yurt (Liaoning Provincial Museum 1998), so no obvious residence sites are left on the surface. A comparison of the composition of the main stone tools can also shed some light (Fig. 7.3). By comparing the proportion of different functional stone tools in several archeological cultural typical sites, it can be seen that Xiaoheyan culture and Xiajiadian upper culture have the least stone unearthed tools, which may be related to people's increased migration frequency and living style. The buildings that can be easily demolished are also a direct reflection of frequent migration. It indirectly reflects the conversion of people's livelihoods. In other words, it relies less on farming than the Hongshan culture and more on hunting and livestock rearing. If farming is the main line of livelihood development, then Xiaoheyan is another form of livelihood specialisation. Moreover, this highly mobile livelihood tended to be less stable than that of the farmers, with whom they needed to exchange essential supplies, which could also explain the impetus for the southward migration of the Xiaoheyan culture—to maintain contact with the farmers and establish a symbiotic relationship, as in the case of the Jiangjialiang Xiaoheyan culture tomb in Yangyuan County (Hebei Provincial Institute of Cultural Relics Research 2001).

From Xinglongwa culture to Hongshan culture, people's mobility has been declining and population density has been increasing. Take Baiyinchanghan site as an example, which contains the remains of Xinglongwa, Zhaobaogou, Hongshan, and Xiaoheyan cultures (Inner Mongolia Autonomous Region Institute of Cultural Relics and Archaeology 2004). The thickness of the braised soil in the hearth shows that the inhabitants of Hongshan culture lived for a longer time. Based on the calculation of site density and the data of several systematic field surveys (Suo 2005), it can be seen that from Xinglongwa culture to Hongshan culture, the site density of Hongshan culture is significantly increased. Even taking into account the longer time span of Hongshan culture, the site density of Hongshan culture is still higher than that of Xinglongwa and Zhabaogou culture in a period of time, which reflects the expansion of population density and scale at this time. The prolongation of settlement time and the increase of population density and scale will inevitably require the prompt of resources that can be provided per

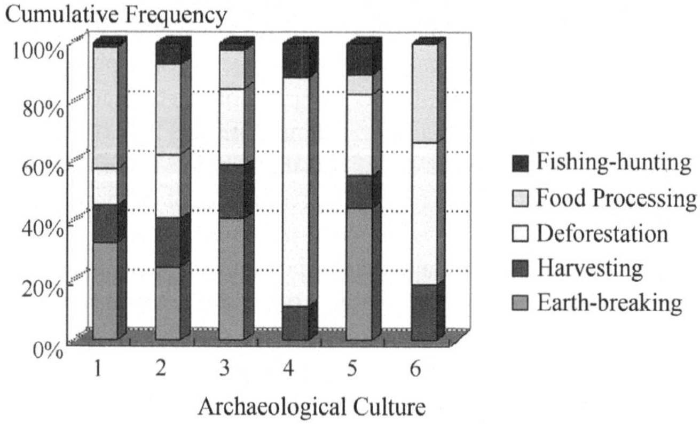

Fig. 7.3 Comparison of the composition of stone tools in six archeological cultural sites[2]

unit area. However, the growth of natural resources is relatively constant and will decline with the increase of human utilisation intensity, so a feasible method is to expand the output of food production. Comparing the lengths of stone shovels of Xinglongwa, Zhaobaogou and Hongshan cultures, among the stone ploughs (stone shovels) excavated from the Baiyinchanghan site, 19 pieces of stone shovels were excavated from the Phase II Class B remains, i.e. the typical Xinglongwa culture, with an average length of 175 mm; while 27 pieces of stone ploughs were excavated from the Phase III Class A remains, i.e. the remains of Zhaobaogou culture, with an average length of 197 mm; only one piece of stone ploughs was excavated from the Phase IV Hongshan culture, with a length of 295 mm. The length of the stone ploughs that can be seen in it has a tendency to increase, indicating that the depth of breaking the soil deepens; and from the shape of the stone shovel, the stone shovels of Zhaobaogou and Hongshan culture are grooved-shaped, which is more convenient for hafting and the sharp angle of the blade, which is conducive to improving the efficiency of breaking the soil. In terms of the geographical location of the settlement, compared with the Xinglongwa culture site, the relative height of Zhaobaogou and Hongshan culture sites decreased and the terraces where they were located were lower (Xia et al. 2000), which indicates that people began to make more use of the forested areas of the river valley, and the expansion of the cultivated area naturally reflects the increase in food production.

During this period, while people's livelihood patterns were differentiated and symbiosis, the internal social groups also began to be differentiated, which was not only reflected in the enhance of the division of labour and the higher degree of specialisation of labor but also reflected in the formation of social classes, which is stabilised through politics and ideology. From a large number of sacrificial relics of Hongshan culture (including rich sacrificial pottery, funerary jade, sacrificial buildings, etc.), it can be seen that a professional division of labour has emerged (Shi Peijun 1989). Widespread sacrificial activities have sanctified the division of social classes, and the sacrificial sites centered on Niuheliang (Liaoning Institute of Cultural Relics and Archaeology 1986) and Dongshanzui (Guo and Zhang 1984) further reflect the stratification of settlements. Stratification among settlements increased, and settlements like Niuheliang became the center of a region. In the lower Xiajiadian culture, an important feature of the site is that it is often located in defensible places, such as a high riverbank with steep rocks and walls. In the vicinity of the Beipiaofengxia site, "Shi Chengzi" (Liaoning Cultural Relics Training Class 1976) with stone walls are commonly found, all belonging to the lower Xiajiadian culture. Obviously, warfare was much more frequent in this period than before. It is worth noting that most of the lower Xiajiadian culture is deeply stratified, complete with farming tools, and the Fengxia site unearthed piles of grains, which are identified as millet and corn, and most of the animal remains were pig bones, followed by sheep bones and cattle bones (Liaoning Cultural Relics Training Class 1976) reflecting a completely settled way of food production and livelihood.

7.4.6 Differentiation and Symbiosis

In the upper Xiajiadian culture, the livelihood patterns in Yan Mountain-Great Wall Transitional Zone changed significantly. The quantity and quality of pottery and stone farming tools in the upper culture of Xiajiadian, as well as the length of settlement time, were not as good as those in the lower Xiajiadian culture, which reflected the change in the livelihood pattern of the upper culture of Xiajiadian, that is, it no longer focused on farming, but turned to livestock raising. The most important factor in this process is the introduction of horses. At present, the conclusion of the research on horse domestication is not clear, and mitochondrial DNA analysis shows a very wide range of origins of horses, which may be of multi-regional origin (Vila 2001), while some suggest an origin in one or two regions of diverse wild horse herds Lister 2001). Domesticated horses appeared in the Yellow River basin in the late Shang Dynasty (Yuan 2003), and the use of domestic horses was already more common in the Upper Xiajiadian culture period. The domestication of the horse gave people mobility that they had never had before, and the range of resources they could use was greatly expanded. This is especially true for hunter-gatherers, whose activities are mostly within a two-hour walking circle when they purely rely on walking (Higgs 1975). Once the population size exceeds the carrying capacity of this range, famine will occur. As a result, it was difficult for hunter-gatherers who relied on foot to form large social groups. (Coastal hunter-gatherers not only had aquatic resources with different seasonal patterns than terrestrial resources but also had the ability to move by boats, allowing large populations to form complex societies such as those of the Northwest Coast Indians of North America.) When horses were available for hunter-gatherers, they provided speed and load capacity far greater than that of humans, allowing them to hunt and gather over a wider range of resources, to use their speed to hunt larger and more dangerous animals, and to transport prey back to central camps. However, it is also because of this that the food-producing Indians in the Great Plains of North America gave up food production and resumed hunting and gathering after owning horses (Binford 2001). As mentioned above, hunter-gatherers formed a symbiotic relationship with food producers after the origin of food production, and this relationship was strengthened when hunter-gatherers adopted horses. It is also particularly noteworthy that the domestication of the horse made it possible, for the first time, for humans to actually have the possibility to fully exploit the grassland environment, which previously was classified as a hunter-gatherer marginal environment along with several environments such as deserts, plateaus, and polar regions because of the relatively low primary productivity (i.e. plant growth) per unit area and the relatively low secondary productivity (i.e. animal growth) that it could support. The low density of resources in this zone and the very high mobility of the fauna made it difficult for humans to sustain themselves by relying on walking, which is one of the reasons why there are fewer Stone Age remains from true steppe areas. The domestication of the horse gave humans a new ecological niche, which is the main reason why the bronze culture of the steppe zone flourished after the domestication of the horse.

From livestock raising to animal husbandry and then to nomadism, people's dependence on domesticated animals keeps deepening. Compared with livestock rearing, livestock rearing means that the social group has to divide the special labour to go out to herd livestock, and the extra labour has to carry out farming; compared with nomadic herding, nomadic herding means that people have completely abandoned farming and rely exclusively on livestock rearing to make a living. Livestock rearing is found in societies where farming is the mainstay, such as the agricultural economy in China's feudal era. Societies with both farming and animal husbandry are more complex, either predominantly farming, as in most European feudal-era economies, or predominantly animal husbandry, as in prehistoric cattle herders in North Africa; that is, different periods of the same society may fluctuate between farming and animal husbandry, as in the Ordos region of the historical period. The emergence of animal husbandry is earlier than the 3200 years before present in the upper Xiajiadian culture (Guangjin 1997), with the burial of sheep mandibles in the middle Zhukaigou period (4000–3800 BP) and pig mandibles in the early period. At the same time, the Qijia culture period (4400–3900 BP) in the Hulu River basin in the northwest decentralised the settlement and livestock farming began to appear (Mo et al. 1993).

During the Upper Xiajiadian culture period, there were well-developed harnesses and even images of stalking and hunting. But a large number of tools suitable for farming were still unearthed, such as in Linxi Dajing (Liaoning Provincial Museum Archaeological Work Team 1983), Xiajiadian (Archaeological Work Team of the Institute of Archaeology, Chinese Academy of Sciences, Inner Mongolia 1974), Zhizhushan (Inner Mongolia Work Team of the Institute of Archaeology, Chinese Academy of Social Sciences 1979), and Ningcheng Nanshangen (Inner Mongolia Work Team of the Institute of Archaeology, Chinese Academy of Sciences 1975), and numerous cellars and evidence of raising other domestic animals, especially pigs, which are not yet nomadic (Wu 2022). Moreover, there are fewer stone shovels in the upper Xiajiadian culture probably because there are more wooden shovels, and stone tools disappeared from the Liaodong Peninsula to western Korea from the Longshan culture period (Hidenori 1997), so fewer stone shovels cannot be regarded as a sign of the decline of farming in the upper Xiajiadian culture. The establishment of nomadic pastoralism actually relied not only on the domestication of horses but also on cattle and sheep, which except for providing meat, could also offer large quantities of dairy products, especially cattle; otherwise, depending only on meat, nomadic pastoralism could not meet the livelihood needs. Furthermore, cattle generally eat only fresh grasses, while sheep eat everything, especially soft-stem bulrush, which are harmful to pasture (Sutton 2004), constituting a good symbiotic relationship. Horses and dogs being domesticated early provide the conditions for large-scale mobile grazing of cattle and sheep. Pig rearing, on the other hand, requires a more sedentary lifestyle and contradicts nomadic livelihoods. In addition, nomadic livelihood was not a completely self-sufficient way of livelihood; nomads obtained some of the necessary means of livelihood such as metal tools, textiles, and food from farmers through exchange (Hidenori 1997), and thus nomadism can be said to be based on farming. In addition to exchange, nomads also obtained their means of livelihood by plundering, as marked by the appearance of bronze weapons in the upper Xiajiadian culture; at the same time, the appearance of the horse also changed the form of war. Armies riding horses and shooting arrows were incomparable in terms of mobility and impact to soldiers on foot in the past. The symbiotic structure of farming and nomads profoundly influenced the later history of China.

After Hongshan culture, the differentiation of livelihood modes in the Yan Mountain-Great Wall Transitional Zone was not only manifested in the emergence of economy and pastoralism, but also in the establishment of a dry farming agriculture system. Yan Mountain-Great Wall Transitional Zone was dominated by the cultivation of setaria italica and panicum miliaceum, especially setaria italica because it is more adapted to the dry and cold environment than panicum miliaceum. Another crop that makes up the dry farming system is Glycine max, which has a similar significance to milk in the dietary structure of Western societies. The Glycine max produces soy milk; Such as milk can be made from cheese, while tofu can be made from Glycine max. The key is that the two have similar nutritional values and occupy similar positions in Eastern and Western societies – both are important sources of protein and fat. Soy products are the most consistent source of protein for dry farmers, allowing them to divert energy from hunting or pastoralising to farming crops, raising livestock for animal power, holiday consumption, and so on. Cultivation of Glycine max and other beans have the function of "land cultivation". In the pre-Qin period, Glycine max and gramineous crops were exchanged or interplanted to maintain soil fertility and improve crop yield (Li 1984). It is through the cultivation of Glycine max that the traditional dry farming system in the north was completed.

Yan Mountain-Great Wall Transitional Zone may be one of the earliest domesticated or cultivated areas of *Glycine max* (Fukuda 1933; Li 1994). At present, there are many controversies about the origin of *Glycine max* (Zhao 2004). The wild ancestor of *Glycine max* may be *Glycine ussuriensis* (Li 1984, p.79) or *G. soja* (Hymowitz 1995), which is distributed over a wide area from the southwest to the northeast of China. T. Hymowitz speculated that the domestication of *Glycine max* probably began around 1100 BC (Hymowitz 1995). The main evidence for the origin of Glycine max in Yan Mountain-Great Wall Transitional Zone is that this region has the largest number of wild and domesticated Glycine max varieties to date, and some of them still retain their original characteristics (Zhao 2004). In addition, it is recorded that Duke Huan of Qi once introduced "soybeans" produced in the northern mountains to the Central Plains for cultivation (Li 1984). If this evidence holds true, we might say that three livelihood systems were established in Yan Mountain-Great Wall Transitional Zone during this period: dry farming, agro-pastoralism (animal husbandry), and nomadic pastoralism, showing a rich diversity.

7.5 Changes in the Environmental Pattern and the Adaptive Transition

We have seen many forms of prehistoric cultural adaptation in Yan Mountain-Great Wall Transitional Zone, usually attributed to climatic changes, because prehistoric cultures were limited in their ability to cope with environmental changes, and so people had to choose to change. There is undoubtedly some truth to this explanation, but it is oversimplified. Since the Late Paleolithic, environmental changes may promote or inhibit cultural development, as in the case of the LGM for the microblade technology, the Younger Dryas Event for the origin of food production, and the Holocene warmest period for Xinglongwa culture, and the coupling between the strongest drought event and the global Old Continent civilisation event in northern China 4000BP (Wu 2000). On the contrary, due to insufficient accumulated temperature, millet farming failed and Hongshan culture migrated southward. The 4800-4200 cal BP cold drying event with global synchronisation was related to the southward migration of Xiaoheyan culture (Jin 2004). However, climate change can be described as climate cycles or fluctuations, and such changes do not make archaeological culture react in the same way every time. What we really have to understand is the initial conditions and historical processes of cultural change. In addition, cultures respond in a variety of ways, including failure (population extinction and cultural disappearance), migration, changes in adaptation, and the development of cultural complexity itself—a corresponding increase in human ability to cope with environmental changes.

According to the map of environmental changes and cultural changes in Yan Mountain-Great Wall Transitional Zone (Fig. 7.4), climate is always fluctuating. Even in the HCO with stable climate, there were two periods of climate fluctuation (Shi et al. 1992), with unstable temperature and a greater variation range of precipitation (Zhang et al. 1992). From the perspective of the current climate of this region, successive drought years occur constantly lasting for 2–3 years even 5 years (Shi 1989). Such short-cycle climatic events are not easy to be reflected in the climatic record, but they are the climatic characteristics of ecotone. The precipitation fluctuation in the early and Middle Holocene was larger than that of the present, with an increase of 30%–40% in the wet period and a decrease of 35%–40% in the dry period (the current annual precipitation variability is 20%–50%) (Shi 1989), which must have a huge impact on prehistoric farming. In the Neolithic, the basic coping strategy was the interlocking of mixed livelihood modes of farming and hunter-gatherer, which fluctuated between farming and hunter-gatherer, and the proportion of farming and hunter-gatherer in livelihood increased alternately. At the same time, depending on the actual resource situation of each region, for example, the northern regions closer to grasslands have more hunting factors, the valleys with better water and heat conditions have more farming factors, and the areas closer to water use more water. In the upper culture of Xiajiadian, animal husbandry livelihood originated and the alternate mode changed to farming and animal husbandry. Later, nomadic economies developed to take full advantage of the steppe environment.

The key factor behind these changes is not climate fluctuations, but the ecological bottleneck of farming. As the population grew, the Neolithic people made use of the river valley, which was the best place for water and heat, and the main vegetation was forest. After the deforestation, crops were planted in this area. As the valley was depleted, there was a shortage of land for further farming. Compared with the agricultural-ecological model in North China (Fig. 7.5), which can gradually expand to plain areas after basin and foothill areas in North China are utilised. Liaohe Plain far from the North China Plain is sandy. During the Holocene, paleosoils only developed four times, representing semi-arid and sub-humid climates, and the vegetation was artemisia grassland or savanna (elm) artemisia grassland (Qiu 1992). There is no doubt that farming risks are huge in this region. On the contrary, grassland resources in this region are very rich and there is a wide space for exploitation. With the domestication of horses, cattle, and sheep, the emergence of animal husbandry, and nomadic economy, people can just exploit and utilise these resources. In short, these changes in cultural adaptation reflect both the adaptation of culture to environmental change and the growth of culture's own complexity.

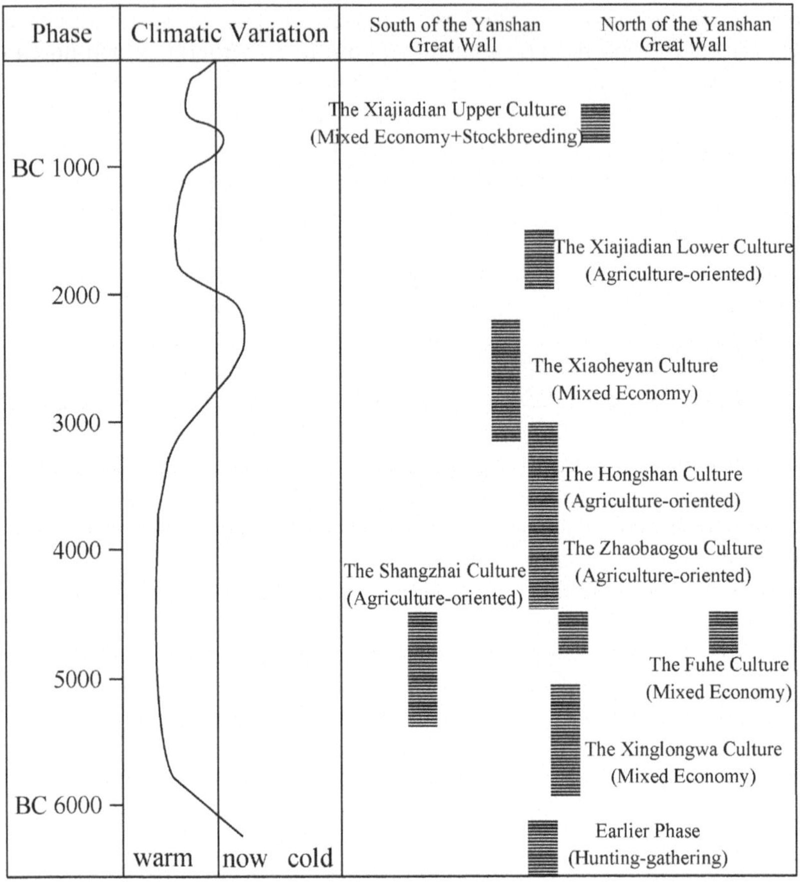

Fig. 7.4 Map of the relationship between climate change and livelihoods represented by archaeological cultures

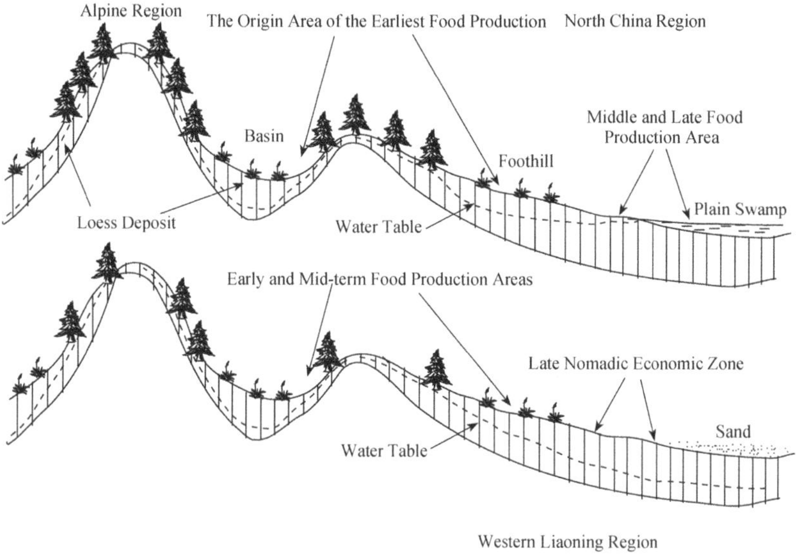

Fig. 7.5 Comparison of agricultural ecological models between North China and western Liaoning

7.6 Conclusion

Ecotone is rich in resources, but sensitive to environmental changes and unstable. Based on the cultural-ecological theory of hunter-gatherers, it can be predicted that hunter-gatherers in ecotone will cope with resource instability by increasing mobility, reserves, intergroup exchange, and expanding diets, and intensifying the use of certain resources with greater frequency and extent than in non-ecotone. Food producers in ecotones, on the other hand, will rely on a variety of livelihoods. At the same time, as the population grows, people seek to expand production; They also make greater use of social resources by developing specialised livelihoods, forming symbiotic relationships with stabilised agricultural groups, or by increasing profits through warfare looting to cope with the risk of resource instability. After sorting out and analysing the archaeological materials in Yan Mountain-Great Wall Transitional Zone, we can find that the most important adaptation strategy chosen by ancient people in the Late Paleolithic was to improve mobility. With the end of the LGM and the beginning of the Holocene, the economy of food production and a warmer and wetter environment created a new ecological space for ancient people, and Xinglongwa culture flourished. Throughout the Neolithic period, the main adaptation strategy in this region was to rely on a diverse mix of livelihoods, in particular retaining a large hunter-gatherer economy to take advantage of the rich diversity of resources in the ecotone and cope with its resource instability. In the upper culture stage of Xiajiadian, with the domestication of horses, the animal husbandry economy originated, and then it further became a nomadic economy. Meantime, dry farming systems in the north were established based on crop cultivation, especially *Glycine max* and livestock raising, and formed symbiotic relationships with nomadic groups. Undoubtedly, environmental changes in this region were constantly influencing people's choices of livelihood, but the real deciding factor in the end was the ecological bottleneck of the development of agriculture in this region, which encouraged people to choose to utilise the richer grassland resources and formed new forms of livelihood.

Notes

1. According to the editor Liu Mingguang's Atlas of China's Natural Geography (2nd edition, China Cartographic Publishing House, 1998, p. 13), the western Liaoning hills possess the broadest distribution zone of loess terraces and loess-covered slopes in China. The thickness of loess cover on these terraces and slopes is relatively smaller compared to Loess Plateaus and loess hills (liang and mao), and due to the vertical distribution characteristics of loess, it is difficult to retain water resources. Therefore, areas with thinner loess cover on loess terraces and slopes often have good vegetation such as forests, while regions with thick loess layers tend to have grassland vegetation. In terms of water resources and soil alone, the western Liaoning hills are undoubtedly suitable for agriculture, and the areas with better water conditions are preferred for early agricultural activities. However, agriculture also requires sufficient heat conditions.
2. (1). the class B remains of the BaiyinChanghan Phase II site; (2). the Zhaobaogou site; (3). the Xishuiquan site; (4). the Danangou site; (5). Beipiao Fengxia site; (6). the upper floor of Xiajiadian site

Referencess

Archaeological Research Team of the Institute of Archaeology, Chinese Academy of Social Sciences, Inner Mongolia, Excavation brief report of the Xinglongwa settlement in Aohanqi, inner mongolia, archaeology, 1997, 1.

Archaeological Research Team of the Institute of Archaeology, Chinese Academy of Social Sciences, and Aohanqi Museum, Investigation of the Xinglonggou neolithic site in Aohanqi, inner mongolia, archaeology, 2000, 9.

Archaeological Work Team of the Institute of Archaeology, Chinese Academy of Sciences, Inner Mongolia, Excavation report of Yaowang temple and Xiajiadian sites in Chifeng city, Acta Archaeologica Sinica, 1974, 1.

Archaeology Teaching and Research Office, Jilin University, The left home mountain neolithic site in Nongan County, Acta Archaeologica Sinica, 1989, 2.

Archaeology Work Team of the Institute of Archaeology, Chinese Academy of Sciences, Inner Mongolia, Excavation brief of the fuhegoumen site in balin left banner, inner mongolia, archaeology, 1964,1.

C. M. Aikens, and T. Higuchi, Prehistory of Japan. Academic Press, New York, 1982. The information on the pottery of the Far East for 10,000 years is based on lectures by Academician Derevyanko at Jilin University.

M. Basgall, Resource intensification among Hunter-Gatherers: acorn economies in prehistoric California. *Research in Economic Anthropology*, 9: 21–52, 1987.

Beijing Institute of Cultural Heritage, Zhenjiangying and Tazhao, Science Press, 1999.

L. R. Binford, In *Pursuit of the past: decoding the archaeological record*. University of California Press, Berkeley, 1983.

L. R. Binford, *Constructing frames of reference: an analytical method for archaeological theory building using ethnographic and environmental data sets*. University of California Press, Berkeley, 2001.

F. Bordes, and D. de Sonneville-DBordes, The significance of variability in paleolithic assemblages. *World Archaeology*, 2: 61–73, 1970.

L. R. Binford, In Pursuit of the past: decoding the archaeological record, University of California Press, Berkeley, 1983. translated by Chen Shengqian, Shanghai Sanlian bookstore Co.Ltd., 2009.

Chen Shengqian, A simulation study of Chinese hunter-gatherers. *Acta Anthropologica Sinica*, 25(01), 2006a.

Chen Shengqian, Adaptive changes of hunter-gatherers during the late upper paleolithic to early holocene transition in China. *Acta Anthropologica Sinica*, 25(03), 2006b.

Chen Shengqian, Adaptation and transition of hunter-gatherers in late pleistocene—Early holocene China. *Acta Anthropologica Sinica*, 25(03), 2006c.

Chen Shengqian, Adaptation, transition, and radiation of late pleistocene humans in Northern China. *Quaternary Sciences*, 26(04), 2006d.

Chen Shengqian, Adaptive changes of prehistoric hunter-gatherers during the pleistocene—Holocene transition in China. Ph. D. Dissertation. Southern Methodist University, Dallas, 2004.

Chen Shengqian, The origin of microblade technology: an ecological and theoretical perspective, archaeological research (issue 7). Science Press, 2008.

Chen Shengqian, The structure of archaeological reasoning, Archaeology, 2007, 10.

Chen Shengqian. The Cultural Perspective of Archaeology, Archaeology, 2009, 10.

R. G. Elston, and P. J. Brantingham, Microlithic technology in Northern Asia: a risk-minimizing strategy of the late paleolithic and early holocene. In R. G. Elston and S. L. Kuhn eds., *Thinking Small: Global Perspective on Microlithization*, pp. 103–116. Archaeological Papers of the American Anthropological Association Number 12, 2002.

Y. Fukuda, Cyto-genetical studies on the wild and cultivated Manchurian soybeans (Glycine L.). *Japanese Journal of Botany*, 6: 489–509, 1933.

Gezidong Excavation Team, Excavation report of the gezidong paleolithic site in liaoning, Journal of Vertebrate Paleontology and Paleoanthropology, 1975, 2.

A. Gilman, Explaining the upper paleolithic revolution. In M. Spriggs ed., *Marxist Perspective in Archaeology*, pp. 1–9. Cambridge University Press, Cambridge, 1984.

P. M. Grootes, et al., Comparison of oxygen isotope records from the GISP2 and GRIP Greenland ice cores. *Nature*, 366: 552–554, 1993.

Guo Dashun, Zhang Keju, Excavation brief of the Hongshan culture architectural complex at Dongshanzui, Kazuo County, Liaoning, Cultural Relics, 1984, 11.

Guo Dashun, Zhang Xingde, *Northeast Culture and the Yan Civilization*, Jiangsu Education Publishing House, 2005.

T. Hymowitz, Soybean. In J. Smartt and N. W. Simmonds eds., *Evolution of Crop Plants*, pp. 413–484. Longman Scientific and Technical, Harlow, 1995.

Heilongjiang Provincial Cultural Relics and Archaeology Work Team, The Xin Kai Liu site in Mishan County, Acta Archaeologica Sinica, 1979, 4.

Hebei Institute of Cultural Relics Research, et al.: Tingsijian microlithic site in Changli, Hebei province, Wenwu Chunqiu, 1992, supplement.

Hebei Provincial Institute of Cultural Relics Research, et al.: Excavation brief report of the Qianxi Xizhai Site in 1988, Wenwu Chunqiu, 1992, Supplement.

Hebei Provincial Institute of Cultural Relics Research, Excavation of the Jiangjialiang neolithic site in Yangyuan county, Hebei, Archaeology, 2001, 2.

I. Hodder, *Postprocessual Archaeology. Advances in Archaeological Method and Theory*, Vol. 8, pp. 1–26. Academic Press, New York, 1985.

E. S. Higgs, *Palaeoeconomy*. Cambridge University Press, Cambridge, 1975.

Inner Mongolia Autonomous Region Institute of Cultural Relics and Archaeology, Baiyinchanghan: excavation report on a neolithic site, Science Press, 2004.

Inner Mongolia Autonomous Region Institute of Cultural Relics and Archaeology, Excavation Brief Report of the Nantaizi site in Keshiketeng Banner, in Collected Works of Cultural Relics and Archaeology in Inner Mongolia, Encyclopedia of China Publishing House, 1994.

Inner Mongolia Work Team of the Institute of Archaeology, Chinese academy of social sciences, excavation of the Zhizhushan site in chifeng, Archaeologica Sinica, 1979, 2.

Inner Mongolia Work Team of the Institute of Archaeology, Chinese academy of sciences, excavation report of the Nanshangen site in Ningcheng, Archaeologica Sinica, 1975, 1.

Institute of Archaeology, Chinese Academy of Social Sciences, *Aohan zhaobaogou—neolithic settlement*, Encyclopedia of China Publishing House, 1997.

Institute of Vertebrate Paleontology and Paleoanthropology, Chinese academy of sciences, Hebei institute of cultural relics, Sifangdong—Hebei's first paleolithic cave site, Wenwu Chunqiu, 1992, Supplement.

Jia Lanpo et al.: Excavation report of the shiyu paleolithic site in Shanxi province, Archaeology, 1972, 1.

Jin Guiyun, Evolution and impact of the holocene climate environment in the Yanshan mountains and the Great Wall Zone, Acta Archaeologica Sinica, 2004,4

Kamura Hidenori, *Residential Patterns of the Neolithic Cultures in the Liao River Basin, in Archaeological Studies in Northeast Asia: sino-Japanese cooperative research report*, Cultural Relics Press, 1997.

F. B. King, R. W. Graham, Effects of ecological and paleoecological patterns on subsistence and paleoenvironmental reconstructions. *American Antiquity*, 46: 128–142, 1981.

J. E. Kutzbach, T. Webb., Conceptual basis for understanding late-quaternary climates. In H. E. Wright eds., *Global Climates since the Last Glacial Maximum*, pp. 5–11. University of Minnesota, Minneapolis, 1993.

Liu Jinxiang, Dong Xinglin, Preliminary study on prehistoric settlement patterns in the Yanshan North and South great wall zone, Cultural Relics, 1997, 8.

A. M. Lister, Tales from the DNA of domestic horse. *Science*, 292: 218–219, 2001.

Li Chaorong et al.: *Excavation Brief of the Paleolithic Site at Wangfujing Dongfang Square*, Beijing, Archaeology, 2000, 9.

Li Fan, Development history of cultivated plants in china, Science Press, 1984.

Liaoning Cultural Relics Training Class, Excavation brief of the Fengxia site in Beipiao County, Liaoning, Spring 1972, Archaeology, 1976, 3.

Liaoning Institute of Cultural Relics and Archaeology, Excavation brief of the Hongshan culture "Goddess Temple" and stone Piled graves in Niuheliang, Liaoning, Cultural Relics, 1986, 8.

Liaoning Provincial Institute of Cultural Relics and Archaeology, Trial Excavation Brief Report of the Cha Hai Neolithic Site in Fuxin, Liaohai Cultural Relics Journal, 1988,1.

Liaoning Provincial Institute of Cultural Relics and Archaeology, Three Excavations at the Cha Hai Site in Fuxin, Liaoning Province, 1987–1990, Cultural Relics, 1994, 11.

Liaoning Provincial Museum Archaeological Work Team, Brief Report on the Trial Excavation of Dajing Ancient Copper Mine in Linsi County, Liaoning, Collection of Cultural Relics, 1983, 7.

Liaoning Provincial Museum, et al.: Discovery of three primitive cultures in Xiaohelian, Aohan banner, Liaoning, Cultural Relics, 1977, 12.

Liaoning Provincial Institute of Cultural Relics and Archaeology, Chifeng Museum, Danan'gou—Post-Hongshan Culture Cemetery Excavation Report, Science Press, 1998.

Lin Yun, *A Reexamination of the Relationship between Shang Culture Bronze Artifacts and Bronze Artifacts in the Northern Region, in Su Bingqi, Collected Essays on Archaeology and Culturec(Volume 1)*, Cultural Relics Press, 1987.

Liu Guoxiang, *Discussion on the Economic Form of the Zhaobaogou Culture and Related Issues, in 21st Century Chinese Archaeology and World Archaeology*, Chinese Academy of Social Sciences Press, 2002.

Liu Jinxian, Dong Xinglin, A brief discussion on the agricultural economy of the Zhaobaogou culture, Archaeology, 1996, 2.

Lü Zun'e, Replication and use of fish darts in the Xiaogushan Xianrendong in Haicheng, Acta Archaeologica Sinica, 1995, 1.

Li Fushan, Study on the origin and evolution of soybeans, Soybean Science, 1994, 13, 1.

P. S. Martin, and D. W. Steadman, Prehistoric extinctions on islands and continents. In R. D. E. MacPhee ed., *Extinctions in Near Time: Causes, Contexts, and Consequences*, pp. 17–55. Kluwer Academic/Plenum, New York, 1999.

R. G. Matson, and G. Coupland, *The Prehistory of the Northwest Coast*. Academic Press, San Diego, 1995.

Man Zhimin, *Migration of the Agro-Pastoral Transition Zone and Subtropical Economic Crop Boundary, in Historical Climate Change in China*, Shandong Science and Technology Press, 1996.

Mo Duowen, Li Fei and Li Shuicheng, Holocene Environmental Evolution in the Huluhu River Basin, Gansu Province, and Its Impact on Human Activities, Acta Geographica Sinica, 1993, 51(01).

Ningxia Institute of Cultural Relics and Archaeology, *Shuidonggou—1980 Excavation Report*, Science Press, 2003.

Okamura Hidenori, *Residential Patterns of the Neolithic Cultures in the Liao River Basin, in Archaeological Research in Northeast Asia—Sino-Japanese Cooperative Research Report*, Cultural Relics Press, 1997.

Okamura Hidenori, Residential *Patterns of Neolithic Cultures in the Liaohe River Basin, in Northeast Asia Archaeological Research—Sino-Japanese Cooperative Research Report*," Cultural Relics Press, 1997.

Ogura Shizuo, Development of early mixed-farming culture in the circum-bohai sea region: centered on the observation of changes in animal assemblages, in archaeological research in northeast asia—sino-japanese cooperative research report, Cultural Relics Press, 1997.

Qiu Shanwen et al.: Ancient soils and holocene environmental changes in the Western Sand Land of Northeast China, Quaternary Sciences, 1992, 3.

Ren Shinan, The discovery and significance of the Xinglongwa culture: a comparison with archaeological cultures in north china during the same period, Archaeology, 1994, 8.

M. C. Steiner et al., Paleolithic population growth pulses evidenced by small animal exploitation. *Science*, 283: 190–194, 1999.

M. D. Sahlins, On the sociology of primitive exchange. In M. Banton ed., *The Relevance of Models for Social Anthropology*, pp.139–236. Association for Social Anthropology, Monograph 1, London, 1965.

J. D. Spector, What this awl means: toward a feminist archaeology. In J. M. Gero and M. W. Conkey eds., *Engendering Archaeology*, pp. 388–406. Basil Blackwell, Cambridge, 1991.

M. Q. Sutton, E. N. Anderson, *Introduction to Cultural Ecology*. Altamira, Walmut Creek, 2004.

G. Shelach, The earliest neolithic cultures of northeast China: recent discoveries and new perspectives on the beginning of agriculture. *Journal of World Prehistory*, 14: 363–413, 2000.

O. Soffer, Storage, sedentism and the Eurasian Paleolithic record. *Antiquity*, 63: 719–732, 1989.

Shenyang Municipal Cultural Relics Management Office, Shenyang imperial palace museum, report on the second excavation of the Xinle site in Shenyang, Acta Archaeologica Sinica, 1985, 2.

Shi Peijun, Precipitation variation and 'fluctuating agriculture and animal husbandry' in the agro-pastoral ecotone of Northern China, Journal of Arid Land Resources and Environment, 1989, 3.

Shi Yafeng et al.: *The Basic Characteristics of The Holocene Optimum And Environment In China*, China Ocean Press, 1992.

Shi Yafeng, Kong Zhaochen, Wang Sumin, et al.: *Basic Characteristics of the Holocene Megathermal Period and Environment in China*," China Ocean Press, 1992.

Suo Xiufen, Human-Environment relationships in the Holocene in the West Liao River Basin, *Research of China's Frontier Archaeology*, Volume 4, Science Press, 2005.

Su Bingqi, Ancient culture, ancient cities, ancient states in Western liaoning—discussing the key points or major topics of current fieldwork, Cultural Relics, 1986, 8.

Sun X. J., Chen Y. S., Palynological records of the last 11, 000 years in China. *Quaternary Science Reviews*, 10: 537–544, 1991.

Tangshan Cultural Relics Management Office, Discovery of paleolithic culture in tangshan region, Wenwu Chunqiu, 1993, 4.

Tian Guangjin, Shi Peijun, Preliminary study on environmental archaeology in the Great Wall Zone of Northern China, Inner Mongolia Cultural Relics and Archaeology, 1997, 2.

B. G. Trigger, *Beyond History: The methods of Prehistory.* Holt, Rinehart and Winston, New York, 1969.

K. C. Tylor, et al., The holocene-younger dryas transition recorded at summit, greenland. *Science*, 278: 825–827, 1997.

C. Vila et al., Widespread origins of domestic horse lineages. *Science*, 291: 474–477, 2001.

Wang Lixin, The cultural pattern and economic form of the western liaoning area during the Xia to warring states period: evolution, Acta Archaeologica Sinica, 2004, 3.

Wang Enlin, New Materials from the Tingsijian microlithic site in Changli, hebei province, Acta Anthropologica Sinica, 1997, 16(01).

Wei Lanying et al.: Preliminary study on climate and environmental changes during the last glacial period in the Beijing area, Quaternary Sciences, 1997, 2.

M. G. Winkler, and Wang P. K., The late-quaternary vegetation and climate of China. In H. E. Wright eds., *Global Climates since the Last Glacial Maximum*, pp. 221–261. University of Minnesota Press, Minneapolis, 1993.

Wang Hong, Paleoenvironment of the holocene shell midden and Oyster Reef in Bohai Bay, Quaternary Sciences, 1996, 2.

Wu En, Some thoughts on early nomadic cultures in the Eurasian Steppe, Archaeologica Sinica, 2022, 4.

Wu Haibin, Guo Zhengtang, Evolution and short-term drought events in the arid regions of Northern China since the last glacial maximum, Quaternary Sciences, 2000, 19(06).

Wu Wenxiang, Liu Dongsheng, The cooling event around 4000 years BP and the birth of Chinese civilization, Quaternary Sciences, 2000, 21, 5.

Xia Zhengkai, Deng Hui, Wu Honglin, Geomorphic background analysis of archaeological cultural evolution in the Xilamulun river basin, inner Mongolia, Acta Geographica Sinica, 2000, 55(03).

Xie Chuanli et al.: Paleogeographic outline of China during the last glacial maximum and its climatic implications, Quaternary Sciences, 1996, 1.

Xu Guangji, *Three Prehistoric Cultures in the Ulgimur River Basin, Collection of Inner Mongolia Cultural Relics and Archaeology*, Volume 1, Encyclopedia of China Publishing House, 1994.

Yan Wenming, The emergence and spread of agriculture in Northeast Asia, Agricultural Archaeology, 1993, 3.

Yan Wenming, The origin of agriculture and the origins of civilization, Science Press, 2000.

Yang Hu, Liu Guoxiang, New achievements in the excavation of the Xinglongwa settlement, China Cultural Relics News, Dec. 26, 1993.

Yuan Jing, *Research on Ancient Domestic Horses in China, in Research on Chinese Prehistoric Archaeology—Collection in Celebration of the 50th Anniversary and 80th Birthday of Mr. Shi Xingbang in Archaeology*, Sanqin Press, 2003.

Zhang Jingming, Microliths from the hongshan culture site at erdaoliang, Balin left banner, inner Mongolia cultural relics and archaeology, 1994, 1. Inner Mongolia autonomous region Institute of cultural relics and archaeology, excavation brief of the erdaoliang hongshan culture site in Balin left banner, Inner Mongolia Cultural Relics and Archaeology, 1994, 1.

Zhang Lansheng et al.: *Holocene Environmental Evolution and Future Hundred-Year Prediction in the Northern Farm-Pastoral Zone of China (Ordos Region), in Holocene Environmental Evolution and Prediction in the Northern Farm-Pastoral Zone of China*, Geology Press, Beijing, 1992.

Zhao Chaohong, Yu Jincheng, Wang Tao, Important discoveries at the early neolithic site in Donghulin, Beijing, China cultural relics news, May 9, 2003. Archaeology and Cultural Heritage College of Peking University, Archaeology Research Center of Peking University, Beijing Institute of Cultural Relics, Prehistoric Sites in Donghulin, Mentougou District, Beijing, Archaeology, 2006, 7.

Zhao Tuanjie, Gai Junyi, Progress in the study of the origin and evolution of cultivated soybeans, Scientia Agricultura Sinica, 2004, 37, 7.

Zhao Zhijun, Exploring new clues for the origin of dryland farming in Northern China, China Cultural Relics News, Nov. 12, 2004.

Zhou Tingru et al.: Climate changes in North China since the last glacial maximum, Journal of Beijing Normal University (Natural Science), 1982, 1.

Zhu Fenghan, Investigation of the Daqin Tala neolithic site in naiman banner, Jilin, Archaeology, 1979, 3.

Zhu Yanping, A preliminary analysis of small mountain-shaped vessels with bird and animal motifs, Archaeology, 1990, 4.

CHAPTER 8

The Initiator but not the Leader: South China

The so-called flexibility means adapting to changes.
—*The Book of Changes*

8.1 Introduction

In the existing studies on the origins of agriculture, it is rare to consider South China as an area in terms of cultural ecology with its own characteristics. More often, it is discussed together with the southwest or the middle and lower reaches of the Yangtze River, which brings obstacle to grasp the unique characteristics of cultural adaptation in South China. From the known archaeological materials, we know that rice agriculture in South China began no earlier than 3000 BC, which is three or four millennia later than the middle and lower reaches of the Yangtze River, but the archaeological materials also show that there were pottery vessels, perforated stone tools, grinding tools and other new features in this area as early as 15,000 years ago. This apparent anomaly is rarely noticed and explained. Why did this region not receive rice agriculture at the same time as the middle and lower reaches of the Yangtze River? The lag in agriculture is often attributed to the tropical climate, where food is available all year round, so farming is unnecessary. But this explanation is undoubtedly too simple. South China is not a single physical geographical unit. It includes Fujian, Guangdong, and Guangxi in the south of Nanling, as well as the southern part of Yunnan, the tropical region of

S. Chen, *Prehistoric Modernization*,
https://doi.org/10.1007/978-981-97-9423-2_8

China. This region includes at least inland mountainous areas, riverside lakes, and coastal areas. Different regions posses different types and richness of resources, and internal differences should be paid attention to. The study of South China also needs to be considered in conjunction with similar environments in Southeast Asia, which can be classified as the same cultural ecological region. We also hope to explore what constitutes cultural adaptation in south China, how it relates to the cultural ecology of the environment, and what kind of changes happened during the Paleolithic-to-Neolithic Transition. In my view, the key problem in South China was that it had been in the Neolithic transition for longer than North China and the middle and lower reaches of the Yangtze River, and that it continued to have a "low level of food production" while agricultural systems had been established in the North and lower reaches of the Yangtze River. This has been fairly well illustrated by the archaeological material found so far, and we need to explain why this is so.

8.2 Hot Topic: Mesolithic in South China

In the second section of Chapter 6, I propose that the concept of "Mesolithic" should not be used in the study of the origins of Chinese agriculture, because it has clear regional characteristics and cultural connotation. But I also put forward that "Mesolithic" is a concept that can be used for reference when studying the transition from the Paleolithic-to-Neolithic in South China and Northeast China, because there is a good comparability between the cultural characteristics of the two regions.[1] The specific meaning and theoretical background of "Mesolithic" have been mentioned above, and will not be repeated here. In the study of Chinese archaeology, South China, Northeast China (An 1956; An 1978) and Northwest China (Gai and Wang 1983) were the earliest regions to adopt the concept of "Mesolithic". However, in the core regions of the origin of Chinese agriculture, the concept is controversial and few people respond to it, which indicates that Chinese scholars are actually aware of the suitability of the concept of "Mesolithic". T. Douglas Price and Chen Chun who from Fudan University have concluded the development process of the concept of "Mesolithic", but Childe does not agree that it constitutes the transition stage from the Paleolithic to the Neolithic. Binford took notice of the small, delicate assemblages of Mesolithic stone tools in Europe and the waterfront environment in which these tools existed, suggesting that they might be related to the use of aquatic resources. His view was further deepened in his later studies (Binford 1968), but without fundamental changes. Similarly, Kozlowski argues that the term "Mesolithic" actually refers to the mode of economic development, rather than to the chronological phase, that is, the hunter-gatherer economy of lowland forests in Europe at the beginning of the Holocene. Marek Zvelebil defines "Mesolithic" as a complex foraging adaptation, noting that "Mesolithic" hunter-gatherer adaptation is different from general hunter-gatherer adaptation.

Among domestic scholars, Wenzhong Pei first classified Inner Mongolia's Jalainur, Shaanxi's "Shayuan Culture" and several discoveries in Guangxi as "Mesolithic", and later denied it due to the reasons of strata; He noted the particularity of Mesolithic sites in Europe and believed that China's Mesolithic culture was not developed. On the other hand, the reason for his rejection of the previous view also shows that he cannot accept the characteristics of the "Mesolithic", which was characterised by the coexistence of microlith culture and pottery, and the coexistence of hammering stone with pottery and grinding stone. What he called "Mesolithic" was more of an era than a special cultural form. Later in 1956, Zhimin An also noted that some sites in Northeast China and Guangxi had characteristics of the Mesolithic. He defined the characteristics of the Mesolithic as that human beings still lived an economic life of gathering, fishing and hunting, and agriculture and animal husbandry did not appear. Stone tools were mainly made by hammering, and typical microliths made by indirect percussion were particularly prevalent, with only a few ground stone tools. His definition is consistent with that of Binford and Kozlowski, emphasising that "Mesolithic" is a special form of adaptation. Later scholars have also noted the dual characteristics of the concept of "Mesolithic": on one hand, as a cultural stage, it was between the Paleolithic hunter-gatherer and Neolithic agricultural production; On the other hand, it has the unique cultural characteristics of Western Europe. Therefore, some scholars have proposed that sites within the last 10,000 years without agriculture can be classified as the Mesolithic (C. Chun 1995, 2000), or that according to the standards of Western Europe, only through systematic field work to find similar findings, can China be sure of the "Mesolithic"—if the latter characteristics is adopted (C. Xingcan 1990).

In fact, what needs to be clarified here is whether the specific meaning of "Mesolithic" as a means of cultural adaptation is found only in

Western Europe? How large is its range of adaptation? In this regard Binford, Kozlowski, and Zwelebier give complementary explanations. Mesolithic culture corresponds to the hunter-gatherer adaptation of aquatic resources. Due to the unique advantages of aquatic resources, this adaptation could support a relatively mobile and even settled hunter-gatherer existence (see Chapter 2), with the emergence of pottery, polished stone, and other features that are usually Neolithic, but they lack the components of agricultural production. As a result of settlement, it formed a population much larger than the average hunter-gatherer group-size settlement, with a higher level of social complexity and even an already stratified social structure. Examples of this adaptation include Western Europe (under the influence of the North Atlantic Drift), Japan (under the influence of Kuroshio), Chile (under the influence of the Peru Current), and the northwest coast of North America, where Marine resources are abundant enough to support settled life. Although Japan is defined by the Jomon Age, it adapted in the same way as the Mesolithic in Western Europe. Therefore, the key to the question of "Mesolithic existence or not" lies in the utilisation of aquatic resources, and whether its benefits are greater than those of agriculture. Of course, the concept is nevertheless of epoch-making significance and it is a stage in the development sequence of a regional culture; whether it was a transition from the Paleolithic to the Neolithic is another matter.

Looking back at the transition from Paleolithic-to-Neolithic in southern China, its characteristics are distinct from those of North China and the middle and lower reaches of the Yangtze River, where agriculture really began much later, but some of the distinctive features of the Neolithic period appear as early, or even earlier. Later development diverged, however, with grain agriculture and animal domestication in North China and the middle and lower reaches of the Yangtze River, and an agricultural system was established during the Middle Neolithic. South China has not developed in this way, but rather until the Late Neolithic, grain agriculture was introduced from the middle and lower reaches of the Yangtze River. From the forking point of cultural development to the real beginning of agriculture, this period lasted five or six thousand years. Such a cultural stage has no comparison in North China and the middle and lower reaches of the Yangtze River. On the contrary, it is very similar to the Mesolithic in Western Europe and the Jomon Age in Japan. It is appropriate to adopt the concept of "Mesolithic" when the concept of "Neolithic" does not apply and has not formed its own concept as

Japanese archaeology has done. As related studies have a long history, the concept of "Mesolithic in South China" can be established. It is roughly equivalent to the transitional stage of the Paleolithic in north China and the middle and lower reaches of the Yangtze River as well as the early and middle Neolithic. Understanding this particularity is very important for us to carry out further research.

8.3 Archaeological Records and Cultural Features of the Mesolithic in South China

Among the Paleolithic archaeological materials in the South China region, the discovery of the end of Pleistocene to the Early Holocene is quite rare, except for the relatively abundant discovery in the early Bose Basin (Table 8.1). The distribution of such materials is striking. On one hand, it seems that human activities in this area are more concentrated and frequent at this stage; on the other hand, its new cultural features stand out, allowing researchers to pay attention. Therefore, in a sense, the South China Mesolithic can be established precisely because of the rich archaeological materials and their characteristics. The main characteristics of prehistoric archaeological materials make the Mesolithic in South China a suitable problem for exploration, which is a favourable condition for research. The disadvantage is the stratigraphy and chronology of the archaeological material. Most of the sites at this early stage are cave sites. These caves were often used for refuge and storage, especially for fertiliser deposit, nitrate production, and stone quarrying, leading to serious damage to cultural heritage. At the same time, the excavation area of these caves is often very small, and part of the work is done early, and the reliability of stratigraphic control is not enough. The 14C dating data of archaeological materials is more controversial (Peking University Department of History, et al. 1982; An 1989; Yuan 1993), mainly due to the older dating of snail shells and calcium board, as well as the "carbonaceous residue" in clay, which makes some materials very old, such as the dating of pottery pieces in Dalongtan Liyuzui, and also raises doubts about materials from other sites. By using a variety of materials, especially charcoal and animal bones, errors can still be ruled out.

In order to understand the development process from the Paleolithic to the Neolithic in South China, we start our discussion with the Late Paleolithic material. The excavated material from this region is fragmentary and scarce, and the only site, Bailiandong in Liuzhou, remains

Table 8.1 Major sites of the pre and post Mesolithic periods in South China

Periods	*Representative sites*
Early Upper Paleolithic in South China (ca.35,000—20000BP)	Baojiyan in Guilin (Wang 1982), Dayan Phase I (Fu et al. 2001; CA, CASS 2010), Liuzhou Bailiandong Phase I and II (Liuzhou Bailiandong Science Museum 2009), Luoshayan in Fengka (Zhang et al. 1994),the fifth cultural layer in Miaoyan (Chen 1999; Zhang 1999), Tiandong Dingmodong (Zeng 1982), the lower cultural layer of Lianhuashan in Zhangzhou (You et al. 1991; Cai 1996)etc
Late Paleolithic or early Mesolithic in South China(ca.20000–11000BP)	Dushi Zai (Qiu 1982), Niulandong Phase I and II (Jin et al. 1998; Yingde City Museum et al. 1999), Dayan Phase II and III, Miaoyan II, III and IV cultural layers, Liyuzui I and II (Liuzhou City Museum et al. 1983; Fu 2004), Yingde Diaozhuyan, Zhuwayan, Xianfoyan, Huangyanmen (Guangdong Provincial Museum 1961b), Huangyandong (Song 1983), Luojiyan (Song 1981), and the upper cultural layer of Lianhuashan

Periods	*Representative sites*
Mesolithic or late Mesolithic in South China (ca.11000–6000BP)	Dayan Phase IV and V, Zengpiyan Phase I to V (CA,CASS 2003), the upper layer of the Liyuzui, Yapu Mountain in Dongxing, Malanju Mountain, and Beijiao Mountain Shell midden (Guangdong Provincial Museum 1961a), Shell middens in Nanning Yong River Coast (Cultural Relics Management Committee of Guangdong Province 1961), Niulandong III, Chenqiao Village in Chaoan, Shiwei Mountain, and Haijiao Mountain,[2] Dingsishan (Guangxi Working Team of the Institute of Archaeology, Chinese Academy of Social Sciences, Cultural Relics Work Team of Guangxi Zhuang Autonomous Region 1997; Guangxi Working Team of the Institute of Archaeology, Chinese Academy of Social Sciences et al. 1998), and Baozitou (Guangxi Working Team of the Institute of Archaeology, Chinese Academy of Social Sciences, et al. 2003), etc
Neolithic in South China (after 6000BP)	Snail Mountain IV, Xiantouling (Shenzhen Museum, Department of Anthropology, Sun Yat-sen University 1990), Keqiutou (Fujian Provincial Museum 1991), Tanshishan (Cultural Relics Management Committee of Fujian Province, Xiamen University 1976; 1983; Fujian Museum 2004), Guye Shell midden (Cui 2007), etc

with clear stratigraphy and chronology and rich archaeological. Only 12 stone artefacts have been excavated from the heavily damaged Baojiyan site in Guilin. 24 stone artefacts unearthed from the second layer of Luoshayan in Fengkai sandstone dating back 22,000 years, which are difficult to accurately grasp the characteristics of stone assemblage. There were considerable number of flint flake tools in the stone assemblages of Bailiandong I and II in Liuzhou. Among the 200 stone artefacts found in the first stage, flint material accounted for 94%, in phase II, out of the 102 stone artefacts, the proportion made of flint was 54.6%, which was obviously different from the gravel stone that was popular for a long time in the Paleolithic in southern China. It is consistent with the trend of miniaturisation and flaking of the stone industry in the Late Paleolithic (Wang 1997). Another site belonging to this era is the Dingmodong in Tiandong County. According to the excavators, the age has a large span from the end of Middle Pleistocene to the late Pleistocene. The cultural layer is the subsurface (from top to bottom), which is brownish yellow loamy sand, most of which is not cemented. Beneath it is a greyish yellow silty clay layer, probably equivalent to the widespread "yellow deposit" in South China. There are only 10 stone artefacts with flint material. The age of the cultural layer probably belongs to the latest stage of the Paleolithic, the same period as that of Dalongtan and Dushizai. Three human teeth were found with dental characteristics similar to caries, which may be related to the frequent consumption of starchy root tubers or sugar-rich wild sugarcane (Zeng 1982), reflecting the relatively simple food characteristics of this period. In general, hunter-gatherers who ate a varied diet had fewer cavities.

The earliest remains belonging to the so-called "Mesolithic" found in South China can be traced back to Pei Wenzhong's collection of chipped gravel stone tools, perforated stone tools, and grinding stones from grey deposits containing snial shells in Baqiao (Cave A), Baxun (cave B), Tengxiang (Cave C) and Guilin (Cave D) in Wuming in 1935, but no pottery or polished stone tools were found. This deposit is followed by an earlier yellow deposit containing stone artefacts, stegodon fauna fossils, and gigantopithecus and anthropoid fossils (Pei 1935). Deposits of two different colours and inclusions are commonly found in the Guangxi and Guangdong sites. In 1985, He Naihan and Qin Shengmin concluded the cultural characteristics of the "Mesolithic in Lingnan" according to the materials of 12 sites in the transition period between Paleolithic and Neolithic times in Lingnan region: ① In strata, the "cultural deposit with shell" belonged to the greyish yellow or greyish brown, which was different from the greyish yellow deposit without shell in the Late Paleolithic; ② In the nature of the fauna, they were all present species, but the possibility of coexistence of one or two extinct species cannot be ruled out. ③ In the nature of cultural relics, there were mainly chipped stone tools, there are a small number of partly polished stone tools or perforated gravel, but there was no pottery and the whole polished stone; ④In the 14C era, it was from about 10,000 years ago; ⑤ In terms of geological time, it belonged to the Early Holocene; ⑥ In economic life, people led a life of fishing, hunting and gathering (He 1985).

He and Qin did not suggest the time of the beginning and termination of this phase. With the discovery of more archaeological and dating materials, the pottery, partially polished and perforated stone tools have been further dated to 15,000 years ago (some dating data are earlier). The so-called "cultural deposit with shell" is also dated much earlier, with the second phase of the Bailiandong site containing a small number of shells dated to 25,000–19,000 years ago. The third phase also contains snail shells and is dated to 20,000–12,000 years ago, and perforated stone tools and grinders appear in this phase. Jiang Yuanjin set the Mesolithic in South China at 20,000–12,000 years ago, and summarised several characteristics, such as "cultural deposit with shell", increased flint small stone components, partially grinding and drilling techniques of stone tools, and pottery (Liuzhou Bailiandong Science Museum 2009).

The intensive appearance of snail shells at the Bailiandong site began in the fourth period, and a ground stone axe was unearthed. The date of this layer is 10,840 ± 580 years ago after correction. Similar discoveries have been made at sites such as Dayan, Miaoyan, and Dushi Zai. According to the current archaeological materials, there are a large number of shells in the site deposit around 12,000 years ago, such as Zengpiyan Phase I, Dingsishan Phase I, Miaoyan Phase III, Dayan Phase II and III, the upper layer of Dushizai, Luojiyan, and nearby Yuchanyan and Sanjiaoyan sites. Therefore, it can be said that the deposit of shells first appeared around 20,000 years ago, but the mass appeared around 12,000 years ago. From this time period until the Late Neolithic, the deposit of shells was common and developed into archaeological remains such as shell middens. In the Middle Neolithic, a large number of shell deposits were found in the sites in South China. It was not until the Late Neolithic, such as the fourth stage of Dingsishan (Work Team of Guangxi Institute of Archaeology, et al. 1998), that the cultural deposit did not contain snail shells. However, this was not a common phenomenon, with obvious regional differences. Coastal areas, such as the Keqiutou cultural site in Fujian (Fujian Provincial Museum 1991) and Chongtang site in Chongzuo, Guangxi (He and Chen 2008) still contain a large amount of shellfish deposit, and there is no sign of agriculture.

Thus, we can divide the Upper Paleolithic to the beginning of the Neolithic into four stages (Table 8.2).

The first stage was the Late Paleolithic, which probably started around 35,000 years ago in South China (Wang 2005). In this stage, the characteristics of stone tools showed a trend of miniaturisation and flaking, and flint was added to stone materials, although gravel still occupied a dominant position. The following animal fossils were all members of the Giant panda—stegodon fauna. Cultural deposit did not contain shells.

The second stage is the Late Upper Paleolithic, or the Early Mesolithic, which is equivalent to the transition period between Palæolithic and Neolithic times in North China and the middle and lower reaches of the Yangtze River, lasting from 20,000 to 12,000 years ago. This was

Table 8.2 Frame and characteristics of prehistoric culture in South China

Periods	*New cultural characteristics*
Early Upper Paleolithic in South China (ca.35,000—20000BP)	①The cultural deposit was yellow without shell, accompanied by fossils of extinct giant panda and members of the stegodon fauna;② Mainly gravel tools; ③ There was a trend of miniaturisation and flaking of stone assemblage. ④ Flint material appeared
Late Paleolithic or Early Mesolithic in South China(ca.20000–11000BP)	①There are small amount of shell-bearing deposits, with limited intensity of human use of aquatic resources; ② pottery; partially polished stone tools; ③perforated ④ gravel; ⑤ Small flint chipping tools
Mesolithic or Late Mesolithic in South China (ca.11000–6000BP)	①Extensive Shell deposit with high intensity of utilisation;②pottery; ③partially to fully polished stone tools; settled site structures;⑤ root tuber remains
Neolithic in South China (after 6000BP)	①The decrease or disappearance of shell-bearing deposits in inland areas; ②the discovery of cereal remains; ③refinement of artefacts;④the predominance of ground stone tools

a period of cultural transition, when Neolithic features such as pottery, partially polished and perforated stone tools began to appear. Cultural deposit began to contain shells.

The third phase is the Mesolithic, or the Late Mesolithic in south China. The period spans from 12,000 to 6000 years ago.[3] Shell deposits occur in large numbers, forming shell-dominated cultural deposits such as shell middens. Archaeological remains such as fully polished stone tools, hard-fired pottery, settled settlements, and burials appeared. It was only this stage that was truly comparable to the Mesolithic in Western Europe, rather than the second stage, which needs to be specifically pointed out and emphasised. Japan's Jomon Period also began at the same time, with a similar assemblage of archaeological remains.

The fourth stage is the arrival of the Neolithic. The earliest the Neolithic can be dated back to 6500 years ago (Jiao 2006) which is the Late Neolithic in this area. In this stage, except for the sites in coastal areas, the shells disappeared or were greatly reduced in the cultural deposit, and there began to be relatively rich rice farming remains, more stone tools were ground, and the degree of refinement of artefacts was improved.

These four periods are marked by three changes, among which the characteristics of the Late Paleolithic period to the Early Mesolithic period had obvious changes, and there were obvious differences in the colour, contents and features of cultural deposit. From the Mesolithic to the Neolithic, the marks were also obvious. Shell deposit disappeared, artefacts were refined, ground stone tools dominated, and pottery types were abundant. From the beginning to the end of Middle Stone Age, the main symbol was the increase of shells in the cultural deposit, and the cultural remains were more rich and complex, but with coherence and strong transitional colour.

The cultural characteristics of the Mesolithic in South China are manifested in the distribution of artefacts and sites. The characteristics of the Early Mesolithic period are relatively simple, and there are more generalisations (He and Qin 1985; Guangxi Liuzhou Bailiandong Science Museum 2009), which will not be repeated here.[4] The main problem is the cultural characteristics of the Late Mesolithic period, which includes the early and middle Neolithic period in this area (An 1978). The typical sites are Zengpiyan and Dingsishan. From the features of the artefacts, the pottery was low-fired and simple, and most of them are round bottom jars. There were few types of ground stone tools, only axes, and adzes, from sharpening the edges to rough grinding throughout. Bone grinding was more exquisite; burials began to be found, with obvious burial styles and burial objects, reflecting the unique customs of the region. Compared with the previous section, the latter part of the Middle Stone Age in South China is still primitive but has been greatly improved.

The most obvious improvement may be in site distribution and structure. The site distribution in the Late Mesolithic period expanded, and the number of shell middens along rivers and coasts began to increase (Guangxi Autonomous Region Cultural Relics Archaeology Training Class, Guangxi Autonomous Region Cultural Relics Work Team 1975), which was a typical feature different from that in the earlier period. Take the sites of Dingsishan culture as an example. Most of them were middles, located in first-level terraces along the banks of Zuo River, You River and Yong River and its tributaries, generally with the back of the mountains and facing the water, choosing the bend of the river or the triangular mouth where the two rivers meet, where fishing and hunting resources are more abundant, reflecting that the choice of site is more restricted by the distribution of fishing and hunting resources. The site piles were dominated by the remains of aquatic animals such as snail and mussel shells and fish bones, and more terrestrial animal remains. The most well-preserved Dingsishan site had an area of 5000 square metres, and its deposit was up to 3 metres thick, showing the characteristics of a settled society with a population larger than that of a general hunter-gatherer society (Chinese Academy of Social Sciences Institute of Archaeology Guangxi Work Team, et al. 1998). This was a leap from the Early Mesolithic cave-dominated small settlements.

The Mesolithic in south China varies from one region to another. The current material shows that Guangxi has the richest finds, especially in the pre-Mesolithic period. Entering the Late Mesolithic, there are significant differences within Guangxi, and there are also two areas (more likely types) in northern Guilin, the southern area is mostly cave sites with more chipped stone tools; the northern area is the upper reaches of the Zi River and Xiang River, and was dominated by hillside sites (Wei 2009). A large amount of charred rice was found in the second and third stages of Xiaojin site in Zi Yuan (Cultural Relics Work Team of Guangxi Zhuang Autonomous Region, Cultural Relics Management Office of Ziyuan County 2004), more rice phytolite was found in the fourth stage of the Zengpiyan site, and there was also some findings in shell midden in Guye, Gao Ming, marking the arrival of agriculture. Bailiandong and Liyuzui sites were developed in central Guangxi. The south of Guangxi area is represented by the Dingsishan, which is mostly distributed on the hillside and terrace, belonging to the shell midden site. The coastal areas of south Guangxi, mainly utilise marine shellfish. There are terraces and a few cave sites in east Guangxi, but they do not contain snail shells, which reflects the particularity of the region. The representatives of the Early Mesolithic in Guangdong are mainly Dushi Zai, Huangyandong, and Niulandong. The sites in the Late Mesolithic are mostly dunes, shell middens and river terraces, and shellfish remains are very common. Taking the shell sites in Zhaoqing as an example, shellfish are mainly in fresh water environment, 20% of which are between salt and fresh water (Guangdong Provincial Museum 1990; C.Yiqin 2009). The findings in Fujian are different. The eastern and southern coastal areas are also dominated by shell middens, such as Keqiutou and Tanshishan, in which semi-crypt houses and public graveyards were found in Tanshishan, and moat was found on the edge of residential areas. The sites of inland mountainous areas in northwest Fujian are mostly mountain sites in intermountain valleys, and there are fewer studies of their livelihood patterns.

The materials of Naminan Cave found in southern Yunnan Jinghong are different. The site was excavated in 1997 and 1998, with a total area of 104 square metres, and two cultural layers were found. The upper cultural layer includes layers 1–4, with layer 4 dating to 13,650 ± 180 years ago (corrected BP); the lower cultural layer includes layers 5 and 6, dating to after 20,000 years ago. Several pieces of coarse sand-tempered shred were unearthed from the upper cultural layer. The stone artefacts combination includes grinding slabs, handstones, large perforated stone tools, and edge-ground stone axes. The ruins have hearth and clam shell pit. Deer faeces and seeds of angiosperms were also found in the site (Yang 2009). Unfortunately, the exact stratification of these finds is not known. These findings are basically the same as those in Guangxi, but grinding slab and handstones are rare in the Mesolithic in Guangxi.

After analysing the concept of the Middle Stone Age and reviewing the archaeological materials from the southern region of China, we have determined the time range and cultural characteristics of the Mesolithic in southern China. It can include two stages, the latter of which is represented by the typical "Mesolithic" in South China, not as a transitional stage of the Paleolithic and Neolithic transition, but the development of cultural branches after this stage. The so-called early Mesolithic in South China is a development stage similar to the transition period between Paleolithic and Neolithic times in North China and the middle and lower reaches of the Yangtze River, both of which contain some features of the Neolithic, such as pottery and edge-ground tools. But the difference is that the cultural relics in South China contain aquatic animal shells, lack of grain agricultural relics, and grinding technology is not common, etc., which represents a different component of resource utilisation.

8.4 Mesolithic Cultural Adaptations in South China

8.4.1 Climate and Environmental Change

Now that we have a general idea of the time–space framework of ancient cultures, we need to know more about how people lived and how their acculturation differed from that of other regions. Why hasn't it entered the agricultural stage like the middle and lower reaches of the Yangtze River? Before discussing cultural adaptation, it is necessary to take a look at the climate change in South China since the LGM. According to the research of Huang Zhenguo and Zhang Weiqiang, during the LGM, there was a cold temperate zone in the north of the Qinling Mountains and Huai River, and the average annual temperature decreased by 7–10 °C, while the middle and lower reaches of the Yangtze River were temperate zone, and decreased by 5–7 °C. However, the variation range of the tropical zone was relatively small, and the response of the tropical region was different. The northern boundary of the central tropical region has shifted five latitudes to the south, while the northern boundary of the northern tropical region has shifted only two latitudes to the south, and the northern boundary of the southern subtropical region has changed the most, shifting six latitudes to the south, that is, the current climate of Guangzhou is similar to that of Changsha. In general, the response of tropical regions to the LGM is much smaller than that of temperate regions. In the direction of longitude, Taiwan Island has the biggest change, with the average annual temperature decreasing by 5–9 °C. Guangxi, Guangdong, and Hainan Island followed with an average annual temperature drop of 3–4 °C. The decrease in southern Fujian and eastern Guangdong is 2–3 °C, while the decrease in southern Yunnan is only 1–2 °C. From east to west, the influence of the Last Glacial Maximum gradually weakened (Huang and Zhang 2000). Studies on multiple environmental indicators in the Pearl River Delta region show that the climate change in the Last Glacial Maximum was not very obvious, with a cooling range of less than 3°C, which was reflected by the Younger Dryas Event, and continued fluctuating warming until 2500 years ago. The Holocene major warm period is difficult to identify, and the warmest period is not in the Middle Holocene, but in the Late Holocene (Zhang and Huang 2005). After the LGM (22,000–15,000 years ago), subtropical evergreen broad-leaved forests dominated by Castanopsis Kawakami grow in South China (Liu and Wang 2004; Liu 2007). As the tropics shifted southward, the temperate zones expanded southward and became more seasonal. Moreover, the sea level decreased during the LGM period, and the distance from the coast of South China expanded, and part of the continental shelf became land. Mangroves were distributed in shallow waters on the front of the continental shelf, and grassland, shrub, and evergreen broad-leaved forest grew in turn on the bare continental shelf from front to back (Liu 2007).

The influence of the warming process after the LGM mainly includes three aspects. First, the tropical range gradually recovers, the tropical forests regain the dominant position, the herbivorous fauna shrinks, hunting resources decrease, and the need to develop new sources of meat. The second is the rise in sea level, which has a direct impact on the location of coastlines, namely the extension of the continental shelf, increasing the opportunity to use marine resources. The third is the influence of monsoon (refer to Chapter 5). The influence of Indian monsoon is manifested in the maximum effective precipitation, which peaked 12,000 years ago. Interestingly, this time coincides with the beginning of the Mesolithic (or Late Mesolithic) in southern China. The Mesolithic in South China was marked by the dependence of aquatic resources (including the appearance of cultural deposit with a large number of shells). The maximum effective precipitation meant abundant river water and increased the possibility of utilising aquatic resources in rivers, which was the expansion of ecological space. In the Early Mesolithic period, the utilisation of aquatic resources had already begun, but it was limited. It wasn't until 12,000 years later that it became an important part of our livelihood. The environmental changes after the LGM created the basic initial conditions for the beginning of the Mesolithic.

8.4.2 *Optimal Foraging Model in South China*

Why does South China turn to aquatic resources? With reference to Fig. 8.1, we categorise the resources available to hunter-gatherers in this area, roughly including big game hunting, small game hunting, aquatic resource exploitation, plant gathering, and root and tuber cultivation (so-called horticulture). Measuring the income scale of these resources, it is no surprise that hunting large animals is the most desired by hunter-gatherers, followed by small animals, then aquatic resources (which provide abundant protein), plant gathering (a more diverse food variety), and finally root tuber cultivation (a relatively simple food supply). Benefits are often obtained in direct proportion to costs, but not all costs are the same. If hunter-gatherers hunt big animals close to their camps, the cost is low; if they are hunting large animals from farther away, it will be costly to have the whole group enjoy the prey, either by bringing the animals back or having someone fetch the food. Here I distinguish two kinds of action costs: action costs between resource patches and action costs within resource patches. From big game hunting to root tuber cultivation, the cost of moving between resource patches is decreasing. The reason is simple. The chance of hunting a big game near a residence is rare, and root tuber cultivation is easy to cultivate near a place of residence. In terms of the cost of action within the resource patch, the trend is the opposite, that is, if they hunt a big animal close to where they live, they will get a lot of meat right away, and the cost is very low. Cultivation of roots and tubers requires labour to open up the land, cultivate, take care of, and harvest. The cycle takes months, and the cost is very high.

What's interesting is where these two trends meet–aquatic resources. When using aquatic resources, the action cost between resource patches and within resource patches is at a medium level. Choosing aquatic resources to use seems to be the best choice to balance the two action costs. In the tropics, species are extremely abundant, but they are more evenly distributed, that is, they don't vary much from patch to patch, and they don't vary much from place to place in the rainforest, unlike in the temperate zone, where the distribution of resources is very zonal. When resource patches are more uniform, the factors that need to be considered in the movement of hunter-gatherer groups are resource levels and operational costs. In heavily forested areas of the tropics, the best way to reduce operational costs is to use rivers and ship back and forth between resource patches, where operational costs are lowest. The proximity to the river and

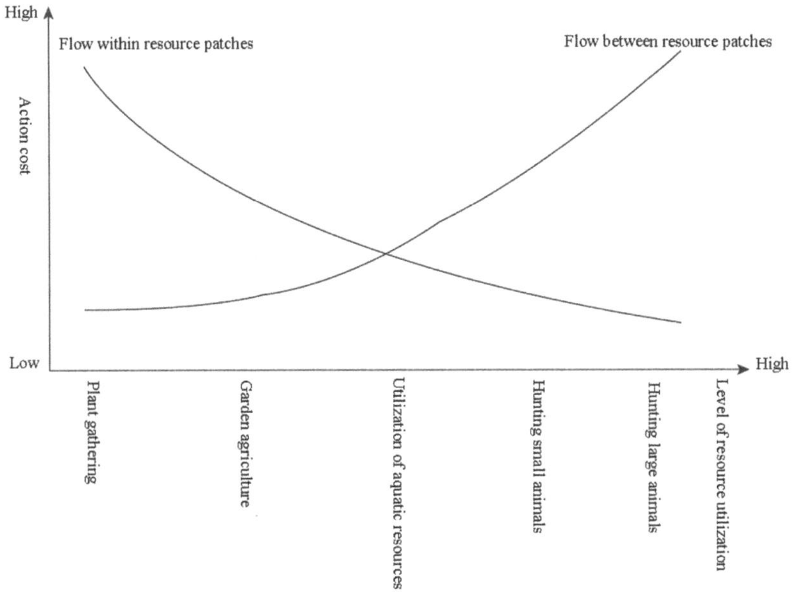

Fig. 8.1 Optimal foraging model in South China

the convenience of using the aquatic resources means that the operation cost within the resource patch is minimised. Once the rowing technology is available and aquatic resources are abundant, aquatic resources will undoubtedly be a very advantageous adaptation strategy in tropical jungle areas. Especially when large animals are scarce and small animals are nimble and expensive to hunt, aquatic resources are the quality resources.

For another thing, although South China belongs to the tropical region, it lies on the edge of it. Different from the tropical rain forest near the equator, it belongs to the tropical monsoon rain forest area with obvious seasonality. In addition, since the LGM, the temperature has decreased, and some areas have been close to subtropical areas. Cultivation of roots and tubers is a good strategy for coping with seasonal poverty. Because they are clonal, cultivation is relatively simple and domestication can occur during unconscious use.

8.4.3 Prehistoric Cultural Ecosystem in South China

According to the natural geographical characteristics of South China, we can distinguish the cultural ecological utilisation zone (Fig. 8.2). The first is the tropical forest zone, which is lush with plants, rich in species, and very high in primary productivity. But one of the fundamental features of a tropical jungle is that the primary productivity is mainly in the canopy, so there are plenty of birds, and there are also good climbers. Under the canopy, there is less sunlight, less productivity, and fewer species of herbivorous or omnivorous animals. The ability of hunter-gatherers to exploit the resources of this region requires access to the canopy. The invention of the bow and arrow was an advantage, whether it was a single bow or a blown arrow, which allowed hunters to hunt canopy birds or climbing animals without having to climb. As for animals under the canopy, it is relatively easy for hunters to capture them through ambushes, though in limited numbers. The jungle is densely covered and difficult to navigate, so the easiest way to travel is by river, which not only saves time to travel through the jungle, but also helps to transport game. Of course, the precondition is that the river is suitable for traffic. The strong current upstream of the river is not conducive to sailing. The Pearl River system in South China, especially in Guangxi and Western Guangdong, has steady and gentle rivers flowing among cascades. Another requirement is to be able to make water transport tools (such as canoes, rafts or bamboo rafts). Bows, arrows, peddles and boats were already available in the last stage of the Paleolithic. The jungle area also contained plant fruits that could be collected and other plant resources (such as arrow wood for making poison). The use of tropical jungle was an indispensable part of the cultural ecosystem of the Mesolithic in South China.

The second zone is the woodland glades, which for some reason (fire, human logging) have fewer trees, more sunlight, and higher surface productivity. This area is the most likely place for agricultural production, but the problem is that removing vegetation from the surface, levelling the land, clearing trees, undergrowth, and removing roots is particularly difficult and requires a lot of labour. Ethnographic materials show that groups using slash-and-burn farming rarely make any effort to remove roots, but leave them in place until they rot (Yin 2008). Furthermore, tropical soils are poor and require more organics, as well as sorbents (e.g. charcoal, animal bones, etc.) to sequester as much organic matter as possible to increase the fertility of the soil, which also requires great

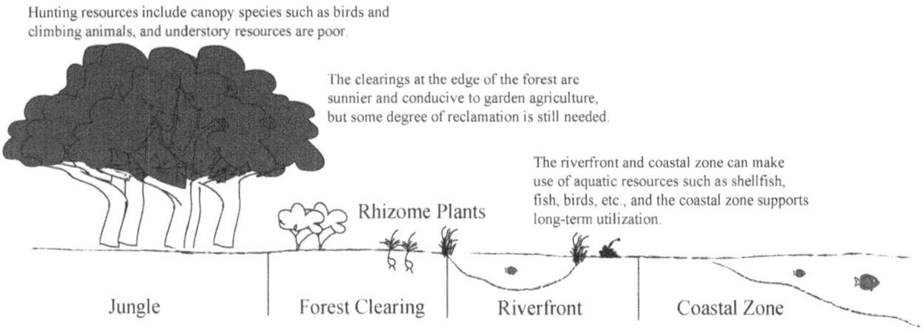

Fig. 8.2 Prehistoric cultural ecosystem in South China

labour. Compared with temperate areas, agricultural production, especially grain cultivation, requires more labour in tropical areas than in temperate areas. It is relatively easy to grow roots and tubers instead of grains. There is no need to clear much vegetation or increase soil fertility, and irrigation is less required. It is relatively easy to use this area to grow roots and tubers rather than grains.

Linked to the above two zones is the riverine zone, which is also rich in resources and diversity. Rivers are not only good for transportation, but also provide fish, snails, and so on. Waterfront lakes and wetlands also attract a large number of birds and other land animals. These animals come from a wide range of sources. Fish may cover the entire basin and even the ocean, birds may be seasonal migratory groups, and land animals may come from both sides of the river. All species are concentrated in the narrow area along the river, forming a resource-rich area. The seasonality of riverfront and lakeside resources can be complementary to terrestrial resources to a certain extent, and the predictability is much better than that of big animal hunting. If the aquatic resources in the riverfront can be utilised, a relatively stable supply of resources can be obtained. Of course, rivers and lakes have abundant water, stable water flow, and relatively rich organic matter, which are the prerequisite for the utilisation of aquatic resources. In areas lacking boats, there are fewer opportunities to use aquatic resources; or even if some lakes are available, they are difficult to be used because of the lack of rowboat transport, such as those on the Qinghai-Tibet Plateau and in Western China. The Pearl River System has a large water volume, stable water flow in the middle and lower reaches, and the Southwest monsoon (Indian Ocean Monsoon) around 12,000 years ago brought the largest effective precipitation, all of which provided favourable conditions for the utilisation of riverside aquatic resources.

The last zone is the coastal zone, where resources are more abundant than riverside and lake areas. The long and narrow coastal area of South China is large. The sea level rise after the end of Last Glacial Maximum led to more zigzag coastline and wider continental shelf area, which was conducive to the utilisation of Marine resources. The source area of Marine resources is wide, and the coastal zone has high productivity and biomass, which is not affected by continental seasonality. The vast wetlands of the coastal zone also attract a large number of birds, as well as land animals. More importantly, the abundance of shellfish and other seafood that could be gathered in the tidal plains greatly expanded the source of protein for people, reducing the need for animal hunting, which was the most mobile livelihood activity and thus reduced the mobility of the entire hunter-gatherer community. The utilisation of marine aquatic resources was one of the main ways of livelihood in South China in prehistoric times, and the rise and fall depended on the changes of coastline (Huang 1996), which affected the available areas and led to the vision of sites in this area.

The four zones in South China contain multiple cultural and ecological niches that allow for different ways of living, and their roles overlap somewhat. The group using the estuary area can exploit the ocean resources and the riverfront resources. Using aquatic resources can provide a source of protein, just like hunting animals. Similarly, root tuber cultivation is a major source of carbohydrates as plant collection. Of course, hunter-gatherer groups can also choose cereal cultivation, which has the same role as root tuber cultivation and plant collection, and animal domestication, which can replace aquatic resource use as well as animal hunting. Which direction offers the best benefit–cost ratio? Which direction of use is the most sustainable? For hunter-gatherers, access to a stable source of food was always required, and root and tuber cultivation and grain agriculture provided a more stable source of plant food than plant gathering. Root and tuber cultivation has an advantage for its cheaper cost. Both domestication and use of aquatic animals and hunting of animals can obtain protein, the first two are undoubtedly more stable and sustainable. Among animal domestication and aquatic resource utilisation, aquatic resource utilisation requires lower input costs and is an activity in which the whole group can participate. Animal domestication not only takes up manpower itself, but more importantly, providing enough food for animals requires other crop production, such as feeding livestock with by-products of root tuber and grain production, which has very high incidental costs. Animal domestication also brings the risk of disease, and animal activity can contaminate habitats and spread disease. Disease pressure is already high in the tropics, and animal domestication greatly exacerbates this pressure.

Based on the analysis of cost–benefit (including risk) and sustainability of resource utilisation, there are four possible acculturation methods in South China: ① In non-coastal riverfront areas, freshwater aquatic resources can be used as well as hunting, gathering and root and tuber cultivation. ② Hunter-gatherers using marine resources in coastal areas may also cultivate root and tuber.③ hunter-gatherers in non-waterfront areas; ④ Hunter-gatherers with root tuber culture in non-waterfront areas. Of the four types, non-waterfront hunter-gatherers have neither aquatic resources to exploit nor root and tuber cultivation to increase the stability of acculturation, and population densities in such areas must be very low; If the population density exceeds a certain threshold, it needs to adopt the reinforcement strategy, which is not a stable adaptation strategy. This was the case in South China before the pre-Mesolithic, when population density was below the threshold and reinforcement strategies such as using aquatic resources or root tuber cultivation were not necessary. During the Mesolithic period, there may have been areas with neither aquatic resources nor root tuber cultivation, such as the Nanling regions, where it was necessary for hunter-gatherers to maintain a low population density. This was a marginal environment for hunter-gatherers. Just as agriculture has become popular, there are still hunter-gatherers living in extreme, marginal environments. These marginal environments cannot be used for agriculture, while hunter-gatherer methods can be used, but the population density needs to be below the threshold. The other three methods all adopt intensive strategies, which can exceed the population density threshold supported by general hunter-gatherers. Among them, the most effective strategy is the utilisation of Marine resources combined with root and tuber cultivation, which is the evolutionary stability strategy in South China. All four strategies were likely to have existed during the Mesolithic period in South China, but it was this evolutionary stability strategy that was most likely to resist rice agriculture for the longest time, as agriculture began to spread along the middle and lower reaches of the Yangtze River.

The reason for the above strategy is that there are many constraints to grain farming in South China, and if these factors are not removed, cereal farming will not be competitive as a livelihood strategy. Basic constraints such as soil conditions require techniques to increase the amount of organic matter in the soil and improve soil fertility. Of course, through burning, using charcoal, grass ash adsorption can solve the problem. The problem then becomes a lot of labour to clear the ground, requiring either effective tools (like metal knives and axes) or a lot of human labour. Population is one of the main constraints for the adoption of grain agriculture in tropical areas (Athens 2003). On one hand, disease pressure limits the total number of population; on the other hand, in order to balance population loss, more births are increased, resulting in a younger population, more shortage of labour force, and slow deposit of technical experience. In addition, stable precipitation is an indispensable basic condition for grain agriculture, because lack of it, rice agriculture encountered obstacles in the process of spreading to Oceania (Dewar 2003, p. 369–388). Stable precipitation is more considered to be the present of Holocene (Richerson et al. 2001), but not synchronised in all regions. The maximum effective precipitation brought by stable monsoon mechanism appeared in South China at 3000 years ago (An 2000) much later than in other regions, which may be another reason for the slower acceptance of rice agriculture.

The above cultural ecology based on hunter-gatherers analyses the possible cultural adaptations used in South China, as well as the constraints, which are undoubtedly rough and framed. What about the actual archaeological findings? Can we piece together a detailed picture of prehistoric adaptation? The current archaeological materials are still quite fragmentary, which is not enough to recover the whole form of cultural adaptation, but it can reveal some signs of adaptation changes, and provide some empirical materials for the derivation of the theory. The changes in the Late Paleolithic in South China are similar to those in the Middle and Lower Yangtze, with the miniaturisation and flaking of the stone tool industry and the introduction of high-quality raw materials such as flint (cf. Chapter 4), reflecting the increased demand for cutting, and it is reasonable to assume that hunting played a stronger role in human livelihoods during this period. With the end of LGM, the Mesolithic in South China began. The most important sign was that human began to use aquatic resources, and snail shells were found in the cultural layer of the site. At the same time, pottery began to appear, along with edge-polished and perforated stone tools. The invention of pottery may have been related to the utilisation of aquatic resources, such as the cooking of snails. Edge-polished and large perforated stone tools may be associated with cutting and digging behaviour, the former reflecting increased investment in settlement construction, and the latter reflecting root tuber utilisation. Of course, more scientific analysis and experimental evidence are needed about the function of these artefacts.

Among the available archaeological materials, the site of Dingmodong in Tiandong, which is considered to belong to the Late Paleolithic, is more likely to belong to the Pre-Mesolithic, and the signs of dental caries on the fossilised human teeth may be related to the consumption of starch-rich plant roots and tubers. It is an interesting phenomenon that deer droppings were found in the Jinghong Naminan Cave. Deer hardly reside in caves in their natural state, and the discovery of deer scat in caves may be evidence of human control of deer. Compared to the Middle and Lower Yangtze transition phase of the Paleolithic, signs of subsistence intensification in South China emerged almost simultaneously, and perhaps even a bit earlier. However, there are two differences that influenced the later trajectories of the two cultural ecological zones: first, the use of aquatic resources, although not much in the pre-Mesolithic period in South China compared to the later period, has already produced differentiation; secondly, the possible beginning of the use of root tubers. Although there is evidence that wild rice may have been distributed in South China during the Mesolithic (Zhao 2003), the ancient people did not make use of it. The possible reason for this is that humans had already begun to exploit the more easily exploitable aquatic animals with root tubers.

Zhao Zhijun has a detailed analysis of root tuber cultivation in South China. The cultivation process of root tuber plants, including taro, is simple. They can reproduce sexually through seeds or asexually through tubers. As long as small buds are left on the leftover tubers, they can continue to sprout and grow next year when external conditions are right. When hunter-gatherers choose a place to live, they often cut down trees and build shelters. At the same time, they discard garbage, including all kinds of food leftovers, such as plant seeds, tubers, etc. These residues are easily sprouted by the cover and fertility of other garbage, and by the sunlight brought through clearing the open space. Hunter-gatherers may have inadvertently established interdependencies with species, known as "moving to produce" (Politis 1996), resulting in "accidental domestication" called by Rindos (Rindos 1984).

Charred tuber remains discovered by flotation at the Zengpiyan site (Zhao 2003), starch granules of taro were found in the surface residue of stone blades, and there is also early indirect evidence of human teeth excavated from the Dingmodong in Tiandong (Lü 2003). Zhao Zhijun proposed that in addition to dry farming in North China and rice agriculture in the middle and lower reaches of the Yangtze River, there are also root and tuber crops in South China (Zhao 2006). There is no doubt that humans may have been able to utilise root tubers of plants as early as the Early Paleolithic, but the question is the intensity of utilisation. Unless such food was consumed regularly, it would not be sufficient to cause dental caries. Evidence from Dingmodong better illustrates the intensity of human utilisation of such plants at that time than direct root tuber plant remains. Equally important as root tubers in the tropics were plant fruits, and tropical fruit trees were numerous, available for long periods of time, and complementary to subsistence.

The emergence of rice agriculture in South China has the character of mutation. Analysis of phytoliths deposited in the first four stages of Dingsishan site shows that rice agriculture has not yet emerged (Chinese Academy of Social Sciences Institute of Archaeology Guangxi Work Team, Guangxi Autonomous Region Cultural Relics Work Team 1997; Chinese Academy of Social Sciences Institute of Archaeology Guangxi Work Team, et al. 1998; Chinese Academy of Social Sciences Institute of Archaeology Guangxi Work Team, et al. 2003; Guangxi Autonomous Region Cultural Relics Archaeology Training Class, Guangxi Autonomous Region Cultural Relics Work Team 1975). This is also proved by the flotation results of Dayan Fifth Stage site (Fu et al. 2011; CA CASS 2010). However, in the cultural deposit of Dingsishan in the fourth stage, a considerable amount of rice phytoliths suddenly appeared, dating back to about 6000 years ago. Ten thousand rice grains were found in the remains of the second and third stages of Xiaojin site in Ziyuan, the age of which was similar to that of the fourth stage of Dingsishan. In stark contrast, no rice remains were found at all in the first phase. The clearest remains of rice agriculture were found in Shixia site in Qu River. A large number of rice grains, husks, and stalks were found in the middle and lower cultural layers (5000–4000 years ago), which were identified as cultivated rice. Interestingly, there may have been wild rice near Zengpiyan site at that time, but no evidence of human utilisation was found. In contrast, phytoliths of rice plants were found in Phase II (11,000–9000 years ago) and Phase III (9000–8000 years ago) cultural deposits in Niulandong, in Yingde Guangdong (Gu et al. 1999), and it is not yet possible to determine whether they belong to cultivated rice or wild rice, but more likely to wild rice. The discovery of Niulandong indicates that there is some variability in the use of plant species in different regions in the same South China region. However, true rice agriculture emerged suddenly and is related to the expansion of rice agriculture in the middle and lower reaches of the Yangtze River. It is not only the proliferation of rice agriculture that is of interest, but also the social complexities that came along with it. Archaeological materials from the Fujian region show that the number of sites increased significantly after the emergence of rice agriculture. Concomitant with rice agriculture may have included the complication of social structures.

Rice agriculture can certainly support higher population densities and requires more human and labour organisation. Rice is more durable than root tubers. More importantly, rice cultivation was accompanied by an "agricultural package" that included not only rice cultivation techniques, but also relatively advanced tools, food processing and storage techniques, social organisation methods, and domesticated animals. In this respect, the spread of rice agriculture is comparable to the spread of modern industrial technology. If a society accepts the technology of another one, it is bound to accept its social culture and even ideological influence at the same time. Technology is connected with many aspects of society. From the interior to the coast, rice agriculture was gradually accepted, relatively early in the interior and later in the coastal areas. It's not only about transmission paths, but about resource sustainability in both zones. Compared with marine resources, freshwater resources are more susceptible to the impact of the terrestrial season, and the resources are not as abundant as marine resources. It has been found in Zengpiyan site that the diameter of shellfish gradually decreases in a day (CA, CASS, 2003), and the enhanced utilisation of freshwater aquatic resources is more likely to affect the reproduction of natural population. That is, while hunter-gatherer livelihoods that combine freshwater resource use with root tuber cultivation can enable local population densities beyond what general hunter-gatherers can support, they are not as sustainable as marine resource, which is why the resource Ziyuan in Xiaojin and Dingsishan sites embraced rice cultivation around 6000 years ago, while Qu

River Shixia site was nearly 1000 years later, and in some areas even later, not fully embracing it until the Bronze Age.

The three signs of the Mesolithic in South China: the beginning; the end and the transition in the middle, respectively represent the start of adaptation and intensification; the beginning and differentiation of rice agriculture. The change 12,000 years ago looks like a transition, but it is actually more historical. It is a turning point in the development of culture, which diverged from the development of the middle and lower reaches of the Yangtze River. In the middle and lower reaches of the Yangtze River, rice was cultivated and an agricultural system centred on rice was developed. South China developed its own evolutionarily stable strategy using aquatic resources (riverside and coastal) and root and tuber cultivation, which kept pace with the middle and lower reaches of the Yangtze River for thousands of years. This difference in development path can be traced back to the Late Paleolithic Period, that is, the cultural adaptation selection at the end of LGM. The initial slight difference was amplified later, leading to the beginning of the Mesolithic in South China. Finally, rice agriculture spread into South China and the Mesolithic ended. The Mesolithic was formed in the middle and lower reaches of the Yangtze River. Obviously, the adaptation in South China is highly regional and depends on local special resource conditions, with few adjustments and changes required by the cultural system of hunter-gatherers. Hunter-gatherer is still the main livelihood mode, and the Cultivation of roots and tubers plays a complementary role, which belongs to what Smith called "low-level food production" combined with aquatic utilisation (Smith 2001, p. 1–43). It is very comparable to the Mesolithic in Western Europe and the Jomon Age in Japan.

8.5 Southeast Asia

It is very necessary to discuss South China in the context of Southeast Asia, not only because South China as a cultural and ecological area is only within the scope of China. If contemporary political boundaries are not taken into account, the scope of South China as a cultural ecological area should include part of Southeast Asia. At the same time, South China is located in the tropical fringe, not the most typical tropical environment. It is sometimes called the transitional tropical, or it is the transition zone from tropical to temperate zone. This zone also includes the northern part of Southeast Asia. Looking at South China from the perspective of

Southeast Asia can give us a comparative perspective and help us to recognise the characteristics of South China. In addition, there is the expansion of materials, looking at South China on a larger spatial scale; Last but not least, the research background and history of Southeast Asian prehistory is different from that of Chinese archaeology, which helps us to understand the process of prehistoric culture in South China from a different angle.

The material on the transition from the Paleolithic to the Neolithic in Southeast Asia is unclear, and can be divided into island and mainland regions on a macro scale (Bellwood 2003). There are differences in the characteristics of stone industry between the two regions. A kind of blade industry is more prevalent in the island region, which is similar to the West Asian region and sometimes called "geometric microliths" (Jia 1978), but different from the microblade technology in North China. The "the Hoabinhian culture" (He and Qin 2009) is popular in mainland China. It is a gravel industry that uses flat, ovate or elongated gravel in rivers to make unifacial or bifacial stone tools. Bifacial processing is dominant in Peninsular Malaysia, while single-sided processing is dominant in other regions. The instrumental combination also includes perforated gravel, whetstone, bone awl, bone needle, and bone knife.[5] At the beginning of the Holocene, edge-ground tools appeared in the Hoabinhian culture stone assemblage, and later pottery appeared, such as the Spirit Cave in Thailand, dating from around 8000 years ago. Most of the sites of the Hoabinhian culture are limestone caves, rock shelter, and shell middens, and quite a few may have been destroyed by the Holocene sea rise. The cultural deposits of the sites all contain large amounts of marine shellfish deposits.

At present, the date of the beginning of the Hoabinhian culture is not clear, and there are many contradictions in the chronology. The Con Moong site found that the Hoabinhian culture overlies the Son Vi culture, but the 14C dating part of it is as late as 12,000 years ago, while the age of the Hoabinhian culture is more than 20,000 years ago. In the Pang Mapha site group in Thailand, the Tham Lod rock house dates from 22,000 to 12,000 years ago, and there are perforated gravels in the stone assembly, which researchers believe belongs to the Hoabinhian culture (CA, CASS 2006). Masa Nishimura also believed that the era of the Hoabinhian culture could be as early as 20,000 years ago, or even earlier, and the end date was 9000–8000 years ago (CA, CASS 2006). Gorman, who excavated the fairy cave in the 1970s, set the age of the Hoabinhian culture at 12,000 to 7500 years ago (Gorman 1972, p. 79–108). Some studies regard Shanwei culture as the local type of the Hoabinhian culture, leading to the expansion of the Hoabinhian culture into a kind of stone industry in the Late Paleolithic and even the Late Pleistocene. Peter Bellwood, a longtime researcher of prehistoric archaeology in Southeast Asia, proposed that around 13,000 years ago, the stone industry in many parts of mainland Southeast Asia gradually entered a typical stage of Hoabinhian culture (Bellwood 2003).

From the somewhat confusing material above, it is still possible to see the beginnings of a prehistoric cultural development sequence in Southeast Asia similar to that of South China, although the material from Southeast Asia is more complex and diverse. It is possible that material from the Hoabinhian culture first appeared around 20,000 years ago, roughly contemporaneous with the pre-Mesolithic period in south China. Around 13,000 years ago, there was an important development. In the north of Vietnam, it is shown as the North Mountain culture. The difference between the North Mountain culture and the Hoabinhian culture is only the edge-ground tools. Nishimura also found a slight difference in the use of raw materials, as well as more bifacial processing tools. The North Mountain culture dates back to 9000–7000 years ago (CA,CASS 2006). Edge-ground tools have also been found in 10,000-year-old strata at the Niah site in Sarawak, reflecting the complexity of the region and suggesting that changes of the Hoabinhian culture were somewhat common in Southeast Asia. After the Hoabinhian culture, the sites of the Duobi Culture (7000–5000 years ago) mostly contained shells and fish bones, and were distributed in lake basins or coastal areas rich in aquatic products, so as to facilitate fishing and collection (Ruan 2006).

During the LGM, sea level in Southeast Asia was about 130 metres lower than it is today, and Sunda Shelf expanded into a vast subcontinent. After the sea level rises, the length of the coastline increased by 46%, increasing the opportunity to exploit Marine resources. At the same time, forests expanded, and native large animals disappeared and miniaturised. The pressure to adapt coincides with the opportunity, and the Hoabinhian culture began to emerge. A typical Hoabinhian culture, as described by Bellwood, took shape around 13,000 years ago, when Southeast Asia entered the Mesolithic (Rindos 1984). The Hoabinhian culture lasted at least until the Middle Holocene. The plant remains found in the Spirit Cave were abundant, including some edible plants, irritants (areca nut), and poisons (white walnut kernel), but no artificial plants were ever cultivated. Remains of rice, dated as late as 5500 years ago, were found in caves in the nearby Jianbanyan Valley Cave.

Rice agriculture in Southeast Asia was formed due to external influences. Charles Higham believes that around 4000 years ago, the Austronesian farmers invaded (Higham 2006) and introduced rice, which had a huge impact on the prehistoric society in Southeast Asia. Studies on the origin of rice agriculture in Thailand have shown that upland rice and rice may have begun to be cultivated simultaneously in Thailand, not as a gradual evolutionary process but as a leap forward (White 1995, p. 37–68). The proliferation of rice-based agriculture has been affecting South Asia, as well as the entire southeast Asian and Oceanian Island Belt (Glover and Higham 1996).

8.6 Conclusion

The concept of "Mesolithic" is applied to the study of the transition of the Neolithic in South China. Through the combing of archaeological materials, we get an interesting cultural chronology sequence, that is, around 20,000 years ago, the beginning of the Early Mesolithic, human adaptation showed some signs of strengthening, but until around 12,000 years ago, the typical South China Mesolithic began to form. This chronological sequence also exists in Southeast Asia, from the beginning of the Hoabinhian culture to the typical the Hoabinhian culture. The beginning of the Mesolithic coincided with the end of the LGM. Rising temperatures, changes in flora and fauna, and rising sea levels reduced the available hunting resources, but gradually made the use of aquatic resources possible. The intensification of the Indian Ocean monsoon 12,000 years ago led to more abundant rivers and much stronger utilisation of aquatic resources. At the same time, prehistoric hunter-gatherers may also have moved from the use of root tubers to cultivation, which combined with aquatic resource use to form an evolutionarily stable strategy in human adaptation in South China, counterbalancing rice agriculture in the middle and lower reaches of the Yangtze River. Around 6000 years ago, rice agriculture expanded into South China, gradually expanding from freshwater aquatic utilisation areas to coastal areas until the Bronze Age and even later, when rice agriculture replaced hunter-gatherer livelihoods that depended on marine resources. During the Late Paleolithic, the differences between South China and the middle and lower reaches of the Yangtze River were not obvious, and the divergent development began in the transitional phase of the Paleolithic and Neolithic, i.e. the Mesolithic in South China. South China developed an alternative mode of food production in conjunction with local resource conditions and developed successful cultural adaptation strategies.

Notes

1. According to the classification in Chinese Archaeology: Neolithic Archaeology Volume.
2. French archaeologist Madeleine Colani excavated 52 caves and rock shelter sites in northern Vietnam from 1924 to 1926. Due to the rapid pace of excavation, the accuracy of stratigraphy has been questioned. In her 1932 report, she described the Hoabinhian culture characterised by unifacial tools, stone hammers, objects with nearly oval cross-sections, a tool assemblage including discoid tools, short axes, almond-shaped stone tools, and a large number of bone artefacts (Praehistorica Asiae Orientalis 1932). See Encyclopedia of Archaeology, entry on Preagricultural Peoples.
3. The second phase of the Xiaojin site has a wide range of radiocarbon dating, and the excavators believe that the upper limit of the age may reach around 6000 years ago. The upper limit of the fourth phase of the Zengpiyan site is also around 6000 years ago.
4. Here, the term "culture" is more akin to the meaning of lithic industry.
5. Due to the earlier warming during the Last Glacial Maximum in Southeast Asia (which is closer to the equator), the beginning of the Middle Neolithic in this region is considered to be approximately one thousand years earlier than in South China.

References

S. Athens, Comments. *Current Anthropology*, 44: 377–378, 2003.

An Zhimin, Early Neolithic 14C chronology and issues in South China, Quaternary Sciences, 1989, 2.

An Zhimin, Several sites of the middle Neolithic age in China, Archaeological Newsletter, 1956, 3

An Zhimin, middle Neolithic remains in Hailar: on the origin and tradition of microliths, Acta Archaeologica Sinica, 1978, 3.

An Zhisheng, The history and variability of the East Asian paleomonsoon climate. *Quaternary Science Reviews*, 19: 171–187, 2000.

L. R. Binford, Post-Pleistocene Adaptation. In S. Binford and L. R. Binford eds., *New Perspectives in Archaeology*, pp. 313-341. Aldine, Chicago, 1968.

P. S. Bellwood, *Prehistoric Southeast Asia, in Cambridge History of Southeast Asia*. Yunnan People's Publishing House, 2003.

CA, CASS et al. *Guilin Zengpiyan*. Cultural Relics Press, 2003.

Chen Chun, On the mesolithic period, acta anthropological sinica, 1995, 14(01).

Chen Chun, Research and reflection on the mesolitic period, Agricultural Archaeology, 2000 (1).

Chen Shilong, *Excavation and research on the Miaoyan cave site in Guilin, in symposium on the middle Neolithic culture and related issues*. Guangdong People's Publishing House, 1999.

Chinese Academy of Social Sciences Institute of Archaeology Guangxi Work Team, et al. Excavation of the Dingshishan site in Yongning County, Guangxi, Archaeology, 1998, 11.

Chinese Academy of Social Sciences Institute of Archaeology Guangxi Work Team, Guangxi autonomous region cultural relics work team, 1996 archaeological survey report on Stone Age in Guangxi, Archaeology, 1997, 10.

Chinese Academy of Social Sciences Institute of Archaeology Guangxi Work Team, et al. Brief report on the excavation of the Baozitou site in Nanning, Guangxi, Archaeology, 2003, 10.

Cui Yong, Significant achievements in the excavation of shell Midden sites in Guye village, Gaoming, Guangdong, Chinese Cultural Relics News, January 12, 2007.

Cultural Relics Management Committee of Guangdong Province, Shell Midden Sites in Chao'an, Guangdong, Archaeology, 1961, 11.

Cai Baoquan, Research and discoveries of Old Stone Age in Southern Fujian, Research on Southeast Asian Archaeology, 1996, 1.

CA, CASS, *Chinese Archaeology: Neolithic Period*. Chinese Academy of Social Sciences Press, 2010.

Chen Yiqin, *60 Years of Chinese archaeology: Guangdong province, in 60 years of Chinese archaeology (1949–2009)*. Cultural Relics Press, 2009.

R. E. Dewar, Rainfall variability and subsistence systems in Southeast Asia and the Western Pacific. *Current Anthropology*, 44: 369–388, 2003.

Fu Xianguo, et al. Re-excavation of the Liyuzui Site in Liuzhou, Chinese Cultural Relics News, Aug.4, 2004.

Fu Xianguo, He Zhanwu, Xiong Zhaoming, Wang Haotian, Initial outline of prehistoric culture in guilin region, chinese cultural relics news, Apr. 4, 2001.

Fujian Provincial Museum, Brief Report on the Excavation of the Keqiaotou Site in Pingtan, Fujian, Archaeology, 1991, 7.

Guangxi Autonomous Region Cultural Relics Archaeology Training Class, Guangxi autonomous region cultural relics work team, Neolithic shell Midden sites in nanning, Guangxi, Archaeology, 1975, 5.

C. F. Gorman, Excavations at spirit cave, north Thailand: some interim interpretations. *Asian Perspectives*, 13: 79–108, 1972.

Gai Pei, Wang Guodao, Excavation Report on the La Yihai middle Neolithic site in the upper yellow river region, Acta Anthropologica Sinica, 1983, 2(01).

Gu Haibin, et al. *Study on rice Phytoliths from the Niulandong site, in symposium on middle neolithic cultures and related issues*. Guangdong People's Publishing House, 1999.

Guangdong Provincial Museum, Shell Midden Site of the New Stone Age in Dongxing, Archaeology, 1961a, 12.

Guangdong Provincial Museum, Gaoyao county cultural bureau, discovery of neolithic shell midden sites in Gaoyao County, Guangdong, Archaeology, 1990, 6.

Guangdong Provincial Museum, The Qingtang site of the new stone age in Wengyuan county, Guangdong, Archaeology, 1961b, 11.

Guangxi Working Team of the Institute of Archaeology, Chinese academy of social sciences, cultural relics work team of Guangxi Zhuang autonomous region, archaeological survey report on Stone Age in Guangxi in 1996, Archaeology, 1997, 10.

Guangxi Working Team of the Institute of Archaeology, Chinese academy of social sciences, et al. Excavation of Dingsishan site in Yongning county, Guangxi, Archaeology, 1998, 11.

Guangxi Working Team of the Institute of Archaeology, Chinese Academy of Social Sciences, et al. Excavation brief of Baozitou site in Nanning, Guangxi, Archaeology, 2003, 10.

Guangxi Liuzhou Bailiandong Science Museum, *Bailiandong in Liuzhou*. Science Press, 2009.

I. C. Glover and C. F. W. Higham, New evidence for early rice cultivation in South, Southeast and East Asia. In D. R. Harris ed., *The Origins and Spread of Agriculture and Pastoralism in Eurasia*, pp. 413–441. Smithsonian Institution Press, Washington, 1996.

He Anyi, Chen Xi, New Discoveries from the Chongtang Shell Midden Site in Chongzuo, Guangxi, in the Neolithic Age, Chinese Cultural Relics News, May 9, 2008. No pottery was found at the Chongtang site, and the excavators estimate its age to be around 5000 years ago.

He Naihan, Qin Shengmin, A preliminary study of the middle Neolithic Age in Lingnan, Acta Anthropologica Sinica, 1985, 4(04).

C. F. W. Higham, *Southern China and Southeast Asia during the Neolithic, in proceedings of the international symposium on prehistoric archaeology in South China and southeast Asia in commemoration of the 30th anniversary of the excavation of the Zengpiyan Site*. Cultural Relics Press, 2006.

Huang Guangqing, Neolithic Archaeology and Paleogeographic environment of the pearl river delta, Acta Geographica Sinica, Vol. 51, No. 6, 1996.

Huang Zhenguo, Zhang Weiqiang, Changes of the last glacial maximum in the tropical region of China, Acta Geographica Sinica, 2000, 55(05).

Jia Lanpo, Characteristics of Chinese Microliths and their tradition, origin, and distribution. *Journal of Vertebrate Paleontology and Paleoanthropology*, 16(02), 1978.

Jiao Tianlong, Chronology of the Neolithic cultures on the coast of Southeast China, in *Proceedings of the International Symposium on Prehistoric Archaeology in South China and Southeast Asia in Commemoration of the 30th Anniversary of the Excavation of the Zengpiyan Site*. Cultural Relics Press, 2006.

Jin Zhiwei, et al. Preliminary report on the Yunniling Niulandong Site in Yingde, Jianghan Archaeology, 1998, 1.

Liu Jinling, Revisiting the vegetation types during the last glacial maximum in South China, Acta Micropalaeontologica Sinica, 2007, 24(01).

Liu Jinling, Wang Weiming, Discussion on the vegetation types during the last glacial maximum in South China, Quaternary Sciences, 2004, 24(02).

Liuzhou Bailiandong Science Museum, *Bailiandong in Liuzhou*. Science Press, 2009.

Liuzhou City Museum, et al. Dalongtan shell Midden site in Liuzhou City, New Stone Age, Archaeology, 1983, 9.

Lü Liedan, *Preliminary analysis of surface residues on stone tools excavated from Zengpiyan, in Guilin Zengpiyan*. Cultural Relics Press, 2003.

G. G. Politis, Moving to produce: Nukak mobility and settlement patterns in Amazonia. *World Archaeology*, 27: 492–511, 1996.

Pei W. C., On a Mesolithic (?) industry of the Caves of Kwangsi. *Bulletin of the Geological Society of China*, 14: 393–412, 1935.

Peking University Department of History, et al. Reliability of carbon-14 dating samples and the dating of Zengpiyan and other sites in limestone areas, Acta Archaeologica Sinica, 1982, 2.

Qiu Licheng, et al. Excavation of the Dushizi Cave site in Yangchun County, Guangdong, Archaeology, 1982, 5.

P. J. Richerson et al., Was agriculture impossible during the Pleistocene but mandatory during the Holocene? A climate change hypothesis. *American Antiquity*, 66: 387–411, 2001.

D. Rindos, *The Origins of Agriculture: An Evolutionary Perspective*. Academic Press, Orlando, 1984.

Ruan Wenhao, The Dong Son culture in Vietnam, in *Proceedings of the International Symposium on Prehistoric Archaeology in South China and Southeast Asia in Commemoration of the 30th Anniversary of the Excavation of the Zengpiyan Site*. Cultural Relics Press, 2006.

Shenzhen Museum, Department of anthropology, Sun Yat-sen University, excavation brief of Xiantouling Sand Dune sites in Dapeng, Shenzhen, Cultural Relics, 1990, 11.

B. D. Smith, Low-level food production. *Journal of Archaeological Research*, 9: 1–43, 2001.

Song Fangyi, et al. Investigation report on caves in Fengkai and Huaiji, Guangdong, Vertebrate Paleontology and Paleoanthropology, 1981, 19(03).

Song Fangyi, et al. The Huangyandong site in Fengkai county, Guangdong, Archaeology, 1983, 1.

Wang Linghong, Ancient human fossils and stone tools discovered in Baojiyan, Guilin, J Acta Anthropologica Sinica, 1982, 1(01).

Wang Youping, *Pleistocene Environment and the Development of Paleolithic Culture in Southern China*. Peking University Press, 1997.

Wang Youping, *The Origins of Ancient Human Culture in China*. Science Press, 2005.

Wei Jiang, *60 Years of Chinese Archaeology: Guangxi Zhuang Autonomous Region, in 60 Years of Chinese Archaeology (1949–2009)*. Cultural Relics Press, 2009.

J. C. White, Modelling the development of early rice agriculture: Ethnoecological perspectives from Northeast Thailand. *Asian Perspective*, 34: 37–68, 1995.

Work Team of Guangxi Institute of Archaeology, et al. Excavation of the Dingsishan site in Yongning County, Guangxi, Archaeology, 1998, 11.

C. Xingcan, Several issues on the middle Neolithic period, Archaeology, 1990, 2.

Yuan Sixun, Difficulties caused by the 14C dating data of early Neolithic period in South China, Archaeology, 1993, 4.

Yang Decong, *60 Years of Chinese Archaeology: Yunnan Province, in 60 Years of Chinese Archaeology (1949–2009)*. Cultural Relics Press, 2009.

Yin Shaoting, *The Vanished Slash-and-Burn Agriculture: Anthropological Perspectives*. Yunnan People's Publishing House, 2008.

Yingde City Museum, et al. Yunniling Niulandong Site in Yingde, in *Prehistoric Archaeological Reports of Yingde*. Guangdong People's Publishing House, 1999.

You Yuzhu, et al. *Prehistoric Culture in Zhangzhou*. Fujian People's Publishing House, 1991.

Zeng Xiangwang, Human fossils and cultural remains of Dingmodong in Tiandong, Guangxi, Archaeology and Cultural Relics, 1982, 2.

Zhang Weiqiang, Huang Zhenguo, New understanding of late quaternary climate fluctuations in the pearl river delta, Tropical Geography, 2005, 25(04).

Zhang Zhenhong, et al. Preliminary report on the first excavation of the Luoshayan Cave site in Fengkai County, Guangdong, Acta Anthropologica Sinica, 1994, 13(04).
Zhao Zhijun, Reconsideration of primitive agriculture in South China, in *prehistoric archaeology in South China and Southeast Asia—Proceedings of the International Academic Symposium Commemorating the 30th Anniversary of Zengpiyan Site Excavation*. Cultural Relics Press, 2006.
Zhao Zhijun, *Study of Plant Remains, in Guilin Zengpiyan*. Cultural Relics Press, 2003.
Zhang Zhenhong, et al. *Study on the animal assemblage of the Miaoyan site in Guilin, in symposium on the middle Neolithic culture and related issues*. Guangdong People's Publishing House, 1999.

CHAPTER 9

The Toughest Hunter-Gatherers: Northeast China

9.1 Introduction

From a modern perspective of physical geography, the western part of the Northeast region is separated from HulunBuir Grassland by Greater Khingan Mountains, and the southern boundary is not clearly defined. It begins at Arxan and goes south along Taoer River Valley and Ulanhot and continues along Zhangwu County, Faku County, Tieling City, and Fushun City—namely along the accumulated temperature line of 3200 °C. Then it proceeds southeastward by way of Kuandian County and stops at Yalu River (Ren 2004). During the prehistoric period, however, the climate has been rather whimsical, making the region of Northeast China hard to define. As far as cultural ecology is concerned, Northeast China is a vague concept which has a fluctuating transition zone with North China and an ecotone with the grassland of Mongolian Plateau. If considering Northeast China in the context of Northeast Asia, it is closely related with Korean peninsula, Japanese archipelago, outer area of Ussuri region, and even the whole Siberia. That is to say, it is where five cultural ecological region—North China, Mongolian Plateau, Siberia, Korean peninsula, and Japanese archipelago—meet. It has four kinds of vegetation: temperate forest steppe, temperate desert steppe, boreal forest and coastal sea; or it is the cooperation-competition region of four traditional living means: farming of North China, pastoral nomadism, hunting-gathering subsistence of forest and fishing-hunting

S. Chen, *Prehistoric Modernization*,
https://doi.org/10.1007/978-981-97-9423-2_9

subsistence of coastal area. With all these features, it follows that Northeast China is sophisticated culturally, which cannot be summarised in one simple model. The best method is to study on a case-by-case basis. In Chapter 7, when the author discusses about North China, the southeastern Inner Mongolia and the southern Liaoning are included; the Western Liaoning Province is discussed as well. Therefore, Northeast China here refers to Jilin, Heilongjiang, and the area east of Inner Mongolia.

Judging from archaeological materials of prehistoric period in Northeast China, we can come to the following conclusions. Firstly, zonal difference is significant. The features of archaeological materials are closely linked with physical geography (Ye 2003; Tang et al. 2007) Secondly, there exist large and coarse chipped stone tools, finely pressured small stone tools based on pressure technology, lithic products by the microblade technology and ground stone tools, which indicates that chipped stone tools cannot serve as a sign of Paleolithic —they may be made in the Neolithic.[1] Thirdly, fishing and hunting tools are sophisticated—this feature can be observed since the Late Upper Paleolithic. And fourthly, agriculture develops very late and rather slow in Northeast China. On the cultural adaption in Northeast China in the Neolithic, Yan Wenming comments:

Generally speaking, the cylinder pottery jar with zigzag pattern is prevalent in southern Northeast China and round bottom jar with fine-toothed comb pattern and round jar with the incised pattern are prevalent in the Western Korean peninsula. The cultures of both regions are characterised by the stable economic culture of production based on farming. The Angangxi culture in northeastern Songnen Plain and Novopetrovka culture and Osinovo culture in the middle reaches of Heilongjiang River where the pottery jar with convex strings is prevalent, suggest a nomadic hunter-gatherer economic culture based on hunting. The Xinkailiu culture in the Sanjiang Plain of China, and the Kongdong culture and Ludenaya culture in the lower reaches of the Heilongjiang River where the Heilongjiang cord-marked pottery is prevalent and the shell midden sites are plentiful, suggest a stable hunter-gatherer economic culture based on fishing. In Japan, the Jomon pottery reflects a stable hunter-gatherer economic culture with an emphasis on both fishing and the collection of plant-based foods. This kind of classification is only relative and there is often overlap in various economic components among different regions. Moreover, with the passage of time, the continuous development and expansion of agricultural economies have led to significant changes in the original economic and cultural patterns. Agriculture was introduced to the Korean Peninsula and the Japanese archipelago through three waves of spreading during the Late Neolithic and Bronze Age (Yan 1993).

In addition to regional difference, we need to explain why such diversity emerges and how it is formed. As we have more archaeological materials, especially those in Western Northeast China, we not only acquire a better understanding of zonal difference but also come to realise that the cultural diversity is beyond what we have thought of. Why agriculture does not take root first in Northeast China? Why Northeast China is slow in picking up agricultural pattern of North China with no significant geographical obstruction? Why Northeast China's fishing-hunting and hunting-gathering lifestyle can last all the way to the historical period? After adopting prehistoric agriculture, what types of agricultural production methods emerged in the Northeast region? What factors influenced the choice of these methods? The fertile black soils in the white mountain and black water only became China's granary with the advent of mechanised agriculture. Why didn't it emerge earlier? The impact of a cold climate on the evolution of cultural mechanisms raises questions that require in-depth exploration, incorporating the mechanisms of cultural evolution, archaeological materials, and multidisciplinary research. Without such comprehensive examination, there is a risk of falling into the deterministic trap of environmental determinism. It is noteworthy that cultural adaptation is the outgrowth of history and is constantly changing. Mechanism of culture evolution cannot replace historical analysis. Backed by the above three aspects, we are in a position to comprehend how cultures evolve in Northeast China.

9.2 Cultural Adaptive Strategies in the Cold Zone

From the perspectives of temperature, precipitation, and terrain, Northeast China can be divided into four transitional zones, as is shown in Fig. 9.1: grassland-forest ecotone between Greater Khingan Mountains and HulunBuir Grassland, Mid-Low Mountain Forest Belt of the Greater Khingan Range, forest steppes and meadow steppes in Songnen Plain, and forested zones in middle and low mountainous area of hills in eastern Northeast China. Historically, nomadic pastoralism is pervasive in HulunBuir Grassland. Ewenki people's way of raising reindeer and Oroqen people's hunting-gathering subsistence survive till modern times in Greater Khingan Mountains, where during Yuan Dynasty this region thrived as the dwelling place for "woodsman" setting it apart from the nomadic tribes residing on the grasslands. Songnen Plain, dotted with lakes and marshes, is the main agricultural region and in it also live people who survive on fishing and hunting—Hezhe people, for example. The southern part of Songnen Plain is different from the northern in that it is higher in height and has low mountains. And forested zones in the eastern Northeast China have a mixed subsistence of both farming and hunting-gathering.

In the period of Han, Wei, and the Southern and Northern Dynasties, historical records state that in Northeast China exist at least three ethnic minorities: Donghu people in the Western region, Yemaek people in the middle, and Sushen people in the southeastern. Even within the same ethnic group, there were noticeable livelihood differences during this period. Some possessed "five cereal" or "grain wheat" while others did not. Some were engaged in nomadic herding of cattle and horses, while others focused on pig farming. Additionally, there were those who specialised in hunting (Zhang 1985). A review on the history of the economy of the nation can give us to understand the diversity in living in Northeast China. Archaeology is concerned with the occurrence of this diversity. One condition shared by all living style in Northeast China is the climate. Northeast China lies between 42° and 53°34′ N and has the coldest weather in China. Because of the continental climate, the cold wave occurs frequently and the temperature can drop to minus 50°C. Winters endure for an extended period, with low temperatures persisting for approximately five to six months. During this time, the soil freezes, the ground is enveloped in a blanket of snow, and rivers freeze over, making agricultural activities impracticable (Ren 2004). Naturally, as the land freezes and becomes covered in snow, and the forest sheds its leaves, the landscape transforms into a vast expanse. The heightened visibility poses challenges for animals attempting to conceal themselves, thereby aiding in hunting, storage, and moving prey (often accomplished by sleds).

Of the four zones, the mountainous area in Greater Khingan Mountains is the highest, then the mountainous and hilly area in the east, and then Songnen Plain. As far as average annual temperature is concerned, it rises from east to west. Because of its high altitude, the mountainous area in Greater Khingan Mountains is colder than HulunBuir Grassland.

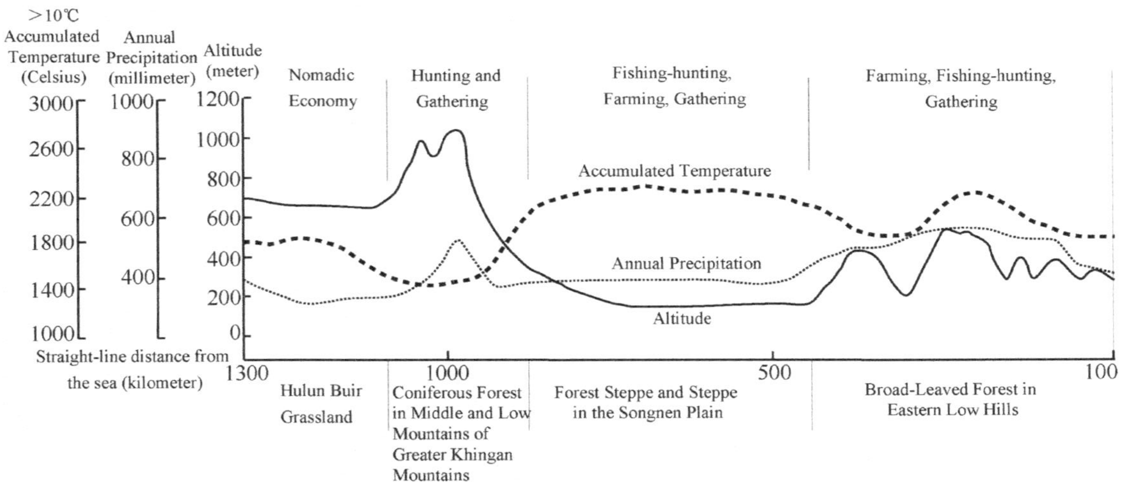

Fig. 9.1 Four cultural and ecological adaptation zones in Northeast China

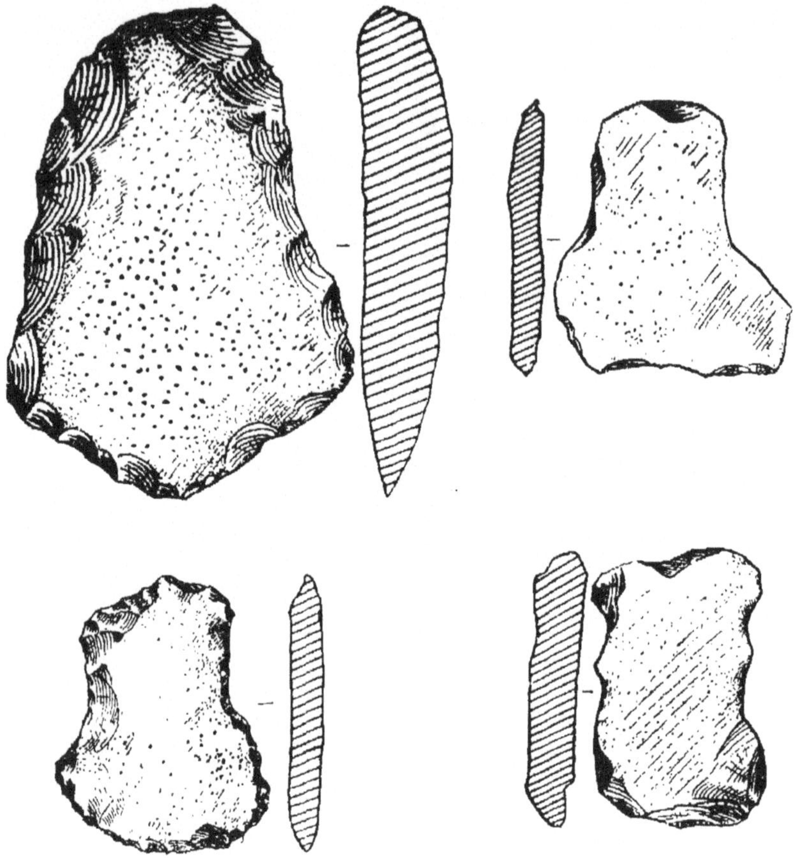

Fig. 9.2 Stone hoe unearthed at the site of Duanliangshan[5]

With an open terrain, stronger winds, higher humidity and more foggy and rainy days, coastal zones in eastern Northeast China are colder in winter than inland regions (Liu 1998). In terms of annual accumulated temperature the Songnen Plain in the central region ranks the highest, followed by the eastern mountainous and hilly areas, especially the small basins in the middle.[2] The HulunBuir Grassland comes next, with the Greater Khingan Mountains having the lowest temperatures in the mountainous regions. Precipitation also exhibits an increasing trend from west to east. The highest precipitation is observed in the eastern mountainous and hilly areas, with slightly lower levels along the coastal regions. Next in line are the Songnen Plain and the HulunBuir Grassland, while the windward slopes of the Greater Khingan Mountains experience slightly greater rainfall.

Seen from the three conditions mentioned above, the most favourable natural conditions for agriculture are found in the Songnen Plain region. Following closely are the eastern mountainous and hilly areas, especially the small basins within. Due to the significantly lower annual accumulated temperature, only half of that in the Songnen Plain, agricultural production is challenging in the Greater Khingan Mountains. The HulunBuir Grassland region, characterised by not only low accumulated temperatures but also arid conditions, is also unfavourable for agriculture.

In the realm of hunting, the most favourable regions are the transitional zones between forests and grasslands. The area where the Western grasslands meet the Greater Khingan Mountains indisputably stands out as the richest hunting ground. Abundant herds of herbivores on the Western grasslands, coupled with deer and medium-sized carnivores at the forest edge, offer a diverse array of hunting opportunities.

Following closely in terms of hunting suitability are the eastern mountainous and hilly areas, where herds of herbivores are somewhat less abundant. The Songnen Plain and the Sanjiang Plain boast numerous lakes, marshes, and winding rivers. During spring, the melting of ice and snow, accompanied by spring floods, transforms these areas into typical wetland regions. Large groups of herbivores find movement challenging in such conditions, making it less suitable for mountainous and forest animals. Nevertheless, these areas teem with fish, birds, and shellfish, particularly migratory fish, making them an extensive source of animal resources. Fish originate from the ocean, and birds migrate from various regions, rendering these areas ideal for fishing and hunting. In unfrozen periods, boats can be utilised for fishing, and during freezing spells, ice can not only be cut for fishing but can facilitate storage. Fishing and hunting emerge as the most favourable livelihoods in the plains and wetland areas.

Animal husbandry, or the raising of domesticated animals, is typically a subsidiary industry in agrarian societies. Animals feed on grain leftovers, and live in or around the settlements. Nomadic pastoralism, on the other hand, relies on grazing livestock for sustenance, with people periodically migrating along with the animals to different pastures. The former requires conditions conducive to agriculture, while the latter necessitates extensive grasslands and suitable animals for herding. Therefore, from a natural conditions perspective, the Songnen Plain and the eastern mountainous and hilly areas are suitable for agriculture and animal husbandry, while the Western grasslands are suitable for nomadic pastoralism.

The various livelihoods mentioned are specific to the current environment in the Northeast region. From the end of the Pleistocene to the middle of the Holocene, there have been phased changes in the environmental conditions of the Northeast region. Nevertheless, comparing different livelihoods under modern environmental conditions can provide valuable insights for assessing prehistoric livelihoods.

9.3 Chronology and Materials

Materials on Northeast China from the Upper Paleolithic to the Early Neolithic are by no means sufficient. Although there are more than 30 Paleolithic sites having been discovered, materials from the Upper Paleolithic period that have undergone systematic excavation and scientific dating are few and far between. From limited materials, we build up a framework. As usual, we will start at the Neolithic and trace back to earlier times. In this way, our research is well-grounded because studies of and materials on the Neolithic are comparatively adequate. From the relatively limited materials available at present, we can establish a preliminary framework of understanding. Similar to the approach mentioned earlier, we start from the Neolithic period, where the dating is clearer, and the materials are more abundant. Step by step, we trace back to the older periods. This progressive method is conducive to grounding our understanding on a more solid foundation of materials and research.

The earliest cultural sequence of the Neolithic period in the Northeast China comprises the Zuojiashan phase I, Xinkailiu culture, and Angangxi culture, dating back to approximately 6000 years ago. This period aligns with the initiation of the Neolithic era in the central and southern regions of Inner Mongolia, roughly corresponding to the Yangshao culture period in the Yellow River Basin and the Zhaobaogou-Hongshan culture period in the northern and southern areas of the Yan Mountains. It signifies a phase of cultural maturity and prosperity during the local Neolithic era and indicates the diffusion of the Neolithic culture. However, at this time, the Neolithic culture in the Northeast region was just beginning to take shape. Considering available materials, I regard the period around

6000 years ago as the Early Neolithic period in the Northeast region, rather than categorising it as the Late Neolithic period (CA, CASS 2010) according to the chronological table of North China. This approach is akin to how one cannot apply the chronology of the Neolithic period in Western Asia to materials from Europe.

Tracing back further, the next chronological marker can be set around 12,000 years ago. The materials identifying this period do not originate from the Northeast China region but rather from the surrounding Far East of Russia and the Japanese archipelago. It is during this time that pottery emerged. Regretfully, they are not found in Northeast China but in the Russian Far East and Japanese archipelago, although breakthroughs are scored in Shuangta site in Baicheng, Jilin (Wang 2012). Another important sign of this phase is that human activity is beginning to enter the lakes and wetlands of Songnen Plain, which means an alteration in people utilisation of natural resources. This crucial transition is closely linked to the climatic changes in the Northeast region from the Pleistocene to the Holocene. Therefore, I designate around 12,000 years ago as the beginning of the transitional phase from the Paleolithic to the Neolithic in the Northeast region. The period spanning from this time until the onset of the Neolithic, around 6000 years later, can be referred to as the "Mesolithic" era in the Northeast China. It bears strong comparability to the Mesolithic period in Europe. Situated at similar latitudes, adjacent to the origins of agriculture but with a later onset of the Neolithic period, the Northeast region also exhibits comparable cultural adaptive features (refer to the next section). The concept of the "Mesolithic" era finds its most applicable region in China in the Northeast.

Studies on the Upper Paleolithic are relatively satisfactory. The microblade technology has been popularised and can be seen from North China and Northeast Asia all the way to Alaska in North America. Current archaeological materials indicate that microblade technology comes into being in Northeast China later than in North China. In fact, the appearance of microblade technology in North China such as Xishi site and Shizitan site can be traced to over 20,000 BP (Wang 2010), whereas in Northeast China 20,000 to 10,000 BP. The excavation of Jinsitai Cave site from 2000 to 2001 discovers three cultural layers, 5 metres deep in total.[3] The lower layer has a date of 30,000 BP. Typical microblade productions are unearthed in the upper layer and produce a date of 10,000 BP. In layer with a date of 20,000 BP, what has been discovered is a flake industry characterised by typical Levallois technology, rather than the microblade technology (Wang 2010). The two excavations of Huma Shibazhan site help us understand their cultural features: like the middle layer of Jinsitai Cave site, the layer of Upper Paleolithic does not have microblade production and the appearance of microblade technology is much later (Zhang 2006). It is noteworthy that there is no evidence that Levallois technology is widespread in Northeast China. It is detected mainly in the area west of Northeast China and occasionally in other sites. Meanwhile, flake tools are common in the eastern part of Northeast China.

Currently, what is most perplexing is the existence of a gravel industry in the Northeast China, and it remains unknown whether its chronology could extend into the Upper Paleolithic period. Examining materials from the Siberian region of Russia, since the Early Paleolithic period, lithic technology has developed along two lines: one involving non-standardised chipping technology and gravel tools, and the other involving discoidal cores and Levallois technology (a standardised method involving prepared cores before chipping) (Derev'anko 1998). The former technique is manifested in some areas as an industry primarily focused on flakes, with a few larger chopping tools; in other regions, it presents itself as a gravel industry utilising large chopping tools and gravel tools resembling handaxes. There are currently several unanswered questions regarding the development of this technological pathway.

First of all, we do not know whether stone tools characterised by flakes and gravel tools are in close relationship with local stone materials—there are no large tools among flake-based ones and if the materials are not good enough, large tools are inevitable. Secondly, we do not know whether large ones and small ones are made by the same group of people. For instance, in the Nianzishan site in Qiqihar, stone materials are collected from the surface and over 300 stone tools, including large tools, refined microcore based on pressure technology, and stone net pendants that resemble those in Neolithic, are unearthed in 25 sites within 10 square kilometres (Gan and Long 1990). Microblade technology users may make larger tools and use them in another place. Another example is the Xiaonanshan site in Raohe County. The mammoth remains in the same layer with stone tools have 14C age from 13,000 ± 60 BP (Yang 1981). To cut the meat off large tools are needed despite developed microblade technology. Moreover, stone tool makers will produce large tools for expedient usage or for ground stone tools. In other words, the lithic assemblages at each location may not represent the predominant lithic technological form at that time. Gravel tools in Northeast China may be made by microblade users as well as by the Neolithic people (Dong 1989). The creators of flaking technology, microblade technology, and stone tools in the Neolithic all have the potential to produce gravel tools. Before confirming the existence of a gravel industry, it is necessary to eliminate the influence of regional material characteristics. A comprehensive understanding of the complete composition of microblade technology assemblages is required, along with knowledge of the Neolithic lithic assemblage and production processes. Surface-collected gravel tools cannot definitively be categorised as belonging to the Paleolithic period. The so-called gravel industry in the Northeast region faces challenges in determining both its chronology and nature.

The relationship between the two branches is another mystery. They exist simultaneously and the distribution of their products is complex. Levallois stone tools are found in the Jinsitai site, Huma Shibazhan site in Northeast China, despite in minority, even in Korean peninsula (Li and C. Quanjia 2008). Technology level and geographic features, which has a bearing on their distribution? The lithic feature in the Early Upper Paleolithic is still vague to us (Table 9.1).

After having a basic understanding of cultural sequence from the Upper Paleolithic to Early Neolithic, we focus on the features of the transition stage between the two ages. From the assemblage of tools, we can ensure that Northeast China Mesolithic is characterised by microblade production including wedge-shaped microcores, microblade, and scrapers. Liang Siyong and Pei Wenzhong are first to take notice of the similarity between microliths in Northeast China and those in Europe and usher in the notion of Mesolithic in Northeast China (Pei 1948). Later Pei Wenzhong realises that microblade technology endures so long that it enters into the historical period and hence denies his previous argument (Pei 1963). On this ground, An Zhimin differentiates the remains in Hailar, see microblade technology without pottery as the feature of Northeast China Mesolithic and believes that they to say the least are made 8000 ~ 9000 BP (An 1978), like those found in Xiachuan site, Lingjing site, Shayuan site, and Hutouliang site. In fact, pottery is found in Hutouliang site and Japan's Jomon period that is very similar to Europe Mesolithic. Therefore, the absence of pottery cannot be counted as a feature of Northeast China Mesolithic. On the contrary, very likely they have come into existence at that time in China because they made at least

Table 9.1 Cultural development sequence, representative sites and features in Northeast China

Time	*Cultural Stage*	*Representative Sites*	*Area*	*Physical Geography*
About 6000 BP	**Early Neolithic:** Zuojiashan site phase I, Xinkailiu culture, Angangxi culture	Zuojiashan site, Yuanbaogou site, Xinkailiu site, Daobeishan site, Xiaolaha site, Huangjiaweizi site, Tengjiagangzi site	Southern of Songnen Plain, Sanjiang Plain, North-western of Songnen Plain	Uniform distribution in lakes and wetlands of the plain
About 12,000—6000 BP	**Mesolithic in Northeast China**	Angangxi site, Dabusu site, Qingshantou site	Songnen Plain	Lakes and wetlands of the plain
About 20,000—12,000 BP	**Upper Paleolithic:** Microblade technology	The upper layer of Jinsitai site, the upper layer of Huma Shibazhan site	More common,the whole Northeast China	Diverse Environment
About 35,000—20,000 BP	**The Early Upper Paleolithic:** Levallois flake industry, small flake industry, gravel industry	The middle layer of Jinsitai site, the lower layer of Huma Shibazhan site, the lower layer of Jinsitai site, Xiaogushan site, Xuetian site, Huadianxianren-dong site, Xiaonanshan site	More common in Western Northeast China, Eastern Northeast China	Mostly in mountainous area

10,000 BP have been founded in neighbouring areas such as the Russian Far East, Japanese archipelago, and North China. Recently pottery made about 10,000 BP are uncovered in Northeast China. Instead, Northeast China during the Mesolithic is likely to have pottery, since pottery has been found in surrounding areas such as the Russian Far East, the Japanese archipelago, and Northern China for tens of thousands of years, pottery dating back 10,000 years has recently been discovered in the Northeast China.

In Songnen Plain lives one the Neolithic culture with the usage of microliths (CA, CASS 2010). It buries in black sub-sandy soil of lakes and wetlands which forms in the Middle Holocene when they dot the ground and plants are all prospering with a warm and humid weather. Taking the microblade combination of Ang'angxi site for example, we can identify that stone materials are chalcedony and jade, and that artefacts include stone arrowheads, projectile point, scrapers, microblades, and blade with alteration on both sides for insetting and adze-shaped chipped stone tools and polished stone adzes. Stone arrowheads have various shapes: flat-bottomed heads, round-bottomed ones, concave-bottomed ones, and tanged point ones. It can be seen that pressure technology is well developed and lithic tend to fix, which are typical of the Neolithic culture (Heilongjiang Provincial Museum 1974). From the end of the Pleistocene to the Middle Holocene, microblade technology is characterised by wedge-shaped microcores. In the Middle Holocene, on the ground of wedge-shaped microcores, cylinder microcores, and conical microcores with pencil-shape are developed. Microblades are unearthed in different cultural layers of Huihe Dam and Hake-Tuanjie site in HulunBuir Grassland, of which the most ancient one could be traced to Liao Dynasty (CA, CASS 2010). It is obvious that microblade technology cannot be counted as the feature of Northeast China's Mesolithic.

In conclusion, Northeast China's Mesolithic is defined not only by certain combination of tools, but also by the change in living. The end of Mesolithic is marked by the appearance of agriculture. It is doubtless that farming is not practised in every region of Northeast China simultaneously. Mesolithic lifeway dose not extinct till modern times—Oroqen people the hunters and Hezhe the fisher-hunters are examples. But we cannot say that China's Mesolithic lasts till modern times. Agriculture will influence neighbouring groups and brings about a mutually beneficial relationship. Oroqen people, though they lead a hunting life, have from farming groups obtained many critical goods and materials from weapons to clothes and by virtue of them grow into professional hunters. With this interrelationship in view, we see the appearance of agriculture as a sign for the end of Northeast China's Mesolithic. On the other hand, agriculture takes a long to develop, and even with it hunting, fishing, and gathering are the main means to earn a living. Different areas have distinct features of their own.

9.4 Diversified Cultural Adaptations in Prehistoric Northeast China

On the ground of previous section, we will settle down to the significance of cultural adaptation of archaeological materials. Based on existing studies and materials, the set of tools is the best way to identify living style. According to the form of tools and ethnological analogy, we can ascertain the overall usage of a tool. Tools for fishing and hunting are distinctly different from those for cultivation. The range, diversity of the remains, and setting of the residential area—the location and structure of the sites are helpful as well. Of course, we want the remains of domesticated plants and animals and detailed analysis of trace elements in the bones of prehistoric people which serve as direct materials for the progress of agriculture, but related researches in Northeast China is far from sufficient. We can only deduce their lifestyle indirectly.

The Xiaogushan site in Haicheng belongs to the Early Upper Paleolithic. Human bones, tens of thousands of lithics most of which are of vein quartz, antler darts with double-row barbs, perforated bone needles, and decorations such as perforated animal teeth and clamshells with inscribed pattern are unearthed. The thermoluminescence dating of the site is 40000 ± 3500 (Liaoning Provincial Institute of Cultural Relics and Archaeology 2009). Experimental archaeology shows that the antler darts can capture fishes as big as 65 cm and can kill a fish effectively even in deep water (Lv 1995). The discovery of Xiaogushan site indicates that in the Upper Paleolithic period of the southern part of the Northeast region, people had already started utilising aquatic resources such as fish and shellfish. The mode represented by Xiaogushan site and the other three represented respectively by Shuidonggou site, Shiyu site, Shandingdong site–Dongfang Square site- Xiaonanhai site are the four adaptive modes in northern China during the Upper Paleolithic (Chen 2006).

In the LGM, Northeast China's climate zone moves southward about 6° in latitude. The location of permafrost and the snow line has shifted from around North Latitude 51° to approximately North Latitude 42° over consecutive years. There has also been a change in the meridional direction, with the dry line moving eastward (Xiao and Hu 1988), roughly by about 7°. Analysis on sporopollen of Maar Lake in Northeast China's Lake Erlongwan signifies that in 29,300–12,600 BP, the forests in this region mainly are cold temperate coniferous forests and birch forests, and the climate turns more cold and dry, especially from

20,600 to 18,700 BP when it happens to be the Last Glacial Maximum. It is not until 12,600 BP when the climate turns gradually from cold and dry to warm and humid and when coniferous and broad-leaved mixed forest becomes dominant that the impact of LGM blows over (Liu et al. 2008). During the LGM, vegetation of Northeast China is steppe-tundra which has no equivalent environment today (Winkler and Wang 1993, p.221–261). Mammathas Coelodonta fauna lives there. Because of a short growing season and low surface productivity, the available plant resources in this region are extremely limited. Hunter-gatherers have to primarily rely on hunting for survival. Yet limited plant resources mean that compared with temperate grassland and savanna there are fewer animals and large herbivores must feed themselves properly in a considerable range. Hunter-gatherers must improve their mobility to catch enough preys. Microblade technology in the Upper Paleolithic comes into being for this purpose. Archaeological materials indicate that the technique first appears in the southern periphery of North China and then spread to Northeast China. Nonetheless, we cannot be sure whether or not human beings have taken up residence in Northeast China and Siberia in the Last Glacial Maximum (Goebel 1999). Mostly likely they have not. When the climate is most unfavourable, they move southward and do not return until it looks up. Northeast China's lagging behind in the microblade technology is possibly due to human's migration spurred by bad weather.

The LGM ends in about 12,600 BP. The climate turns gradually from cold and dry to warm and humid; palaeomonsoon brings higher precipitation and the maximum precipitation (effective precipitation) occurs (An 2000). With more water, aquatic resources bring new hope. It is beyond possibility to make use of them in steppe-tundra of the Last Glacial Maximum. Northeast China's maximum precipitation (effective precipitation) comes earlier than North China and the middle and lower reaches of Yangtze River, which lies the foundation for the utilisation of aquatic resources such as fishing-hunting, shell collecting, and bird catching. The evidence for exploiting water resources in the Early Upper Paleolithic has been found in Xiaogushan site. The access to those resources marks the beginning of Northeast China's Mesolithic.

This utilisation means that Mesolithic sites of Northeast China's will reach to Songnen Plain and coastal areas. Coastal zone in the outer area of Ussuri region is too cold for utilising marine resources. The warmer inland river is conducive to aquatic resources utilisation and so it is. Typical Mesolithic sites with microblades in majority—Angangxi site (Daxingtun site), Dabusu site (Dong 1989), and qingshantou site (Li 1984)—are all located in the hillocks of lake and wetlands in Songnen Plain. No earlier Paleolithic remains are found in these areas. Hunter-gatherers in the Mesolithic expand their living space to these areas where they can make use of water resources effectively and they hunt if necessary. Unfortunately, materials on means of living are extremely scarce. Cultural remains found in the Daxingtun site, Ang'angxi County which is about ca. 11,000 BP, includes ashes, burned bones, and various artefacts (Huang 1984). Artefacts made mainly of chalcedony, agate, and flint. Flakes, most of which are blades, are small. The existence of microcores signifies microblade technology. Sporopollen analysis of cultural layers reveals a sparse forest grassland full of Artemisia and chenopodium album plants which, instead of being warm and humid with lakes and wetlands here and there and prospering plants as it is in the Middle Holocene, is cold and dry. The layer manifests that people live along the lake shore or riverbank. Burned bones and ashes involve broken animal bones, none of which belongs to a carnivore. Mammal bones consist of large herbivores such as wild ox horse, as well as small animals such as hare, pika, hamster, and vole.

We still do not know whether people of the Ang'angxi site has an access to aquatic resources or not, but we know that they are able to do so. Their prey includes agile small animals such as hare and pika, which is the result of the shortage of resources. With the end of LGM, large herbivores like mammoth, woolly rhinoceros, and *Bos primigenius* either die out, or live in a limited area. Two choices are offered by Northeast China's Mesolithic: food production and water resource. The extinction of large herbivore in Northeast China coincides with its maximum precipitation. Northeast China naturally chooses the second choice. It is of epoch-making significance for people to enter into Songnen Plain for aquatic resource.

The beginning of the Northeast farming era is often marked by the appearance of specific agricultural tools, with the use of farming tools being a significant indicators. Existing archaeological materials provide valuable evidence, and the types of agricultural tools, including chopping tools, digging tools, soil cultivator, intermediate cultivation tools, reaping tools, grinding tools, among others, serve as useful indicators. Yet many of them may be multifunctional. For instance, chopping axe can be used to shape the wood as well as for building a house; Lei—a

Chinese plough—can be used for gathering as well; besides domesticated crops, undomesticated ones and nuts can be put under the stone grinding slabs and handstones. We need to find a tool solely for agriculture as the indicator for farming. The "stone hoe" discovered in the Northeast China is a suitable indicator. Despite its name "hoe", it is not a digging tool but rather a soil cultivator. Observations of usewear experimental research, artefact design analysis, and ethnographic comparisons all indicate that this tool is more likely used for soil cultivation rather than digging (Chen et al. 2014). Furthermore, the existence of a group of tools is a strong proof for agriculture, even though they may be multifunctional. The discovery of tools for grinding, reaping, digging, and cutting speaks louder than the discovery of certain tool.

The Xiduanliang Mountain site of county Dongfeng, Jilin province is representative in the variety of farming tools. Its phase II (ca.5000 BC) contains stone axes, stone spades, stone grinding slabs, and handstones, but seldom contains microliths, bone artefacts, stone or pottery net pendants, which is rather distinctive (Jilin Provincial Institute of Cultural Relics and Archaeology 1991). Stone axes contain chipped ones and polished ones. The chipped stone axes have thin groove, making it more likely to be the stone hoe mentioned earlier. The stone spades are all chipped and may belong to the hoe as well (see Fig. 9.2). Phase I does not have those spades, but does have chipped and polished stone axes, stone grinding slabs, handstones, and stone knives. The chipped axes are different from the polished ones in design and resemble to the spade of phase II. The shape of its stone knife is similar to those found at the Baiyinchanghan site of the Xinglongwa culture (Inner Mongolia Autonomous Region Institute of Cultural Relics and Archaeology 2004), appearing elongated.[4] It differs from perforated or concave-sided stone knives used for harvesting. It is likely a tool for clearing weeds and shrubs during land cultivation (Li 1954). With respect to farming tools, stone tools of the Xiduanliang Mountain site Phrase II are not so developed as the Xinglongwa culture. It can be concluded that agriculture there is still at an early stage.

Yinggeling site of Ningan county, Heilongjiang Province is another site where stone hoes are found comparatively prodigiously. In the lower layer three kinds of chipped hoes are found: long-handled, short-handled, and grooved-shaped. There was also an antler hoe featuring a Fang Qiong (the hole on axe for installing a handle). Each side of the Qiong is pierced with a hole for reinforcement, indicating a refined antler digging tool (Heilongjiang Provincial Institute of Cultural Relics and Archaeology 1981) (Fig. 9.3). In the Beishachang site in Yabuli town of Shangzhi County of Heilongjiang province, grooved-shaped stone hoes, polished spades, stone grinding slabs, and handstones are discovered. The lower layer of Yinggeling site and Beishachang site (Heilongjiang Provincial Institute of Cultural Relics and Archaeology 1988) can be traced back to circa 5000 BP, so can Jingudshan site in Longjing, Jilin province (Yanbian Museum 1991), and Longxingcheng site (Jilin Provincial Institute of Cultural Relics and Archaeology et al. 1998). Hoe-shaped tools with a handle are discovered in the Jingudshan site and large grooved-shaped stone hoes in the Longxingcheng site. Therefore, it can be concluded that primitive farming occurs in Northeast China in about 5000 BP.

The Xiduanliang Mountain site locates in the hilly areas of Songliao watershed. The pattern represented by Yinggeling site and Beishachang site can be found around hilly and mountainous areas in eastern Heilongjiang. Jingu-Xingcheng culture locates in the transitional zone of a hilly area and valley of Changbai Mountain. They all share in the lack of aquatic resource.

The Zuojiashan site in Nong'an County in Songnen Plain is earlier, with its archaeological remains divided into three phases. Phase I contains only 5 lithic products—stone axe, stone grinding slab and handstone, blade and jade—and 14 bone antler tools—needle, awl, spade, arrowhead, drill, antler spear, chisel, harpin, tube, and shuttle-shaped tool; no typical fishing and hunting tool is found; a big volume of pottery is excavated; one house in remnant and one ash pit are located. Fishing darts are recognised in phase III (Archaeology Teaching and Research Office, Jilin University 1989). Animal remains in phase I indicate domesticated dogs and pigs. Fish and mollusk remains are perceived as well. Judging by the remains, deer, domesticated pig and wild boar are the main meat provider (the proportion of domesticated pig and wild boar stay the same during the three period) (Chen 1993). Due to pig breeding at Zuojiashan, a significant number of pig bones were also discovered at the same cultural site of Yuanbaogou site (Jilin Provincial Institute of Cultural Relics and Archaeology 1989). In broad sense, agriculture has come into existence. During Zuojiashan's existence, people may plant grains, though in limited scale, because the stone grinding slab and handstone has been invented and fishing and hunting is undeveloped. Lifeways are rather complex, including hunting, gathering, fishing, farming, and

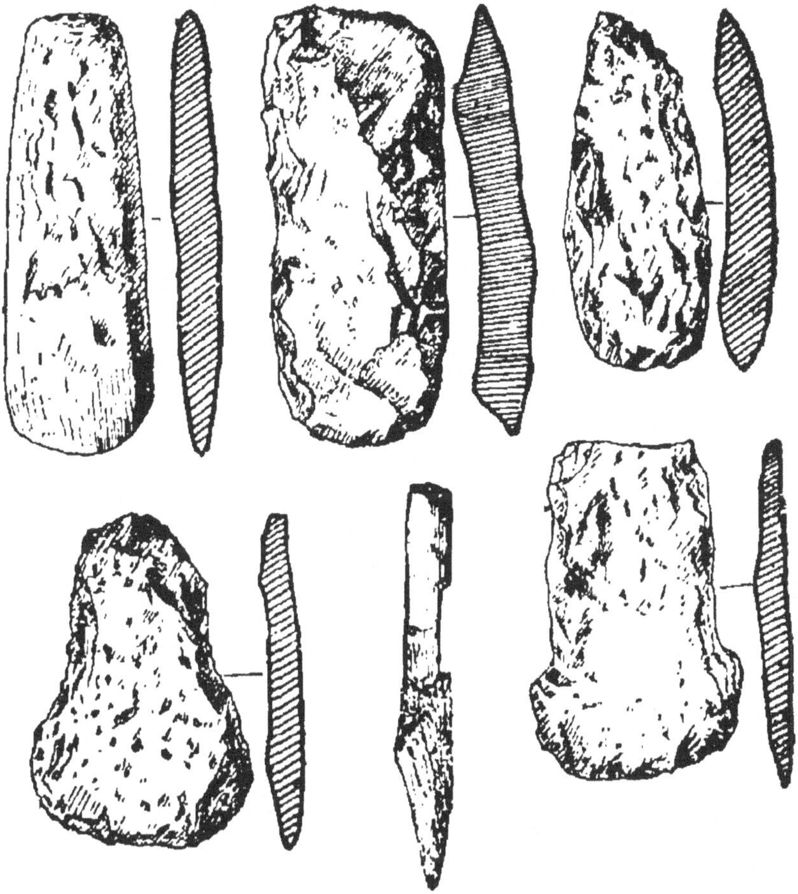

Fig. 9.3 Stone hoe unearthed at the site of Yinggeling[6]

husbandry and therefore farming is not as important as that during Xiduanliang Mountain existence. Grooved-shaped spades (Jin 2009) of the middle and late Neolithic are reported in the Yangcaogou site in Yitong County, Xingshan site, Yaohongzuizi site, and Xiaojiatun site, which bear witness to the development of agriculture.

An'gangxi culture at the heart of Songnen Plain focuses more on fishing and hunting with its sites along the floodplain of Nenjiang which rises 10 to 15 metres above the ground. High floodplain is safer when flood comes near the lakes and wetlands where fishing and hunting are convenient (Heilongjiang Provincial Museum 1960). Microblade technology is well developed. Polished and chipped lithic implements are in the minority. Chipped tools include net pendant and adze-shaped tools; polished adze and chisel, but no tools for digging, cultivating, grinding, and farming. Bone tools which are common include spear head, arrowhead, and dart. Animal remains include frog, fish, pig (we do not know if it is domesticated), bird, deer, rabbit, and dog (Heilongjiang Provincial Museum 1974, 1961; Liang 1959). Archaeological investigations conducted in the 1950s along Nenjiang show that animal bones like fish and bird are very prevalent (Heilongjiang Provincial Museum 1960). The Huangjiaweizi site in Zhenlai County shares in this feature. A layer of fish bone of Neolithic, 10 cm in depth, is discovered in the ash pit. The set of tools, animal remains and the location of ruins are all similar to Ang'angxi's fishing-hunting subsistence. Therefore, Zhao Binfu calls it as "fishing-hunting culture in Neolithic"(Zhao 2007). Angangxi culture goes along Nenjiang. Remains with kindred features can be traced in midstream area of the Nenjiang and Songhua Rivers in Heilongjiang, the Yaojingzi site in Changling County (Jilin Provincial Institute of Cultural Relics and Archaeology et al. 1992) near Zuojiashan site, Chuanzijing site of Qianan county, Daqingzui site, Erqingzui site of Dehui County (Jin 2009). They are all located at a higher place and near lakes and wetlands. Animal remains such as fish, bird, and shell take a great volume in their remains. These sites reflect different levels of dependence on fishing and hunting. The further a place is from the heart of Songnen Plain, the more unimportant the fishing and hunting are.

The Xinkailiu site in Mishan locates in Sanjiang Plain. 32 graves and 10 fish cellars are found in the lakes and wetlands area between Xinakai Lake and Xiaoxingkai Lake. There are two cultural phases in total. The researchers believe that these two phases may be the two stages of the same culture since their dates are not far apart. Digging, cultivating, grain grinding tools are completely vacant. Grinding tools there are too small to smash grains. Nevertheless, fishing and hunting tools are prodigiously found, including fishing dart, fork, hook, and projectile point. Ethnographic materials suggest that the fish cellar is for stocking fresh fish. Patterns on the pottery are shaped like fish scale, lozenge, and wave, which vividly reflect the theme of life (Heilongjiang Provincial Institute of Cultural Relics and Archaeology 1979). According to archaeological materials of the Xinkailiu site, its dependence on fishing and hunting is much higher than Songnen Plain.

Hierarchy can be discerned in the Xinkailiu site. "Grave M3, M7, M20, and M31 follow the secondary burial. Burial objects are rare. Some graves have only one or two pottery wares or a few artefacts, and some do not even have any at all. Yet graves M6, M3 and M7 have many" (Heilongjiang Provincial Institute of Cultural Relics and Archaeology 1979). Tombs of jade are discovered in the Xiaonansha site in Raohe County that exists about the same period with Xinkailiu site. The grave is large in size and rich in burial objects. Judging by the number of jade artefacts, the tomb owner enjoys a prominent position (Li and Gao 1996) Hierarchy can also be observed in Wokenhada Cave tomb in Yilan County (Li 1954) and Bashan tomb in Baicheng city (Jilin Provincial Institute of Cultural Relics and Archaeology 1988). It is worth noting that these burial sites all belong to prehistoric communities that relied on fishing and hunting. This, along with ethnological support, proves that a complex society may occur even in a fishing-hunting-gathering group. Social differentiation is by no means unique to agricultural communities and may happen in a hunting-gathering community (Yesner 1980). To our surprise, Xiduanliang Mountain in the same period which depends on farming is less stratified.

From the Bronze Age all the way to the Iron Age, people in Songnen Plain and Sanjiang Plain live by fishing and hunting. The same is true of the Xiaolaha site in Zhaoyuan County (Heilongjiang Provincial Institute of Cultural Relics and Archaeology, Department of Archaeology, Jilin University 1997, 1998), which is similar to the Angangxi culture and is the late stage of early Neolithic site. When it comes to hilly and mountainous areas where fishing and hunting is hard to carry on, farming takes root. The lower layer of Yinggeling Mountain is one example.

In Western Northeast China, the most ancient cultural layer of Huihe Dam and Hake-Tuanjie site in HulunBuir Grassland dates back to 7000 BP. Extremely refined microcores, microblades, and arrowheads are reported. In Phase I of Huihe Dam, over 7000 items, mostly microblade artefacts, 602 pottery sherds, 3656 animal bones, 1 residential site, 2 bonfire sites, 1 ash pit of animal bones are discovered. In the pit 15 animals are distinguished, including herbivore (donkey, horse, cattle, Mongolian gazelle, antelope, sheep, and rabbit), middle and small carnivore, bird, fish, and rodent which are common local animals (CA,CASS 2010). The absence of deer indicates that people focus on utilising animals on grassland and nearby wetland. Phase I of Hake-Tuanjie site is similar to that of Huihe Dam, except for the broken hadnstones in a few large artefacts and stone pottery pad. The pad signifies that the pottery are local products. The residential site of Huihe Dam is like a local basement(cellar)—Diyinzi—which is put in use in winter. This, together with the pottery and remains, demonstrates that people here take up a long residence of about one year or longer. In other words, they settle here. They live on broad spectrum hunting and perhaps limited grains (Yanbian 1991). They certainly fish, but fishing tools and fish remains are not as many as those sites in Songnen Plain and Sanjiang Plain.

In conclusion, prehistoric people in Northeast China survive on a variety of means such as hunting, gathering, fishing, farming and husbandry. Hunting and gathering enjoy the longest preference and hunting is the most suitable in Northeast China. In the Upper Paleolithic, fishing-hunting subsistence occurs and in Mesolithic prevails among the people who enter into lakes and wetlands. In about 6000 BP, the Neolithic begins. Animal raising and agriculture in a limited scale come into being but from mountainous and hilly areas to lakes and wetlands, food production witnesses an incessant fall in significance. In about 5000 BP, archaeologically comparatively convincing evidence for agricultural production is found in mountainous and hilly areas in eastern Northeast China. The establishment of agriculture waits until Han Dynasty (Heilongjiang Provincial Institute of Cultural Relics and Archaeology, Department of Archaeology, Jilin University 2001). Fishing and hunting remain the main means of survival from the Bronze Age all the way to the Iron Age. In the grassland in Western Northeast China, a broad spectrum of subsistence is widespread with hunting growing increasingly important. Hunting phases out when nomadic pastoralism appears. In a word, while taking advantage of nearby resources, prehistoric people in Northeast China find the best means of survival by making a full use of local resources, the consequence of which is a very rich and diverse cultural adaptation. From another point of view, instead of choosing agriculture as the sole support for life as North China and middle and lower reaches of Yangtze River do, Northeast China adopts a clear unity in the prehistoric cultural adaptation in the Northeast China, that is, a mixed way of livelihood and different areas have a bias. Why is it so? Why not adopt food production? How could it exist under the impact of agriculture? Next section will solve these questions.

9.5 Spreading of Agriculture and Changes of the Forager Society

Northeast China's environment is favourable to farming. Its temperate and monsoonal climate ensures that farming period takes 190 to 220 days in a year (daily average temperature over 5°C). Although the accumulated temperature is low, in summer both daylight time and sunshine time are long and the effective temperature is high. It can last for 4 to 5 months, which meets the growing conditions of general crops. In winter, coldness slows down the decomposition of organism and therefore fertile black soil and chernozem covers the ground. Songnen Plain is full of thick and fertile black soil. Permafrost holds back the water so that the ground is always humid. Because surface water is prevented by the permafrost, lateral erosion is significant and vast river valley and flat and shallow terrace are formed, which are favourable to massive farming. Snow melts in spring, moisturising the earth. Concentrated summer rains together with heat and productive soil create an ideal cradle for agriculture (Ren 2004).

Northeast China do need the support of agriculture. Because growing season is short, more food is needed for a long winter, which can be provided by crops. Furthermore, grains are easier to stock. In fact, even hunter-gatherers in this region need to make a preparation for winter. Coldness is conducive to the preservation of food. Compared with tropics, temperate zones have fewer kinds of animals, but each kind is in large numbers. It is possible that people run a herd of deer or cattle into a lake or down the cliff and gain much more meat than they need. The excess meat has to be handled. According to Hayden's argument, surplus production brings about social complexity in hunting-gathering groups (Hayden 1992; 1998). That is to say, hunting-gathering groups with high social complexity have a larger demand for surplus production. With this in view, Northeast China is equipped with the social mechanism for agriculture. But archaeological materials show that agriculture originates earlier in North China than in Northeast China. More than this, it develops at a slow speed and hunting-gathering and fishing-hunting are more popular. This is against natural conditions and cultural mechanism. Why?

Natural environment of Northeast China is in favour of agriculture under the condition that there are enough labour forces. To cultivate forest steppes and meadows, much effort is needed. Drainage facilities are required to evacuate snow water and rainwater, which also needs much labour. Moreover, because of the short growing season, crops must have a short growth cycle, and sowing and harvesting must be taken effectively. The former requires agricultural knowledge and the latter demands a high efficiency. However, population growth in cold regions is slow in that inadequate caloric intakes put a restriction on pregnancy rate and raising a baby costs more mentally and physically than in tropical regions. Cold weather combined with a seasonal lack of food pulls down the survival rate of babies. Hence the population density of Northeast China is rather low in prehistoric time (Goebel 1999). During the LGM, population density here falls to the lowest and some places become uninhabited. The density does not turn up until dark clouds roll away, but even though it turns up, it is still lower than that of North China. Population density is the engine as well as precondition for agriculture.

Without enough labour forces, the land could not be cultivated, especially where the soil is solid and humid and the plants thrive as in Northeast China. North China turns into a plain where plants flourish and lakes and wetlands star the ground in the Neolithic after agriculture develops a certain amount of time. Valleys in southern and northern areas of Yanshan Mountains are not cultivated until the middle and late Neolithic. In the Early Neolithic, people there farm on the loess slopes. Except for the constriction of labour forces and plants, Northeast China's climate in spring is unstable with strong wind. Arid wind, sandstorm, and frost tend to occur. The risk of farming particularly high in Western Songnen Plain where the snow is thin and ground is sandy. According to ethngraphic records, Oroqen people do engage in food production, but due to restricted labour force, the result is unsatisfactory. All in all, Northeast China is indeed suitable for agriculture, but for mature one instead of the one that just comes into being.

Hunting is more appealing to low population density. On the one hand, it consumes fewer resources; on the other, after the end of LGM, people return to find more available hunting resources. The forest steppe transitional zone of Northeast China is the most advantageous location for hunting. Therefore, unlike people in North China and the middle and lower reaches of Yangtze River who find hunting resources unable to support them between the end of Pleistocene and the beginning of Holocene, people in Northeast China live comfortably. This also contributes to the lagging behind of Northeast China in farming.

However, this does not mean that cultural adaption does not take place in Northeast China during the Pleistocene-Holocene transition. It is implied by the occurrence of pottery and people's migration to Songnen Plain. In about 12,000 BP, people turn to aquatic resources. This is significant. The occurrence of pottery and the utilisation of water resources are like two friends—each is supporting as well as supported by the other (Rice 1999). They coordinate with social complexity. It has been explained that there are many migratory fishes from the sea. Because they can capture a great volume of food during their migration, people settle down there, which prompts a complex society. Aquatic resources to some degree can replace farming—it can support more people and provide a relatively immobile life. This is achieved in much smaller area than farming, which reduces the conflict over territory. Aquatic resources as distinct from agriculture do not demand large labour forces, intensive labour or suitable plants, nor is it as risky as farming. Like hunting and gathering, it depends on natural resources and like agriculture, it can bring an immobile life and increase social complexity by enlarging population. No wonder why the utilisation of aquatic resources is more popular in Northeast China.

In North China's Mesolithic, basic agricultural production occurs in Northeast China, yet fishing-hunting is still the order of the day. This is not only because it can provide protein, fat and leather (Hezhe people make clothes out of fish skin), but also because it leaves enough time for farming. In winter when it is too cold for plants to grow, fishing-hunting shines. Coldness makes food preservation easier. In the meantime, hunting becomes painless in that snow greatly slows preys down, and sledges become available; the frozen lake and wetland further improve hunting efficiency. People could hunt and fish in the long winter and farm in spring, summer and autumn. Therefore, fishing and hunting have always been the first choice. This is decided by the climate. Even in spring before farming begins, people could survive on gathering, not to mention the time left in autumn after the harvest. Restricted farming time makes fishing, hunting, and gathering possible. This mixed subsistence characterises Northeast China's Neolithic. Different place decides the nature of livelihood: in mountainous areas, people are all hunters; in lakes and wetlands of Songnen Plain, fisher-hunters; in southern Songnen Plain where fishing and hunting are less satisfactory, farmers.

When farming began in Northeast China, husbandry originated. Domesticated horses were introduced in the Bronze Age. Husbandry can provide not only meat but also animal power. With husbandry on the side, both means of survival and cultural adaption in Northeast China are enriched. The horse immensely increases the hunting space and efficiency, which in the meanwhile weaken the reliance on farming. With horse, metal weapon, hunting technique, and complex society, Northeast China's cultural adaption is competitive. People there become brilliant worriers, which explains why their numerous invasions on the central plains of China. Complex society makes plunder profitable. In the competition with Northeast China, agricultural society by no means holds the upper hand.

Northeast China resists agricultural society of North China with mixed and intricate means of survival. It can draw on different resources in different seasons and regions. Social institution and value system (faith, morality, and prestige evaluation system) based on it are divergent to that in North China. Animism is more important to them and it is believed that the Shaman can interact with the sky and earth. As social complexity goes up, people in the upper level are unwilling to exchange their value system, which is harmful to agricultural development as well.

9.6 Conclusion

This chapter first analysis the cultural and natural conditions in Northeast China—which can be divided into four zones, and discusses the cultural adaptation of farming, gathering, and hunting in this land of cold. Then on the ground of concrete archaeological materials, it establishes the idea of "the Mesolithic of Northeast China", a cultural development stage which can be compared with Europe's Mesolithic and Japan's Jomon period. It makes clear the feature of cultural adaption of Northeast China's Mesolithic: various means of survival such as farming, hunting, fishing, and gathering are involved with different focus on different means in different areas. On this ground Northeast China's complex society is organised and it tremendously holds back the spread of agriculture from North China.

Notes

1. I have conducted a survey of stone tools on the Dalin River Terrace in Kazi Township, Liaoning Province, and found a large number of chipped stone tools. These terraces often coincide with Hongshan culture sites, suggesting that these widely distributed chipped stone tools are more likely products of the Neolithic rather than the Paleolithic.
2. The excavation materials do not mention the discovery of large stone tools, only a few mentions of chipped large stone tools. Although grinding tools are present, their quantity is much smaller compared to the hilly areas in the eastern and southern parts.
3. Dating of the Xishi site is based on discussions with Professor Wang Youping; reference for the Shizitan site is Shi Jinming and Song Yanhua, Excavation Report on the Ninth Location of the Shizitan Site in Shanxi, Archaeology, 2010, 10.
4. Recently, we have studied the stone knives unearthed from the Baiyinchanghan site, including sickles and elongated stone knives. Experimental archaeology has shown that elongated stone knives are not as effective as sickles in cutting grass, but they are more efficient when it comes to felling plant stems like reeds.
5. According to the original report, Figures 1 and 3 are stone shovels, 2 is a stone axe, and 4 is a stone tool.
6. According to the original report, Figures 1, 2 and 3 are stone axes, 4 and 6 are stone hoes, and 6 is an antler hoe.

References

An Zhimin, Middle stone age remains in Hailar: on the origin and tradition of microliths, *Acta Archaeologica Sinica*, 1978, 3.

An Zhisheng, The history and variability of the East Asian paleomonsoon climate. *Quaternary Science Reviews*, 19: 171–187, 2000.

Archaeology Teaching and Research Office, Jilin university, Neolithic Sites at Zuojiashan, Nongan, Acta Archaeologica Sinica, 1989, 2.

CA, CASS, *Chinese archaeology: Neolithic archaeology volume*. China Social Sciences Press, 2010.

CA, CASS et al. *Hake site: archaeological excavation report 2003–2008*. Cultural Relics Press, 2010.

Chen Quanjia, *Animal bone identification and traces study at the Zuojiashan site, Nongan, in collected works on agriculture and fruit*. Zhishi Publishing House, 1993.

Chen Shengqian et al. Functional study of stone hoes unearthed from the Dashanqian site, Kalaqin, Inner Mongoli, Acta Anthropologica Sinica, forthcoming. stone hoes in the Northeast region are mostly in a grooved shape, identical to those unearthed from the Dashanqian site. however, to draw definitive conclusions, further analysis of the use of traces of stone hoes from various sites in the Northeast region is needed. apart from this, the research in the other three aspects is consistent, 2014.

Chen Shengqian, Adaptation, transition, and radiation of late Pleistocene humans in Northern China. *Quaternary Sciences*, 2006, 26(04).

A. P. Derev'anko eds., *The Paleolithic of Siberia: New Discoveries and Interpretations*. University of Illinois Press, Urbana, 1998.

Dong Zhu'an, The Microliths of Dabusu, Acta Anthropologica Sinica, 1989, 8(01).

Gan Zhigeng, Li Long, Stone Tools Discovered in Nianzishan Area, Qiqihar City, Northern Cultural Relics, 1990, 3.

B. Hayden, Nimrods, piscators, pluckers, and planters: the emergence of food production. *Journal of Anthropological Archaeology*, 9: 31–69, 1992.

B. Hayden, Practical and prestige technologies: the evolution of material systems. *Journal of Archaeological Method and Theory*, 5: 1–53, 1998.

T. Goebel, Pleistocene human colonisation of Siberia and peopling of the Americas: An ecological approach. *Evolutionary Anthropology*, 8: 208–227, 1999.

Heilongjiang Provincial Institute of Cultural Relics and Archaeology, Yinggeling site in Ning'an County, Heilongjiang, Archaeology, 1981, 6.

Heilongjiang Provincial Institute of Cultural Relics and Archaeology, Brief report on the clearance of Yabuli Neolithic site in Shangzhi county, Heilongjiang, Northern Cultural Relics, 1988, 1.

Heilongjiang Provincial Institute of Cultural Relics and Archaeology, Xinkailiu site in Mishan County, Acta Archaeologica Sinica, 1979, 4.

Heilongjiang Provincial Institute of Cultural Relics and Archaeology, Department of archaeology, Jilin University, Excavation brief of Xiaolaha site in Zhaoyuan county, Heilongjiang, Northern cultural relics, 1997, 1

Heilongjiang Provincial Institute of Cultural Relics and Archaeology, Department of archaeology, Jilin University , Excavation Report of Xiaolaha Site in Zhaoyuan County, Heilongjiang, Acta Archaeologica Sinica, 1998, 1.

Heilongjiang Provincial Institute of Cultural Relics and Archaeology, Department of Archaeology, Jilin University, River mouth, and revitalization: excavation report of Lianhua reservoir in Mudanjiang (part 1), Science Press, 2001.

Heilongjiang Provincial Museum, Archaeological survey brief of the left bank of Nen River downstream, Archaeology, 1960, 4.
Heilongjiang Provincial Museum, Investigation of Angangxi Neolithic site, Archaeology, 1974, 2.
Heilongjiang Provincial Museum, Survey of Microlithic culture sites along the Nen River, Archaeology, 1961, 10.
Huang Weiwen et al. The Paleolithic of Angangxi, Heilongjiang, Acta Anthropologica Sinica, 1984, 3(03).
Inner Mongolia Autonomous Region Institute of Cultural Relics and Archaeology, Baiyinchanghan: Neolithic site excavation report, Science Press, 2004.
Institute of Archaeology, Chinese Academy of Social Sciences, et al. Hake site: archaeological excavation report 2003–2008, Cultural Relics Press, 2010.
Jilin Provincial Institute of Cultural Relics and Archaeology, et al. Excavation brief of Xingcheng site in Helong, Jilin Province, Northern Cultural Relics Journal, 1998, 1.
Jilin Provincial Institute of Cultural Relics and Archaeology, et al. Yaojingzi Neolithic site in Changling county, Jilin, Archaeology, 1992, 8.
Jilin Provincial Institute of Cultural Relics and Archaeology, Excavation of Neolithic sites in Xiduanliangshan, Dongfeng county, Jilin, Archaeology, 1991, 4.
Jilin Provincial Institute of Cultural Relics and Archaeology, Excavation of Yuanbaogou Neolithic site in Nongan County, Jilin, Archaeology, 1989, 2.
Jilin Provincial Institute of Cultural Relics and Archaeology, Excavation brief of Bashan Tomb in Baicheng, Jilin, Archaeology, 1988, 12.
Jin Xudong, et al. 60 years of archaeology in China: Jilin province, in Zhongguo Kaogu 60 Nian (60 years of Chinese archaeology, 1949–2009), Cultural Relics Press, 2009.
Liang Siyong, Angangxi prehistoric site, in Liang Siyong's collection of archaeological papers, Science Press, 1959.
Li Wenxin, Caves in Yilan Wokenhada, Acta Archaeologica Sinica, 1954, 7.
Li Xikun et al. The discovery and significance of Qingtoushan people and Qian Guo people in Jilin, Jilin Geology, 1984, 3.
Li Yingkui, Gao Bo, Neolithic Tombs in Xiaonanshan, Raohe county, Heilongjiang, Archaeology, 1996, 2.
Li Youqian, Chen Quanjia, Preliminary study on Paleolithic materials and industrial types in the Korean Peninsula: also on several understandings of Paleolithic research in the Eastern part of Jilin province, research of China's Frontier Archaeology, 2008, 7.
Liaoning Provincial Institute of Cultural Relics and Archaeology, Xiaogushan: comprehensive study of prehistoric cave sites in Haicheng, Liaoning, Science Press, 2009.

Liu Mingguang (ed.) *Atlas of China's physical geography*, China Cartographic Publishing House, 1998.

Liu Yuying, Zhang Shuqin, Liu Jiaqi, et al. Palynological records of vegetation and environmental changes in late Pleistocene Ma'er Lake, Erlongwan, Northeast China, Acta Micropalaeontologica Sinica, 2008, 25(03).

Lv Zun'e, Replication and use of fish spearheads from Xianrendong cave, Haicheng, Acta Archaeologica Sinica, 1995, 1.

Pei Wenzhong, *Chinese Stone Age (3rd edition)*. China Youth Publishing Group, 1963.

Pei Wenzhong, *Research on China's Prehistoric Period*. The Commercial Press, 1948.

Ren Meie, *Outline of China's Physical Geography (Revised 3rd Edition)*. The Commercial Press, 2004.

P. M. Rice, On the origin of pottery. *Journal of Archaeological Method and Theory*, 6: 1–54, 1999.

Tang Zhuowei, et al. Human activities and climate environment in jilin province, in Liu Jialin (ed.), *Research on the Historical Evolution of Natural Environment and the Impact of Human Activities in Northeast China*, Science Press, 2007.

Wang Lixin, et al. Discovery of pottery from the Shuangta site in Baicheng, Jilin province, dating to around Ten thousand Years ago, Chinese Cultural Relics News, Sept. 14, 2012.

Wang Xiaokun, et al. Excavation report on the Jinsitai cave site in inner Mongolia, Acta Anthropologica Sinica, 2010, 29(01).

M. G. Winkler, P. K. Wang, The late-quaternary vegetation and climate of China.In H. E. Wright eds., *Global Climates Since the Last Glacial Maximum*, pp. 221–261. University of Minnesota Press,Minneapolis, 1993.

Xiao Ronghuan, Hu Jianbin, Characteristics of climate and geomorphology in Northeast China since the last glacial maximum. *Journal of Glaciology and Geocryology*, 1988, 10(02).

Yan Wenming, The origin and spread of agriculture in Northeast Asia, Agricultural Archaeology, 1993, 3.

Yanbian Museum, Brief report on the clearance of Jingu Neolithic site in Longjing county, Jilin, Northern Cultural Relics, 1991, 1.

Yang Dashan, Newly discovered Paleolithic sites in Xiaonanshan, Raohe county, Northern Cultural Relics, inaugural issue, 1981.

Ye Qixiao, Research on cultural remains of the Paleolithic in Heilongjiang province, Research of China's Frontier Archaeology, 2003, 2.

D. R. Yesner, Maritime hunter-gatherers: ecology and prehistory. *Current Anthropology*, 21: 727–750, 1980.

Zhang Boquan, *The Local History of Northeast China*, Jilin University Press, 1985.

Zhang Xiaoling, et al. New materials and dating of the Shibaizhan site in Heilongjiang, Acta Anthropologica Sinica, 2006, 25(02).
Zhao Binfu, Research on the livelihood patterns of the Neolithic in the Nen River Valley, Archaeology, 2007, 11.

CHAPTER 10

The Last Hunter-Gatherers: Southwest China

> Man from the state of Song wants to sell hats to the Yue tribes but in vain because the Yue people were hairless and wore tattoos all over their bodies.
>
> —*Zhuangzi*

10.1 Presentation of Questions

At present, Southwest China has the most colourful and best-preserved ethnic culture. There are the most diverse forms of marriage and 25 ethnic minorities distributed here. Undoubtedly, it's closely related to its history. Can it be traced back to the Prehistoric Age? According to the archaeological materials, lithic assemblages of the Upper Paleolithic in this area are extremely diverse. In Southwest China, the Neolithic began late and lasted short, and soon was replaced by the Bronze Age and the Iron Age; what is puzzling is that the use of chipped stone tools has lasted until modern times, which shows the relatively poor agricultural development in this area. In contrast, there are some near "abnormal" developments in some areas, such as pottery and settlement-like structures while there is no similar development in the nearby areas. Why? During the process of entering the agricultural era in North China and the middle and lower reaches of the Yangtze River, what changes have taken place in Southwest China when the North China and the middle and lower reaches of the

S. Chen, *Prehistoric Modernization*,
https://doi.org/10.1007/978-981-97-9423-2_10

Yangtze River entered the agricultural era? What enlightenment can the unique cultural process in Southwest China provide for us to understand the cultural mechanism of prehistoric hunter-gatherer lifeway? The unique process of prehistoric culture in Southwest China needs to be explained, but the research related to this area is still not in depth. I will combine the theories of cultural ecology, behavioural ecology, and archaeological materials to analyse the characteristics and evolution process of prehistoric cultural adaptation in Southwest China, and explain the reasons for its uniqueness. Our research on Southwest China is rarely carried out independently, most of which is discussed together with South China and even the whole southern region (Zhang and Hong 2009), or combined with the Tibetan Plateau to form a greater southwest region (Tong 1986). Ignoring the above problems is not conducive to understanding the particularity of Southwest China.

10.2 Cultural-Ecological Conditions in Southwest China

Before discussing Southwest China, it is necessary to make it clear what it refers to and why it is determined. Geographically, Southwest China is the area in the southeast of the Tibetan Plateau and the west of the Guizhou Plateau, almost including Yunnan Province and the southwest of Sichuan Province. In terms of cultural ecology, it mainly includes Yunnan Province, Guizhou Province, and Sichuan Province, as well as the northwest region of Guangxi Province, excluding the tropical region of Yunnan Province. Southwest China can be roughly divided into three physical geographical units: Sichuan Basin, the high mountains and valleys on the eastern edge of the Tibetan Plateau, and the Yunnan-Guizhou Plateau. In fact, there are significant differences among the three places, but their common characteristic is isolation. Because of the blocked geographical location, they have poor transportation systems. At the same time, Sichuan Basin is the only one that has a relatively large area within the three units, and the other two are composed of many small units. Being cut into small pieces by rivers flowing in different directions, the Yunnan-Guizhou Plateau is undulating with a broken original surface. The same is true for the eastern edge of the Tibetan Plateau where mountains and valleys crisscross. Therefore, they have small cultural geographical units (except Sichuan Basin) (Archaeology Teaching and Research Office of the History Department, Peking University, Sichuan

Provincial Museum 1983) usually composed of small basins (dams). Between different geographical units, there is complex terrain, and significantly diversified species, climate and soil, which is so-called "ten miles in different days". The mutually divided resource patches constitute the geographical units of cultural adaptation with significant differences.[1]

The so-called cultural ecological condition refers to the ecological condition that exists relative to and interact with cultural adaptation. The cultural ecological condition is a relative objective existence, and its state depends not only on the natural condition, but also on human cultural adaptation. Human cultural adaptation is constantly evolving, with the trend of continuous development of technology, increasing complexity of society (of course, there is also the collapse of complexity), and continuous population growth (with reference to Chapter 1). Subject to natural ecological condition, the cultural adaptation has its own evolution mechanism, and can affect or even transform natural ecological condition. What we need to emphasise is that the cultural ecological condition suitable for hunter-gatherers is not necessarily suitable for agricultural production groups, because there are significant differences in the way their cultural adaptation interacts with natural conditions. Therefore, it is necessary to distinguish them when investigating the cultural ecological condition in Southwest China. Let's start with the familiar part—agricultural ecological condition.

Because of the serious erosion, the Yunnan-Guizhou Plateau is characterised by Karst landform, showing the features of poor soil and abundant rocks. Due to the river diving into the ground, the surface water is insufficient, making it inconvenient for people to use (Fig. 10.1). At the same time, the rugged terrain and sparse plains cause less arable land and difficult irrigation. In addition, the low temperature in summer here is not conducive to the growth of rice which is suitable for growing in the zone where rain and heat are in the same period. Especially in rainy and foggy Guizhou Province, the short sunshine time is not suitable for developing agriculture. More seriously, Once the native forest in the limestone area is destroyed, water resources will deteriorate rapidly, and the forest recovery will be difficult. Instead, the forest will be replaced by tantlery shrubs and grass slopes that are heliophilous, tolerant of drought, and barren soil. Now, almost 80% of the vegetation in Guizhou Province has the above characteristics. The vegetation on Qianling Mountain in Guiyang is well-preserved, which shows that its original natural condition is relatively superior. The soil under the evergreen broad-leaved forest is black lime soil, and the topsoil is thick and soft (Ren 2004). However, this kind of environment is relatively fragile. Deforestation and development of agriculture have resulted in deterioration of ecological condition, hindering the further development of agriculture.

Since ancient times, Yunnan-Guizhou area has been regarded as "a malaria-infested place". From the perspective of modern science, it is the area with the humid and hot environment easy to cause infectious diseases, especially falciparum malaria. This kind of environment leads to a low average life expectancy. To maintain the population size, it is necessary to increase the birth rate, leading to a low average age. That is, a low proportion of the working population and a high proportion of the young people who need to be raised. Accordingly, each child will receive less care, which is not beneficial to accumulate (less elderly people) and spreading (more young children) knowledge as well as experience in the community, hindering the cultural development. Population is one of the prerequisites for agriculture. Only when the necessary population size

Fig. 10.1 Karst mountain landscape in Guizhou

is reached can labour-intensive agricultural production be adopted. Both land reclamation and harvest require more labour force. Under the pressure of disease, the small population size also restricts the development of agriculture in Yunnan-Guizhou region. Of course, this pressure is not only applied to farmers, but also to hunter-gatherers. However, it has a greater impact on the former. Garbage accumulation, water pollution, and more concentrated population will all lead to the occurrence and spread of diseases.

Although Southwest China is not suitable for farmers, it is relatively suitable for hunter-gatherers that subsist through nomadic foraging. The favourable environment must include rich and diverse resources (animal and plant food, stone, water, fuel, etc.), as well as geographical condition conducive to human shift. Hunter-gatherers can easily use these resources. The traffic in Southwest China is inconvenient, but the rich plant resources make up for its weakness to a large extent. Southwest China has complex terrain, including low-altitude valleys and dams, as well as high mountains, especially at the edge of the Tibetan Plateau. The mountains run north and south, which is conducive to the deepening of the moist and hot air flow. The valley becomes the channel for animals to migrate and plants to spread. The area's vertical distribution creates a particularly complicated habitat for animal and plant evolution, making it home to the greatest diversity of species in all of China. Southwest China is less impacted by the cold stream because of the mountain barrier, resulting in less climate change and greater stability. These are environmental conditions conducive to hunter-gatherers.

10.3 Adaptation: Perspective from Human Behavioural Ecology

Under the current environmental conditions, how can hunter-gatherers survive in Southwest China? We can use the basic principles of behavioural ecology to derive the basic characteristics of cultural behaviour, providing a reference for archaeological research. Behavioural ecology studies the significance of behaviour to adaptation, mainly from the perspective of evolution and ecology, the most prominent of which is the optimal foraging theory (MacArthur and Pianka 1966). It uses models such as diet width, patch choice, central foraging, and linear program analysis to discuss the behaviour selection of human beings in certain environment (Winterhalder 2001). This theory is very enlightening in analysing the changes of adaptation of hunter-gatherer society. It may be helpful for us to understand the unique characteristics and process of cultural adaptation in Southwest China by analysing the optimal strategies of human adaptation in Southwest China with the theory of behavioural ecology.

Patching is a significant aspect of the biological environment in southwest China (numerous dams and valleys). The traffic is inconvenient, and the various resource patches are clearly separated. The marginal value principle of behavioural ecology shows that. Firstly, the marginal utility of utilisation of any resource patch decreases: the longer the utilisation time, the lower the income. Secondly, there is an optimal utilisation time length: the income will be maximised when the utilisation time length is appropriate. Thirdly, the optimal foragers stay longer in high-quality resource patches than in inferior ones. Fourthly, the longer the travel time between resource patches, the longer the foragers stay in the resource patches. Fifthly, if the quality of the whole environment is poor, the foragers' stay time in a resource patch will also be correspondingly prolonged. Lastly, if the foragers can evaluate the quality of the resource patch habitat, they will be more likely to take extreme strategies: stay in high-quality resource patches for longer time, while in poorer quality resource patches for a shorter time (Zhang et al. 2000).

On the other hand, the degree of diversity of habitat resources also affects human choices. The high diversity of resources means that more diverse ecological space can be used, enriching the diversity of cultural ecology and means of livelihood. More diversified habitat resources means more choices and the more stable livelihood system. If there is no subsistence pressure, foragers usually choose to use the best quality resources, which is easy to form a more specialised adaptation (Bucking et al. 2003). In an unstable environment, they must adopt a generalised strategy to decrease risks instead of relying on only one type of resources. However, Southwest China has a good environmental stability and a high degree of resource diversity.

Based on the above principles and the cultural ecological condition of Southwest China, we can further make the following reasoning for the adaptation of hunting-gathering in this region.

Reasoning 1: If the living environment is convenient for migration (such as boats and horses), the flow between resource patches will be more frequent, and the range of activities of groups will be larger. We will see similar archaeological relics distributed in a larger range. On the contrary, if migration is difficult, the mobility between resource patches will decrease, and we will see archaeological remains with limited distribution and different characteristics. Secondly, if the living environment is more diversified, the ways of human beings to utilise resources will also increase, and we will also see various archaeological remains. Moreover, if diversity exists in a relatively stable and closed environment with little external influence or interference, it will form a more specialised adaptation in the resource patch, rather than a universal utilisation strategy to minimise risks.

Reasoning 2: If there are more diverse and stable resources in a living environment, it is easy to form diverse and specialised culture to adapt to life, and it will maintain the existing adaptation mode for a long time due to less external influence. This trend will be further strengthened when the distribution of resource patches is patchy and the migration between different patches are relatively difficult. The hunter-gatherers living in different resource patches will choose to stay in their living areas for a long time, especially them living in the patches with better resource conditions, and will not be easily abandoned; If the population pressure increases, this trend will also strengthen, because once you leave, you may never find a suitable place. If different groups try to maintain their own territory, they will form diverse and conservative cultural adaptation. In material record, it will be "cultural tradition" (stable and unique heritage characteristics) in a small range (within the resource patch) of long-term.

Reasoning 3: In Southwest China (except Sichuan Basin), agricultural production will bring serious ecological problems. Agricultural ecological conditions in this place is poor and the rich resources here are conducive to hunting-gathering, which makes it more competitive than agricultural production. It can be predicted that hunting-gathering here will probably last longer. If the population increases, it may spread to the more marginal and smaller resource patches, because it is impossible to spread to the eastern agricultural areas. The conditions for agricultural production in these areas are worse, and the hunting-gathering livelihoods will be preserved for a longer time in these areas. When agriculture spreads into Southwest China, it can be predicted that those resource patches with better agricultural conditions will accept agriculture earlier, and those resource patches on the edge will still maintain hunting-gathering livelihoods.

The livelihood diversity of hunters and gatherers based on the simulation of modern weather station data (with reference to Fig. 3.7 in Chapter 3) also shows that the level of livelihood diversity in Southwest

China is relatively high (Due to the lack of fishing and hunting resources, it is not as good as the southeast coastal zone). The vertically distributed mountains enable hunter-gatherers to use resources of different heights; at the same time, the diversified species also provide a variety of resources, which need to be used in different ecological niches and seasons.

The ethnological study of Cashdan has also proved that there are strong associations between the diversity in ethnic groups and ecological system (Cashan 2001). She analysed the materials related to the simple population distribution in the world and found that the population diversity is straightly relevant with the productivity of the ecosystem, especially the key variables such as annual rainfall and temperature. However, the diversity in ethnic groups is directly related to the other three variables: climate stability, disease pressure, and habitat diversity. Areas with low diversity in ethnic groups have poor climate stability, low disease pressure, and low habitat diversity while those with high diversity in ethnic groups have high climate stability, high disease pressure, and high habitat diversity. Her research is consistent with the deduction of behavioural ecology. The ethnographic materials in Southwest China also confirmed her hypothesis.

The diversity of habitats also affects the marriage patterns of human society. Generally speaking, the forms of marriage in human society can be divided into monogamy and polygamy (polygamy, polyandry, and mixed marriage). From the perspective of behavioural ecology, monogamy is more likely to form in areas with evenly distributed resources. In situations with patchy resource distribution, it is more likely to form polygamy, and the party with resources is more likely to dominate the form of marriage. The conditions among different resource patches in the southwest region are significantly different. The ethnographic materials also reveal the extremely diverse marriage patterns in Southwest China. This complex state is more likely to be related to the resource conditions, rather than the remains of a certain stage of human social development. And it is not universal, such as the "matriarchal society" remnants of Mosuo of the Naxi nationality.

The closed geographical environment in Southwest China has exacerbated this differentiation. It is difficult to communicate with the outside world, and the traffic between different resource patches is inconvenient. Therefore, different groups are isolated and closed. Due to the lack of effective means of transportation (such as boats, livestock, roads, etc.), the degree of isolation between groups will increase with the barrier of the terrain. In contrast, the Northeast and South China, and the middle and lower reaches of the Yangtze River can all rely on boats. The terrain in North China and the grassland and Gobi areas are flat and convenient for human activities while Southwest China is characterised by close and isolation. The stability of the environment is also conducive to the stability of cultural adaptation and the emergence of specialised adaptation. The cost of stability and specialisation means that other options are excluded. In a relatively closed environment, it can naturally last for a long time.

10.4 Cultural Sequence of the Archaeological Materials

In order to understand the transformation process of the Neolithic, we need to establish at least a cultural sequence from the Upper Paleolithic to the peak of the Neolithic like other regions. However, this is not easy to do in Southwest China. Among the archaeological materials, the materials related to the Neolithic have been relatively deficient. In recent years, there has been significant progress, but it is not as rich as North China and the middle and lower reaches of the Yangtze River, which are sufficient to establish a complete cultural sequence. At present, the total amount of materials is relatively limited. Compared with thousands of sites in the Yangshao culture period in North China, the exploration of Neolithic sites in Southwest China is still in its early stages. On the other hand, the reason is that archaeological materials are extremely diversified, and the features of different sites vary significantly, resulting in difficult induction of cultural sequence. Based on the particularity of the Neolithic materials in Southwest China, the main archaeological discoveries in Southwest China will be sorted out in chronological order to establish a basic chronological framework and summarise the representative cultural characteristics existing in different ages and regions.

Southwest China has a wide range of Karst terrain and numerous caves, which is a good place for Paleolithic archaeology. It has also been one of the central areas of archaeological research of the Paleolithic in China. After discovering the Guanyindong site in Bijie in 1964, Guizhou Province has found more than 60 Paleolithic sites, 15 of which have been excavated. There are few sites with clear absolute dating data, and the age distribution is inconsistent. With reference to Table 9.1, there are most sites in the transition period of the Neolithic and the Paleolithic (It can be called "Post-Paleolithic Age" in Yunnan-Guizhou region). This is a favourable research condition which can provide rich research materials for us to explore the cultural changes before and after the origin of agriculture. Here, I divide Southwest China into three cultural and geographical units: Yunnan-Guizhou Plateau, the eastern edge of the Tibetan Plateau, and Sichuan Basin, which is consistent with the previous division of physical geography. Through the archaeological analyses, we can also see whether this classification is reasonable enough.

The materials of the Paleolithic in the Yunnan-Guizhou Plateau area began in the Guanyindong site in Bijie, Western Guizhou Province. The early Paleolithic lithic assemblages are represented by flakes. The stone artefacts have steep edges, various types, and lack of stable shape for repair and processing (Li and Wen 1986), which is different from the gravel chopping tools industry popular in the middle and lower reaches of the Yangtze River. Another characteristic of the Paleolithic in this area can be traced back to the Xiaohuidong site in Shuicheng county, which is famous for its discovery of ridged-hammer bipolar flaking method. This technology has been considered as a typical feature of the Paleolithic in Southwest China, and has continued to the production of stone artefacts in the Neolithic. In the Upper Paleolithic materials, taking the Baiyanjiaodong site as an example, its lithic assemblage is mainly flaking tools (more than 90%) and many scrapers with steep blade antlers by refined processing from the back to the plane of fracture. During the Upper Paleolithic in North China, the common endscrapers and backed knives are rare, and the carving tools, concave scrapers and points are not developed. The gravel tools are mainly chopping tools (Li and Cai 1986, 2). The radiocarbon dating of the Baiyanjiaodong site is 12800 ± 200 years ago (the third layer) and 14,600 ± 200 years ago (the fifth layer). The Caohai site in Weining County, belongs to the later stage of the Upper Pleistocene, and the lithic assemblage also adopts steep and two-way processing, which is more consistent with the Guanyindong site (Wu et al. 1983). Zhangkoudong in Yiliang County, Yunnan Province is about 15,000 C14 BP. The lithic assemblage mainly includes small and medium-sized stone artefacts made of flint raw materials, and also adopts traditional techniques in Yunnan-Guizhou Plateau such as steep processing and ridged-hammer bipolar flaking method (Hu 1995).

At present, the best materials in the eastern edge of the Tibetan Plateau are from the Dahe site in Fuyuan County, Yunnan Province. The AMS and U-series dating of it are both between 30,000 and 50,000 years. Its lithic assemblage is famous for containing Mousterian stones as well as the Levallois technology, and of course it also includes ridged-hammer bipolar flaking method (Ji et al. 2006). Stone artefacts similar to Mousterian stones are also found at NO.2 Longtanshan site in Kunming (with a history of 30,500 ± 800 years) (Qiu et al. 1985). It is said that there are also stone artefacts similar to Levallois technology at the Dadong site in Pan County, Guizhou Province (Huang et al. 1997). This technology was first discovered at the Shuidonggou site in Lingwu City, Ningxia Hui Autonomous Reign, and later Jinsitai site in eastern Inner Mongolia in Hetao area, and Shibazhan site in Huma County, Heilongjiang Province. Its distribution is consistent with the transition zone in physical geography in China from northeast to southwest, and it is also a crescent cultural distribution zone from northeast to southwest from the Neolithic to historical period (Tong 1986).

Another important evidence of the existence of this cultural zone is the distribution of microblade technology, which is mainly popular in North China and Northeast Asia. In North China, its southern boundary does not cross the Qinling-Huaihe line. However, in Southwest China, it extends to Yunnan Province. The earliest Fulin site in Hanyuan County, Sichuan Province has 10% of pseudo-blade and the size of stone artefacts is small (Zhang 1977) which is close to that of microblades. Many sites in the transition period of the Neolithic have been found, such as the Zhongzipu site in Guangyuan and Yanyundong site in Beichuan County (the shape of stone axes unearthed together is relatively regular, close to the ground stone axes in the Neolithic, so it is included in this stage (Sichuan Provincial Institute of Cultural Relics and Archaeology 2006). Stone artefacts of microblade technology were further found in Neolithic sites. However, they were mainly found in the eastern edge of the Tibetan Plateau, and occasionally distributed in the eastern part of Southwest China (Chongqing Cultural Relics and Archaeological Research Institute 2009).[2] Through the distribution of lithics similar to Mousterian stone artefacts, Levallois technology, and microblade technology, we can roughly distinguish the eastern edge of the Tibetan Plateau and Yunnan-Guizhou Plateau (part of Yunnan Province).

There are few discoveries in the Upper Paleolithic period in Sichuan Basin. The radiocarbon dating of Zhangertang site in Tongliang County is more than 20,000 years ago. Its lithic assemblage is mainly large stone artefacts, in which end choppers are in the majority. There are also large triangular points similar to those at Dingcun site. The stone artefacts are mostly steeply processed (Li and Zhang 1981), which is significantly different from the Fulin site. Later materials show that there are 20 stone artefacts that may have been transported, with unconspicuous artificial characteristics unearthed at the Liyuqiao site in Ziyang (Archaeology Teaching and Research Office of the History Department, Peking University 1983).

The eastern margin of the Tibetan Plateau has a strong consistency, while the other two regions are different in a lithic assemblage of Paleolithic sites, such as the Fulin site, Zhangertang site, Baiyanjiaodong site, and Ma'anshan site. There are two reasons for this. On the one hand, the shape of stone artefacts is largely restricted by the unique local raw materials. On the other hand, the characteristic of Paleolithic stone artefacts in Southwest China are highly diverse. In short, it is difficult to generalise. Corresponding to the diversity is ridged-hammer bipolar flaking method, steep processing, and so on in this area, which has a very good continuity, extending from the early and middle Paleolithic to the Neolithic. Of course, ridged-hammer bipolar flaking method is not unique in Southwest China. It may be related to the "throwing against anvil technique", which Gao Xing called "Yangtze technology" (Gao et al. 2008). It is also found at No. 2 Yuzui Paleolithic site in Yun County, Hubei Province, marked by some flat chipped gravels (Chen et al. 2014). This technology undoubtedly reflects the characteristics of local raw materials. It is most fully represented in Southwest China, especially in the Yunnan-Guizhou Plateau, forming a technological tradition. The reason why it is called the technological tradition is that ridged-hammer bipolar flaking method is not the only method to produce flakes, and the hammering method can produce, too. It is precisely because this cannot be fully explained in function, so it formed the stone technological tradition in the Upper Paleolithic and later stages in Southwest China.

There is no obvious boundary between the Upper Paleolithic and the transition period of the Neolithic in Southwest China. Setting a time limit is not because there has been an obvious cultural change which is difficult to find from the current archaeological materials. It is determined based on the time frame of the surrounding areas and is set for comparison which needs researchers to pay special attention. The lack of a clear beginning of the Neolithic transition period is also one of the key characteristics of Southwest China.

There are many sites belonging to the transition period of the Neolithic, and the lithic assemblages vary in characteristics. The Chuandong site in Puding County, is only a few kilometres away from the Baiyanjiaodong site. The stone technology of the upper cultural belt of Chuandong site in Puding County and Baiyanjiaodong site is mainly ridged-hammer bipolar flaking method. There are many large stone artefacts and bone tools of complex types. The blade antler of the scraper is relatively sharp, mostly around 60°, which is different from the Paleolithic relics of the same era and later. According to the radiocarbon dating, the lower part of the Chuandong site is about 16,000 years old. The middle part is 9610 ± 100 years old (the sixth layer) with few remains. The upper part (layers 3, 4, and 5) has existed for more than 8000 years. It can be seen that the shallower the layer, the older the age (Zhang 1995). According to the radiocarbon dating, the Maomao Cave site existed 8820 ± 130 years ago. There are a large number of points and scrapers (with sharp edges) with stable shape and refined processing of the lithic assemblage. The types of points are diverse. This site maintains the traditional technology in this area, and nearly 80% of flakes are produced by ridged-hammer bipolar flaking method, and most of the tool repairs are processed towards the plane of fracture (Cao 1982). Similar discoveries have been made at the Laojiangdi site in Xingyi, and a small number of polished stone tools were unearthed (Zhang et al. 2008).

From the current materials, the implements made of bone and antler seem to be an important feature of this period. Polished bone tools have been found at the Ma'anshan site with a history of about 15,000 years (Zhang 1988).[3] The Chuandong site is famous for its numerous and exquisite bone tools (Zhang 1995; Mao and Cao 2012). There are abundant polished bone awls, bone needles, bone arrowheads, antler awls, antler shovels, antler spears, etc. unearthed at the Tangzigou site in Baoshan, Yunnan Province(with a history of about 7000 years in the radiocarbon dating) (Ji et al. 2004). Similar discoveries have also been made at the Maomao Cave site in Xingyi (Cao 1982). The Daerwo site in Kaiyang County, with a history of 22,000–3000 years in the radiocarbon dating, is divided into 18 layers. More than 5 layers of it contain pottery shreds, polished stone tools, and bone tools. Chipped stone tools, bone tools and animal fossils were unearthed below the fifth layer. Chipped bone tools and aggregates account for 98% of all excavated materials. There are 10 antler shovels unearthed at the Hongtudong site in Puding County with a history of 8000 years (Wang et al. 2009). There may also be more than 630 stone artefacts and 16 bone tools unearthed from the Laolongdong site in Ershan, Yunnan Province, which may belong to the same period (Zhang 1994).

The pottery that appeared in this period was first seen in the area adjacent to Guangxi Province. The typical representative is the Guanyindong site in Anlong County, which is located in the upper reaches of the Pearl River. More than 50, 000 pieces (Zhang 2003)[4] of chipped stone tools, polished stone tools, bone tools, pottery shreds, human and animal remains were unearthed, as well as fire-related relics. The excavators conducted the radiocarbon dating on the samples collected from the four consecutive cultural layers in the middle of the site accumulation during the trial excavation in 1986, and obtained several consecutive absolute dating data from 9970–7080 years ago. The pottery shreds were unearthed at the layer with a history of 8000 years, and there were stone axes and stone adzes in the few polished stone tools. Chipped lithic assemblage is mainly flakes (Cai and Wang 1998). As the loess deposit the lower layer of the Feihushan in Pingba County has a history of 12,000 years. Although it is comparable to that of South China, there is a lack of pottery and polished stone tools. Instead, there are excavated flint stone tools, bone and antler tools, fire-related relics, animal fossils, etc. However, the axe-shaped tools are similar to those of the polished stone axe, which may be related to be in the transition period of the Neolithic (Liu 1996). Column holes in building are also found in Baoshan Tangzigou which shows that there may be some form of building here (Zhang 1992). The discovery of this period is still lacking in Sichuan Basin. The eastern edge of the Tibetan Plateau has been mentioned above, and it is mainly characterised by stone artefacts of microblade technology. In general, the southwest region in this period is characterised by bone and antler tools, stone artefacts and pottery similar to those in the Neolithic, as well as relatively fixed residence, of which pottery and fixed residence may appear relatively late.

The beginning of the Neolithic in Southwest China is marked by the emergence of pottery groups, settlement structures, and polished lithic assemblages. The starting age is about 5000 years ago, thousands of years later than the surrounding areas. Although earlier sites such as the Guanyindong site in Anlong County and Tangzigou site in Baoshan have similar findings, their characteristics have not constituted a complete implement assemblage. Moreover, they are located at the edge of Southwest China where can be classified as either South China or Southwest

China, and are rarely found in the central region of Southwest China. The Neolithic in Southwest China is influenced by culture from three regions: South China, the middle reaches of the Yangtze River, and Northwest China, which are not only reflected in the style of pottery, but also in the characteristics of stone artefacts, such as the shouldered stone tool (Wang 1989). In a sense, the Neolithic in Southwest China is the product of the foreign cultures. The Neolithic cultural characteristics appeared earlier in the marginal regions and later in the central regions.

The Neolithic cultural sequence of the whole Sichuan Basin is divided into four sub-regions: the hilly area in northern Sichuan, the Three Gorges area in eastern Sichuan, the high-mountain and valley area in northwestern Sichuan, and the centre of Sichuan Basin, which are respectively affected by the Lijiacun culture in northern China, the Daxi culture in the east of the Three Gorges, and the Majiayao culture in Northwest China. Neolithic discoveries in western and northern Sichuan began at the Daditou site in Hanyuan County (Sichuan Provincial Institute of Cultural Relics and Archaeology, et al. 2006), Jiangweicheng site in Wenchuan County (Sichuan Provincial Institute of Cultural Relics and Archaeology, et al. 2006), and the Yingpanshan site in Mao County (Chengdu Institute of Cultural Relics and Archaeology, et al. 2000). In recent years, it has been discovered that there are dozens of polished stone axes and more than 200 stone net pendants and housing sites (Liu 2011) are unearthed at the Jiaohuayan site in the Xiangjiaba reservoir area. Similarly, there are stone artefacts of microblade technology and housing sites (Lei 2011), all have a history of 5000 years unearthed at the Guiyuanqiao site in Shifang. Later, there are multibay housing sites found at the Maiping site where the polished stone tools are more developed (CA, CASS 2006; Sichuan Provincial Institute of Cultural Relics and Archaeology, et al. 2008). The Neolithic here can be at least as late as the late Shang Dynasty in the Central Plains, such as the Neolithic site in Lizhou (Lishou Site Joint Archaeological Excavation Team 1980). The eastern edge of Sichuan Province is obviously influenced by the early Neolithic culture in the middle reaches of the Yangtze River, and the Neolithic remains are found much earlier than the central area. The Neolithic culture centred on the Chengdu Plain emerged relatively late, and developed rapidly from 4000 to 4500 years ago. It forms the "Baodun Culture" with local characteristics, and city sites appear, rapidly beginning the process of social complexity (Zhang 2003).

In recent years, the discovery of archaeological remains from the Neolithic to the Shang and Zhou Dynasties in Guizhou has increased significantly. There have been attempts to study in different regions which have formed their own systems with obvious differences. Taking western and northwestern Guizhou as an example, they can only be classified as a mixed cultural area. "Although the regional peculiarities of each local region are also quite evident, the influence of numerous cultural variables is its common feature from the perspective of the overall characteristic and vast regional scope of culture". Taking Jigongshan in Weining County and Qingchang in Bije, as examples, the distance between the two places is not very far from the perspective of geographical units, and the geomorphic units are also very similar, but the cultural characteristics are quite different. It seems that this region also presents a kind of situation of "large-scale mixed settlement" and "small-scale clustered settlement" (Wu 2011). Similar to the complex cultural pattern of Guizhou Province, the Neolithic sites in Yunnan Province can be divided into three geographical types: riverside terraces, riverside caves, and lakeside seashells, while in terms of cultural characteristics, they can be classified into 8, 9 or even 11 types (Xiao 2001) with extremely complex patterns. This constitutes another particularly obvious feature of the Neolithic in Southwest China.

The characteristics parallel with the late beginning of the Neolithic and the complex regional pattern are as follows: the Neolithic in Southwest China retains the local ancient technological tradition, such as ridged-hammer bipolar flaking method of the entire Paleolithic. There is also a remnant of microblade technology belonging to the Paleolithic. After entering the Neolithic, this technique rapidly disappeared in North China, and remained in the Sichuan-Yunnan corridor in the southwest. Some areas remained it until the Shang and Zhou Dynasties (Table 10.1).

10.5 Diversified Adaptations in Southwest China

10.5.1 The Origin of Agriculture in Southwest China

In theory, agriculture in Southwest China may begin to develop in the Neolithic, but archaeological evidence cannot fully support this assertion. This is due to insufficient archaeological materials as well as the complex and diverse regional conditions here. The analysis from the perspective of archaeological culture supports that the eastern part of Sichuan Basin first accepted rice farming introduced from the middle and lower reaches

Table 10.1 Space and Temporal Framework and Typical Sites of Prehistoric Times in Southwest China

Time	*Stage*	*Typical Sites*		
		Yunnan-Guizhou Plateau	*Sichuan Basin*	*East edge of Tibetan Plateau*
15,000–35,000 years ago	The Late Paleolithic	Caohai site; Maludong site	Zhangertang site, Liyuqiao site	Dahe site, Fulin site, Huilongwan site
5000–15,000 years ago	Transitional stage between the Paleolithic and the Neolithic	Baiyanjiaodong site, Maomao Cave site, Feihushan site, Daerwo site, Guanyindong site, Zhangkoudong site		Zhongzipu site, Tangzigou site, Yanyundong site
	The Neolithic The Bronze Age	Zhongshui site, Jigongshan site, Dadunzi site	Shizishan site, Yingpanshan site, Baodun site, Zhangjiapo site	Maiping site, Jiangweicheng site, Lizhou site, Maojiakan site

of the Yangtze River, and then people from the northwest moved and brought millet farming to Southwest China. Rice replaced other crops as the primary agricultural crop after adaptation (Sun 2009). From the perspective of stone artefacts, in the areas south of Dabashan and on both sides of the Jialing River, taking the Zhangjiapo site as an example, lithic assemblage is mainly axe, adze and chisel (Sichuan Working Team of the Institute of Archaeology, Chinese Academy of Social Sciences and the Cultural Relics Management Office of Guangyuan City, Sichuan 1991), among which the so-called "jue" (an Chinese earth-digging agricultural tool in China). "jue" is more likely to be the reuse of abandoned stone axes in terms of form. Stone pestle unearthed together does not have the use characteristics of pestle, and the large scar marks indicate that it may have been used to hit hard objects. The same is true for rectangular stone knives with short edges which do not have the characteristics of stone knives for harvesting (Fig. 10.2). In general, the Zhangjiapo site lacks tools for farming and grain processing. Similarly, axes, adzes, chisels, spears, whetstones are unearthed from the Dashuidong Neolithic site in Jiangyou, but there are no agricultural tools (Hu 2005). Stone pestle, grindstones, and tools similar to grain processing tools are unearthed from the Jiangweicheng site in Wenchuan County, with a history of 5000 years (Sichuan Institute of Cultural Relics and Archaeology, et al. 2006). But there is only one real stone pestle (The cross section of the other is like a stone axe). The centre diameter of a grindstone is only 6–7 cm and it is not a stone grinding slab. The site also has microblade artefacts, indicating the existence of hunting economy.

A large number of microblade artefacts and a small amount of ground stone tools—adzes and perforated stone knives have been found at the Zhongzipu site in Guangyuan, Sichuan Province, which has a history of 6000 to 7000 years. The excavators believe that they may belong to different periods (Sichuan Working Team of the Institute of Archaeology, Chinese Academy of Social Sciences 1991). However, it is quite possible that they belong to the same period, because microblade artefacts and ground stone tools are possible to exist at the same time. If this conjecture is correct, farming may already exist at the Zhongzipu site, which

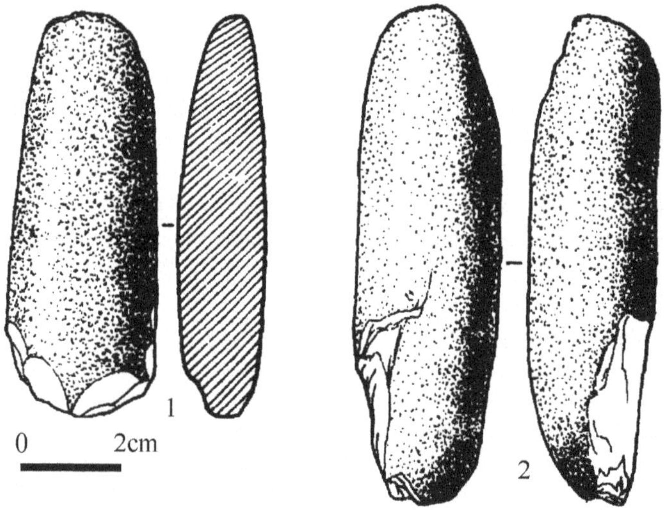

Fig. 10.2 Stone adze and stone pestle at the Zhangjiapo site[5]

can be well proved by perforated stone knives used for harvesting. Similarly, ground stone tools and microblade stone tools, such as stone axes, perforated stone knives, stone adzes, net pendants, etc. It were unearthed at the Daditou site in Hanyuan County, Suchuan Province. The excavators speculated that these indicate the existence of fishing, hunting, and farming activities around 4000 years ago (Sichuan Institute of Cultural Relics and Archaeology, et al. 2006). After the investigation and trial excavation of the Maiping site and Majiashan site in Hanyuan County of the Neolithic to the Shang and Zhou Dynasties, it was found that there was a lack of agricultural tools in lithic assemblage, and the ground stone tools were mainly axes and adzes. There are microblade stone tools and chipped stone tools unearthed there (CA, CASS et al. 2006). After the systematic excavation of Maiping site, fishing and hunting tools such as net pendants and arrowheads are unearthed, and perforated stone knives are also discovered (Sichuan Institute of Cultural Relics and Archaeology, et al. 2008). There are also many stone knives (12 pieces) unearthed at the Maojiakan Neolithic site in Dechang County, Liangshan Yi Autonomous Prefecture, Sichuan Province. Other productions are mainly axes and adzes, as well as stone arrowheads and microblade artefacts (Sichuan Institute of Cultural Relics and Archaeology, Liangshan Yi Autonomous Prefecture Museum 2007). The Lizhou site in Xichang similar to Dadunzi site where have abundant perforated stone knives, which equivalently existed in the late Shang Dynasty (Lizhou Site Joint Archaeological Excavation Team 1980). In general, with the popularisation of agriculture, the number of perforated stone knives is increasing. On the other hand, the sites in Sichuan Basin reflected that the agricultural development is gradual and preliminary, because other agricultural tools and grain processing tools are scarce (Of course, it may also be related to the easier availability of wooden tools) except for perforated stone knives. However, the discovery of microblade stone tools as well as fishing and hunting tools shows that the livelihood composition at that time is still mixed, and agriculture has not taken an absolute leading position.

Yunnan Province is home to wild rice, but the cultivated rice found in Yunnan Province existed in a relatively late age. Beiqiu site in the Dianchi Lake area (Huang and Zhao 1959; Yunnan Provincial Cultural Relics Working Team 1961), Baiyang site in Binchuan county (Yunnan Provincial Museum 1981), Dadunzi site in Yuanmou County (Yunnan Provincial Museum 1977), and Haimenkou site in Jianchuan county have a history of 4260, 3770, 3210 and 3115 years relatively (Xiao 1995). Another important evidence from the environmental change materials and the magnetic susceptibility analysis of the sediments of Erhai Lake suggests that farming began around 2470 BC (Zhang et al. 2000). According to the analysis of multiple environmental indicators of sedimentary cores, high-resolution analysis and accurate dating, the new study shows that human began to cut down forests as early as 4400 BC, and gradually expanded from the low-lying evergreen broad-leaved forest to the higher hemlock forest. The pollen of higher percentages from gramineae (including rice) reflects the emergence of nearshore farming. Plantago, Labiate, and other plants indicate the existence of animal husbandry, and other weeds, such as Artemisia, Chenopodium, also indirectly indicate the existence of human activities (These weeds usually grow on the roadside, near residential settlements and in abandoned fields). The evidence of pollen deposition is about 1500 years earlier than the archaeological evidence. The analysis of lake sediments also supports this point. Deforestation leads to soil erosion and increase of coarse matter. The analysis of environmental indicators also showed that in the Western Han Dynasty, farming agriculture began to develop extensively due to large-scale migration (Shen et al. 2004). Yunnan Province has long been regarded as the main origin area of rice farming, but there is no evidence to support it in materials in archaeology and environmental change.

The direct evidence of the origin of agriculture in Guizhou Province is the discovery of rice remains. In 1995 and 2002, abundant rice remains were found at the Zhongshui site in Weining County. The radiocarbon dating shows those remains have a history of 3120 ± 65 years (Guizhou Institute of Cultural Relics and Archaeology, et al. 2006). Later, more than 6500 grains were collected from 120 ash pits at the Jigongshan site in Weining County (Guizhou Institute of Cultural Relics and Archaeology, et al. 2005). The rice is short and fat, which may be upland rice, a variety suitable for growing in plateau. The excavation of stone knife, sickle, pestle, and mortar further confirms the existence of farming (Zhang et al. 2008). The ground stone tools unearthed at the Qingchang Neolithic site in Bijie are mainly adzes (38 pieces), axes (9 pieces), and chisels (4 pieces), all of which are fully polished, as well as shovels (2 pieces, only the blade was slightly sharpened) (Guizhou Provincial Museum 1987). Typical agricultural and grain processing tools are still scarce. The same is true for the ground stone tools in other places in Guizhou Province. From the outline of stone hoes rarely found, the single-sided blade is very obvious, still in the shape of stone adzes, and stone knives are also rare (Li 1982; Li and Wang 1984). At the Laopodi sites group in Liuzhi Special Zone, the lower layer is characterised by sand-tempered brown pottery tools with round bottom and ground stone tools, and the original soil surface has architectural relics such as bases of buildings and fences, the age of which is not clear. There are only polished stone axes, stone anvils, and chopping tools in stone artefacts, and only bone knives, bone shovels, and spinning wheels in bone tools, lacking agricultural tools (Wang 2006). The materials related to Guizhou Province show that its agriculture began later than that of Sichuan Province and Yunnan Province, which indicate characteristics of the Bronze Age rather than the Neolithic.

10.5.2 Diversified Cultural Adaptations of Hunter-Gatherers and Response to Agricultural Proliferation

Based on the limited archaeological materials at present, it can be seen how the cultural adaptation of Southwest China has changed since the Upper Paleolithic. In the archaeological materials related to the Paleolithic in Guizhou Province, stone artefacts are often processed in a steep way, which is obviously not used for slaughtering activities, but more likely related to the processing of organic tools such as bamboo, wood, and bones. Flakes are mostly processed towards the splitting surface, which is due to the characteristics of gravel stones. The smooth back of flakes can reduce the scraping resistance in use and are convenient to process organic tools. Organic tools have good elasticity and are not easy to be damaged, but not as efficient as stones. If used as a hunting tool, it is more suitable for hunting some smaller animals (Bleed 1986; Brown 2005). Most of the sites with a history of about 10,000 years in Guizhou Province have bone tools. To some extent, it reflects that hunting small animals began to play a more important role. At the same time, the antler shovel used for excavation reflects the utilisation of rhizome plants with high starch content which needs a larger labour input. The rhizome plants can be regarded as evidence of intensified utilisation of resources. Of course, there are many sharp scraping tools at Chuandong site and Maomao Cave site, which not only reflect the distinctions between different sites, but also indicate the differences between regions.

Levallois technology and other stone artefacts similar to Mousterian stone artefacts, which appeared in the eastern edge of the Tibetan Plateau in Sichuan Province and Yunnan Province in the Upper Paleolithic, reflect the adaptation of relative mobility in hunting. Prepared stone cores and finely prepared stone artefacts are designed to improve the efficiency of hunting. The microblade technology popular later further supported the view of hunting dependence. Standardised production of blades of implements is very conducive to the replacement of composite tool blades, which is the adaptation of hunting. Before the general promotion of agriculture, even in the early stage of the formation of agriculture, hunting in this area still played an important role in livelihood. I mentioned earlier that there is an ecotone from northeast to southwest in China. The northern and southern areas of Yanshan Mountain is one part of the ecotone, and the eastern edge of the Tibetan Plateau belongs to the southern part of it. The advantage of adaptation in the ecotone is the diverse hunting resources, and the disadvantage is the unstable resource conditions. Therefore, the effective adaptation strategy is to adopt a diversified and mixed livelihood form, in which hunting plays an important role.

There are few materials related to the central part of Sichuan Basin. The stone artefacts in Tongliang County, Sichuan Province are characterised by large chopping tools and triangular points. The lithic assemblage at the Taohuaxi site is also dominated by large tools (Chongqing Natural Museum 1992), which shows a means of livelihood that focuses on plant gathering. After entering the Holocene, the large rivers make Sichuan Basin embrace relatively rich fishing and hunting resources, while Guizhou Plateau region is upstream of rivers and has few fishing and hunting resources. Therefore, the utilisation of aquatic resources in Sichuan Basin and its surrounding areas has become a part of livelihood. Of course, it is not as important as South China and Northeast China, and also inferior to the middle and lower reaches of the Yangtze River. We know little about the adaptation of hunting-gathering culture in Southwest China at the end of Upper Pleistocene and for a period of time after the end of Upper Pleistocene, and there is much speculation in the conclusion. In general, it can be concluded that Sichuan Basin is an area dominated by gathering while the western part of Sichuan Province and Yunnan Province is by hunting, and the Guizhou Plateau is more likely to be a mixed economic mode with hunting preference (The primary productivity of the plateau is not as high as that of Sichuan Basin), which is basically consistent with the conclusion of the simulation study. At this time, it is worth noting that interesting signs of intensified utilisation of resources can be seen in Yunnan-Guizhou Plateau.

The abundant remains of animals and plants unearthed at the Tangzigou site in Baoshan, Yunnan Province, at the transitional stage between the Neolithic and Paleolithic, reflect a highly diverse resource environment. The animals studied include invertebrates (snails, mussels, snails, etc.), fish, birds, mammals, etc. However, the lithic assemblage more reflects the function of plant gathering and processing, such as many single-sided hammers (Ji et al. 2004). The typical remains in Guizhou Province at this time is the Guanyindong site in Anlong County. The overall cultural characteristic is very similar to the remains of the Mesolithic in South China. It is a livelihood way that emphasises the use of extensive resources, especially aquatic resources. There are tens of thousands pieces of animal bones unearthed from the site, including more than 20 species and genera, of which there are 4–5 species of invertebrates such as snails and mussels. Among all animal remains, snail shells accounts for a large proportion, and there are differences among cultural layers. The number of snail shells is a gradual increase from bottom to top (Cai and Wang 1998). However, such remains are not commonly found in Guizhou Province. Because it is adjacent to Guangxi Province in geography, it is more appropriate to include it in South China than in Guizhou Province.

The climate environment in Southwest China is relatively stable among all cultural and ecological regions since the Upper Pleistocene, which can be compared with that in South China. South China is located in the tropics, so it is not as affected as the temperate zone. Due to the barrier of surrounding mountains and plateaus, the impact of the climate in the ice age in Southwest China is smaller than that in the same latitude region. In addition, it is also a region less affected by extreme natural events such as cold waves, typhoons, and floods, and the climate environment is relatively stable. After the end of the Late Pleistocene, it was first affected by the southwest monsoon. The climate changed from cold and wet to warm and wet, and the climate improvement after the end of the ice age was accepted earlier than that in North China and the middle and lower reaches of the Yangtze River. According to the environmental data of Erhai Lake, from 11,000 BC to 6400 BC, the lake expanded due to the strengthening of the southwest monsoon. However, in the Holocene warm period, the decrease of effective humidity in the basin caused the lake surface to decrease (Zhang et al. 1998; Zhang et al.

2000), which is significantly different from the warmer and wetter climate in other cultural and ecological areas.

In theory, a stable environment is undoubtedly conducive to hunting-gathering, which will reduce the risk of feeding. At the same time, the diverse resources are also beneficial to reduce the risk of feeding. Moreover, the relatively closed and independent basin environment also decrease the competition between groups. The above analysis in behavioural ecology supports the highly diverse adaptation in Southwest China. The archaeological materials related to the Neolithic is abundant. There are 11 cultural types in Yunnan Province in the Neolithic. Those types are different not only in pottery assemblage, but also in the combination of tool assemblage, reflecting that there may also be differences in livelihood composition between different regions (Xiao 2001). In the Neolithic of Sichuan Basin, there are also cultural areas such as the Western Sichuan Plain, the northern Sichuan hills, the upper reaches of the Minjiang River, the upper reaches of the Dadu River, the middle reaches of the Dadu River (Hanyuan Basin), the Anning River basin, the southern Sichuan region, and the Jialing River basin, which are also very complex in livelihood composition. This is especially true in Guizhou Province. The stable, diverse and relatively closed environment leads to the specialisation of cultural adaptation and forms the regional culture with unique characteristics. In a sense, adaptation specialisation is highly successful. Therefore, it does not need to change. As a result, it makes a conservative impression on people.

When agriculture spread to Yunnan-Guizhou-Sichuan, Sichuan Basin quickly accepted agricultural production, and the process of social complexity was rapid. The major reason is related to the impact on it. At that time, the middle reaches of the Yangtze River and the Northwest China began social differentiation. From this point of view, the acceptance of agriculture in Southwest China is accompanied by complex social factors, that is to say, this process may be accompanied by the strong implementation of external factors. In Yunnan-Guizhou region, agricultural groups first spread into some large basins with surface runoff or lakes (Fig. 10.3), and when they occupied them, the hunter-gatherers were squeezed into smaller basins. This process continued and the agricultural groups occupied all the areas where agriculture could be carried out finally. The hunter-gatherers had to live in the most marginal environment. These areas can hardly carry out agricultural production and are not the best places for hunting-gathering. The "backward ethnic groups"

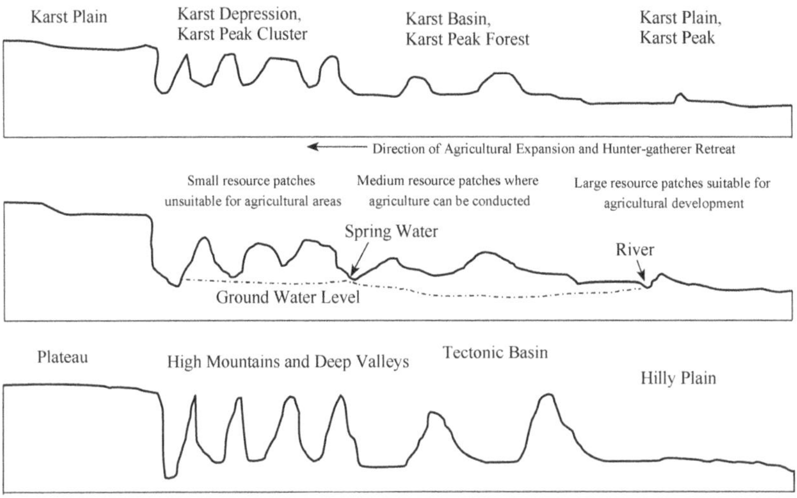

Fig. 10.3 The spread of agriculture in the Southwest and the response of hunter-gatherers

we see in ethnology are the groups forced to live in these marginal environments.

10.5.3 Guizhou Province in Post-Paleolithic

The characteristics of Guizhou Province in the Neolithic are still not very clear. At present, the so-called Neolithic remains are mostly equivalent to the Shang and Zhou Dynasties in the Central Plains. The age of the Neolithic cultural layer of the earlier Feihushan site has a history of 4120 ± 90 years (Peking University Radiocarbon Laboratory 1984). The Neolithic remains found in the North-Pan River Basin are also mainly stone artefacts, which are mostly smashed by sharp edges and less polished. The "Neolithic Age" in the whole Guizhou Province is more like a theory, lacking sufficient material proof. Zhang Senshui first noticed the particularity of the Paleolithic culture in Guizhou Province and put forward the concept of "Epipaleolithic" (Zhang et al. 2000, pp. 51–66). However, at the same time, although the main discussion is about the materials of Yunnan-Guizhou region, he used the concept to refer to the transition stage between the Paleolithic and the Neolithic in the whole

China. Here I prefer to use "Post-Paleolithic" to describe the period from the Upper Paleolithic to the Shang and Zhou Dynasties in Guizhou Province, which is equivalent to the Neolithic in other regions. There is no typical Neolithic Age in Guizhou Province.

The lack of Neolithic in Guizhou Province can be explained from the perspective of cultural ecology. This is not a suitable area for prehistoric agriculture. As previously analysed, once agriculture began to develop, deforestation and land reclamation in this area will lead to the loss of fertile soil and the formation of unfavourable conditions for agriculture. On the contrary, under the conditions of hunting-gathering, human beings have relatively little interference with natural vegetation which is still accumulating soil fertility. Developing agriculture in an area unsuitable for agriculture will inevitably lead to deterioration of the ecological condition. There is no doubt that hunter-gatherers are not willing to adopt this adaptation strategy. As we have seen before, there are signs of intensified utilisation of some resources in the materials related to the sites in Guizhou Province in the early Holocene. But it is not suitable for the extensive operation of the early agricultural type of "slash-and-burn cultivation". Then, the Neolithic in Guizhou Province has not really started. The agriculture in Guizhou Province is foreign, which is produced through the strong implementation of the complex society. That is why the agriculture in Guizhou Province only began in the Shang and Zhou dynasties. However, the history shows that the extension of agriculture in Guizhou Province has not been successful, and its agriculture has been relatively underdeveloped compared with other regions, resulting in the impression of poverty in the history of it. However, this area is actually a living area suitable for hunter-gatherers. Just as the coastal area is the most advantageous area in the industrial era, the plain area of large rivers is advantageous in the agricultural era, and the area with a stable environment and diverse resources is advantageous in the hunting-gathering era. Guizhou is such an advantageous area. "Every dog has its day". Different ways of livelihood require different cultural and ecological conditions. Improper use will cause difficulties in livelihood.

Many caves and rock shelter sites have been found in Guizhou, which are often regarded as relics of the Paleolithic, and quite a few of them may belong to the Epipaleolithic (Equivalent to the Neolithic in North China and the middle and lower reaches of the Yangtze River). During this period, there has been livelihood pressure in Guizhou Province, and human beings have begun to adapt widely, and expand the range significantly, that is, from large resource patches to smaller resource patches (Smaller basins have also begun to be occupied by hunter-gatherers). A difficult problem has emerged: If there is no external agricultural influence, will Guizhou Province begin to develop agriculture? So far, agricultural production is the only strategy to solve the contradiction between population and food.

Notes

1. The term cultural geographical unit refers to a geographical space in prehistoric times that could be easily traversed on foot and had relatively similar environmental conditions.
2. Microblade artefacts made during the Neolithic to Shang-Zhou periods were unearthed from the stratigraphic layers at Qingyuan Site in Youyang County, Chongqing. The estimated age of the Neolithic remains is approximately 4800 to 4600 years ago, approaching the Bronze Age of the Central Plains. Microblade technology is extremely rare in this region, suggesting a possible relationship with population migration during the process of social complexity. See Chongqing Cultural Relics and Archaeological Research Institute, et al. (2009).
3. Based on uranium-series dating of the third layer's deer tooth artefacts, the Ma'anshan site was determined to be approximately 18,000 ± 1000 years old. Radiocarbon dating of the broken bones unearthed from the same layer yielded an age of approximately 15,100 ± 1500 years. See Zhang (1988).
4. It is said to have more than 200,000 artefacts. See Zhang (2003).
5. 1. stone adze; 2. stone pestle.

References

A. Buckling et al., Adaptation limits diversification of experimental bacterial populations. *Science*, 302: 2107–2109, 2003.

A. R. Brown, *The Andaman Islanders*, translated by Liang Yue, Guangxi Normal University Press, 2005.

Archaeology Teaching and Research Office of the History Department, Peking University, Sichuan Provincial Museum, Excavation Report on the Liyuqiao Paleolithic Site in Ziyang, Sichuan. *Acta Archaeologica Sinica*, 1983, 3.

B. Winterhalder, The behavioural ecology of hunter-gatherers. In C. Panter-Brick, R. H. Lyton, and P. Rowley-Conwy eds., *Hunter-Gatherers: An Interdisciplinary Perspective*, pp. 12–38. University of Cambridge Press, Cambridge, 2001.

CA, CASS, et al. Excavation Brief of the Maiping Village and Majiashan Sites in Hanyuan County, Sichuan. *Sichuan Cultural Relics*, 2006, 2.

Cai Huijin, Wang Xinjin, Initial Excavation and Significance of the Guanyindong Site in Anlong. *Guizhou Culture and History*, 1998, 6 and10.

Cai Huiyang, Wang Xinjin, The First Excavation and Significance of Guanyindong Site in Anlong. *Guizhou Cultural and Historical Collection*, 1998, 6 (10).

Cao Zetian, Research on the Maomao Cave Paleolithic Site. *Journal of Vertebrate Paleontology and Paleoanthropology*, 1982a, 20 (02).

Cao Zetian, Study of Bone and Antler Artifacts from Maomao Cave. *Acta Anthropologica Sinica*, 1982b, 1 (01).

Chen Shengqian et al. Excavation Report on the Yuzui Paleolithic Site2 in Yun County, Hubei. *Acta Anthropologica Sinica*, 2014, 33 (01).

Chengdu Institute of Cultural Relics and Archaeology, et al. Excavation Report of the Yingpanshan Site in Mao County, Sichuan. In Discoveries in Chengdu in 2000. Science Press, 2000.

Chongqing Cultural Relics and Archaeological Research Institute, et al. *Youyang Qingyuan*. Science Press, 2009.

Chongqing Natural Museum, Paleolithic Site at Taohuaxi. *Acta Anthropologica Sinica*, 1992, 11 (02).

E. Cashan, Ethnic diversity and its environmental determinants: effects of climate, pathogens, and habitat diversity. *American Anthropologist*, 103: 968–991, 2001.

Gao Xing et al. Excavation Report on the Ranjialukou Paleolithic Site in 2005. *Acta Anthropologica Sinica*, 2008, 27 (01).

Guizhou Institute of Cultural Relics and Archaeology, et al., Excavation Brief of the Jigongshan Site in Weining County, Guizhou in 2004. *Archaeology*, 2005, 3.

Guizhou Institute of Cultural Relics and Archaeology, et al., Shangzhou Period Site at Wujia Daping in Weining County, Guizhou. *Archaeology*, 2006, 8.

Guizhou Provincial Museum, Qingchang Neolithic Site Survey in Bijie County, Guizhou. *Archaeology*, 1987, 9.

Hu Changyu, Discovery of New Stone Age Cave Sites in Jiangyou, Sichuan. *Chinese Cultural Relics News*, Nov. 30, 2005.

Hu Shaojin, Paleolithic Discoveries in Zhangkoudong, Jiuxiang, Yiliang County, Kunming. *Acta Anthropologica Sinica*, 1995, 14 (01).

Huang Weiwen et al., Stone Tool Industry at Dadong, Guizhou. *Acta Anthropologica Sinica*, 1997, 16 (03).

Huang Zhanyue, Zhao Xueqian, Investigation of the Dianchi Lake East Bank Neolithic Site in Yunnan. *Archaeology*, 1959, 4.

Ji Xueping et al. Discovery of Stone Artifacts with Mousterian Culture Characteristics in Daheshui, Fuyuan, Yunnan. *Chinese Cultural Relics News*, Aug. 18, 2006.

Ji Xueping, et al. Excavation Report of Tangzigou Site in Baoshan, Yunnan in 2003. In *Proceedings of the 9th Chinese Conference on Vertebrate Paleontology*. China Ocean Press, 2004.

Lei Yu, Guiyuanqiao Site from the Neolithic Period to the Song Dynasty in Shifang City. Archaeology Yearbook of China 2010. Cultural Relics Press, 2011.

Li Xuanmin, Zhang Senshui, Research on the Tongliang Paleolithic Culture. *Journal of Vertebrate Paleontology and Paleoanthropology*, 1981, 19 (04).

Li Yanxian, Cai Huiyang, Paleolithic Site at Baiyanjiaodong in Puding, Guizhou. *Acta Anthropologica Sinica*, 1986, 5 (02).

Li Yanxian, Wen Benheng, Guanyindong: Cultural Site of the Early Paleolithicin Western Guizhou, Cultural Relics Press, 1986.

Li Yanyuan, Wan Guangyun, Trial excavation and preliminary study of Feihushan Cave Site. *Prehistory*, 1984, 3.

Li Yanyuan, Guizhou's Neolithic Period and the Feihushan Cave Site. *Guizhou Social Sciences*, 1982, 4.

Liu Enyuan, Preliminary exploration of prehistoric culture in Puding and Pingba, Guizhou. *Cultural Relics in Southern China*, 1996, 3.

Liu Zhiyan, The Jiaohuayan Neolithic Period and Warring States, Han, Ming, and Qing Dynasty Sites in Pingshan County. Archaeology Yearbook of China 2010. Cultural Relics Press, 2011.

Lizhou Site Joint Archaeological Excavation Team, Lizhou Neolithic Site in Xichang, Sichuan. *Acta Archaeologica Sinica*, 1980, 4.

Mao Yongqin, Cao Zetian, Preliminary Study of Ground Bone Artifacts Discovered at the Chuandong Site in Guizhou in 1979. *Acta Anthropologica Sinica*, 2012, 31 (04).

P. Bleed, The optimal design of hunting weapons: maintainability or reliability. *American Antiquity*, 51: 737–747, 1986.

Peking University Radiocarbon Laboratory, Radiocarbon Dating Report (VI), Cultural Relics, 1984, 4.

Preliminary Study of the Prehistoric Site of Chuandong (Excavated in 1981). *Acta Anthropologica Sinica*, 1995, 14 (02).

Qiu Zhonglang et al. Fossil and Paleolithic Artifacts from Longtanshan, Chenggong, Kunming. *Acta Anthropologica Sinica*, 1985, 4 (03).

R. H. MacArthur and E. R. Pianka, On optimal use of a patchy environment. *American Naturalist*, 100: 603–609, 1966.

Ren Meie, *Outline of China's Natural Geography* (Revised 3rd Edition), the Commercial Press, 2004.

Secondary Processing at the Baiyanjiaodong in Guizhou, Jianghan Archaeology, 1986, 2.

Shen Ji et al. Lake sediment records of climate change and human activities in the Erhai Lake Basin Since the Holocene. *Science in China Series D: Earth Sciences*, 2004, 34(02).

Sichuan Institute of Cultural Relics and Archaeology, et al., Daditou Neolithic Site in Hanyuan County, Sichuan. *Cultural Relics*, 2006, 2.

Sichuan Institute of Cultural Relics and Archaeology, et al. Excavation Brief of the Jiangweicheng Neolithic Site in Wenchuan County, Sichuan. *Archaeology*, No. 11, 2006.

Sichuan Institute of Cultural Relics and Archaeology, et al. Excavation of the Maiping Neolithic Site in Hanyuan County, Sichuan in 2007. *Archaeology*, 2008, 7.

Sichuan Institute of Cultural Relics and Archaeology, Liangshan Yi Autonomous Prefecture Museum, Excavation Brief of the Maojiakan Neolithic Site in Dechang County, Sichuan. *Sichuan Cultural Relics*, 2007, 1.

Sichuan Provincial Institute of Cultural Relics and Archaeology, Excavation Report on the Yanyundong Paleolithic Site in Beichuan County, Sichuan. *Sichuan Cultural Relics*, 2006, 6.

Sichuan Provincial Institute of Cultural Relics and Archaeology, et al. Daditou Neolithic Site in Hanyuan, Sichuan, Cultural Relics, 2006, 2.

Sichuan Provincial Institute of Cultural Relics and Archaeology, et al. Excavation Report of the Jiangweicheng Neolithic Site in Wenchuan County, Sichuan. *Archaeology*, 2006, 11.

Sichuan Working Team of the Institute of Archaeology, Chinese Academy of Social Sciences and the Cultural Relics Management Office of Guangyuan City, Sichuan, Investigation and Excavation of the Zhangjiapo Neolithic Site in Guangyuan City, Sichuan. *Archaeology*, 1991, 9.

Sichuan Working Team of the Institute of Archaeology, Chinese Academy of Social Sciences, Microliths at Zhongzipu, Guangyuan City, Sichuan. *Archaeology*, 1991, 4.

Study on the Stone Tool Types of Baiyanjiaodong. *Acta Anthropologica Sinica*, 1986, 5 (04).

Sun Hua, Evolution of prehistoric grain types in the Sichuan Basin: information mainly derived from archaeological cultural interactions. *Journal of Chinese Culture*, Nov. 2009.

Tong Enzheng, A discussion on the cultural transmission belt in the shape of a crescent from Northeast to Southwest China. In *Cultural Relics Press Editorial Department, Collection of Cultural Relics and Archaeology Essays*. Cultural Relics Press, 1986.

Tong Enzheng, The Paleolithic in Southwest China. In Research on Chinese Archaeology: Essays in Honor of Mr. Xia Nai's 50th Anniversary. *Archaeology*. Cultural Relics Press, 1986.

Wang Haiping, Research on shouldered stone tools in Southwest China. *Sichuan Cultural Relics*, 1989, 2.

Wang Hongguang, New discoveries and new understanding in Guizhou Archaeology. *Archaeology*, 2006, 8.

Wang Xinjin, et al. 60 years of archaeology in China: Guizhou Province. In *60 Years of Archaeology in China (1949–2009)*. Cultural Relics Press, 2009.

Wu Maolin et al. New discoveries of Paleolithic in Guizhou Province. *Acta Anthropologica Sinica*, 1983, 2 (04).

Wu Xiaohua, Discoveries and zoning of Neolithic to Shang-Zhou Period Cultural Relics on the Guizhou Plateau in recent years. *Sichuan Cultural Relics*, 2011, 11.

Xiao Minghua, An overview of archaeology in Yunnan. *Archaeology*, 2001, 11.

Xiao Minghua, Early Bronze Age Site of Haimenkou in Jianchuan, Yunnan. *Archaeology*, 1995, 5.

Yunnan Provincial Cultural Relics Working Team, Brief Investigation Report of Neolithic Sites around Dianchi Lake in Yunnan. *Archaeology*, 1961, 1.

Yunnan Provincial Museum, Dadunzi Neolithic Site in Yunnan. *Archaeology*, 1977, 1.

Yunnan Provincial Museum, Baiyang Village Site in Binyang, Yunnan. *Acta Archaeologica Sinica*, 1981, 3.

Zhang Chi, Hong Xiaochun, The time of agricultural emergence and related issues in South China and Southwest China. *Cultural Relics in Southern China*, 2009, 3.

Zhang Dayong et al., Research on Theoretical Ecology, Higher Education Press and Springer Group, 2000.

Zhang Dingfu, Research on the Guanyindong Site in Anlong. *Guizhou Culture and History*, 2003, 3.

Zhang Gaike et al., Discovery of a Neolithic Site in Xingyi, Guizhou. *Chinese Cultural Relics News*, Sept.12, 2008.

Zhang Senshui, The Fulin Culture. *Journal of Vertebrate Paleontology and Paleoanthropology*, 1977, 15(01).

Zhang Senshui, Excavation Report of the Ma'anshan Paleolithic Site. *Acta Anthropologica Sinica*, 1988a, 7 (01).

Zhang Senshui, Some Issues Regarding the Late Paleolithic Culture in Guizhou. In *Collection of Essays Commemorating the 30th Anniversary of the Discovery of Maba Human Fossil*. Cultural Relics Press, 1988b.

Zhang Senshui, Preliminary Study of the Chuandong Prehistoric Site (Excavated in 1981). *Acta Anthropologica Sinica*, 1995, 14 (02).

Zhang Senshui, The Epipaleolithic in China. *Journal of East Asian Archaeology*, 1: 51–66, 2000.

Zhang Xingyong (Ed.), *Prehistoric Archaeology of Baoshan*, Yunnan Science and Technology Press, 1992.

Zhang Xingyong, Discovery of Paleolithic Artifacts in Laolongdong, Eshan County, Yunnan. *Acta Anthropologica Sinica*, 1994, 13 (02).

Zhang Zhenke et al., Analysis of Lake Sediment Records of Human Activities in the Erhai Lake Basin, Yunnan, Since the Holocene. *Acta Geographica Sinica*, 2000, 55 (01).

Zhang Zhenke, et al., Stable Carbon Isotope Records of Climate Evolution in the Erhai Lake Area, Yunnan, in the Last 8kaBP, Marine Geology and Quaternary Geology, 1998, 18 (03).

Zhao Dianzeng, Overview of Ancient Cultures in Sichuan. *Journal of Chinese Culture*, 2003, 2.

Zhao Xiaofan, Discovery of early rice remains and harvesting and processing of grains in Guizhou. *Ancient and Modern Agriculture*, 2008, 2.

CHAPTER 11

Who Were the First Inhabitants in the Qinghai-Tibet Plateau?

11.1 Introduction

The Qinghai-Tibet Plateau is an extreme environment for human survival. People travelling to The Qinghai-Tibet Plateau from lower altitudes often face shortness of breath and difficulty walking, let alone toiling or running. However, ancient humans have adapted to the plateau, both physically and culturally.

With regard to the theme of "prehistoric modernisation", three questions are central to the region: ①Who first colonised the Qinghai-Tibet Plateau, the hunter-gatherers, or the early food producers? ②Why did humans come to the Qinghai-Tibet Plateau, given its very harsh environment? How could humans successfully use the resources here? ③How did ancient humans successfully adapt to the Qinghai-Tibet Plateau, and what developmental process did they undergo? What archaeological material evidence is available? Current research generally assumes that hunter-gatherer populations had successfully established adaptation strategies on the Qinghai-Tibet Plateau during the Late Pleistocene, or at least during the Upper Paleolithic.

In contrast, my view is that the successful colonisation of the Qinghai-Tibet Plateau was conditioned by the origin of agriculture, and the earliest Qinghai-Tibetans would have been food producers, not hunter-gatherers. I will analyse who actually colonised The Tibetan Plateau from both theoretical and material perspectives, and discuss the formation process of cultural adaptation of the plateau.

S. Chen, *Prehistoric Modernization*,
https://doi.org/10.1007/978-981-97-9423-2_11

11.2 Cultural and Ecological Conditions of the Qinghai-Tibet Plateau

11.2.1 Basic Physical Geographic Conditions

The Qinghai-Tibet Plateau, as it is commonly referred to, is a vast area that includes plateaus over 4000 metres above sea level and numerous parallel high-mountain ranges, making it an extreme environment for human existence. Covering a total area of about 2.2 million square kilometres, the Qinghai-Tibet Plateau is an area that experienced rapid uplift, rising in some places by as much as 500 metres in 10,000 years. There are mountain glaciers and many lakes on the plateau, while the hinterland of the plateau is very dry and most of the lakes are saltwater lakes. The soils of the plateau are primitive, and most of the soils contain more than 30% gravel. The vegetation gradually transitions from alpine grassland and alpine meadow to alpine desert from southeast to northwest. The coverage of grassland is 20–50%, and although it does not produce much grass, the area is vast and conducive to grazing. In addition, the alpine meadow vegetation at 4500–5000 metres and above, although short and monotonous, can still be used as summer pasture. There are forests in southeastern Qinghai, northwestern Sichuan, southeastern Tibet, and the Himalayas, with good grasslands near the forest line (Ren 2004). According to the forest edge effect (Ling and Braham 1981), this zone has relatively better resource conditions and is more conducive to animal and human survival. Animals on the Qinghai-Tibet Plateau are poor, even insects are scarce, and the animals which can be hunted include wild yaks, Tibetan antelopes, Tibetan donkeys, rock sheep, plateau rabbits, etc. They are fast-moving and not easy to hunt, and the low density of animal resources and fragile ecosystems can easily lead to overhunting of prey. The southeastern edge of the Qinghai-Tibet Plateau is relatively richer in species.

The Qinghai-Tibet Plateau is not a single physiographic unit, and includes at least several physiographic units such as the Yarlung Tsangpo River valley (the Qiangtang Plateau, the Qaidam Basin, the Qinghai Lake Basin, and the high-mountain valley areas on the eastern edge (Ren 2004) mong them, the Yarlung Zangbo River valley is 3700–3900 metres above sea level, and the high mountains to the north and south confront each other, making it difficult to cross. The water and heat conditions inside the valley are relatively good, and it is the best area for agriculture on the

Tibetan Plateau. The Qiangtang plateau is 4300–5000 metres above sea level, where it is arid and barren, and only a few areas with good water and heat conditions have grass growing. The Qaidam Basin is 2600–3000 metres above sea level and is very arid and barren. Deserts, salt lakes, and gobi are widely spread here. The edges of the basin are dotted with oases because of the ice and snow melt water. The Qinghai Lake Basin, at an altitude of about 3200 metres, has vast grasslands and relatively few survival challenges. The high-mountain valley area on the eastern edge of the Qinghai-Tibet Plateau widens from south to north and is influenced by the vertical distribution of vegetation, where resources are rich and diverse but difficult to access.

11.2.2 Cultural and Ecological Conditions of Agriculture: Is the Qinghai-Tibet Plateau Suitable for Food Producers?

Whatever the physical geography, it is always relative to certain cultural conditions. The aforementioned natural geographical conditions of the Qinghai-Tibet Plateau are actually relevant to modern cultural conditions, specifically in terms of agricultural and pastoral production. In order to explore the theme of prehistoric modernisation, we need to understand two aspects: on the one hand, the impact of the Qinghai-Tibet Plateau on food producers, which is currently better known to us as the impact on agriculture and pastoralism; and on the other hand, the impact on hunter-gatherers. Knowing the impact of these two aspects will help us to understand the transition from hunter-gatherer to food production—the prehistoric modernisation process.

The impact of physical geography is usually both positive and negative. The former may attract certain livelihoods or make it difficult to change; the latter may be a hindering constraint, but may also be a pressure that drives cultural change. In short, there is no simple correspondence between physical geography and cultural adaptation, and each case needs to be analysed on its own merits. On the negative side, due to the high altitude, the heat conditions on the Qinghai-Tibet Plateau are unfavourable. The regions with accumulated temperatures greater than or equal to 10°C are significantly lower in areas at the same latitude in comparison to subtropical lowlands. For example, the accumulated temperature of Gyantse County (28°55'N, 4040 m above sea level) is only 1482°C, and it is lower than the indicator of 1700°C at the southern boundary of the cold temperate zone, which is unfavourable to the growth of plants. Additionally, these regions experience large temperature variations, making them susceptible to frost damage. Secondly, the plateau has strong radiation and the airflow is strongly disturbed by the terrain, so thunderstorms and hailstorms are particularly frequent, which can easily lead to catastrophic weather. In addition, the plateau is under the control of high altitude western airflow all year round, with strong and frequent winds and dry climate.

On the positive side, the Qinghai-Tibet Plateau is rich in sunshine, with strong solar radiation and more ultraviolet and infrared parts of the spectrum, both of which are conducive to plant cultivation. Crops such as highland barley can therefore be grown as high as 4900 metres. The annual temperature difference is small, and the winter is not very cold, for example, the average winter temperature in Lhasa is minus 2.3°C. Rainfall and high temperature seasons are synchronised on the Qinghai-Tibet Plateau, and evaporation is low, making the climate humid during the growing season. Soil mineral loss is low, which is conducive to crop growth. The daily difference in temperature is great, and the low night temperature reduces the consumption of nutrients in crops and facilitates the synthesis of crop carbohydrates. Therefore, the crop fruits are large and the agricultural yield is quite high (Ren 2004). Weighing the advantages and disadvantages, it should be said that the Qinghai-Tibet Plateau is still favourable for agricultural production, especially when compared with hunter-gatherer livelihoods.

11.2.3 Cultural and Ecological Conditions for Hunter-Gatherers: Can Hunter-Gatherers Survive on the Qinghai-Tibet Plateau?

(1) The cultural ecological principles adapted by hunter-gatherers are the theoretical basis for understanding human colonisation of the Qinghai-Tibet Plateau.

The current research on human colonisation of the Qinghai-Tibet Plateau is largely centred on archaeological materials. This kind of research that relies entirely on archaeology is very unfavourable. The blind reliance on materials has basically deprived archaeological research of the ability to self-correct. If the material itself is faulty, then the opinion based on it will be inaccurate. Questionable material does exist.

All Paleolithic archaeological material about the Qinghai-Tibet Plateau rarely has accurate stratigraphic evidence to simply correspond the stylistic features of stone tools to the age. For example, chipped stone tools belong to the Paleolithic and stone artifactss of microblade technology to the Mesolithic, while the material is completely unreliable. In addition, archaeology formation processes are basically not considered, so no credible chronological framework has been established so far. It is obviously unconvincing to explore the Paleolithic or NPT transition on the Qinghai-Tibet Plateau when there is no basic reliable spatial and temporal framework. From another point of view, archaeologists would turn to hunter-gatherer studies. Although it began as an ethnographic study of hunter-gatherers in modern times, it studies the cultural adaptation mechanisms of hunter-gatherers. That is, both ancient hunter-gatherers and modern ones, as long as they live by hunting and gathering, are necessarily subject to certain basic cultural adaptation mechanisms (naturally grown plant and animal resources through mobile foraging), which do not differs according to time and place. Looking at the cultural adaptation mechanisms of hunter-gatherers, we then turn to the question of the earliest pioneers of the Qinghai-Tibet Plateau, which can at least theoretically answer the question of whether this area was suitable for hunter-gatherer survival and to what extent it could be exploited by hunter-gatherers. Theoretical studies can narrow the scope of our exploration of the problem. If the theories are not even tenable, then ambiguous archaeological material alone will not solve the problem. Hunter-gatherer studies are the theoretical cornerstone for understanding issues related to the earliest Qinghai-Tibetan topographers. Unfortunately, little attention has been paid to this point in either domestic or foreign research to date.

With regard to hunter-gatherers, we can be certain that hunter-gatherers depend on mobile ways of foraging for natural resources for their livelihoods, as has been common throughout time. But we also know that there are certain exceptions, such as hunter-gatherers who used aquatic resources are able to settle to some degree. This commonality determines that its survival must be constrained by two factors: one is the mobility of humans, that is, the range that can be covered within a certain period of time. If relying on walking, and without the help of horses and oars, the range of foraging must be limited. Meanwhile, we can also be certain that human mobility will be further limited by the rugged terrain and oxygen-deprived plateaus. The second is the density of natural resources. It is obvious that if the resources are abundant, then humans need a small scope of mobile foraging; if the resources are scarce and sparsely distributed, then humans need a greater scope of foraging in order to meet their basic needs. If the resources are scarce beyond what the hunter-gatherer's mobility can cover, then the hunter-gatherer cannot survive. The idea that hunter-gatherers can use any habitat is an untested assumption and a misguided notion informed by the technology of the industrial-agricultural era—the idea that hunter-gatherers can produce the resources they need just like the inhabitants of the industrial-agricultural era. They are people who depended on naturally growing plants and animals for their livelihoods and are subject to the most basic biological laws that required them to find food in order to survive.

(2) Is the land productivity on the Qinghai-Tibet Plateau sufficient for hunter-gatherers to survive?

The net primary productivity of the Qinghai-Tibet Plateau is low because of the low total accumulated temperature. Its distribution decreases from southeast to northwest, and this trend is generally consistent with the hydrothermal gradient. The net primary productivity is lowest in winter (December to February of the following year), accounting for only 1.3% of the year. And it is highest in summer (June to August), accounting for 80% of the year (Zhou et al. 2004). Net Above-Ground Productivity implies that the secondary productivity (animal density) that the area can support is also limited. The seasonality of productivity corresponds to the seasonality of food resources. How much population an area can carry is not determined by the season of greatest resource abundance, but by the season of greatest resource deprivation. Although it should be taken into account that hunter-gatherers can store resources, it is undoubtedly very difficult to supply the other resource-poor ten months with just two months of the peak growing season. Of course, it is still possible to utilise the Qinghai-Tibet Plateau[1] only during the summer and fall seasons, that is, to utilise this period of the most abundant resources seasonally and then not to involve in the plateau for the rest of the year.

According to the hunter-gatherer simulation study, assuming the climate is as it is today, except for the eastern edge areas, large areas such as the highland hinterland and the northwestern desert belt are all unsuitable for hunter-gatherer habitation. This is because the productivity of this area is too low. And with the mobility that depends on walking is not enough to take advantage of the sparsely distributed resources. If the climatic and environmental conditions are harsher than they are now, the possibilities of utilisation will be reduced, even in the season when the resources are most abundant will also be shortened, and the scope for hunter-gatherers to utilise the eastern edge area may be reduced. In summary, the chances of hunter-gatherers surviving in the Qinghai-Tibet Plateau hinterland are slim to none.

There is another noteworthy phenomenon in the simulation study. Although the Qinghai-Tibet Plateau hinterland is very unsuitable for hunter-gatherers' living, the eastern edge of the Qinghai-Tibet Plateau and extending to the southwest is the best zone for the hunter-gatherer adaptation. According to this model, we can deduce that the last area where hunter-gatherers disappeared could be located here. Since this area has undulating terrain, it will have more surface space compared to the same area. In addition, the vertical distribution of resources is also conducive to the acquisition of diverse resources. The ethnographic material also shows that this region has indeed kept the greatest sector of the hunter-gatherer economy. On the one hand, this strong contrast makes us negate the view that the hinterland of the Qinghai-Tibet Plateau is suitable for hunter-gatherer survival. On the other hand, we have to affirm that the eastern edge area has a suitable environment for hunter-gatherer habitat. Therefore, it is wrong to simply say that the Qinghai-Tibet Plateau is suitable or unsuitable for hunter-gatherer survival, and it needs to be viewed in two regions. We can predict the results of archaeological material finds: the highland hinterland will have a small amount of hunter-gatherer archaeological material. In contrast, the edge of the plateau surroundings will reveal the archaeological material of hunter-gatherers with very late age.

(3) Challenges to human adaptation in highland environments

If humans are to live at altitudes above 2500 metres for long periods of time, they will encounter problems such as low oxygen, night cold, low biomass, and high neonatal mortality (Baker 1969), which will require a series of cultural, behavioural, and physiological changes. People from lower altitudes living at higher altitudes need to reduce their activity level and increase their caloric intake to adapt to the cold natural conditions.

People may also experience certain disease risks, including various respiratory infections and altitude sicknesses. From a physiological point of view, the above problems can be avoided only by being born on the highlands. Otherwise, they can only be improved by means of cultural adaptation. This includes minimising the distance and frequency of migration and choosing the best habitat to reduce the range of access to food. If one is going to hunt, then he needs to take longer, better preparation of supplies (stocking up on supplies needed for the journey in advance), greater transport capacity (for example, horses, oxen, vehicles with wheels, etc.), and carry out more ambush hunting rather than chasing. It is even more indispensable to be able to make fire at any time and to be skilled in the processing of leather clothing (Aldenderfer 2003).

Highland environments have low average annual temperatures. Human survival requires high caloric content. One of the signs of successful adaptation to the plateau is the people are able to establish a diet structure required by the plateau environment, such as the high-calorie foods consumed by modern Tibetans, for example ghee (yak butter), tsampa (Zanba), and beef. With the absence of agriculture, it is necessary to rely on hunting to obtain high-calorie food. The problem that immediately follows is the ability of hunter-gatherers to obtain sufficient meat resources through hunting. The thin air of the highland environment obviously limits human movement, making long-distance walking and running difficult. While we may occasionally observe herds of antelopes or even yaks, they are distributed over vast areas, and the average secondary productivity per unit area is very low. In other words, it takes a very wide geographical area for hunter-gatherers in order to be able to hunt enough animals. They need to be more mobile and cover as large an area as possible when seeking food. However, the thin oxygen of the highlands will limit human mobility. It is highly doubtful that hunter-gatherers can successfully survive on the Qinghai-Tibet Plateau without this contradiction being resolved. If hunter-gatherers have to utilise the resources of the region, then it is undoubtedly more appropriate to choose the summer months when food is more abundant.

The Qinghai-Tibet Plateau hinterland, with its extreme environment, infertile land, changing climate, unpredictable and extremely fragile environment, requires highly reliable and flexible livelihood strategies. Among the adaptation strategies of hunter-gatherers, those usually used for extreme environments include improving mobility, expanding diet, increasing reserves, intensifying the use of certain resources (e.g.

plant seeds), promoting sharing and exchange, controlling population size, inventing more sophisticated technologies, using more stable food resources (e.g. aquatic resources), and developing religious rituals to strengthen social identity. But the only ways among them that are applicable to the Qinghai-Tibet Plateau region are more complex and flexible technologies, increasing reserves, and strengthening utilisation. Of these strategies, mobility is the central element, and it is precisely this factor is limited by the Qinghai-Tibet Plateau environment. An effective means of transport requires reliance on the domestication of animals, which is not owned by the hunter-gatherers. Therefore, it is very difficult to develop a highly reliable and flexible hunter-gatherer strategy on the Qinghai-Tibet Plateau. Among the "cultural adaptation strategies" available to hunter-gatherers, there are no necessary strategies to meet the challenges of plateau adaptation.

The eastern edge of the Qinghai-Tibet Plateau is relatively rich in resources. There are high mountains and valleys interspersed with great ecological diversity and rich resources. On the one hand, there is a need for intra-group division of labour (Chen et al. 1990) to exploit the diverse resources; on the other hand, it is basically the headwater area of rivers with rugged terrain. Some of the roads are only seasonally accessible (snowbound in winter), which is not conducive to movement. Therefore, a combination of tools that can facilitate mobility is needed. In comparison to the highland hinterland, this area is lower in elevation and certainly has much better conditions for movement. However, the eastern fringe region is severely deprived of flat land suitable for cultivation and its steep terrain makes it prone to soil erosion after the destruction of vegetation. Therefore, among the two livelihoods of hunting and gathering and food production, this area is actually more suitable for hunting and gathering than food production.

In short, the Qinghai-Tibet Plateau is not suitable for the survival of hunter-gatherers. If hunter-gatherers need to take advantage of the Tibetan Plateau, the first strategy is to utilise it seasonally, and to penetrate deep into the plateau during the summer months; the other strategy is to make use of the eastern edge of the Qinghai-Tibet Plateau. In summary, the best option for hunter-gatherers is to live on the eastern edge of the Qinghai-Tibet Plateau and to access it seasonally when the climatic conditions are suitable to exploit certain resources. It is only possible for people to exploit the Qinghai-Tibet Plateau hinterland if they can master the food production technologies.

11.3 Paleoenvironment

The present environment of the Qinghai-Tibet Plateau is not suitable for hunter-gatherers, but was it suitable in the Neolithic? There is no evidence that humans existed on the Qinghai-Tibet Plateau before the Upper Paleolithic. Moreover, during the Upper Paleolithic (later than 35,000 years before present), the altitude of the Qinghai-Tibet Plateau was already comparable to that of today. The limitation of mobility for hunter-gatherers by the thin oxygen was already present, which is the most important variable in the ancient environment of the Qinghai-Tibet Plateau. That is, the Qinghai-Tibet Plateau was equally unsuitable for highly mobile hunter-gatherers during the Upper Paleolithic. Furthermore, the Pleistocene climate environment is more variable than the rare stable climate of the Holocene (Richerdson et al. 2001), which will increase the survival risk of hunter-gatherers on the Qinghai-Tibet Plateau.

Earlier studies have shown that the climate trends on the Qinghai-Tibet Plateau over 40,000 years include a wet period for 25,000 years, a dry and cold climate for 25,000–10,000 years, a dry and warm climate for 10,000–7000 years, and a wet and warm (suitable) climate for 7,000–36,000 years (Kezao et al. 1990). New studies have showed a high temperature heavy precipitation event 25,000 years ago (Shi et al. 2002). Analysis of lake boreholes, peat profile, and pollen from seven sites on the Qinghai-Tibet Plateau showed that 25,000–15,000 years ago, the climate was dry and cold in southeastern Tibet, Ruoergai County in Western Sichuan, Qaidam in Qinghai, and Kekexili region, and the average annual temperature was about 6 °C lower than today. The vegetation is mainly desert grassland, and the forests in the Qinghai Lake area retreated to grassland vegetation. Only in the eastern and southern edges of the Qinghai-Tibet Plateau are forests distributed (Tang et al. 1998). Climate simulation shows a 13–2 °C decrease in mean annual temperature during the Last glacial period on the Qinghai-Tibet Plateau, accompanied by a more arid environment (Liu et al. 2002). The Qaidam Basin and the Qinghai Lake Basin were extremely arid 15,000–9000 years ago. And the eastern edge of the plateau became warmer and wetter from 13,000 years ago due to the influence of monsoons (Lehmkuhl and Haselein 2000). the Holocene Climatic Optimum in the Qinghai Lake Basin peaked at

6700 years before present (Liu et al. 2002). The studies of paleoenvironment show that the best use of the Qinghai-Tibet Plateau by humans was during HCO, approximately 7000 years ago.

Another time when humans may have entered the Qinghai-Tibet Plateau is 25,000 years ago, the interglacial period of the Last glacial period (53,000–25,000 years ago), when nearly half of the Qinghai-Tibet Plateau was forested (Shen et al. 1996). Although the climate was not as stable as the Holocene at that time, it was much better than the LGM Period. Therefore, even if humans were able to enter the Qinghai-Tibet Plateau 25,000 years ago, they would have had to leave at the time of the Last glacial period, just as they did in the Siberia region (Geobel 1999). During the interglacial period of the Last glacial period, the Mu Us Desert and its margins were successively discovered with the sites of Salawusu and Shuidonggou, indicating that this marginal environment was once suitable for human habitation. However, it is noteworthy that the entire American region was not colonised by humans at this time (Gao et al. 2008), and there was still a huge living space to expand. The space for hunter-gatherers in Eurasia was at its peak, and the marginal environments that have now become deserts were available for human use. In short, there was no need for humans to enter the Qinghai-Tibet Plateau at this time. And at the height of the Last glacial period, the Qinghai-Tibet Plateau environment is even more extreme. Even if humans wish to enter, there are not enough resources available.

The Younger Dryas Event impact hypothesis is another challenge. Current evidence suggests that most of it on the Qinghai-Tibet Plateau occurred between 11,000 and 10,000 years ago. Due to its huge area and special geographical features, the Qinghai-Tibet Plateau will be more sensitive to climatic and environmental changes, and its ecosystem is also more fragile. So the impact of the Younger Dryas Event impact hypothesis on the Qinghai-Tibet Plateau is more obvious (Shen et al. 1996). The prolonged cooling event, spanning over a thousand years, has rendered the already low productivity even more barren. The amplified environmental effects have made this region even more unsuitable for human habitation. Even if any humans ventured into the Qinghai-Tibet Plateau, they would undoubtedly have struggled to survive. This time period is usually considered to belong to the Mesolithic, and many sites on the Qinghai-Tibet Plateau are classified as this stage, which is clearly unreasonable from the point of view of environmental conditions.

11.4 Neolithic Archaeological Record

Before answering the question of who colonised the Qinghai-Tibet Plateau, why they approached the plateau, and how adaptation was formed, it is necessary to review the available archaeological material to understand what extent it can answer the above questions. I will still start with the known and relatively clear and late material and gradually progress to less clear material and times, with the aim of gaining knowledge of the process from the known to the unknown. The sites that have been systematically excavated on the Qinghai-Tibet Plateau include the Karub (kha-rub) site in Chamdo, the site at Jiaritang (lcags-ri-thang) in Damxung County, the site at Qugong(chu-gong) in Lhasa, the site of Changguogou (phreng-sgo) in Gongga County, as well as the site at Xiaoenda (dngul-mdav) in Chamdo, the site of Yunxing, Jumu in Linzhi, and several sites in Medog County. I will first focus on these few clearly dated and important finds.

The Site at Jiaritang in Damxung County. This is one of the sites that have been systematically excavated. The Karub relics in Chamdo, the site at Qugong in Lhasa, the relics site of Changguogou in Gongga County and this site are all cornerstones of the prehistoric era for the Qinghai-Tibet region. The site is located in the connection between the nomadic area of the northern Tibetan grasslands and the agricultural area of the Lhasa Valley, at an altitude of 4234 metres. After three excavations, a total of 2900 square metres were uncovered. One hearth and one ash pit, and two stone pavement remains were found. The hearth is composed of three stones. A total of 2800 artefacts were collected and excavated, and the lithic assemblage consisted mainly of products based on microblade technique, including pencil-shaped microcores and prismatic microcores. There are few ground stone tools (including perforated stone balls and pitted grinding stones on one side. The pottery sherds are sparse and fragmented, with jars predominant. Carbon-14 was used to date it to 3200 BC–2900 BC. Researchers believe that its subsistence base is nomadism and hunting, which is different from the Karub and Qugong cultures. This site belongs to a nomadic site at a high altitude (Tibet Autonomous Region Cultural Relics Bureau, et al. 2005).

The Site of Changguogou in Gongga County. This site is located on the first terrace of the seasonal river in the Changguogo Gorge valley of the Yarlung Tsangpo River basin, with an altitude of 3570 metres and surrounded by high mountains on three sides. More than one thousand cultural artefacts were obtained through investigation and small-scale test excavations. The lithic assemblages consist of 257 pieces of general ground stone tools and 587 pieces of microblade artefacts (conical and semi-conical microcores, microblades, and debris), with the stone artefacts found in a concentrated area of more than 20 square metres, which may be the site of stone tool production. Six pieces of polished stone tools, including stone knives, stone mortars, grinding slabs, handstones, drilled round tools, and denticulates (Binford 2001). In addition to these, 162 pieces of pottery sherds were found. Plant remains such as millet and highland barley were also found at the site. Using radiocarbon dating, the age of the cultural artefacts can be determined. The animal bone sample yielded a radiocarbon date of 2896 ± 99 years before present, while the charcoal sample yielded a radiocarbon date of 3044 ± 102 years before present. These dates correspond to the early Western Zhou period in the Central Plains (Archaeological Team of the Institute of Archaeology, Chinese Academy of Social Sciences, Tibet Work Group, Tibet Autonomous Region Cultural Relics Management Committee 1999). The date is closer to that of the Lhasa Qugong site, but there are more components of microblade artefacts in the stone assemblage.

Karub Ruins. The site is located on a terrace at the confluence of the Karuob River, a tributary of the Lancang River, in Changdu County. Situated at an altitude of 3100 metres, it lies in the high mountain canyon region of the eastern Qinghai-Tibet Plateau. The valley's shady slopes are covered with forests. The site has been excavated twice, uncovering a total of 1,800 square metres. And it is divided into two phases: early and late (the early phase is also divided into two sections: former and later). From the early to the late period, its stone tool assemblage shows an increase in microblade stone tools, and a significant decrease in the number of ground stone tool. The pottery assemblage also exhibits similar characteristics, with the late period showing simpler vessel shapes and decorations, and the absence of painted pottery. From the style of the building, there are three styles in the early period. But only one kind of stone wall semi-cave house is seen in the late period. Judging from the function of stone tools, ground stone tools are mainly axes, adzes, chisels and knives, and ground perforated stone knives are abundant in form. The chipped stone tools are shovel-shaped tools, hoe-shaped tools, plough-shaped tools, etc., which may be earth-breaking tools. The ground stone tools are more developed in the early period, probably, mainly agriculture, and in the late period agriculture declined and the proportion of hunting increased. Karub Ruins has revealed uncharred millet husks and possible remains of domesticated pigs. These findings date back approximately 5000 to 4000 years ago, corresponding to the Longshan period in the Central Plains (Tibet Autonomous Region Cultural Relics Management Committee, Department of History, Sichuan University 1985).

The Site of Qugong in Lhasa. The site is located in the northern suburbs of Lhasa at the foot of the Sulawesi Mountain on the northern edge of the Lhasa Valley, with an altitude of 3680 metres. After three excavations, the total excavated area of the site and the burial area is 3187.5 square metres. Fifty-eight burials, rituals, ash pits, and other remains were found. More than 10,000 relics were unearthed, including mostly stone tools, pottery, bone tools, bronze tools and a large number of animal bones. The lithic assemblages (12,000 pieces in total) are mainly made of chipped stone tools, including microblade artefacts. The ground stone tools (25 pieces) are few, mainly combs, knives (1 piece), tooth sickle (1 piece), adzes, arrowheads, and perforated stones. A total of 382 pieces of grinding tools were also found. Animal bones show that yaks, sheep, and domesticated dogs were raised in early times. The other animals are wild animals, with deer predominating. The date is 2000 BC–1500 BC (Institute of Archaeology, Chinese Academy of Social Sciences, Tibet Autonomous Region Cultural Relics Bureau 1999).

Through more systematic field archaeological work, the above four sites have covered the basic features of Neolithic remains dating from 5000–3000 years ago. A particularly striking feature is the prevalence of microblade artefacts. They are not the remains of the Mesolithic, but of the Neolithic. Certainly, the microblade artefacts coexist with pottery at these sites. At present, we can be sure that the Neolithic of the Qinghai-Tibet Plateau still retains the microblade technology, which is dated as late as the historical period of the Central Plains. Secondly, its pottery and ground tools are not very well developed, but there are already crop harvesting tools such as stone knives and denticulate sickles. Furthermore, Karub Ruins exhibits notable characteristics of a coarser and less refined culture in its later period compared to its earlier phase. This observation is highly significant and deserves attention.

Given the available material, we can determine that the earliest evidence of the Neolithic on the Qinghai-Tibet Plateau is only 5000 years old.

11.5 The Earliest Qinghai-Tibetans

11.5.1 The Human Remains of the Late Pleistocene on the Qinghai-Tibet Plateau

1) The main findings

For a long time, the Paleolithic remains on the Qinghai-Tibet Plateau were considered to be: the Sure site at Dingri County in southern Tibet (Zhang 1976), Zhulele at Shenzha County in northern Tibet (An et al. 1979),[2] Geting at Bange County (Qian et al. 1988), Zhabu and Duogeze at Ritu County in the Ali region (Liu 1986), Sanchakou at Kekecili in Qinghai (Qiu 1958), and Xiaochaidan in the Qaidam Basin (Huang et al. 1987). The excavated objects from the Leng Lake and Xiaochaidan sites in the Qaidam Basin are surface collections. The stratigraphy of the excavated stone tools was later re-evaluated and dated to 30,000 and 37,000 years ago, respectively. Since the association of the excavated material with the stratigraphy was not obtained from excavation, the reliability of the dating is questionable. The material from the other sites was surface-collected with no stratigraphic evidence (Table 11.1).

In recent years, new sites have been discovered in the Qinghai Basin, located on the edge of the Qinghai-Tibet Plateau (Gao et al. 2008; Yi et al. 2011). Among them, Heima River, Jiangxi Ditch, and the back of ditch 001 sites have undergone small-scale excavations. Heima River and Jiangxi Ditch sites have been subjected to radiocarbon dating based on ash piles from hearths (Brantingham and Gao 2006, pp. 387–414). The dating results are relatively reliable, but not earlier than 15,000 years ago at the earliest (corrected dating). Importantly, these sites are characterised by lower elevations and are situated in the peripheral regions of the Qinghai-Tibet Plateau.

Nineteen hands and footprints are found in the travertine at the Chusang site near Yangbajain. The palm prints are photoluminescently dated to 22,000 years ago (Zhang and Li 2002, pp. 161–163). In addition, a hearth is also found nearby. In recent years, the Paleolithic site at Selincuo on the northern Qinghai-Tibet Plateau is considered to be evidence of the Late Pleistocene human activity (Yuan et al. 2007). Besides, thirteen lithic sites are also found around Zhongba in southwest Tibet, and some of them are also considered to belong to the Paleolithic (Zhang et al. 2006).

Table 11.1 Typical Paleolithic Sites on the Qinghai-Tibet Plateau

Sites	*Altitude (a.s.l.m)*	*Typical remains*	*Date of identification*
Su Re	4500	Flake tools	Middle and Late Paleolithic
Ge Ting	4663	Siliceous flake tools	Late Paleolithic or Neolithic
Duo Geze	4830	Flint, jasper, agate and other stone pieces	Late Paleolithic
Zha Bu	4600	Same as above	Late Paleolithic or Mesolithic
Zhu Lele	4800	Microblade stone artefacts and flake tools	Late Paleolithic or Early Mesolithic
Se Lincuo	4600	Flint flake tools, microblade cores	Before the Last Glacial Maximum
Xiao Chaidan	3170	Collection of 160 lithics	30,000 years BP(?)
Chu Sang	4200	19 hand and footprints, hearths	22,000 years old BP
Leng Lake 1	2804	Levallois artefacts	37,000 years old BP(?)
Hei Ma Lake 1	3210	Microblade artefacts	Carbon-14 dated about 10,000 BC
Jiangxi Ditch 1	3300	Microblade artefacts, ash heap	Carbon-14 dated about 12,000 BC
Xiadacuo	4350	Handaxes, thin-bladed axes, etc	Early Paleolithic period
Luhuo, Kangding	3556	Handaxes	Early Paleolithic period

The earliest evidence of human activity on the Qinghai-Tibet Plateau comes from the Xiadacuo lakeshore in the Ali region, where typical handaxes are collected (Huo 1994).[3] Before that, Mr. Tong Enzheng has studied the two handaxes donated by the American J. H. Edgar in the 1930s. These two handaxes are respectively found in the area of Luhuo and Kangding (Tong 1989).

The Qinghai-Tibet Plateau region is an alpine zone, making field archaeology very difficult. Therefore, most of the current major discoveries are made by surface surveys and most of the discoverers are geological surveyors, and the surface collection is not systematic. This also makes the accuracy and representativeness of the archaeological materials less than ideal. The excavated (and only small-scale test excavations) sites are located on the edge of the Qinghai-Tibet Plateau, that is, around the Qinghai Lake basin. No Paleolithic sites have been systematically excavated in the hinterland of the Qinghai-Tibet Plateau. This makes the discussion of the Paleolithic transition on the Qinghai-Tibet Plateau rather vague (Tang 2011). The first issue to address is whether the Qinghai-Tibet Plateau had an existence during the Paleolithic. It is necessary to examine whether the current materials and research methods are sufficient to support the notion of a Paleolithic presence on the Qinghai-Tibet Plateau. Once these questions are resolved, it would then be possible to consider the periodization of the Paleolithic and the transition from the Paleolithic to the Neolithic.

2) The dating issue: the technology and typology are unreliable in determining the date of stone tools

The first thing that needs to be determined in order to explore the earliest evidence of human activity on the Qinghai-Tibet Plateau is the dating of lithic remains. However, the technology and typology currently used are largely unreliable in determining the date of stone tools. The primary reason for this is that the products of one lithic technology are very diverse.[4] The production process for ground stone tools includes blank knapping, which produces many flakes, broken chunks, debris, and semi-finished products (which for some reason are not further processed into finished products). If the site itself is a lithic manufacturing place, it is likely that no ground stone tool would have coexisted (the semi-finished products for further grinding and processing are taken away), and only purely chipped stone artefacts would have survived (Qian et al. 1988). If only these flakes and semi-finished products are collected, it is likely that they will be considered as Paleolithic products. Moreover, the technique of chipping is not unique to the Paleolithic, as there is no shortage of chipped stone tools from the Neolithic and even more recent times. It is not impossible that chipped stone tools were distributed separately, and it is entirely reasonable that there were chipped stone tools belonging to the Neolithic period in stone tool manufacturing sites as well as in sites with special functions.[5] It is possible that isolated handaxes, such as those found at Luhuo and Kangding, served as semi-finished products for Neolithic stone tools.

A similar problem exists for the microblade technology artefacts. The iconic artefacts include microblade, microcores, prepared stone cores, and the typical by-products of the production of the microblade process (crest blade, sled-shaped flake). As for the by-products formed in the process of making blanks of bifacial tools during the prepared stone cores, they are all chipped stone pieces with characteristics that are not typical.

If they were found separately, it is likely that they would be considered to belong to the type of flake industry and would be regarded as typical Palaeolithic remains distinct from the microblade technology artefacts. But they are really nothing more than the by-product of the microblade technology stone assemblage. Moreover, because the prepared processing of microblade cores resulted in generally small flakes, some of them may indeed have been used for repair. Thus the understanding similar to that of the North China Small Stone Tool is concluded.

At the same time, these pieces can easily be considered as Levallois pieces because of the repair marks all over the back and the presence of prepared platforms. In addition, the microblade technology, as a technology suitable for mobility, and its lithic are very commonly used outside of the central site. So it is entirely possible that they did not coexist with pottery and ground stone tool, even if the users were people from Neolithic period. The microblade technology lasted from its origin to its disappearance from the Upper Paleolithic all the way into the historical period. This technique disappeared in the Central Plains in the Early Neolithic, but persisted in the surrounding areas. It is not possible to judge their age based on the shape alone. For example, in the Qiangtang area, the microblade artefacts were unearthed along with the ground stone tool, and researchers concluded that they belonged to the Early Neolithic sites by the similarity of the technical types of stone artefacts to those in northern China, which is obviously unreliable (Wang et al. 2009). This is evidenced by the prevalence of microblade artefacts at the Neolithic sites (5000–3000 years ago) on the Qinghai-Tibet Plateau.

In areas where there is no systematic field archaeological work (survey and excavation), the reliance on a few under-represented collected lithic and their dating based on technical types is undoubtedly not very credible. The technological types of stone tools lasted for a very long time. The technology of chipping and the microblade technology lasted throughout the Paleolithic, Neolithic, and historical periods. It is not theoretically valid to answer whether there was a Paleolithic period on the Qinghai-Tibet Plateau with this.

3) The discovery and interpretation of microblade technology on the Qinghai-Tibet Plateau

Theoretically, there is no good reason to suggest that the microblade technology artefacts are Paleolithic remains. In terms of actual archaeological findings, there is also insufficient evidence to support the existence of microblade technology in the Paleolithic of the Qinghai-Tibet Plateau. To date, a considerable number of sites containing microblade technology artefacts have been discovered on the Qinghai-Tibet Plateau: 42 sites were counted in 1989 (Duan 1989); there were more than 80 sites by 1994 (Tang 1999). During the construction of the Qinghai-Tibet Railway, 27 more lithic remains sites were discovered along the route (Tibetan Autonomous Region Cultural Relics Bureau, et al. 2005),[6] including the Jiaritang site. From the available material, none of the sites have conclusive stratigraphic and chronological evidence to suggest that the microblade stone artefacts belong to the Paleolithic. The so-called slice stone artefacts are also questionable, as they could have been a by-product of the microblade technology. It is therefore questionable whether the flake tools are independent lithic assemblages in areas where microblade stone artifacts are widespread. This is because the chipped flakes collected at the surface can never be equated with old stone tools. Some scholars believe that the presence of Levallois technology exists in the Zhu Lele site and Geting site (Qian et al. 1988), but this also cannot be equated with the Paleolithic technology. As the prepared flakes existed during the processing of blanks (bifacial tools) of microblade core, which is similar to the Levallois technology.

The remains of microblade stone artefacts have been widely found in the Yarlung Tsangpo River basin. Either they are simply distributed, or coexist with flakes and stone tools, or coexist with ground stone tool and chipped stone tools. Some scholars believe that its distribution is related to the physical geographic features of the region, such as the first one is mostly found on the plateau surface, and the last one is mostly found in the high-mountain valley area. It is originally assumed that the microblade stone artefacts in southern Tibet is less typical than in northern Tibet, but new findings show that the difference is not as great as originally thought.

Typical conical microcores were also found around Zhongba region in southwestern Tibet (Zhang et al. 2006). A large number of microblade stone artefacts found in the Qiangtang basin of northern Tibet are also very delicate (Wang et al. 2009). The microblade stone artefacts in the high-mountain valley region of eastern Tibet are represented by the Karub site and Xiaoenda site in Chamdo with similar technical techniques and very late ages. The earliest dates of these microblade stone artefacts remains found so far may be comparable to the HCO. Associated with different economic forms, they may have belonged to both hunter-gatherers and food producers (crop growers or livestock groups). The HCO on the Qinghai-Tibet Plateau was warm and humid, with adequate lake recharge and many freshwater lakes, which favoured human survival at this time (Liu 1986).

At present, the earliest microblade materials in the Qinghai-Tibet Plateau region are all found in the northeastern margin. "microblade debitage" have been found at the Heima River, Jiangxi Ditch and its back sites, but no microblade cores have been discovered. The distributional characteristics of the microblade stone artefacts remains are closely related to the nature of the technique. The standardised stone blade was lightweight and easily replaced, making it ideal for a highly mobile subsistence practice. Its presence on the Qinghai-Tibet Plateau suggests that the population living here retains at least part of a highly mobile form of livelihood, namely hunting. Even agricultural groups probably had seasonal hunting activities to supplement their livelihood. As highly mobile hunting groups, they generally did not carry heavy implements such as pottery and polished stone tools. So the fact that certain microblade stone artefacts did not coexist with pottery and polished stone tools does not mean that they are not from the same period. In addition, hunting groups usually repaired tools and used some flakes while waiting for animals at lookout points, animal slaughterhouses or temporary places for overnight stays (Binford 2009).[7] And it may not leave remains with typical microblade features such as microblade and microblade cores, thus resulting in flakes and stone tool remains.

In short, the available material does not support the idea that the microblade technology artefacts are Paleolithic remains. These artefacts more likely to be evidence of human activity after the HCO.

4) Are there any Late Pleistocene sites on the Qinghai-Tibet Plateau?

With the exception of the Late Pleistocene microblade stone artefacts remains on the margins of the Qinghai-Tibet Plateau, the Paleolithic sites found on the Qinghai-Tibet Plateau are questionable. The problems of the Leng Lake and Xiaochaidan sites have been mentioned earlier, and the theoretical and archaeological material perspectives have also been discussed in terms of microblade technology artefacts and possible associated lithic remains. In other words, we do not yet have conclusive evidence of Paleolithic lithic materials. The hand and footprints found in the Chusang site were dated just before the Last Glacial Maximum Period, and no accompanying lithic were found. It is also unknown whether the hearths were contemporary with the palm prints. Curiously, instead of using the more mature Carbon-14 dating technique to date the hearth material, the method of photoluminescence was used, which makes Chusang's findings less convincing.

In 2007, Yuan Baoyin and others reported the discovery of Paleolithic remains from the ancient lakeside terrace (4600 m above sea level) on the southeastern shore of Selincuo, northern Tibet, dating the terrace to 40,000–30,000 years ago. They pointed out that the artefacts that were found showed a strong European Middle Paleolithic cultural style in terms of technology and typology, suggesting that the emergence of early humans in northern Tibet might be related to the Late Pleistocene transcontinental wave of early human migrations (Yuan et al. 2007). However, Carbon-14 dating is inverted, with modern lakeside carbonate "pebbles" dating to only 6360 years ago. In addition, the age of the terraces does not equal the age of the lithic artefacts, so the question of the age of the lithic remains is still unresolved. It is noteworthy that its excavated artefacts contain a wedge-shaped microcore. Its presence would suggest that this site is probably not Late Pleistocene in age, but rather a relatively recent product of the Holocene. Because this style of stone artefact is widespread in the Qinghai-Tibet Plateau region, the remains with clear absolute dates are quite recent. This is also consistent with the age of the microblade technology that persisted in the Northeast China and Inner Mongolian grassland regions.

The most difficult to deny the evidence of Early Pleistocene human activity on the Qinghai-Tibet Plateau is the hand axe remains found on the lakeshore of Xiadachuo in Ali region. The handaxes excavated from the Luhuo and Kangding areas are not typical enough (Lü 2011), or will also negate the evidence of human activities. While the handaxes excavated in Xiadacuo have typical features. At the same time, scrapers, points, chopping tools were also found, and the surface of the stone tools were severely weathered, but there was no trace of water carrying and washing (Huo 1994). The typical hand axe is not an Upper Paleolithic artefact. If this finding is true, then it is reasonable to assume that the Qinghai-Tibet Plateau was occupied by humans as early as the Early Paleolithic. At that time, the Qinghai-Tibet Plateau was not as high as it is now (4350 m above sea level at the Xiadacuo site) (Tang et al. 1998), and humans may have used it during the season when food resources were more abundant. The question may not lie in whether early humans in the Paleolithic accidentally ventured into the Qinghai-Tibet Plateau, but rather whether there was sustained human habitation in the region, especially during the peak of the LGM and the subsequent the Younger Dryas Event.

11.5.2 The Origin of Ancient Humans on the Qinghai-Tibet Plateau

Studies on the origin of ancient humans on the Qinghai-Tibet Plateau have gathered contributions from multiple disciplines, including physical anthropology, linguistics, molecular biology and archaeology. On the one hand, it answers the origin of the earliest Qinghai-Tibetans, and on the other hand, it answers the time when ancient humans entered the Qinghai-Tibet Plateau. The physical anthropological characteristics of contemporary Tibetans show two types: the southern type (group A) and the Kam type (group B). The former is closer to Yunnanese, Assamese, and Southern Mongolians, while the latter is closer to North Chinese. However, there is no absolute boundary between the two (Han and Zhang 1991), and there may also be a mixture of western and northern inhabitants (Lin 1981).[8] The human skeletons excavated from the site at Qugong in Lhasa still retain undifferentiated features that are similar to both of two types (CA, CASS, and Tibet Autonomous Region Cultural Relics Bureau 1999).

Molecular biology studies that do not rely on archaeological materials provide another clue to information on the origin of the earliest Qinghai-Tibetans. A. Torroni and other suggested that Tibetans originated from North Asia and Siberia based on mitochondrial DNA analysis (Torroni et al. 1994, pp. 189–199), and the study of the Y chromosome affirmed that Tibetans originated from North Asia (Qian et al. 2000). The Y chromosomes of 31 members of Sino-Tibetan language groups were analysed by Su Bing and others which showed a very strong linkage with each other. Given that the westward and southward migrations of Tibetan-Burmese groups were influenced by the bottleneck effect, it is inferred that the groups that invented agriculture after 10,000 years ago in the middle and upper Yellow River were the ancestors of Tibetans. The divergence of language may have occurred 5000 to 6000 years ago. And the earliest Tibetans probably entered from the east through the Tibet-Burma corridor (Su et al. 2000). It has also been suggested that the earliest population in Tibet arose 5000–4000 years ago (Sforza 1998). Van Driem, combining linguistic, demographic, and archaeological analyses, proposed that the homeland of the Sino-Tibetan-Burmese people was in the area of Shanxi and Shaanxi provinces. Around 6500 B.C., the Sino-Tibetan-Burmese group expanded into the Gansu and Qinghai regions and then entered Tibet (Driem 2002, pp. 233–249). Although different scholars differ on the chronology of human entry into the Qinghai-Tibet Plateau hinterland, the time is basically within the Neolithic period, rather than the Paleolithic.

There is also some evidence on lithic archaeological materials. Mark Aldenderfer and Zhang Yinong noted that there are two distinct styles of lithic culture in Tibet, with the northern part characterised by microblade technology and the southern part characterised by small flakes (Aldenderfer and Zhang 2004). The recognition of the two lithic craft traditions has also been proposed in earlier studies. It is consistent with the characterisation of physical anthropology studies and supports the theory that the Tibetan population originated from both North and Southwest China (Tang 1999; Duan 1989). However, the "small flake" lithic tools from southern Tibet are more likely to be an irregular microblade production (Zhang et al. 2006), which does not differ much from the northern part of the country, as shown by the findings from the northern site of Zhongba. The origin of lithic traditions and populations is still an unresolved theoretical issue (Wang et al. 2009), and different lithic technology traditions do not equal different population origins. From the perspective of climatic conditions, there do exist two periods of time suitable for human migration, respectively 6000–3600 and 3000–2200 years before present (Wang and Xiong 2004), when the climate was relatively warm and humid.

Currently archaeologists largely ignore the bioarchaeological evidence, as if such studies are less practical than the discovery of archaeological material. In fact, the more reliable archaeological evidence available does not contradict the bioarchaeological view, and both support the idea that the earliest Qinghai-Tibetans would have migrated to the Qinghai-Tibet Plateau in the Holocene. There is no direct continuity between the Paleolithic era and the Neolithic era on the Qinghai-Tibet Plateau.

11.5.3 The Time and Way of Human Access to the Qinghai-Tibet Plateau

Bioarchaeological studies have answered the timing, and even the direction, of human entry into the Qinghai-Tibet Plateau. But it does not answer the reason why humans entered this barren region. "Barrenness" is a relative and historical concept, and whether an area is "barren" or not is relative to the way it is used. The desert of Saudi Arabia is very barren, but it is rich in oil. In a modern economy sense, it is also very productive. What is meant by barren is that it is not suitable for agriculture and plant growth. As pointed out earlier, although the Qinghai-Tibet Plateau is barren, it is rich in sunshine. As long as the appropriate soil is available, it will be suitable for agriculture. But it is indeed unfavourable to hunter-gatherers, with low primary productivity (few plants) and even lower secondary productivity (monotonous and scarce animals), and the plateau environment is not conducive to human action. So for hunter-gatherers, the Tibetan Plateau is a barren place. But why did hunter-gatherers still enter the Qinghai-Tibet Plateau?

Brantingham and Gao Xing proposed a colonisation model of the Qinghai-Tibet Plateau, suggesting that some hunter-gatherer groups were forced to move to the Qinghai-Tibet Plateau to survive due to "competitive exclusion" caused by the squeeze of agricultural populations. This is similar to the Southwest region mentioned earlier, where hunter-gatherer groups squeezed by agricultural groups have to live on smaller and smaller land resources. They suggest that hunter-gatherers may have gradually entered the Qinghai-Tibet Plateau in "three steps". Firstly, they may have survived as early as 30,000 years ago in areas with altitudes of 3000–4000 metres. Secondly, during the period 30,000–15,000 years ago, they were mainly used the plateau margins seasonally. Finally, after 8200 years before present, hunter-gatherers chose to enter the Qinghai-Tibet Plateau due to the competitive squeeze from agricultural groups at lower altitudes, which coincided with the beginning of the HCO Period (Brantingham et al. 2003; Brantingham and Gao 2006).

However, my viewpoint is different. I believe that the earliest pioneers of the Qinghai-Tibet Plateau were food producers, rather than hunter-gatherers. The time of entry into the Qinghai-Tibet Plateau was also relatively late. It was in the Middle Neolithic of the eastern region, and after the primitive agricultural ecosystem was more mature, when agricultural populations began to expand into the Qinghai-Tibet Plateau. The above article explains why the earliest pioneers on the Qinghai-Tibet Plateau were not hunter-gatherers in terms of cultural and ecological conditions, palaeoenvironment, archaeological materials, and bioarchaeology in the Qinghai-Tibet Plateau region. This is very different from the southwest region. In the southwest, after being ostracised by agricultural groups, hunter-gatherers chose to retreat to the high-mountain valley areas where agriculture was not suitable for production and continued to engage in hunting and gathering. From the cultural and ecological conditions and ethnographic materials, we know that the southwest is a suitable environment for hunter-gatherers to survive. As far as the Qinghai-Tibet Plateau region is concerned, why would hunter-gatherers give up the eastern living area suitable for hunting and gathering, and enter a completely unsuitable area for survival? If the earliest Qinghai-Tibetan inhabitants were agricultural groups, everything would be logical. Indeed, the interior of the Qinghai-Tibet Plateau hinterland is more suitable for sedentary agricultural production and a stable pastoral economy, rather than a highly mobile hunting and gathering livelihood.

Moreover, there is another issue that needs to be addressed. That is, how did the agricultural groups enter the Qinghai-Tibet Plateau through the southwest zone, where it is suitable for hunting and gathering. A possible explanation is that primitive agriculture matured in the Middle Neolithic. And this "package" included not only domesticated crops and livestock, but also the corresponding tools, social organisation, and religious rituals, which already constituted a cultural ecosystem. While these groups were expanding, the neighbouring hunter-gatherer groups partially accepted their influence, either through food exchange, the spread of tools and technology, or intermarriage. In other words, although people in the southwest still retain their hunter-gatherer-based livelihoods, they are no longer a population completely unknown to agricultural production. Some of the regions with more suitable environmental conditions for agriculture may have carried out crop cultivation and livestock domestication. In the process of spreading agricultural groups to the Qinghai-Tibet Plateau, the influence of millet farming coming from the Shan-Shaan region was even greater. And this will be discussed in the following section.

According to ethnographic records, there are many discoveries of hunter-gatherers in marginal environments. For example, the hunter-gatherer groups in the Kalahari and Western Australian deserts. In addition, hunter-gatherer groups have been found in polar environments, and they still exist in tropical rainforests. However, we must be clear that the Kalahari is not a simple desert. It is rich in species, even though seasonally distinct. In addition, most of the plants in the desert margins have well-developed rhizomes that are suitable for gathering in order to store water and nutrients, but this area is clearly not suitable for crop agriculture. The Australian Aborigines live in the western desert zone are also associated with the displacement of European colonists. As for the polar environment, there lived mainly groups that depended on marine mammals for hunting, and groups that depended on reindeer in a semi-domesticated state. The tropical rainforest zone hunter-gatherers may also be associated with the later colonial history. Moreover, they accepted metal tools and maintained exchange relationships with agricultural groups. They are more like specialised "forest experts", which is similar to the Oroqen people who live by hunting and gathering. So far there is no real hunter-gatherer living in the grassland environment found in ethnography (Binford 2001).

The dominant environmental type on the Qinghai-Tibet Plateau is the grassland. The mobility of hunter-gatherers is not sufficient to access the highly mobile animal resources here, especially during food-poor seasons. More importantly, and as I have repeatedly emphasised, it is the high plateau environment that imposes constraints on the mobility of hunter-gatherers. In short, hunter-gatherers' cultural adaptation mechanisms limit their ability to exploit the Qinghai-Tibet Plateau. It would clearly be a violation of cultural ecology principles for hunter-gatherers to abandon an area better suited to their livelihood style and enter this dangerous land. It would be like trying to farm in the desert or develop industry in an inland region with limited transportation. These crazy initiatives may have existed before, but they all ended in failure.

11.6 Establishment of the Cultural Adaptation Pattern in the Qinghai-Tibet Plateau

11.6.1 The Successful Adaptation to the Qinghai-Tibet Plateau

Before exploring the establishment of cultural adaptations on the Qinghai-Tibet Plateau, it is necessary to first learn about the known adaptations on there. The adaptation based on modern transportation, industrial and commercial production and exchange, and medical technology security occurred in recent decades is not very helpful in exploring this issue. *The Survey of Tibetan Social History* provides the valuable material that predates the development of modern industry and commerce, showing that Tibetans have established successful cultural adaptations on the plateau.[9] And we know that such an adaptation can be traced back to the Neolithic on the Qinghai-Tibet Plateau. The most important feature of this approach is the symbiotic relationship between agriculture and animal husbandry. Some areas are predominantly agricultural, some are totally dependent on livestock, and some are mixed. There is a constant exchange between the agricultural and pastoral areas. The agricultural areas provide food, tools, fruits, and vegetables needed by the pastoral areas. And the herding areas exchange salt, alkali, hides and skins, and dairy products. This interchange is necessary for both sides, and interestingly, it is the agricultural areas that are dominant in the exchange relationship, which reflects the fundamental importance of agriculture in the adaptation of the Qinghai-Tibet Plateau (Yuan et al. 2007).

The high plateau and alpine environment requires a high-calorie food supply. Tibetan ancestors grew highland barley, raised horses, cows, sheep, donkeys, and other animals, and developed tsampa, ghee, milk dregs, dried meat, and other foods that are suitable for plateau life, ensuring the basic need for calories. The low oxygen environment of the plateau limits people's ability to move, but animals can pack, tug, and ride. The yak, known as the "boat of the plateau", is the main force for carrying goods on the plateau, along with horses, donkeys, mules and cows, etc. The use of domestic animals greatly eases the labour intensity, and at the same time, greatly improves the range of activities of people, successfully solving the restrictions of the plateau environment on human mobility.

Tibetan society has also developed a religious form that allows people to minimise their material needs and place their hopes in the afterlife. We do not know how long this ideology dates back to, but it is fair to say that it has relieved the pressure of material life in the harsh highland environment. Obviously, even with the help of agriculture and animal husbandry, and even with the low population density of this region, it is still a harsh condition for human survival compared to other cultural ecological zones. Furthermore, even though the Tibetan people living on the plateau have built up a physical adaptation to the low oxygen, low air pressure, low temperature and strong radiation environment. However, there are still challenges associated with adapting to the high altitude of the plateau, such as high infant mortality rates and a higher prevalence of congenital heart diseases. These are considered costs of successfully adapting to the high altitude environment.

11.6.2 *Changes in Adaptation Patterns on the Neolithic Qinghai-Tibet Plateau*

Archaeological evidence to date has supported the successful establishment of human cultural adaptation on the Qinghai-Tibet Plateau during the agricultural era. The traditional view on the origin of agriculture on the Qinghai-Tibet Plateau, however, is that it spread from northern China. Mike Aldenderford and Zhang Yinong criticise the study of agricultural diffusion theory, arguing that it does not take into account the local development, especially the formation process of the unique local nomadic economy (Aldenderfer and Zhang 2004). The Qinghai-Tibet Plateau may be one of the domestication sites of highland barley (barley). Fieldwork to Tibet and the western part of Sichuan confirmed that the southwestern region is the distribution area of wild barley.[10] The simultaneous presence of several types here represents several evolutionary stages of domestication process. This may be one of the centres of origin of cultivated barley (Shao et al. 1975).

From the current paleoethnobotany evidence, Fu Daxiong identified millet and highland barley among the crop remains excavated from the Changguogou site,[11] and a large number of charred grains of highland barley seeds were mixed with a few charred grains of wheat seeds. Based on the comparison of rinsed and screened specimens, Fu believes that highland barley was the main crop at that time, but millet also occupies an important position. Interestingly, only a large amount of single "millet" was found at the earlier Karub site, and there was no highland barley. The later site of Donghuishan at Minle County in Gansu province

(1700 m above sea level, dated 5000–3770 years ago) has found coexisting remains of highland barley, wheat and millet. Charred grains such as millet and sorghum were also found at the Dongshanhui site, which included almost all the major crops of the Yellow River basin (Li et al. 1989). Archaeological materials support the view that agriculture in Tibet was preceded by millet and followed by barley, and since we know that northern China was the origin of millet crop, it is not untrue to suggest that millet agriculture on the Qinghai-Tibet Plateau came from northern China.

In addition to the characteristic that millet appeared early and barley appeared late, another major adaptation changes also took place in the Neolithic of the Qinghai-Tibet Plateau, namely the origin of the nomadic economy. There are significant differences in the characteristics of the early and late remains from the Karub site in Chamdo. Is this a change in adaptation due to climate change, or a change in site function? Since Karub was a culture characterised by sedentary farming, where people lived in the centre of the site and did not move as often as hunter-gatherer or nomadic groups, thus leaving behind site types with different seasons or different functions. So there is no suggestion that the function of the site has changed. The early and late changes at the Karub site in Chamdo are around 4000 years before the present, when climatic deterioration events did occur and are thought to have some relationship with the origin of civilization (Wu et al. 2001). The early and late changes at the Karub site are more likely to be related to the decline of the agricultural economy. In the face of the deteriorating climatic conditions, the eastern fringe, which was not very suitable for agriculture, turned to herding domesticated animals.[12] By increasing mobility, relying more on animal meat and its by-products, and developing exchanges with agricultural groups, the degree of cultural adaptation was increased and the risk and uncertainty of food acquisition was reduced. More microblade artefacts, more simple dwelling sites, and pottery types in the late cultural features of the Karub site reflect the increased mobility of the group. However, the domesticated animals excavated from the site are mainly pigs, which are distinctly different from the later nomadic economy based on horses, cattle, and sheep. Thus, the remains from the late Karub site represent a strategy of adapting to risk by diversifying operations, which also laid the groundwork for the development of a nomadic economy in the Qinghai-Tibet region in the future.

The changes exhibited at the Karub site in Chamdo are not unique. At the Zongri site at Tongde county in Qinghai Province (5000–4000 years ago, roughly contemporaneous with the Majiayao culture), the pottery assemblage also shows a trend of constant decline. The regression of pottery is usually interrelated with the development of nomadic economies. The upper Xiajiadian culture, where a nomadic economy emerged in the north and south of Yan Mountains, also shows a marked decline in the pottery assemblage compared to the lower Xiajiadian culture, where farming was the mainstay. However, stable isotope analysis of human bones excavated from the Zongri site shows that the diet of the early inhabitants was all based on millet and corn, which was increasing from the early to the late period. In contrast, the proportion of nitrogen isotopes reflecting a meat diet was, on the contrary, decreasing (Cui et al. 2006). The Zongri site, located in an area suitable for agriculture, reflects the differential forms of livelihood between regions on the Qinghai-Tibet Plateau, even during a period of climatic deterioration, which was the basis for the later symbiotic relationship between farming and animal husbandry. We can infer from the differential responses seen at the Karub and Zongri sites that typical farming-pastoral symbiosis began to originate on the Qinghai-Tibet Plateau around 4000 years ago. However, a true nomadic economy requires a combination of domesticated animals such as horses, cattle (yaks and bulls), and sheep (goats and jumuck). At present, we do not know when these animals began to be utilised in the Qinghai-Tibet Plateau region, so the origin of the nomadic economy stated here is still quite rudimentary.

The origin of nomadic economy does not mean the decline of agricultural economy. What it replaced was a mixed economy of animal husbandry, farming, hunting, and gathering. Farming shifted to the river valleys where water and heat resources were better, and the use of domesticated animals made hunting much less necessary and enhanced people's ability to settle. We do not yet know when people began to use domestic animals. But there is no doubt that the use of animals for carrying, hauling, and riding will greatly enhance people's ability to utilise the Qinghai-Tibet Plateau. At the same time, the formation of the nomadic economy facilitated the use of the vast pasture resources of the Qinghai-Tibet Plateau, and further expanded the space for people to utilise resources. The division of labour and cooperation between farming and nomadic groups also improved the stability of prehistoric livelihood systems in the Qinghai-Tibet Plateau region. Simultaneously practising agriculture and pastoralism within a single community can be more efficient, especially when agricultural climatic conditions become more adverse.

11.6.3 An Earlier Possibility

Are there other possibilities for human adaptation on the Qinghai-Tibet Plateau? Such as, hunter-gatherers, which is discussed previously from several perspectives, including cultural ecological theories of hunter-gatherers, archaeological materials, bioarchaeology, and paleoenvironment. It is very difficult for hunter-gatherers to establish a successful adaptation when the Qinghai-Tibet Plateau has been lifted as a whole to near its present height. It may be a bit extreme to say that humans would never have ascended the Qinghai-Tibet Plateau, but from the hunter-gatherer perspective, living on the Qinghai-Tibet Plateau was never a good choice unless there were some more extreme factors that compelled them to choose this extreme environment. At present, we know that there may have been population pressure on the eve of the origin of agriculture after 10,000 years ago, and that hunter-gatherers with enhanced mobility (such as using microblade technology) may have entered the Qinghai-Tibet Plateau. Archaeological evidence suggests that humans were already living on the margins of the Qinghai-Tibet Plateau at this time. Beyond this window of time, earlier possibilities are not sufficiently justified. Why use a marginal environment when there were better areas to exploit.

It is known from materials from other regions that after the origin of agriculture, the spread to the agricultural marginal environment was a result of the maturation of primitive agriculture. That is, it began after the formation as an agroecosystem. The origins of agriculture in South China, Southwest China, and Northeast China all occurred in North China and the middle and lower reaches of the Yangtze River after the Middle Neolithic. The primitive agricultural system required the development of a whole set of farming techniques, including not only tools, species, and farming methods, but also social organisation and infrastructure. Only when the primitive agricultural system was perfected as a whole was it possible to overcome the obstacles to agricultural production in marginal environments. And it was also the best time for human beings to start utilising the Qinghai-Tibet Plateau.

Notes

1. Based on the illustrations in the report, it could actually be a stone shovel.
2. The study includes 16 locations, all of which were surface collections. The microcores include wedge-shaped, conical, cylindrical, and other forms, indicating a mature technology.
3. Considering that the Nielamu area has risen 500 metres within ten thousand years, if the age of Xiadacuo belongs to the Early Paleolithic, there is sufficient reason to believe that the Qinghai-Tibet Plateau at that time did not pose such a serious challenge to human adaptation as it does now.
4. The dispersal of modern humans from Africa to various parts of the world does not show clear evidence of a lithic technology tradition. It remains uncertain whether there is a problem with the Out-of-Africa hypothesis or with considering lithic technology traditions as a method for studying populations. Currently, it appears that the latter issue may be more significant.
5. The large stone tool manufacturing sites at Mount Dagudui and Emao Kou in Xiangfen, Shanxi, discovered a large number of unfinished products and flakes from the Neolithic period. At first, they were mistakenly identified as Paleolithic stone tools.
6. The report states three types of stone tools: microliths, small to medium-sized flake tools, and a combination of microliths and flake tools. In fact, they may all belong to the category of microblade technology because the production process of microblade technology includes flakes, and flake tools would also be used together with microblade technology. Therefore, the claim of the existence of three types of stone tools is not reliable.
7. In Chapter 6, "Hunters in the Environment," discusses the behavioural characteristics of hunter-gatherers at different types of activity sites.
8. Lin believes that the characteristics of the Linzhi human fossils show the influence of the Europoid and South Asian races.
9. Survey of Tibetan Social History, a six-volume series, is a collection of surveys conducted on different regional groups. It primarily focuses on social relationships and also mentions their production conditions.

10. This forms the basis for diverse forms of marital relationships in the southwestern region, where practices such as polyandry and walking marriages are associated with its unique resource distribution conditions.
11. The report does not mention what the dating material is. If it is charcoal, there may be a possibility of an earlier dating due to the issue of dead wood (trees that have died a long time before being utilised by humans). Schiffer has noted that in dry and sparsely populated areas, dead wood may remain unused for hundreds or even thousands of years. See Schiffer (1987). The discrepancy in the dating data between animal bones and charcoal dating at the Changguo site indicates this point.
12. There is a time difference in the utilisation of animals and plants. For example, animals are fattest during autumn, so if hunting, one may choose autumn instead of summer. This is the time difference.

References

A. Torroni et al., Mitochondrial DNA analysis in Tibet—implications for the origin of the Tibetan population and its adaptation to high altitude. *American Journal of Physical Anthropology*, 92: 189–199, 1994.

An Zhimin, et al. Paleolithic and Microlithic Industries in Northern Tibet. *Archaeology*, 1979, 6.

Archaeological Team of the Institute of Archaeology, Chinese Academy of Social Sciences, Tibet Work Group, Tibet Autonomous Region Cultural Relics Management Committee. Neolithic Sites in Changguo, Gongga County, Tibet. *Archaeology*, 1999, 4.

B. Su et al., Y chromosome haplotypes reveal prehistorical migrations to the Himalayas. *Human Genetics*, 107: 582–590, 2000.

Brantingham, P. J. et al., Speculation on the timing and nature of Late Pleistocene hunter-gatherer colonisation of the Tibetan Plateau. *Chinese Science Bulletin*, 48: 1510–1516, 2003.

CA, CASS, and Tibet Autonomous Region Cultural Relics Bureau, Lhasa Qu Gong, Encyclopedia of China Publishing House, 1999.

Chen Kezao, et al. Climate change on the Qinghai-Tibet Plateau in the past 40,000 years. *Quaternary Sciences*, 1990, 1.

Cui Yaping et al. Stable isotope analysis of human bones from the Zongri Site. *Quaternary Sciences*, 2006, 26 (04).

D. D. Zhang and S. H. Li, Optical dating of Tibetan human hand-and footprints: An implication for the palaeoenvironment of the last glaciation of the Tibetan Plateau. *Geophysical Research Letters*, 29: 161–163, 2002.

Duan Qingbo, Microlithic remains in Tibet. *Archaeology and Cultural Relics*, 1989, 5.

F. B. King and R. W. Braham, Effects of ecological and paleoecological patterns on subsistence and paleoenvironmental reconstructions. *American Antiquity*, 46: 128–142, 1981.

G. van Driem, Tibeto-Burman phylogeny and prehistory: Languages, material culture, and genes. In Examining the Farming/Language Dispersal Hypothesis, edited by P. Bellwood and C. Renfrew, pp. 233–249. McDonald Institute for Archaeological Research, University of Cambridge, 2002.

Gao Xing, et al. Late pleistocene human remains and survival patterns in the marginal area of the Qinghai-Tibet Plateau. *Quaternary Sciences*, 2008, 28 (06).

Han Kangxin, Zhang Jun, Anthropological Characteristics of the Tibetan Ethnic Group and their Ethnic Origins, Wenbo, 1991, 6.

Huang Weiwen, et al. Paleolithic Sites in Xiaochaidamu Lake, Qinghai. In *Proceedings of the China-Australia Quaternary Symposium*. Science Press, 1987.

Huo Wei, Discovery of Paleolithic relics on the shore of Lake Xiada in Ali. *China Tibet*, 1994, 6.

Institute of Archaeology, Chinese Academy of Social Sciences, Tibet Autonomous Region Cultural Relics Bureau. Qugong in Lhas. Encyclopedia of China Publishing House, 1999.

L. R. Binford, *Constructing Frames of Reference: An Analytical Method for Archaeological Theory Building Using Ethnographic and Environmental Data Sets*. University of California Press, Berkeley, 2001.

L. R. Binford, In pursuit of the past, translated by Chen Shengqian, Shanghai Sanlian Bookstore Co.Ltd., 2009.

Lehmkuhl, F. and F. Haselein, Quaternary paleoenvironmental change on the Tibetan Plateau and adjacent areas (Western China and Western Mongolia). *Quaternary International*, 65/66: 121–145, 2000.

Li Fan et al. New discovery of ancient agricultural remains at the Donghuishan Neolithic Site in Minle County, Gansu Province. *Agricultural Archaeology*, 1989, 1.

Lin Yipu, Ancient human remains discovered in the Linzhi Village in Tagong, Tibet. *Journal of Vertebrate Paleontology and Paleoanthropology*, 1981, 3.

Liu Xingqi et al., A 16,000-year pollen record of Qinghai Lake and its paleoclimate and paleoenvironment. *Chinese Science Bulletin*, 47: 1931–1936, 2002.

Liu Zechun et al. The Paleolithic of Dogezê and Zhabu in the Tibetan Plateau: a discussion on the influence of Plateau Paleoenvironment on the distribution of Stone Tool Cultures. *Archaeology*, 1986, 4.

Liu, J. et al., Palaeoclimate simulation of 21ka for the Tibetan Plateau and eastern Asia. *Climate Dynamics*, 19: 575–583, 2002.

Lü Hongliang, Reconsideration of the Paleolithic Age in Tibet: taking the Xiadacuo Northeast Shore Site in Ali as the Center. *Archaeology*, 2011, 3.

M. Aldenderfer and Y. Zhang, The prehistory of the Tibetan Plateau to the seventh century A. D.: perspectives and research from China and the west since 1950. *Journal of World Prehistory*, 18: 1–55, 2004.

M. Aldenderfer, Montane Foragers: Asana and the South-Central Andean Archaic. University of Iowa Press, 1998. Moving up in the world. *American Scientist*, 91: 542–549, 2003.

M. S. Schiffer, Formation processes of the archaeological record. University of Utah Press, 1987.

P. Cavalli-Sforza, The Chinese human genome diversity project. *PNAS*, 95: 11501–11503, 1998.

P. J. Brantingham and X. Gao, Peopling of the northern Tibetan Plateau. *World Archaeology*, 38: 387–414, 2006.

P. J. Richerdson et al., Was agriculture impossible during the Pleistocene but mandatory during the Holocene? A climate change hypothesis. *Americian Antiquity*, 66: 387–411, 2001.

P. T. Baker, Human adaptation to high altitude. *Science*, 163: 1149–1156, 1969.

Qian Fang et al. Preliminary Observations on the Stone Tools of Geting Sites in Northern Tibet. *Acta Anthropologica Sinica*, 1988, 7 (01).

Qiu Zhonglang, Discovery of Paleolithic Sites in the Tibetan Plateau. *Journal of Vertebrate Paleontology*, 1958, 2 (2–3).

Ren Meie, *Outline of China's Physical Geography* (Revised 3rd Edition). The Commercial Press, 2004.

Shao Quanren, Li Changsen, Basang Ciren, Origin and evolution of cultivated barley: wild barley in Tibet and Western Sichuan, China. *Acta Genetica Sinica*, 1975, 2 (02).

Shen Yongping, et al. Climate and environment of the younger Dryas impact hypothesis in the Qinghai-Tibet Plateau. *Journal of Glaciology and Geocryology*, 1996, 18 (03).

Shi Yafeng, et al. Characteristics, Impacts, and causes of high-temperature and heavy precipitation events on the Qinghai-Tibet Plateau and its surrounding areas during 40–30kaBP. *Journal of Lake Sciences*, 2002, 14 (01).

T. Geobel, Pleistocene human colonisation of Siberia and peopling of the Americas: an ecological approach. *Evolutionary Anthropology*, 8: 208–227, 1999.

Tang Huisheng, A brief discussion on the Paleolithic and Microlithic tools on the Qinghai-Tibet Plateau. *Archaeology*, 1999, 5.

Tang Huisheng, Archaeological culture and economic form of the Late Paleolithic to Early Neolithic on the Qinghai-Tibet Plateau. *Acta Archaeologica Sinica*, 2011, 4.

Tang Lingyu, et al. Pollen evidence of the climatic optimum during the last glacial maximum in the Eastern Qinghai-Tibet Plateau. *Journal of Glaciology and Geocryology*, 1998, 20 (02).

Tibet Autonomous Region Cultural Relics Bureau, et al. *Field Archaeological Report on the Tibetan Section of the Qinghai-Tibet Railway*. Science Press, 2005.

Tibet Autonomous Region Cultural Relics Management Committee, Department of History, Sichuan University. Karub in Chamdo. Cultural Relics Press, 1985.

Tong Enzheng, Handaxes on the Tibetan Plateau. *Archaeology*, 1989, 9.

Wang Jianlin, Xiong Wei, Migration of ancient humans in Tibet Since the Late Pleistocene and its relationship with climate change. *Acta Geographica Sinica*, 2004, 59 (02).

Wang Yong et al. Preliminary study on the stone artifacts from the Gemuco Basin in the Central Qiangtang, Northern Tibet. *Geological Bulletin of China*, 2009, 28 (09).

Wu Wenxiang, Liu Dongsheng, Cooling events around 4000 BP and the birth of Chinese Civilization. *Quaternary Sciences*, 2001, 21 (05).

Xia Zhengkai, Yang Xiaoyan, Preliminary study on abnormal flooding events around 4 ka BP in Northern China. *Quaternary Sciences*, 2003, 23 (06).

Y. Qian et al., Multiple origins of Tibetan Y chromosome. *Human Genetics*, 106: 453–454, 2000.

Yi Mingjie, et al. Investigation and excavation report of prehistoric sites in the marginal area of the Qinghai-Tibet Plateau in 2009. *Acta Anthropologica Sinica*, 2011, 30 (02).

Yuan Baoyin, Huang Weiwen, Zhang Dian, New evidence of Late Pleistocene human activities on the Northern Tibetan Plateau. *Chinese Science Bulletin*, 2007, 52 (13).

Zhang Senshui, Paleolithic discoveries in Dingri, Tibet. In Scientific Survey Report of the Mount Everest Region (1966–1968)—Quaternary Geology. Science Press, 1976.

Zhang Zhenli et al. Discovery and research on ancient human activity relics in the Zhongba Area, Southwest Tibet. *Guizhou Geology*, 2006, 23 (01).

Zhou Caiping, et al. Net primary productivity of major ecosystems on the Qinghai-Tibet Plateau. *Acta Geographica Sinica*, 2004, 59 (01).

CHAPTER 12

Conclusion: Answered and Unanswered Questions

The chapters that came before this one cover the origins of Chinese agriculture from about four different angles. The first section provides a theoretical explanation based on the notion of cultural ecology, for why agriculture started and why it did not in some areas. The archaeological evidence is not directly relevant to this explanation. The origin of agriculture has traditionally been viewed by archaeologists theoretically. It is simple to observe that theoretical research is always a precursor to issue solving, despite the fact that material-centred archaeologists may not acknowledge this. This is true if one looks at the history of the study of agricultural origins and considers the starting point of archaeologists' research.

The second section simulates potential cultural adaptations of hunter-gatherers in various cultural ecological areas in China using current data in the weather station based on the cultural ecological principles of hunter-gatherers. Similar to a contemporary nuclear weapon simulation exercise, the outcomes serve as a reference even though test explosions are no longer permitted. As archaeologists cannot travel back 10,000 years in prehistory, we can partly understand the cultural responses of various cultural ecological areas based on simulations, which is difficult to grasp through other methods.

In the third section, it starts to focus on specific archaeological and palaeoenvironmental materials, tracing the preparation of cultural adaptations prior to the origin of agriculture and borrowing the contemporary

S. Chen, *Prehistoric Modernization*,
https://doi.org/10.1007/978-981-97-9423-2_12

concept of development economics—"resource endowment structure"—in attempt to explain that the origin of agriculture was not a "cultural mutation" without any accumulation. Food production has always been a part of hunter-gatherer cultural systems, developing into the agriculture only when the circumstances were favourable. Just as the exchange has always been a part of human behaviour, it wasn't until modernisation that it took on the form of a market economy and became the backbone of contemporary society.

The fourth section, the book's main body, concentrates on the investigation of archaeological materials and classification of previously discovered archaeological artefacts. The most surprising thing to me was that I discovered that different cultural ecological zones had their own unique cultural adaptations. Some regions had developed agriculture, others had initially attempted but later gave up, others were hesitated between agriculture and hunter-gatherer lifeways, and still others discovered innovative ways to deal with their livelihoods. From this, we can understand the cultural adaptability of the various prehistoric Chinese regions and sense the multiple historical trajectories.

Finally, I'd like to summarise and consolidate the information from the earlier chapters and try to integrate them together to form a more concise thought. It can be founded some errors in this course. As the author and first reader of this book, I must present which aspects of the issue I have made some headway on and other aspects I have not been able to address.

12.1 Origins of Agriculture from the Perspective of Modernisation

One of archaeologists' key objectives stands for the reconstruction of prehistory accounting for more than 99% of human history, which only can be made by archaeologists. This book's major goal is to try to piece together a crucial period of Chinese history that took place roughly 10,000 years ago. There is no doubt that this is a daring attempt because archaeological materials are always fragmentary and must be repeatedly assembled before a complete history can be obtained. Archaeological materials are discovered by chance and altered by cultural and natural processes, and thus do not always accurately reflect the true history. In addition, unlike historical writings, which immediately spread information, archaeological evidences are material and must be carefully evaluated by archaeologists in order to be understood. Therefore, the work of reconstructing prehistory faces many obstructs. Historians who are not familiar with archaeology will lack a strong material foundation and comprehension of prehistoric human life.

Certainly, they will be deeply inspired by the connection of prehistory with history after having a firm grasp of the archaeological evidence. *Guns, Germs, and Steel: The Fate of Human Society*, Jared Diamond's Pulitzer Prize-winning masterpiece in 1998, appears a remarkable effort in this direction (Diamond et al. 2000). Archaeologists should become familiar with this macro-historical viewpoint. History can be studied from the present back into the past or from the past into the present, the latter is often used by archaeologists. For this case, the book is titled by *Prehistoric Modernisation*, which aims to integrate the reconstruction of prehistory with knowledge of contemporary societal evolution. This work still centres on the archaeological research but tends to an empirical study than Diamond, who has a more macroscopic perspective.

In the millions of years of human prehistory, the origin of agriculture is only seen as one segment or point. However, this point is of great significance because it directly relates to the beginning of later civilizations and the future of various societies, which is the main subject of Diamond's work. It is undeniable that the big events of a person's life are essential to their development, just as every day in a person's life is necessary. Similar to this, despite the lengthy human history, it is crucial to understand the changes that have occurred at significant turning points.

I divide the evolution of humans into three periods based on humans' ability to gather food safely. At that time, people engaged in the hunting and gathering, agricultural production, and industrial and commercial production. As the old saying "Food is the people's most basic need" implies that survival is people's main issue. There are differences in the three accesses to food security. People would be mobile during the hunter-gatherer stage in order to search for food. They moved to find and acquire food, as well as gathering information about food distribution. Hunter-gatherers would be mobile to find friends and mates in a vast area, and complete the race's reproduction. People start to cultivate and raise animals during the agricultural production phase. They created their own food instead of going on the hunt for natural resources; food production was crucial at this point. Modern society is featured by industrial and commercial production where people didn't produce food rather than trading them in order to acquire enriched goods and services from the market. The market is most fundamental during this time.

I called these three factors as the “order parameters” of each historical stage in the previous section which have a significant impact on other societal aspects. Five dimensions can be used to analyse the three stages of human society (Table 12.1).

Given few places filled with sufficient natural resources for a human group to use continuously, humans could only live on mobility for food during the hunter-gatherer era. The scale of socially organised groups must be mostly maintained at nomadic levels due to limited (with few left) and uncontrolled resources. There exists an equivalent relationship or a difference in prestige among these groups. The information is often charged by Shamans, military commanders, and senior members of the community. The common mindset at this time was one of complete reliance on natural resources for survival.

In the age of agriculture, people had to grow more food on scarce land resources. Land was the most valuable resource for production, and settlement was primarily an agricultural social structure (nomadism was a derivative mode of production after agriculture emerged, and there was a dependent relationship with agricultural communities). More complicated social organisation is needed to regulate production surplus and land, which is the most valuable productive resource. Centralization was the most prevalent form of government in this society, due to the elite of aristocratic scholar-bureaucrats that dominated this society’s governing class. The agricultural output at the period could only support a small and disconnected ruling class; if the class grew too large, the social load would increase and the society would be on the verge of disintegrating (Tainter 2010). It is currently in the optimal state of heavenly and human harmony. People should follow the social order and not disobey the laws of nature, just as crops are in line with the heavenly periods.

In contrast, the market’s exchanges and incentives improved both the productivity and land use efficiency in the industrial and commercial era. More and more people moved to reside in cities to lower the cost of exchange. A large number of professional class has been formed through the division of labour in modern society, which is frequently referred to the “middle class” or the mass elite. The spirit of contract is thriving in this exchange-based society, and social politics tends towards democracy and the rule of law. It is perfect to keep an ecological (interdependent and interconstrianed) and science-based relationships between human and the environment.

Table 12.1 Comparison of human social characteristics in three historical stages

	The hunting and gathering phase	*The agricultural production phase*	*The industrial and commercial production phase*
Way of resource acquisition	Mobile for searching food	Self-sufficient life style	Industrial production and market exchange
	Mobility-oriented	Food production-oriented	Market economy-oriented
Social existence form	Nomadic communities	Rural society	Urban society
Social organisation system	A society that values equality or prestige	Centralised society	Democratic and law-governed society
One who controls information and transmission	Shamans, military commanders, and senior members of the community	Noble and scholar-bureaucrat class	Mass and elite
The spiritual relationship between humans and the environment	Everything in life has its spirituality	Unity of Heaven and Man	Ecological Science

It is useful to understand prehistory and contemporary China's modernisation process in five aspects. For instance, Chinese academic fields lay much stress on the growth of productive forces, an idea taken from Marxist theory, to evaluate the process of current modernisation (Luo 2009). But what does productivity actually mean? Social production is so extensive that it can be categorised under productivity. In that way, the notion is overly broad and vague and does not assist with our understanding of the modernisation process.

We can observe from this macro-historical comparison that the main determinants of productivity differ at various historical junctures. Mobile foraging is the main means for higher productivity that in turn mostly depend on the mobility in the hunter-gatherer stage; In the agricultural production stage, the productivity means the method of creating food. Food production is vital and its basic factors are species, land, labour, and their organisation. In the industrial and commercial age, productivity also involves the generation of goods and services, but its core is the market where human exchange products with each other to support social operation. The analysis of the "order parameters" reveals that the key of modernisation process is commercialization, to be more exact, the market, rather than the industrialization we are familiar with. This conclusion is drawn through macro-historical analysis instead of a study of modern society. In other words, if productivity can be used to classify the historical development of humans (the fundamental goal of productivity is to meet humans' most basic food needs and ensure that they have adequate, abundant, and stable access to food), then contemporary modernisation should be achieved through the market, with industry and commerce being the only specific means.

From the perspective of modernisation, the highly diversified global social developments are encompassed within a comparative framework. This point may not be appreciated by relativists or supported by zeitgeist that promotes diversity. However, we have to recognise that there is a unity in human historical evolution, which paradoxically unites with diversity. The only way to break the impasse is on a case-by-case basis, and this dialectical stance is easily regarded as unprincipled eclecticism. The issue of the origins of agriculture overlaps with the concept of prehistoric modernisation, combining two characters of uniformity and diversity. Chapter 6 goes into considerable detail about the fundamental patterns of Chinese agricultural origins, and it discusses the pervasive influence of prehistoric agriculture in relation to other cultural ecological zones.

Indeed, this process is highly complex and spans thousands of years, which is much longer than the timespan of modern society measured in hundreds of years. For instance, early evidence of Neolithic culture in South China pointed to a shift towards agricultural production; however, as the Last Glacial Period ended and the tropical climate returned, prehistoric agriculture in this area became less desirable and more challenging to develop. Agriculture was not gradually introduced to this region until the middle and late Neolithic period in the middle and lower reaches of the Yangtze River. Only once the agricultural system was upgraded did it become attractive and have the drive to spread (as the population grew, more land was required). This is similar to the spread of the Industrial Revolution. In reality, the West has entered into the modernisation since the Renaissance, but a robust capability for extension and dissemination didn't fully emerge until the second revolution.

Differential regional development is highly correlated with regional cultural processes and natural environmental factors. The prehistoric agriculture first emerged in the North China and the middle and lower reaches of Yangtze River can be attributed not only to environmental factors following the end of the Last Glacial Period, but also to the resource endowment structure created by cultural development prior to the origin of agriculture. Because agriculture first appeared when hunter-gatherers were less mobile or even keep settled. Hunter-gatherers had a difficult time gathering natural resources for their limited mobility. In terms of what reduces or eliminates mobility, issues like population growth, the evolution of ownership ideologies, and a change in resource utilisation (such as the intense use of small seed plants) are all contributing factors. The adaptive changing characteristics of other cultural ecological zones are frequently disregarded and are typically thought to be the result of environmental limits, in contrast to the great attention given to the subject of agricultural origins. Such a crude perspective ignores the local development because these regions are not conducive for the growth of agriculture, thereby the agricultural origins do not exist. Each area has a distinct development strategy. Aquatic resource became popular in South and Northeast China, and this new utilisation helped to relieve the pressure on pre-agricultural lives. The Yan Mountains' northern and southern regions alternate between agricultural and hunter-gatherer growth patterns, sometimes emphasising agriculture and other times hunting and gathering in response to climatic shifts. Thus there are variations even within the same civilization. The characteristic of dry farming and nomadic economy emerged with the mature of species conditions. The Southwest region has had the longest duration of hunting and gathering livelihoods due to its suitability for these activities. Even though agriculture represents a more "advanced" form of production, it was not always the best option for all prehistoric regions, nor was it always a viable option. However, the history of agricultural spread cannot be discounted despite the presence of the distinctive adaptations in the area. All of China's cultural ecological zones had agriculture (including nomadism) by the Late Neolithic or, at the latest, the Bronze Age. It is obvious that agricultural production has advantages over hunter-gatherer lifestyles. It can feed a greater population and produce more left food on less land, which realised the specialised division of labour and food security, people who were not involved in agricultural production such as artisans, soldiers, priests, and bureaucrats began to appear. This greatly expanded societal complexity, which in turn made agricultural communities more suited to growing. When faced with agricultural communities, hunter-gatherers were undoubtedly at a disadvantage because of the scale of the population, the rigid social structure, and the availability of food, making it impossible for them to sustain a protracted battle. Germs are lethal weapons of agricultural communities. High-density populations tend to settle in specific site in which garbage and excrement are abandoned, water are always contaminated, and domestic animals like rats and cockroaches can carry and spread infectious agents. In contrast, hunter-gatherers are rarely bothered by garbage problems, and there is less risk of domestic animal-transmitted diseases. This distinction has developed stronger immunity in agricultural communities. However, it was fatal for them when they spread diseases to hunter-gatherers who lacked immunity. This dynamic has repeatedly occurred throughout the history of Western colonial expansion.

Agricultural groups benefited from prehistoric modernisation in that it gave them a sense of cultural superiority. In anthropological materials, it is typical to find criticisms of agricultural populations' attitudes towards hunter-gatherers, such as their sloth. In addition, hunter-gatherers and nomadic groups also face more difficulties in exchange goods, because agricultural societies are self-sufficient and do not have a strong need for exotic goods. The same is true of marriage and employment relationships; in essence, women and labour are exported by hunter-gatherers. The relationship is similar to that of agricultural communities and industrial and commercial societies. It can be seen from the urban–rural gap in

China. Prehistoric modernisation faced significant difficulties. People in agricultural society are more favourable to develop inaccessible marginal regions, such as the Qinghai-Tibet Plateau, desert oases, and limitless grasslands (mainly nomadic populations). Furthermore, hunter-gatherers were restricted to areas with forests and high-mountain valleys. Even in this location, they were still reliant on farmers for food, clothing, and other necessities like metal tools. In a sense, they are specialised hunters in an agricultural social group that do not live in complete isolation but rather in an agrarian cultural system (Diamond 2000). Modernisation in prehistoric times was also widespread and ubiquitous. No society in the agricultural global system was exempt from its effects; nonetheless, it was longer than the current modernisation process. No matter how glorious and profound the traditional history, we must accept and embrace modernisation in today's China. Instead of fighting against modernisation, our task is to embrace its benefits to create a more effective model for us. The same is true of history!

12.2 Origins of Agriculture from the Culture-Ecological Perspective of the Hunter-Gatherers

This book's characteristics or contribution stands that the research mostly depends on the cultural ecology of hunter-gatherers. The study of the Neolithic era's archaeological remains has long been the main focus of research on the origins of agriculture, and more efforts have gone into finding pertinent artefacts. There are few explanations for why and how agricultural origins might have happened, why some regions lack agricultural origins, and how these regions dealt with the problem. Even if there are, the solutions are frequently taken from theoretical findings pertaining to Western archaeology, lacking Chinese sources and Chinese viewpoint. However, this book is an attempt, somewhat interesting, and perhaps holds some significance.

From this point, it is first required to define the nature of agricultural origins and the potential role of archaeologists. Archaeologists need to investigate the process of the origin of agriculture because it entails a significant change in the stage of development of human cultural adaption. To solve the problem, we must first know the distinction between agricultural producers and hunters-gatherers. Since most people are familiar with agricultural societies, our main challenge is to clear about the cultural adaptation system of hunter-gatherers, particularly when they face threats to their livelihoods and resource scarcity.

This demonstrates that food production has been a long-standing adaption in hunter-gatherer subsistence practices. That is, hunter-gatherers are not totally ignorant of food production, thus they are aware of how to increase the diversity of natural species by using them more extensively as well as how to turn inedible components, such small-sized seeds, into edible ones. The fact that agriculture did not develop sooner, however, shows that more extreme circumstances were required for its development. As for me, the origin of agriculture was a comprehensive and emergent reaction of the hunter-gatherer cultural system in the Paleolithic. The presence of beginning conditions and the building up of resource endowment structures inside the cultural system are prerequisites for the system's qualitative evolution. To put it another way, hunter-gatherers had to adopt the approach of food collectors and make the associated preparations in tool technology, site development, reserve technology, and even social organisation.

The rationality of the underlying assumptions and the consistency of deductive reasoning make it impossible to provide a theoretical explanation for how agriculture first developed. Theoretical research would otherwise become unsupported supposition. Anyone can take what they already know for granted; it's just pointless to do like this. The theoretical explorations and intensive ethnographic studies conducted by earlier researchers like Julian Steward, Lewis Binford, and other pioneering work form the foundation of the cultural ecology of hunter-gatherers. Deeper thought of the mobility of hunter-gatherers is essential to understand their way of life. I regard the key factor as the "order parameters" of the hunter-gatherer cultural system, which determines the hunter-gatherer's range of access to food resources, as well as their chances of survival. People will face the risk of survival if they cannot find enough food resources within the foraging range. Binford lays more stress on population change. He contends that excessive population growth causes the shift of cultural systems because hunter-gatherers have relatively stable cultural mechanisms (Binford 2001). However, population scale closer to the starting conditions has no decisive impact on hunter-gatherer cultures. The market economy must be used to explain the formation of modern industrial and commercial societies rather than population growth or other considerations. My theoretical research has mobility as its central tenet. The advent of agriculture is characterised by a sharp decline in population movement, that is, the rise of sedentism, though there may still have been some seasonal hunting among the groups. It's crucial to stress that an extreme viewpoint cannot explain the decline in mobility, such as presuming that if people do not settle down, they will have mobile lives. Mobility is linearly, or more correctly, periodically reduced. Hunter-gatherer's mobility started to decline in the Late Upper Paleolithic, and by the Middle Neolithic, the sedentary culture had fully developed. Since then, it has endured for tens of thousands of years. The origin of agriculture is also a process of gradual agricultural maturation. Therefore, in the early stages of agricultural origin, primitive agriculture was still in its infancy, and the population may have still maintained some degree of mobility, especially before the agricultural system was established. Crop cultivation started fairly early in the Mexican highlands, but there weren't enough suitable domesticated animals to create an agricultural system that synchronised crop production and livestock breeding. At last, there was always certain mobility among the agricultural population (Hard and Merrill 1992, pp. 601–620). We cannot consider this aspect of early agricultural growth to be the final form of prehistoric agriculture and must thus consider the paradoxical relationship between agriculture and mobility.

Research using simulations can also make advantage of the hunter-gatherer culture's ecological principles. It should be noted that this type of study is not just simple ethnic archaeology. While the information about hunter-gatherers comes from ethnography, Binford's cultural ecology model is based on sound principles. Anyone who relies on hunting and gathering for a living and obtains food from nature must be subject to nature's supply that is measurable indeed, as was previously mentioned. The simulations' findings surprised everyone. I never intended for the simulations to match the archaeological evidence, and that was not the fundamental purpose. Instead, my goal is to discover differences between the simulations and the evidence in the beginning, but the simulations were surprisingly accurate in their predictions of the archaeological evidence. It's a little unexpected, but why is that the case? What should be said to clarify this? One argument is that the differences between the present climates and that of 10,000 years ago had no bearing on the cultural systems of hunter-gatherers (the simulations are based on data from current weather stations). The region of North China is still a temperate climate (perhaps a little bit colder than now), neither a plateau nor a cold zone. The middle and lower reaches of the Yangtze River are also in a temperate region, neither turning tropical nor becoming a cold zone. The second reason is that the cultural ecology model of hunter-gatherers has passed the test of archaeological evidence, giving us a chance to examine the origins of agriculture without reliance on these kinds of resources. I am only a beneficiary of the theoretical research that Binford conducted from a fresh viewpoint, which has unquestionably made a significant contribution to our study of the history of agriculture. From the perspective of hunter-gatherers, the study of the origins of agriculture begins with the Paleolithic where the agriculture first be founded. What do we observe during the tracing course? We can see that in the core region of China's agricultural origin, North China and the middle and lower reaches of the Yangtze River, some new features of reduced mobility and enhanced resource utilisation began to emerge in the Late Upper Paleolithic period before the origin of agriculture, that is, after the Last Glacial Maximum and before the beginning of the Neolithic. It's interesting to note that a similar trait also developed simultaneously (or even a little earlier) in South China. Other regions undoubtedly responded, just in different ways and not in any way at all. The use of aquatic resources was considerably more constrained than in South China, where fewer shellfish could be collected, and it evolved simultaneously in Northeast China, along with the use of fishing and hunting resources. The Paleolithic always shows evidences of the beginning of agriculture that are extremely sparse, sometimes even obscure. Fortunately, in recent years, sites like Yuchanyan and Xianrendong in the middle and lower reaches of the Yangtze River, as well as archaeological sites like Zhuannian, Donghulin, Nanzhuangtou, and Lijiagou in North China have given us extremely valuable materials. The artefacts from the numerous Early Neolithic sites are linked to them and fill in any remaining gaps. As a result, by contrasting five stages—the Early and Late Paleolithic, the NPT period, and the Early and Middle Neolithic period—we may trace the development of agricultural origins step by step. Each stage, from germination to the creation of the prehistoric agricultural environment, has new characteristics and massive development. This tracking has allowed us to know the full history of how prehistoric agriculture first appeared in China. We already have a fairly clear structure, despite the fact that the specifics are not yet complete, especially now that the framework has been improved to five more distinct stages.

With the hunter-gatherer perspective, we also need to examine the adaptive migrations made by hunter-gatherers in regions without agricultural origins. Thought-provoking changes have occurred in the northern and southern Yanshan Mountains, South China, Northeast China, Southwest China, and the Tibetan Plateau in Qinghai province. Agriculture is also required to survive in the fragile ecological ecotone, and the north and south of the Yanshan Mountains are near to the core regions where agriculture first appeared. Due to the excellent hunting conditions in this area, prehistoric people opted for a mixed economy that prioritises both agricultural production and hunting and gathering. But this tactic is unstable. Hunting and gathering are incompatible with agricultural production; while the latter demands mobility but also labour input, the former requires settlement and a significant amount of workforce. Due to the fluctuating environment, the richest and most diverse prehistoric cultural artefacts have been left in this region, and some materials remain difficult to categorise, like the Fuhe and Xiaohexi cultures, as well as other cultures that lack necessary resources for settlement, including the Xiaoheyan culture and a particular stage of the Hongshan culture. In South China, agriculture appeared early but being abandoned for a while, it reappeared after a few thousand years. If we don't delve into this tough process of agricultural origins, it can be a little confusing. The explanation for this is simple: In the beginning, hunting and gathering were much more alluring than agriculture, particularly in South China where aquatic resources could be used and possibly equipped with gardening for cultivating root plants. Therefore, it is unreasonable to engage in labour-intensive and insecure crop agriculture. As the most powerful hunter-gatherer, people in the Northeast region gradually incorporated limited agricultural production into the fishing and hunting economy and built a complex society around it. Due to its relative isolation, the southwest is a good choice for hunter-gatherers. Although the agricultural conditions in this area are not ideal, the diverse resources and environment result in an extremely diverse cultural adaptation. Every stage of agricultural dispersion is obstructed, and agriculture steadily pushes hunter-gatherer areas to the margins in a slow but cruel way. There is little interaction between hunter-gatherers and the Tibetan Plateau. The earliest colonisers may have been food growers, but hunter-gatherers continued to subsist in the primitive agricultural system until a nomadic economy developed.

The viewpoint which I studied is theory-oriented, and its assumption is the hunter-gatherer cultural ecology. Even in the west, this concept is not universally recognised, and Binford has always promoted and practised this component of the study. I make an effort to research the beginnings of agriculture in ancient China using this theory in order to carry forward it, but I do not anticipate being able to resolve this issue. Instead, I have a desire to broaden the scope of our ongoing investigation into the origins of agriculture in China.

12.3 Direct or Indirect Archaeological Evidence

Whether it be theoretical or critical inquiry, the study of the origins of agriculture requires proof from archaeological materials. Archaeological materials must be explored for all information because they cannot speak for themselves. Scientific analysis is based on specific concepts to infer changes in the population's movement at the period, such as in the examination of human bone isotopes or stone tool combinations. There are direct and indirect distinctions even though they are all part of the archaeological data. It can also be separated into hard evidence and soft evidence according to the reliability of the underlying ideas.

The biochemical analysis of the isotopes in human bone provides the most direct evidence of all the archaeological material evidence. There is never too much evidence in archaeological artefacts for archaeologists. Archaeologists look for enriched and multidisciplinary evidences in order to avoid the limitations of single viewpoint. The archaeological evidence of diet is the most convinced. There is evidence that both people and domesticated canines eat millet throughout the Early Neolithic like the time period of the Dadiwan site, suggesting that domesticated crops did not become the staple food of human diets until the Middle to Late Neolithic. The isotope analysis of human bones still has several issues to be resolved. The first is the scarcity of materials, particularly those from the Upper Paleolithic to the NPT phase. Human bone materials are found at a very small number of sites which includes the Donghulin site. It is unrealistic to build a complete change sequence. Accuracy is the second problem. Although the species of C3 and C4 plants clearly differ from one another, we are unaware of all the plant species that ancient humans consumed. Even the Neolithic human bone materials that have been discovered lack a complete reference sequence, which refers to the isotopic variations in human bones brought on by various dietary habits. As a result, the present insufficient research may only be utilised as indirect evidence and cannot be used as direct one.

Another method for direct evidence is to find plant and animal remains to determine whether they have been domesticated. When the domestication characteristics of the species are well-established, this aspect of the evidence is typically not a block for the Middle Neolithic sites, but earlier periods—particularly the transitional period between the Paleolithic and Neolithic periods—pose a challenge. We often read that certain species possess both wild and early domestication qualities at the transitional stage of domestication. This description is ambiguous for what level of domestication has been attained and what role did these creatures play in historical human society? These are unknown to us. People have preferred to use the flotation techniques in plant archaeology recently. We can learn something new by observing the shape and abundance of plant megafossils like seeds and microfossils like phytoliths, which are currently the most valued evidence in the investigation of the origins of agriculture in China. We cannot ensure that these plants were used by people at that time, due to the complexity of site evolution. For instance, the strata were created after the site was abandoned, and it's possible that the items on the surface of the living space were also accumulated afterwards. Furthermore, paleoethnobotany research makes it hard to ascertain the degree to which people relied on domesticated crops during the early stages of agricultural origins. Rice may have been used by humans in the middle and lower Yangtze River as early as the late Paleolithic, or around 20,000 years ago, according to recent paleoethnobotany research. In the Yuchanyan and Xianrendong sites, there are already indications of the domestication of rice forms. There is more evidence from northern China even if it comes a little later.

Archaeologists favour evidence that consists of archaeological remnants, which are representations of the Neolithic era, such as pottery, polished artefacts and settlements. They separately refer to pottery vessels and polished stone tools rather than pottery artworks and decorative items. It wasn't until the Neolithic that the technology for making pottery and grinding stone tools started to be widely used in daily life. The Neolithic settlement stands out from Paleolithic sites due to its well-organised buildings, caves, tombs, and various assemblages of artefacts. Archaeologists distinguish between the Neolithic and Paleolithic based on the distinctive characteristics of the assemblage of relics, and this sort of evidence is rich and varied, with recognisable features and definite systems. Archaeologists have amassed a substantial amount of archaeological material as a result of their extensive fieldwork, thus such conclusions are neither isolated nor straightforward. However, there is no conclusive proof that agriculture originated during the Neolithic. Ceramic jars, polished stone tools, and finished settlements cannot prove that there was formerly agricultural activity. Examples of periods that exhibit the aforementioned traits include the Jomon Age in Japan, the Mesolithic in Northwest Europe, and the Neolithic in Northeast China, represented by the Xin Kailiu culture, all of which lack any indication of an agricultural basis. It would be contrary to most archaeologists' common sense if the Neolithic cultural traits did not always signify the beginning of primitive agriculture. The previous article outlined the reason why this special circumstance and promoted the adoption of the term "Mesolithic" to associate with the hunting and gathering civilization that depended on water resources. This is a cultural adaption similar to early agriculture. Common archaeological evidence alone cannot pinpoint the beginning of agriculture in regions with poor agricultural conditions and abundant aquatic resources.

There is indirect evidence in addition to the direct evidence already presented. As people started to produce their own food instead of just consuming it, the relationship between humans and the environment fundamentally changed with the advent of agriculture. Although food is still a result of natural growth, humans have begun to control the environment and processes of growth. It involves both the artificial selection of species' traits and the creation of artificial environmental factors. Vegetation disturbance by humans is the most obvious indicator. Agriculture did not first cause of human interference with the natural vegetation; rather, it started with the use of fire. Fire was a powerful tool for clearing vegetation after people mastered its application method. With the advent of agriculture, human usage of fire in constrained regions expanded significantly, and the local plant species also changed. Both the Kuk site in New Guinea and the sediment from the Erhai Lake in Yunnan Province provide proof of major deliberate disturbance with natural vegetation, which may have something with agricultural origins.[1] This indirect evidence ultimately still needs to be corroborated by direct evidence, just as the Kuk site is being questioned. The oldest forms of agriculture may not have significantly disturbed the ecosystem, and artificial interference is not the only reason for vegetation replacement. It can be difficult to attribute all vegetation changes to people.

All of the aforementioned factors must definitely be supported by evidence when examining the origins of agriculture, and this book expands extra two sorts of indirect evidence. One comes from the study of stone tools, and the other from the study of site processes. My expertise in Paleolithic archaeological training, which has evolved into a feature of this study in archaeological material analysis, is primarily responsible for the formation of these two types of evidence. The analysis of Neolithic stone tools currently concentrates on determining their purposes. Chinese Paleolithic archaeology now places a lot of emphasis on the function estimation achieved through trace analysis. Because of the chance to examine first-hand materials and conduct experimental archaeology research, I was able to infer the purpose of some stone tools and then make assumptions about the farming practices used at the time in my study of the adaptation and change of prehistoric culture in the northern and southern regions of Yanshan. It is evident that there was a progressive shift from slope cultivation to valley cultivation by comparing the stone shovels at the Xinglongwa culture at the Baiyinchanghan site, grooved-shaped stone shovel with pointed blade at the Zhaobaogou culture, stone sickle at the Hongshan culture. This idea is further supported by research on the stone hoes and shovels used by the Xiajiadian Lower culture. However, the analysis of chipping stone tools is the article's main focus. As the agriculture first emerged in the Paleolithic, it is crucial to figure out the changes in the technology and assemblages of stone tools since the Upper Paleolithic. The study of microblade technology is one of them, and it is important because it is a symbol stone tool technology from the Late Upper Paleolithic of North China, and its rise and fall are closely tied to the beginnings of agriculture. Prehistoric hunter-gatherers used the microblade technology to increase mobility. It reflects the peak of hunter-gatherer movement and chipping technology of stone tools. After their rapid growth, however, prehistoric humans in North China made a disruptive change in adaptability, drastically lowering their level of mobility and eventually turning towards settlement. The most fundamental materials are the stone artefacts found in Paleolithic sites. We can further observe the variations in mobility by the combination features of stone products at various phases, such as those at the Lijiagou site. Even though it is indirect, we may undoubtedly get useful knowledge about agricultural sprouts from simple stone tools.

The analysis and exploration of the archaeological site process (I came up with the phrase “archaeological sitology”) begins with human habitation and continues with usage, abandonment, and modification before ending with research and utilisation. In terms of dwelling choice, site layout, and disposal techniques, hunter-gatherers and agricultural producers differ significantly. Hunter-gatherers’ adaptive strategies underwent a substantial transformation, particularly in the Late Upper Paleolithic. After their mobility shrank, the change process was easily discernible through site analysis. The performance is especially noticeable in North China, where site groups for the technology of microblades include Shizitan, Xiachuan, Hutouliang, Shayuan, and Malingshan. Thousands of fire marks at the enormous Shizitan Localty S29 site show that it has frequently been inhabited by prehistorical hunter-gathers. The construction has no difference and the combination of discovered artefacts is simpler than sites like Donghulin, Zhuannian, and others from the NPT period, depicting the highly mobile lifestyle. The material of the site combination was substantially richer during the NPT period, showing a significantly reduced level of people mobility (partially settled). This indicates the beginning of agriculture as well as the transition of ancient humans from highly mobile to less mobile lifestyles. Without food production, it is impossible to access enough food resources on a smaller scale as mobility declines. A complete sedentism gradually takes shape. The Early Neolithic sites, such as those at Cishan and Baiyinchanghan, revealed that there may have still been a lot of mobility at that time based on the disposal techniques used there. The abundance of carefully preserved artefacts at the site, including intact tools and grain cellars, suggests that people originally plan to leave for a while—possibly seasonally—in order to make the most of specific unique resources. This kind of abandonment was less frequent than it was in the Early Middle and Late Neolithic sites. The study of site process demonstrates that settlement is a slow process, which is also compatible with how early agriculture developed.

We require known direct evidence, indirect evidence, and extended evidence. Archaeologists need to compare and integrate different types of evidence in addition to creating new evidence in order to answer questions concerning the beginnings of agriculture. The more abundant the evidence is, the broader and more deep the questions we can answer. We can grasp more intricate procedures, geographical variation, and the causes of agricultural origin except for providing answers to the fundamental forms of agricultural origin. Although it seems to be a macro-historical theory from the title, positivism is still present in this study's analysis of the archaeological evidence.

12.4 Unanswered Question: Difficulties of *Touwujianren* in Archaeology

As a key question, the origins of agriculture depend on the level of archaeological progress, specifically the approaches that may be applied in theory, methods, and practice. These limitations on our understanding are precisely what we study the history of agriculture and conduct archaeological investigation. It's necessary to review and consider the strategies we use. This job preserves the past while paving the way for the future. Any research is just a small piece of knowledge, whether it is significant or not; whether it is accurate or not. I'll discuss the challenges I faced on five different levels, from simple to complex and from concrete to abstract. The issue is no longer trivial, thus I expect that future study can resolve or outperform them. With different emphasis on each level, they are all parts of the archaeological process of judging humans through materials.

In any case, archaeological materials are needed to study the history of agriculture. The proverb "Even the cleverest housewife cannot cook a meal without rice" is true. The absence of archaeological evidence makes it extremely difficult to research the beginnings of agriculture in some areas. The omission might have been noted by readers of the text. Although I included the northwest region in the cultural ecology region, I did not talk about it at the end. The scarcity of materials is a primary cause (Institute of Chinese Borderland Studies, CASS 2008). One element of the issue is the sporadic Paleolithic-to-Neolithic findings in the vast Xinjiang region, and the other is the quality of the archaeological materials. The most fundamental characteristic of archaeological materials possesses the precise spatiotemporal correlations, or the layers and ages. While the latter includes relative and absolute ages, the former includes horizontal distribution (also known as the distribution within the site and the area) and vertical distribution. Archaeological materials will lose their most fundamental characteristics if the surface-collected materials take away from the original stratigraphic relationships. If a systematic surface survey has been conducted, this deficiency can be somewhat made up for. The quality of such archaeological materials is difficult to guarantee, and it is undoubtedly very difficult to draw trustworthy conclusions from them if only randomly collected from the surface based on the interests of investigators. The vast majority of archaeological findings in the northwest region, with the exception of a few occasional ones, were collected from various eras (meaning that the techniques utilised for each collection may vary). Chinese and foreign researchers have also looked into the Paleolithic archaeology in Xinjiang in recent years, however, there are still few items with distinct time and spatial linkages available. One of the key ideas is that chipped stone tools are not equivalent to those from the Paleolithic (the stone tools from the Paleolithic). Neolithic, Bronze Age, and even earlier these stone tools exist today. The province of Guizhou continued to use chipped stone tools in the 1950s. Chipped stone tools are frequently considered to be Paleolithic tools (Hard and Merrill 1992) in the exploration and research of caves in Guizhou province, however their true age may be fairly recent. Stone tools of the microblade technology have a distinct making method, which are frequently encountered in China's northwest. Its date ranges from roughly 20,000 years ago to the prehistoric era, meaning it could be Paleolithic, Neolithic, or prehistoric artefacts. It is now hard to reliably estimate the age based on the characteristics of stone tool technology kinds, including the microblade technique. This is because the archaeological typology of stone tools is still undeveloped. The limitations imposed by the conditions of the archaeological material are clear, and they are also the main issues that need to be addressed in the next archaeological studies.

Coastal adaptation hasn't been thoroughly studied yet, despite the fact that it's a crucial component of hunting and gathering livelihoods. For instance, hunter-gatherers had established a settled way of life during the Mesolithic in northwest Europe, the Jomon Period in Japan, and the Paloma and Chilca periods on the Peruvian coast.[2] The continental shelf where ancient humans might have lived was submerged as a result of sea level rise in North China and the middle and lower reaches of the Yangtze River between the end of the Late Pleistocene and the beginning of the Early Holocene, making it challenging to find materials about coastal adaptation. At the same time, prehistoric agriculture developed here, so the coastal adaptation has a minor cultural impact. It may also be related to the greater lack of coastal marine resources in China compared to the aforementioned regions, in addition to the substantial influence of the agrarian cultural system. Neolithic shell midden sites can be found in Liaoning province's eastern and southern regions, the Jiaodong Peninsula, Fujian province, and the Guangdong and Guangxi area (Institute of Chinese Borderland Studies, CASS 1999), but they are all very recent and have been influenced by the agricultural economy. This issue can be investigated further if the data are more plentiful and thorough, particularly for the coastal sites that date from the Upper Paleolithic to the Neolithic.

The second level, which is frequently disregarded, has to do with how archaeological materials are made. Despite paying attention to this aspect, the book does not offer a thorough analysis. On the one hand, because it deviates from the focus of this book; On the other hand, it also stems from insufficient information and analytical approaches. The manner in which archaeological materials were created has a direct bearing on their dependability and the profundity of our judgments. Only the most basic conclusion can be supported by rough materials, whereas precise and detailed materials lead to a deeper understanding. The definition of "living floor" is frequently questioned. This issue existed in both Paleolithic and Neolithic settlements. What vertical distance should be taken into account while determining the living floor range? What happens if the initial surface is not level? As a result, some individuals consider it to be a strip of residential areas in archaeological (Schiffer 1987), following a question of "How long did this strip form?" In the Paleolithic, the abandonment process was mostly rapid, so the cultural accumulation was relatively less disturbed; while the houses of the Neolithic may have been gradually abandoned, so the process is quite lengthy. As a result, it's possible that the structure and combination of remains inside the dwellings have been fluctuated, and perhaps that it wasn't done by the same group or at the same time. The remnants found on a living floor, however, are typically seen to be the most reliable indicator of the people's way of life at the time. How well, for instance, do the plant remnants in the strata reflect the use of plants by residents during that time? Pollen and phytolith discovered in the stratum cannot symbolise historical behaviour if they are discarded deposits. Cultural artefacts changed by cultural and natural processes are known as archaeological materials. The transformation can impact the completeness and expression of materials, and not all remains or relics can be retained or entirely disappeared. The formation process not only results in the loss of information but also adds new information, such as some ecofacts. Overall, the analysis is relatively insufficient, which has an effect on the precision of our reasoning and leaves some undiscovered threats.

The third level is the specific concept of "*Touwujianren* (seeing humans through materials)", which refers to the investigation of archaeological artefacts in order to comprehend the behaviour of ancient people. The use of theories, methodologies, and materials combines this specific archaeological research procedure. From a theoretical standpoint, archaeologists make conclusions about ancient people based on "reason" in all cases. What is the "reason" in this case? I initially think it as logic. From a concrete perspective, it is essential to adhere to the most fundamental tenets of logic, such as the law of identity and the law of excluded middle, otherwise it will lead to a logical self-contradiction and irrationality. From an abstract perspective, archaeologists infer using the tenets of induction, deduction, and analogy. Archaeologists frequently use the induction method since no one has ever experienced prehistoric times and they don't have a basic understanding of them. Therefore, it is inevitable to gain the initial comprehension through induction. A comparison based on resemblance is called an analogy. It is typically a technique utilised in fields like experimental archaeology and ethnoarchaeology. Analogical reasoning is often doubted for the comparability between the present situation and prehistoric times. Analogies are nevertheless essential, just as metaphors cannot be eliminated from our language. Deductive reasoning is perhaps the most challenging, which requires a premise first and relies on more general principles to infer, Is there a set of reliable archaeology principles? Some answers are negative because no one has ever lived in prehistory. Binford, though, responded positively. For instance, humans rely on naturally growing animals and plants for a living, regardless of periods and places under the hunter-gathering lifeway; How many hunters and gatherers (population limit) that a place support can also be roughly counted based on the number of animals and plants that a region can produce (a field of natural science research). He studied hundreds of hunter-gatherer groups and some people believe this as ethnoarchaeology. In fact, what he seeks is the cultural ecology principle of hunting and gathering livelihood, and then inferring based on the established principles, which is called as deductive reasoning.

With a methodological perspective, archaeologists have actually developed multiple ways to infer human behaviour from archaeological materials, including the analysis of stone tools, pottery, tombs, settlement patterns, spatial analysis within the settlement, zooarchaeology and paleoethnobotany, sitecatchment, artefact source, landscape, etc. If we continue to list them all, the list will get very long. In order to comprehend the human behaviours in the past, archaeologists investigate material relics, target diverse artefacts, and use numerous disciplines to get knowledge from archaeological materials. For most people, the study of numerous categories is the most active and quickly developing branch of archaeology. Every phase of research depends on the formulation of fundamental principles and it serves as the basis for subsequent conclusions. Researchers have discovered that human mobility is directly tied to the standardisation and refinement of stone tools, taking stone tool analysis as an example. These tools are incredibly lightweight, and their blades produced regularly are convenient for replacement and suitable for different situations. It unquestionably requires extremely delicate and hard raw materials as well as cutting-edge stone processing technology, thus the raw materials and labour is highly priced. It is cost-ineffective to produce such tools unless there are unique requirements. Stone tool researchers thereby acquired a key concept of stone tool making and used it to guess the lifestyle of ancient people based on the fundamental tenets of economics (the principle of opportunity cost). For instance, several microblades were discovered in the Western Liaoning region's Xinglongwa and Hongshan cultural sites in the Neolithic period. I've come to the conclusion that these societies have at least a few highly mobile populations, indicating that hunting still plays a role in their way of life.

All theories and methods are based on archaeological findings, which form the basis for inferring human behaviour through materials. It is essentially regarded as an approach. Although almost all of the methods that archaeologists have developed can be used to study the origins of agriculture, each scholar is constrained by their own knowledge background and is rarely proficient in using them, typically only referring to the results of various methods. The constraints of a particular method or the premise of the argument are frequently unclear due to a lack of process comprehension. Just like stone tool analysis which I am relatively skilled at, no matter how exact the analysis approach is, it must rely on the initial definition and classification of stone tools. The difference among them leads to various conclusions. Without any doubt, interdisciplinary cooperation is necessary for the study of the origins of agriculture, but this should go beyond simply combining research reports from various disciplines and instead goes towards integration. It is the right course for us to take despite its barriers.

The success of this study, "*Touwujianren* (seeing humans through materials)", depends not only on reliable sources of information and strict methodology, but also on the soundness of theoretical tenets, which is also the fourth problem I have mentioned: limitations of the theory. This book focuses on the history of agriculture from the view of hunter-gatherer cultural ecology. The logic of this principle determines both its traits and the basis for its foundation. There is an unresolved issue of the adaptation of hunter-gatherers to grassland environment. The tropical savanna is a very productive habitat that receives nourishment from volcanic ash. It is home to a variety of animals and a hot climate, without the need of heating. Human ancestors did, in fact, inhabit tropical savannas in Africa. But can they make use of the Eurasian Savannah's environment? These pristine grasslands resemble the harsh environments that prehistoric humans encountered, such as polar areas, plateaus, deserts, and tropical rain forests. Mobility poses the biggest challenge due to the high mobility of large herbivore groups and the relatively small range that humans can cover by walking hunting. Approaching to the prey in wide terrain counts a difficult task, and catching them is more challenging for people. Another issue relates to the fuel, particularly in the winter when a lot of fuel is needed, but grasslands cannot supply these. The best option lies near to forests and grasslands, taking advantage of their complementary resources. According to research on the cultural ecology of hunter-gatherers, early people has no capacity to adjust to an environment without trees in an open grassland, and it was much harder in a desert. The theory of cultural ecology discloses the most common circumstance in an ideal state yet. In fact, grasslands and trees are mostly dispersed in an irregular pattern, and not all deserts are lifeless. For instance, the Atacama Desert in South America is as barren as the moon, yet the Kalahair Desert where Bushmen reside, is abundant in plant and animal resources. It is obviously dangerous to judge the feasibility of human cultural adaptation exclusively on generic qualities without knowledge of the details of prehistoric ecological conditions.

The human behaviour patterns contribute significantly to see humans in materials for archaeology's quest. Whether agriculture was first used for "eating" or "inviting people to eat" is one of the most essential questions in the study of the origins of agriculture. The former holds that humans developed agriculture as a pressure-response adaptation and a need to solute survival issues. As for the latter, there is no reason to support the notion that humans experienced a survival crisis around ten thousand years ago. As a result, agriculture is associated with internal competition in human civilization, in other words, it means "inviting people to eat" and achieving higher social standing through feasting. The latter is more relevant to the problem of human reproduction because only the "strongest" are able to have children. The so-called "strongest" among people is also a fluid idea. It may refer to those individuals and societies that could consistently produce food resources about 10,000 years ago. The activity of "inviting people to eat" has a social significance because competition exists not only among individuals but also within the entire society. Human survival and reproduction, two major things, can be made simultaneously. The true purposes of agriculture include both for "eating" and "inviting people to eat". But at the moment academic research both domestically and globally still sees the two as opposed. There are deeper causes for this disagreement, if we think the beginning of agriculture as a reaction to pressure, then human adaptation is passive; however, if we consider the beginning of agriculture as an human active pursuit, then we believe that agriculture is the result of human initiative, which in the end is a conflict between materialism and idealism. In terms of people and things, who comes first and then? The presence of things is meaningless in the absence of people, and human existence is impossible in the absence of objects. Two fundamental philosophical pillars are involved in the study of the history of agriculture. Should we approach the study of human behaviour from a material perspective, or is it the opposite? Perhaps we choose one to study, adopt a compromise attitude, or express a general position, there are both benefits and drawbacks. And then we should transcend the opposition of dualism, but has this entered a new cycle?

The final stage is that when performing archaeological study, we must reflect on ourselves. What role does archaeology play in interpreting human behaviour through artefacts? Only by realising that "the swellhead will be punished" when doing archaeological study can the archaeologists engage in scientific research with a diligent and humble attitude. Whether or not the origins of agriculture count a meaningful topic, and if so, what does it mean? The origin of agriculture in China in this book includes a modernisation perspective that is consistent with China's ongoing modernisation theme. Modernisation, however, is a concept that is quickly losing relevance. Modernity is being broken into pieces by the rising postmodernist trend. Laws, structures, patterns, and so forth are all being deconstructed as the grand narrative of history is being abandoned. Will my system of early Chinese agricultural processes be destroyed? But I am aware that postmodernism's fragments ultimately stem from the development of modernity. There wouldn't be any significant postmodernist accomplishments without modernity.

Everything will pass because that is the only rule of history, so why should I feel regret?

Notes

1. The beginning of the Neolithic Age in the Yunnan-Guizhou region was relatively late, making it difficult to adopt the stage division used in North China and the middle and lower reaches of the Yangtze River. For the purpose of comparison, a "Late Paleolithic" period was established, corresponding to the early and middle stages of the Neolithic Age in North China and the middle and lower reaches of the Yangtze River.
2. The so-called isolated hunter-gatherer groups are often sensationalised by the media. For example, there was a news report on October 25, 2010, about an isolated hunter-gatherer group living in the jungles of Peru, and a researcher was even killed during an investigation. In reality, no human population can survive without intermarriage with the outside world. The notion of isolation is simply unknown to people living in cities, but nearby agricultural communities are well aware of them.

References

Institute of Chinese Borderland Studies, CASS compiled, Environmental Archaeology of Shell Midden sites in the Jiaodong Peninsula, social sciences academic press(CHINA),1999.

Institute of Chinese Borderland Studies, CASS compiled, Stone Age and Bronze Age in Xinjiang, Cultural Relics Press, 2008.

J. A. Tainter, The Collapse of Complex Societies, translated by Shao Xudong, Hainan Publishing House, 2010.

J. M. Diamond, *Guns, Germs, and Steel: The Fates of Human Societies*, translated by Xie Yanguang, Shanghai Translation Publishing House, 2000.

L. R. Binford, Constructing Frames of Reference: An Analytical Method for Archaeological Theory Building Using Ethnographic and Environmental Data Sets. University of California Press, Berkeley, 2001.

Luo Rongqu, A New Discussion on Modernization: The Modernization Process of the World and China (Revised and Enlarged Edition), the Commercial Press, 2009.
M. B. Schiffer, The Formation Processes of the Archaeological Record. University of Utah, Salt Lake City, 1987.
R. J. Hard and W. L. Merrill, Mobile agriculturalists and the emergence of sedentism: Perspectives from Northern Mexico. American Anthropologist, 94: 601–620, 1992.

Glossary

Chinese	*English*
食谱	diet
细石叶技术	microblade technology
地表净生产力	Net Above-Ground Productivity
生态交错带	ecotone
聚落形态	settlement pattern
鱼叉	harpoon
石制品	lithic artefacts
细石核	microblade core/microcore
细石叶	microblade
尖状器	point
新仙女木事件	the Younger Dryas Event
磨盘	grinding slab
磨棒	handstone
磨制石器	ground stone tool
石网坠	stone net pendant
锄形器	hoe-shaped tool
石斧	stone axe
研磨器	grinder
细石器	microlith
灶膛/火塘	hearth
窖穴	storing pit
磨制技术	grinding technology
灰坑	ash pit
石耜	stone *si*
磨制工具	grinding tool

(continued)

S. Chen, *Prehistoric Modernization*,
https://doi.org/10.1007/978-981-97-9423-2

(continued)

Chinese	*English*
螺壳	snail shell
砾石石器	gravel tool
打制石器	chipped stone tool
穿孔砾石	perforated gravel
贝丘	shell midden
壕沟	moat
植硅石	phytolith
勒瓦娄哇技术	Levallois technology
砍砸器	chopper/chopping tool
石片工业	flake industry
刮削器	scraper
锛形器	adze-shaped tool
压制技术	pressure technology
广谱	broad spectrum
琢背石刀	backed knife
大三棱尖状器	large triangular point
石片	flake
鹿角棒	antler rod
粮食堆积	grain pile
根块茎农业	root and tuber agriculture
族属	ethnicity
圆头刮削器	round-headed scraper
石镞	arrowhead
骨制品	bone artefact
石叶疤	blade scar
融冻褶皱、冰楔	the thawing fold and ice wedge
陶容器	ceramic vessel
石容器	stone vessel
灶火	stove
陶器纹	pottery decoration
夹砂陶	sand-tempered pottery
碳化黍	charred millet
鱼窖	fish cellar
骨鱼镖	bone dart
石铲	stone shovel
人骨13C分析	13C isotope analysis on human bones
挖掘棒	digging rod
加重石	weighting stone
石穿孔器	stone perforator
石锛	stone adze
石凿	stone chisel
杵	pestle

(continued)

(continued)

Chinese	*English*
小石棒	small stone rod
臼	mortar
饼形器	pie-shaped stone artefact
石球	spheroid
石刀	stone knife
斧形器	axe-shaped tool
玉器	jade
凹弦纹	concave string pattern
筒形罐	cylinder jar
石质破土工具	stone earth-breaking tool
祭祀用陶器	sacrificial pottery
炭化粟米	charred millet and rice
水稻植硅石	rice phytolite
夹砂粗陶片	coarse sand-tempered pottery shreds
大型穿孔器	large perforated tool
刃部磨光石斧	edge-polished axe
蚌壳堆积坑	pit filled with shells
磨刃石器	edge-ground tool
舟楫技术	rowing technology
刀耕火种	slash-and-burn farming
金属刀斧	metal knife and axe
螺蚌	snail shell
砾石工业	gravel industry
扁平、卵形或长条形的砾石	flat, ovate or elongated gravel
骨锥	bone awl
骨针	bone needle
骨刀	bone knife
石灰岩	limestone
小型石器	small-sized stone tool
细石叶工艺产品	products based on microblade technique
之字纹筒形陶罐	zigzag pattern cylindrical pot
篦纹圆底罐	round bottom jar with fine-toothed comb pattern
刻划纹圆底罐	round bottom jar with incised pattern
烧制的非容器陶土块	baked clay for nonvessel pottery
绳纹陶器	Jomon pottery
勒瓦娄哇打片技术	Levallois flaking technology
砾石工具	gravel tool
手斧	handaxe
石片工具	flake tool
压制细石核	microcore based on pressure technology
猛犸象化石	mammoth fossil

(continued)

(continued)

Chinese	*English*
新石器时代	the Neolithic
旧石器时代	the Paleolithic
晚更新世末	end of the Late Pleistocene
全新世之初	Initial Holocene
新石器时代革命	the Neolithic Revolution
前陶新石器时代	the Pre-Pottery Neolithic
全新世大暖期	the Holocene Climatic Optimum (HCO)
末次盛冰期	the Last Glacial Maximum(LGM)
新石器时代早期	the Early Neolithic
东北中石器时代	the Mesolithic in Northeast China
旧石器时代晚期晚段	the Late Upper Paleolithic
旧石器时代晚期早段	the Early Upper Paleolithic
旧石器时代晚期	the Upper Paleolithic
中石器时代	the Mesolithic
碧玉	green nephrite
投枪头	projectile point
圆柱形细石核	cylindrical microcore
脉石英	vein quartz
双排倒刺鹿角鱼鳔	antler darts with double-row barbs
穿孔骨针	perforated bone needle
带刻纹的蚌壳	clamshell with inscribed pattern
热释光时代	the chronology date based on thermoluminescence
光释光	Optically Stimulated Luminescence (OSL)
实验考古	experimental archaeology
烧骨	burned bone
玉髓/石髓	chalcedony
玛瑙	agate
燧石	flint
打制石锄	chipped stone hoe
鹿角锄	antler hoe
方銎	square adze
角制掘土工具	antler -made digging tool
有柄锄形器	hoe-shaped tool with a handle
角矛	antler spear
笄	hairpin
管	tube
梭形器	shuttle-shaped tool
骨器	bone tool
鱼骨层	layers of fish bone

(continued)

(continued)

Chinese	*English*
玉器墓	tomb with jade
青铜时代	the Bronze Age
铁器时代	the Iron Age
陶垫	pottery pad
石器组合	lithic assemblage
最佳寻食理论	the Optimal Foraging Theory (OFT)
斑块选择	patch choice
中心地采食与线性程序分析	central place foraging and linear program analysis
文化序列	cultural sequence
锐棱砸击法	the ridged-hammer bipolar flaking technology (RHBF)
劈裂面	the plane of fracture
边刮器	side scraper
凹缺刮器	concave scraper
陡向加工	direct percussion
复向加工	bipolar technique
燧石原料	flint raw material
AMS 与铀系测定年代	AMS and U-series dating
端刃砍砸器	end chopper
打片砾石	chipped gravel
锤击法	percussion flaking
刮削器的刃角	the blade antler of the scraper
骨角工具	bone and antler tool
磨制骨器	polished bone tool
磨制骨锥	polished bone awl
骨镞	bone arrowhead
角锥	antler awl
角铲	antler shovel
陶片	pottery sherds
打制骨器	chipped bone tool
骨料	bone material
用火遗迹	fire-use feature
磨制石器组合	polished lithic assemblage
打制石器组合	chipped lithic assemblage
黄土堆积	loess deposit
燧石石器	flint tool
磨光石斧	polished stone axe
陶器群	pottery group
肩石器	shouldered stone tool
城址	city site
滨河阶地	riverside terrace

(continued)

(continued)

Chinese	English
滨河洞穴	riverside cave
滨湖贝丘	lakeside shell midden
镢	(jue) a big spade
穿孔石刀	perforated stone knife
夹砂褐陶圜底器	sand-tempered brown pottery with circular base
石砧	stone anvil
纺轮	spinning wheel
竹木骨	bamboo-wood-bone
骨尖状器	bone point
穿孔贝壳	perforated shell
绑柄	hafting
木标枪	wooden spear
石制矛头标枪	stone-tipped spear
骨质标枪	bone spear
石器制造场遗址	lithic workshop site
储藏坑	storage pit
大石片砍斫器	large flake chopping tool
船底形刮削器	boat-shaped scraper
拇指盖状刮削器	thumbnail-shaped scraper
石器压制修理技术	pressure-flaking retouch technology
锥形石核	conical core
楔形石核	wedge-shaped core
铅笔头细石核	pencil-shaped microcore
半锥形石核	semi-conical core
柱锥形细石核	columnar-conical microblade core
肩尖状器	shouldered point
两面器石核	bifacial core
除皮石片	debitage flake
两面器修理石片	retouch flake from biface
雕刻器技术	burin technology
软锤修理	Soft hammer retouch
热处理	heat treatment
棱柱状石核技术	prismatic core technology for removing blades
石叶剥制技术	blade flaking technology
嵌刃石刀	embedded-blade stone knife
薄刃斧	cleaver
细小燧石器工业	small flint lithic industry
箭杆	arrow shaft
嵌刃标枪	blade-embedded spear

(continued)

(continued)

Chinese	*English*
射弋工具	projectile tool
标枪头	spearhead
似细石叶	pseudo-microblade
小石叶	bladelet
琢背石叶工业	backed blade industry
两极小石叶	bipolar bladelet
两极技术	bipolar technology
几何形细石器技术	geometric microlithic technology
古环境	paleoenvironment
泥炭剖面	peat section
石臼	stone mortar
钻孔圆形器	drilled round tool
齿刃器	denticulates
磨制穿孔石刀	grinding perforated stone knife
铲形器	shovel-shaped tool
犁形器	plough-shaped tool
励石	whetstones
破土工具	earth-breaking tool
铜器	bronze tool
梳形器	comb-shaped tool
齿镰	tooth sickle
硅质岩石片石器	siliceous flake tool
灰堆	ash midden
鸡冠状的第一剥片	crest blade
雪橇形的剥片	sled-shaped flake
预制台面	prepared platform
勒瓦娄哇石片	levallois flake
预制石片	prepared flake
锥状细石核	cone-shaped microcore
序参量	order parameter
亚腰形尖刃石铲	grooved-shaped stone shovel
考古类型学	archaeological typology
居住面	living floor/surface
民族考古、实验考古	ethnoarchaeology experimental archaeology
民族志	ethnography
环境考古	environmental archaeology
人口考古	demographic archaeology
植物考古	paleoethnobotany
动物考古	zooarchaeology
古食谱	paleo-diet
贾湖人骨材料	human bones at the Jiahu site

(continued)

(continued)

Chinese	*English*
陶盂	*yu* pottery
燧石细小石器	tiny flint tool
夹炭陶	charcoal-tempered pottery
农业边缘区起源理论	Marginal Zone Hypothesis
放射性碳测年	radiocarbon dating
播撒媒介	dispersal agent
广谱适应	broad spectrum adaptation
竞争性的宴飨	competitive feasting
过程考古学	processual archaeology
考古学认识论	archaeological epistemology
奥杜威技术	Oldowan technology
阿舍利技术	Acheulean technology
莫斯特技术	Mousterian technology
石叶技术	blade technology
采食者策略	forager strategy
采集经济	gathering economy
渔猎经济	FGH
培高田地农业	raised field agriculture
培土农业	raised-bed planting
石片碎屑	debitage flakes
摩擦技术	Friction technique
最佳栖息地	the best habitats
古季风	Palaeomonsoon
碳块	charcoal chunk
直口筒形器物	straight-sided cylindrical pottery
间断似绳纹	interrupted rope-like motifs
刻划纹	incised floral motifs
大敞口瓶	the cup with a large open-mouth
双耳罐	double-ear jars
大平底盘	large flat-bottomed plate
镂空圈足盘	hollowed plate with ring foot
圜底	circular base,
多角沿陶器	polygonal edged potteries
颜料块	pigment chunks
鸵鸟蛋壳穿孔装饰品	ostrich egg shells perforated ornaments
木棒(柄)	wooden stick (handle)
石英砂岩	quartz sandstone
断块	broken chunks
断片	broken flakes
石圈	stone circle
动物骨骼碎片	animal skeletal remains
土坑竖穴墓	vertical pits

(continued)

(continued)

Chinese	English
仰身直肢式墓葬	supine with straight limbs
仰身屈肢式墓葬	supine with flexed limbs
硅质泥岩	siliceous mudstone
凝灰岩	tuff
角页岩	hornfels
灰岩	limestone
人头骨	human skulls
肢骨	limb bones
灶坑	cooking pit
果核	plant kernels
非容器陶土块	non-container clay chunks
穿孔蚌器	perforated mussels
燧石镰刀	flint scythes
食物残片	food debris
随葬玉器	funerary jade
剑齿象动物群化石	stegodon fauna fossils
巨猿与猿人化石	gigantopithecus and anthropoid fossils
大熊猫 – 剑齿象动物群化石	Giant panda—stegodon fauna fossils
含介壳 的文化堆积	cultural deposit with shell
圜底罐	round bottom jars
随葬品	burial objects
和平文化	the Hoabinhian culture
盘状石核	discoidal cores
预制石核	prepared cores
铅笔头形的圆锥形细石核	conical microcores with pencil-shape
建筑柱洞	Column holes in building
彩绘陶	painted pottery
石墙半地穴房屋	stone wall semi-cave house
筑丘种植	mounding cultivation
陶雕像	ceramic figurine
封泥	clay sealing
陶轮	pottery wheel
陶币	pottery token
陶珠	pottery bead
几何形细石叶	geometric microblade
共生驯化	symbiotic domestication
燧石细小石器组合	flint assemblage
磨刃石锛	edge-polished stone adze
素面夹砂陶片	plain sand-tempered pottery shreds
粗夹砂陶	coarse sand-tempered pottery
大龙谭鲤鱼嘴	the Liyuzui site in Dalong Tan
柳州白莲洞	the Bailiandong in Liuzhou

(continued)

(continued)

Chinese	*English*
东湖林和转年遗址	the Donghulin and Zhuannian site
宁夏灵武水洞沟遗址	the Shuidonggou site in Lingwu, Ningxia
萨拉乌苏遗址	the Salawusu site
小南海遗址	the Xiaonanhai site
王府井东方广场	the Dongfang square in Wangfujing
道县玉蟾岩	the Yuchanyan aite in Dao County
大地湾遗址	the Dadiwan site
彭头山遗址	the Pengtoushan site
上山	the Shangshan site
胡家屋场遗址	the Hujiawuchang site
籍箕滩	the Dianjitan site
玉田孟家泉	the Mengjiaquan in Yutian
滦县的东灰山	the Donghuishan in Luan County
凌源西八间房	the Bajianfang in the Lingyuanxi
泥河湾盆地	the Nihewan Basin
东寨	the Dongzhai site
上宅文化	the Shangzhai Culture
镇江营一期	the Zhenjiangying Phase I
磁山文化	the Cishan Culture
北福地甲类遗存	the Beifudi class A remains
后岗文化	the Hougang Culture
小河西文化	the Xiaohexi Culture
白音长汗与查海遗址	the site of Baiyin Changhan and Chahai
迁西东寨、西寨	the Dongzhai and Xizhai in Qianxi
白音长汗遗址二期乙类遗存	the class B remains of the BaiyinChanghan Phase II site
南台子	the Nantaizi
兴隆沟	the Xinglonggou
赵宝沟文化	the Zhabaogou Culture
红山文化	the Hongshan Culture
沈阳新乐下层文化	the ShenyangXinle Lower Culture
富河文化	the Fuhe Culture
大沁他拉、二道梁遗址	the Dachintala and Erdaoliang sites
沟门、金龟山	the Goumen and Jinguishan sites
夏家店上层文化	the Xiajiadian Upper Culture
阳原姜家梁小河沿文化	the Jiangjialiang Xiaoheyan Culture
农安左家山一期文化	the Nong'an Zujiashan PhaseI Culture
密山新开流文化	the Mishan Xinkailiu Culture
资源晓锦遗址	the Xiaojin site in Ziyuan
黄岩洞	the Huangyan dong
牛栏洞	the Niulan dong

(continued)

(continued)

Chinese	English
肇庆蚬壳洲遗址	the shell site in Zhaoqing
壳坵头、昙石山	the Keqiutou and Tan Shishan sites
山岗遗址	the Shangang site
景洪娜咪囡洞	the Jinghong Nami dong
泰国精灵洞	the Spirit Dong in Thailand
昆门洞遗址	the Con Moong dong site
山韦文化	the Shanwei Culture
泰国邦马帕地点群	the Pang Mapha site group in Thailand
谭罗德岩厦	the Tham Lod rockshelter
北山文化	the North Mountain Culture
沙捞越的尼阿遗址	the Niah site in Sarawak
建班延山谷洞穴	the Jianbanyan Valley Cave
昂昂溪文化	the An'gangxi Culture
新彼得罗夫卡文化	the Novopetrovka Culture
奥西诺沃文化	the Osinovo Culture
鲁德纳亚文化	the Rudnaya Culture
仰韶文化	the Yangshao Culture
吉林白城双塔遗址	the Shuangta site in Baicheng Jilin Province
西施、柿子滩遗址	the Xishi and Shizitan sites
金斯太洞穴遗址	the Jinsitai Cave site
呼玛十八站	the Shibazhan site in Huma county
齐齐哈尔碾子山遗址	the Nianzishan site in Qiqihar
饶河小南山遗址	the Xiao Nanshan site in Raohe County
刀背山遗址	the Daobeishan site
小拉哈	the Xiaolaha site
黄家围子	the Huangjiaweizi type
滕家岗子	the Tengjigangzi type
元宝沟	the Yuanbaogou site
大布苏	the Dabusu site
青山头	the Qingshantou site
金斯太上层	the Upper Layer of Jinsitai Cave
呼玛十八站上层	the Upper Layer of Shibazhan in Huma
仙人洞	the Xianrendong site
下川	the Xiachuan site
灵井	the Lingjing site
沙苑	the Shayuan site
虎头梁	the Hutouliang site
辉河水坝与哈克-团结遗址	the Huihe Dam and Hake-Tuanjie site
海城小孤山遗址	the Xiaogushan site in Haicheng city
吉林东丰西断梁山遗址	the Duanliangshan site in Dongfeng, Jilin

(continued)

(continued)

Chinese	*English*
黑龙江宁安莺歌岭	the Yinggeling site in Ningan, Heilongjiang
金谷-形城文化	the Jingu-Xingcheng Culture
农安左家山遗址	the Zuojiashan site in Nong'an County
伊通羊草沟	the Yangcaogou in Yitong County
杏山	the Xingshan site
腰红嘴子	the Yaohongzuizi site
肖家屯	the Xiaojiatun site
密山新开流遗址	the Xiankailiu site in Mishan
依兰倭肯哈达洞穴墓地	the Wokenhada Cave tomb in Yilan County
白城靶山墓地	the Bashan tomb in Baicheng
毕节观音洞	the Guanyindong site in Bijie
水城硝灰洞	the Xiaohuidong site in Shuicheng Coutny
白岩脚洞遗址	the Baiyanjiaodong site in Puding County
威宁草海遗址	the Caohai site in Weining County
云南宜良张口洞	the Zhangkoudong in Yiliang County, Kunming, Yunnan Province
云南富源大河遗址	the Dahe site in Fuyuan Coutny, Qujing, Yunnan Province
昆明龙潭山第二地点	the NO.2 Longtanshan site in Kunming, Yunnan Province
贵州盘县大洞	the Dadong site in Pan County, Liupanshui, Guizhou Province
河套地区、内蒙古东部的金斯太遗址	the Jinsitai site in eastern Inner Mongolia in Hetao area
富林遗址	the Fulin site
广元的中子铺	the Zhongzipu site in Guangyuan
北川烟云洞	the Yanyundong site in Beichuan County
铜梁张二塘遗址	the Zhangertang site in Tongliang County
丁村遗址	the Dingcun site
资阳鲤鱼桥遗址	the Liyuqiao site in Ziyang
马鞍山遗址	the Ma'anshan site
湖北郧县余嘴2号旧石器地点	the No. 2 Yuzui Paleolithic site in Yun County, Hubei Province
穿洞遗址	the Chuandong site in Puding County
兴义猫猫洞遗址	the Maomaodong site in Xingyi
兴义老江底遗址	the Laojiangdi site in Xingyi

(continued)

(continued)

Chinese	*English*
云南保山塘子沟遗址	the Tangzigou site in Baoshan, Yunnan Province
开阳县打儿窝岩厦遗址	the Daerwo site in KaiyangCounty
红土洞遗址	the Hongtudong site
云南峨山老洞洞遗址	the Laolongdong site in Yuxi, Yunnan Province
安龙观音洞遗址	the Guanyindong site in Anlong County
汉源大地头	the Daditou site in Hanyuan County
汶川姜维城	the Jiangweicheng site in Wenchuan County
茂县营盘山	the Yingpanshan site in Mao County
向家坝库区叫化岩遗址	the Jiaohuayan site in the Xiangjiaba reservoir area
什邡桂圆桥遗址	the guiyuanqiao site in Shifang
多开间的房址	multibay house site
汉源麦坪遗址	the Maiping site inHanyuan County
宝墩文化	the Baodun Culture
威宁中水鸡公山	the Jigongshan in Weining County
毕节青场	the Qingchang site in Bije
马鹿洞遗址	the Maludong site
飞虎山遗址	the Feihushan site
打儿窝遗址	the Daerwo site
张口东遗址	the Zhangkoudong site
中水遗址	the Zhongshui site
鸡公山遗址	the Jigongshan site
大墩子遗址	the Dadunzi site
狮子山遗址	the Shizishan site
营盘山遗址	the Yingpanshan site in Mao County
宝墩遗址	the Baodun site
张家坡遗址	the Zhangjiapo site
回龙湾遗址	the Huilongwan site
礼州遗址	the Lizhou site
毛家坎遗址	the Maojiakan site
江油大水洞新石器时代遗址	the Dashuidong Neolithic site in Dakang Town, Jiangyou
麻家山遗址	the Majiashan site
滇池区域贝丘遗址	the Beiqiu site in the Dianchi Lake area
宾川白羊村遗址	the Baiyang site in Binchuan
剑川海门口遗址	the Haimenkou site in Jianchuan
毕节青场新石器时代遗址	the Qingchang Neolithic site in Bijie
六枝老坡底遗址群	the Laopodi sites group in Liuzhi
波克塔克提克遗址	the Boker Tachtit site

(continued)

(continued)

Chinese	English
卡萨阿科尔遗址	the Ksar' Akil site
许家窑遗址	the Xujiayao site
刘家岔	the Liujiacha site
下川富益河	the Xiachuan Fuyihe site
西白马营遗址	the Xibaimaying site
柿子滩遗址	the Shizitan site
大岗	the Dagang site
于家沟	the Yujia Gou site
薛关	the Xueguan site
柴寺	the Chaisi site
大贤庄	the Daxianzhuang site
黑龙潭	the Heilongtan site
小空山上洞	the Upper Cave of Xiaokongshan
织机洞	the Zhijidong site
塔水河	the Tashui River site
西沟	the Xigou site
龙牙洞	the Longyadong site
泥河湾盆地的马鞍山遗址	the Mahanshan site in Nihewan Basin
头马坊遗址	the Toumafang site
山西襄汾柴寺 (或称丁家沟)	the Chaisi site (also known as Dingjiagou site) in Shanxi Province
陕西大荔沙苑	the Shayuan site in Dali, Shaanxi Province
乌鸦山	the Wuyashan site
燕耳洞	the Yan'erdong site
十里岗	the Shiligang site
樟脑洞	the Zhangnaodong site
华龙洞	the Hualongdong site
打鼓岭	the Daguoling site
神农架地区的犀牛洞遗址	the Xiniudong site in the Shennongjia Forestry District
贵州观音洞	the Guanyindong in Guizhou Province
万年仙人洞	the Xianrendong in Wanxian
潕水文化类群	the Wushui Culture Group
澧水文化类群	the Lishui Culture Group
二卵石	the Erluanshi site
岩屋滩	the Yanwutan site
长坪	the Changping site
虎爪山	the Huzhaoshan site
鸡公垱	the Jigongdang site
大圣庙	the Dashengmiao site
石家坪	the Shijiaping site

(continued)

(continued)

Chinese	*English*
胡家堡	the Hujiabao site
谢家堡	the Xiejiabao site
王家山	the Wangjiashan site
临澧竹马	the Zhuma site in Linli County
三山岛	the Sanshandao site
张家营	the Zhangjiaying site
石龙头	the Shilongtou site
九道河	the Jiudao River site
毛竹山	the Maozhushan site
水阳江	the Shuiyang River site
官山	the Guanshan site
石鼓后山坡	the back slope of Shigushan site
余嘴	the Yuzui site
丁村 77: 01 地点	the Locality 77:01, Dingcun site
河南新密西施遗址	the Xishi site in Xinmi, Henan Province
内蒙古东部金斯太遗址	the Jinsitai Site located in the eastern part of Inner Mongolia
江孜	Gyangzê county
昌都卡若	Karub ruins in Chamdo
居木	Jumu in Linzhi
藏南定日县的苏热	Sure at Dingri County in southern Tibet
藏北申扎县的珠乐勒	Zhulele at Shenzha County in northern Tibet
班戈县的各听	Geting at Bange County
阿里地区日土县的扎布、多格则	Zhabu and Duogeze at Ritu County in the Ali region
青海境内可可西里的三岔口	Sanchakou at Kekecili in Qinghai
柴达木盆地的小柴旦	Xiaochaidan in the Qaidam Basin
柴达木盆地的冷湖	Leng Lake in the Qaidam Basin
黑马河	Heima River
江西沟	the Jiangxigou site
沟后001地点	the back of ditch 001 site
羊八井	the Yangpaijing town
楚桑遗址	the Chusang site
藏北高原的色林错旧石器遗址	the Paleolithic site at Selincuo on the northern Tibetan Plateau
藏西南仲巴一带	the Zhongba in southwest Tibet
阿里地区夏达错湖滨	the Xiadacuo lakeshore in the Ali region
日塘遗址	the Jiaritang site
珠乐勒与各听遗址	the Zhu Lele and Ge Ting site
藏北色林错东南岸古湖滨阶地	the ancient lakeside terrace on the southeastern shore of Selincuo, northern Tibet

(continued)

(continued)

Chinese	English
阿里夏达错湖滨	the lakeshore of XiadachuoinAli region
南部类型 (A 组) 与卡姆类型 (B 组)	the southern type (group A) and the Kam type (group B)
甘肃民乐东灰山遗址	the Donghuishan site at Minle County in Gansu Province
青海同德宗日遗址	the Zongri site at Tongde County in Qinghai Province
马家窑文化	the Majiayao Culture
华北南庄头、李家沟遗址	the Nanzhuangtou and Lijiagou sites
新开流文化	the Xin Kailiu Culture
新几内亚的库克遗址	the Kuk site in New Guinea
虎头梁、沙苑、马凌山细石叶工艺遗址	the site groups for the microblade technology of Hutouliang, Shayuan, and Malingshan
磁山遗址	the Cishan site
秘鲁海岸的帕洛玛与奇尔卡时期	the Paloma and Chilca periods on the Peruvian coast
胶东半岛	Jiaodong Peninsula
老官台文化	the Laoguantai Culture
后李	the Houli site
北辛	the Beixin site
皂市下层	the Lower Zaoshi Culture
城背溪	the Chengbeixi Culture
裴李岗文化的贾湖类型	the Jiahu type of the Pei Ligang Culture
小山口一期文化	the Xiaoshankou Phase I Culture
白家文化	the Baijia Culture
李家村	the Lijiacun site
大地湾一期	the Dadiwan Phase I Culture
前仰韶文化遗址	the pre-Yangshao Culture site
陶寺文化	the Taosi Culture
后李文化小荆山遗址	the Xiaojingshan site of the Houli Culture
月庄遗址	the Yuezhuang site
姜寨遗址	the Jiangzhai site
西坡遗址	the Xipo site
大溪文化	the Daxi Culture
城头山城址	the Chengtoushan site
浦江上山遗址	the Shangshan site in Pujiang
嵊州小黄山遗址	the Xiaohuangshan site in Chengzhou
阳原马鞍山遗址	the Ma’anshan site in Yangyuan
河姆渡遗址	the Hemudu site
柱坑类遗迹	the pillar pit site
桂林庙岩遗址	the Miaoyan site in Guilin

(continued)

(continued)

Chinese	*English*
宜川龙王辿遗址	the Longwangshan site in Yichuan
柿子滩遗址 S1 地点	the Shizitan S1 site
S9DH 01用火遗址	the S9DH 01 charcoal block from the Songhuo site
扁扁洞遗址	the Bianbiandong site
淅川坑南遗址	the Kangnan site in Xichuan
华垱	the Huadang site
宋家岗	the Songjiagang site
新密李家沟遗址	the Lijiagou site in Xinmi
新几内亚高地大河谷	the Grand Valley of the Baliem River in the highlands of New Guinea
尼罗河谷	Nile Valley
伊拉克耶莫遗址	the Jarmo site located
阿布·休莱拉遗址	the Abu Hureyra site
纳布塔遗址	the Nabta site
纳吐芬文化	the Natufian Culture
耶利哥遗址	the Jericho site
几何形克巴兰文化	the Geometric Kebaran Culture
安马拉哈遗址	the Ain Mallaha site
阶地遗址	the Jiedi site
奥哈罗遗址	the Ohallo II site
纳布塔遗址	the Nabta site
白莲洞二期	the Bailiandong Phase II
顶蛳山 (一期)	the Dingsishan site Phase I
阳春独石仔	the Dushizai site in Yangchun
封开黄岩洞	the Huangyandong in Fengkai
英德牛栏洞	the Niulandong in Yingde
新乐下层文化	the Lower Xinle Culture
水洞沟12 地点	the locality 12 at the Shuidonggou site
鸽子山	the Gezishan site
班村遗址	the Bancun site
豹子头遗址	the Baozitou site

The manufacturer's authorised representative in the EU is Springer Nature Customer Service Centre GmbH, Europaplatz 3, 69115 Heidelberg, Germany. If you have any concerns regarding our products, please contact ProductSafety@springernature.com

Printed and bound by CPI Group (UK) Ltd, Croydon, CR0 4YY

01/10/2026

02225924-0013